Algebraic Geodesy and Geoinformatics

Second Edition

Joseph L. Awange · Erik W. Grafarend ·
Béla Paláncz · Piroska Zaletnyik

Algebraic Geodesy and Geoinformatics

Second Edition

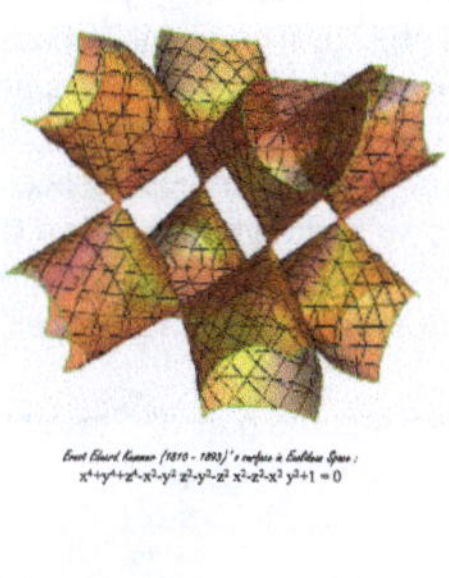

Ernst Eduard Kummer (1810 - 1893)'s surface in Euklidean Space:
$x^4+y^4+z^4-x^2-y^2z^2-y^2-z^2x^2-z^2-x^2y^2+1=0$

Springer

Prof. Dr.-Ing. Joseph L. Awange
Department of Spatial Sciences
Faculty of Science and Engineering
Curtin University of Technology
GPO Box U1987
Perth, WA 6845
Australia
J.awange@curtin.edu.au
joseph.awange@gmail.com

Prof. Dr.-Ing. Erik W. Grafarend
Universität Stuttgart
Geodätisches Institut
Geschwister-Scholl-Str. 24 D
70174 Stuttgart
Germany
erik.grafarend@gis.uni-stuttgart.de

Prof.Dr. Béla Paláncz
Budapest University of
Technology & Economics
Dept. Photogrammetry & Geoinformatics
1521 Budapest
Hungary
palancz@epito.bme.hu

Dr. Piroska Zaletnyik
Budapest University of
Technology & Economics
Dept. Geodesy & Surveying
Müegyetem rkp. 3
1111 Budapest
Hungary

Additional material to this book can be downloaded from http://extra.springer.com

ISBN 978-3-642-43113-5 ISBN 978-3-642-12124-1 (eBook)
DOI 10.1007/978-3-642-12124-1
Springer Heidelberg Dordrecht London New York

Printed on acid-free paper

Springer is part of Springer Science+Business Media (www.springer.com)

Preface to the first edition

While preparing and teaching 'Introduction to Geodesy I and II' to undergraduate students at Stuttgart University, we noticed a gap which motivated the writing of the present book: Almost every topic that we taught required some skills in algebra, and in particular, computer algebra! From positioning to transformation problems inherent in geodesy and geoinformatics, knowledge of algebra and application of computer algebra software were required. In preparing this book therefore, we have attempted to put together basic concepts of *abstract algebra* which underpin the techniques for solving algebraic problems. Algebraic computational algorithms useful for solving problems which require exact solutions to nonlinear systems of equations are presented and tested on various problems. Though the present book focuses mainly on the two fields, the concepts and techniques presented herein are nonetheless applicable to other fields where algebraic computational problems might be encountered. In Engineering for example, network densification and robotics apply resection and intersection techniques which require algebraic solutions.

Solution of nonlinear systems of equations is an indispensable task in almost all geosciences such as geodesy, geoinformatics, geophysics (just to mention but a few) as well as robotics. These equations which require exact solutions underpin the operations of ranging, resection, intersection and other techniques that are normally used. Examples of problems that require exact solutions include;

- three-dimensional resection problem for determining positions and orientation of sensors, e.g., camera, theodolites, robots, scanners etc.,
- coordinate transformation to match shapes and sizes of points in different systems,
- mapping from topography to reference ellipsoid and,
- analytical determination of refraction angles in GPS meteorology.

The difficulty in solving explicitly these nonlinear systems of equations has led practitioners and researchers to adopt approximate numerical procedures; which often have to do with linearization, approximate starting values, iterations and sometimes require a lot of computational time. In-order to offer solutions to the challenges posed by nonlinear systems of equations, this book provides in a pioneering work, the application of *ring* and *polynomial theories*, *Groebner basis*, *polynomial resultants*, *Gauss-Jacobi combinatorial* and *Procrustes algorithms.* Users faced with algebraic computational problems are thus provided with algebraic tools that are not only a MUST, but essential and have been out of reach. For these users, most algebraic books at their disposal have unfortunately been written in mathematical formulations suitable to mathematicians. We strive to simplify the algebraic notions and provide examples where necessary to enhance easier understanding.

For those in mathematical fields such as applied algebra, symbolic computations and application of mathematics to geosciences etc., the book provides some practical examples of application of mathematical concepts. Several geodetic and geoinformatics problems are solved in the book using methods of abstract algebra and multidimensional scaling. These examples might be of interest to some mathematicians.

Chapter 1 introduces the book and provides a general outlook on the main challenges that call for algebraic computational approaches. It is a motivation for those who would wish to perform analytical solutions. Chapter 2 presents the basic concepts of ring theory relevant for those readers who are unfamiliar with abstract algebra and therefore prepare them for latter chapters which require knowledge of ring axioms. Number concept from operational point of view is presented. It is illustrated how the various sets of natural numbers $\mathbb{N}$, integers $\mathbb{Z}$, quotients $\mathbb{Q}$, real numbers $\mathbb{R}$, complex numbers $\mathbb{C}$ and quaternions $\mathbb{H}$ are vital for daily operations. The chapter then presents the concept of ring theory. Chapter 3 looks at the basics of polynomial theory; the main object used by the algebraic algorithms that will be discussed in the book. The basics of polynomials are recaptured for readers who wish to refreshen their memory on the basics of algebraic operations. Starting with the definition of polynomials, Chap. 3 expounds on the concept of polynomial rings thus linking it to the number ring theory presented in Chap. 2. Indeed, the theorem developed in the chapter enables the solution of nonlinear systems of equations that can be converted into (algebraic) polynomials.

Having presented the basics in Chaps. 2 and 3, Chaps. 4, 5, 6 and 7 present algorithms which offer algebraic solutions to nonlinear systems of equations. They present theories of the procedures starting with the basic concepts and showing how they are developed to algorithms for solving different problems. Chapters 4, 5 and 6 are based on *polynomial ring theory* and offer an in-depth look at the basics of *Groebner basis*, *polynomial resultants* and *Gauss-Jacobi combinatorial algorithms*. Using these algorithms, users can develop their own codes to solve problems requiring exact solutions.

In Chap. 7, the Global Positioning System (GPS) and the Local Positioning Systems (LPS) that form the operational basis are presented. The concepts of local datum choice of types $\mathbb{E}^*$ and $\mathbb{F}^*$ are elaborated and the relationship between Local Reference Frame $\mathbb{F}^*$ and the global reference frame $\mathbb{F}^\bullet$, together with the resulting observational equations are presented. The test network "Stuttgart Central" in Germany that we use to test the algorithms of Chaps. 4, 5 and 6 is also presented in this chapter. Chapter 8 deviates from the polynomial approaches to present a linear algebraic (analytical) approach of Procrustes that has found application in fields such as *medicine* for gene recognition and *sociology* for crime mapping. The chapter presents only the partial Procrustes algorithm. The technique is presented as an efficient tool for solving algebraically *the three-dimensional orientation problem* and the *determination of vertical deflection.*

From Chaps. 9 to 15, various computational problems of algebraic nature are solved. Chapter 9 looks at the ranging problem and considers both the GPS pseudo-range observations and ranging within the LPS systems, e.g., using EDMs. The chapter presents a complete algebraic solution starting with the simple planar case to the three-dimensional ranging in both closed and overdetermined forms. Critical conditions where the solutions fail are also presented. Chapter 10 considers the Gauss ellipsoidal coordinates and applies the algebraic technique of Groebner basis to map topographic points onto the reference ellipsoid. The example based on the baltic sea level project is presented. Chapters 11 and 12 consider the problems of resection and intersection respectively.

Chapter 13 discusses a modern and relatively new area in geodesy; the GPS meteorology. The chapter presents the theory of GPS meteorology and discusses both the space borne and ground based types of GPS meteorology. The ability of applying the algebraic techniques to derive refraction angles from GPS signals is presented. Chapter 14 presents an algebraic deterministic version to outlier problem thus deviating from the statistical approaches that have been widely publicized. Chapter 15 introduces the 7-parameter datum transformation problem commonly encountered in practice and presents the general Procrustes algorithm. Since this is an extension of the partial Procrustes algorithm presented in Chap. 8, it is referred to as Procrustes algorithm II. The chapter further presents an algebraic solution of the transformation problem using Groebner basis and Gauss-Jacobi combinatorial algorithms. The book is completed in Chap. 16 by presenting an overview of modern computer algebra systems that may be of use to geodesists and geoinformatists.

Many thanks to Prof. B. Buchberger for his positive comments on our Groebner basis solutions, Prof. D. Manocha who discussed the resultant approach, Prof. D. Cox who also provided much insight in his two books on rings, fields and algebraic geometry and Prof. W. Keller of Stuttgart University Germany, whose door was always open for discussions. We sincerely thank Dr. J. Skidmore for granting us permission to use the Procrustes 'magic bed' and related materials from Mythweb.com. Thanks to Dr. J. Smith (editor of Survey Review), Dr. S. J. Gordon and Dr. D. D. Lichti for granting us permission to use the scanner resection figures appearing in Chap. 12. We are also grateful to Chapman and Hall Press for granting us permission to use Fig. 9.2 where malarial parasites are identified using Procrustes. Special thanks to Prof. I. L. Dryden for permitting us to refer to his work and all the help. Many thanks to Ms F. Wild for preparing Figs. 13.9 and 14.7. We acknowledge your efforts and valuable time. Special thanks to Prof. A. Kleusberg of Stuttgart University Germany, Prof. T. Tsuda of Radio Center for Space and Atmosphere, Kyoto University Japan, Dr. J. Wickert of GeoForschungsZentrum Potsdam (GFZ) Germany and Dr. A. Steiner of the Institute of Meteorology and Geophysics, University of Graz, Austria for the support in terms of literature and discussions on Chap. 15. The data used in Chap. 13 were provided by GeoForschungsZentrum Potsdam (GFZ). For these, the authors express their utmost appreciation.

The first author also wishes to express his utmost sincere thanks to Prof. S. Takemoto and Prof. Y. Fukuda of Department of Geophysics, Kyoto University Japan for hosting him during the period of September 2002 to September 2004. In particular Chap. 13 was prepared under the supervision and guidance of Prof. Y. Fukuda: Your ideas, suggestions and motivation enriched the book. For these, we say "*arigato gozaimashita*" – Japanese equivalent to *thank you very much*. The first author's stay at Kyoto University was supported by Japan Society of Promotion of Science (JSPS): The author is very grateful for this support. The first author is grateful to his wife Mrs. *Naomi Awange* and his two daughters *Lucy* and *Ruth* who always brightened him up with their cheerful faces. Your support, especially family time that I denied you in-order to prepare this book is greatly acknowledged. Naomi, thanks for carefully reading the book and correcting typographical errors. However, the authors take full responsibility of any typographical error. Last but not least, the second author wants to thank his wife *Ulrike Grafarend*, his daughter *Birgit* and his son *Jens* for all support over these many years they were following him at various places around the Globe.

Kyoto (Japan) and Stuttgart (Germany)
September 2004

Joseph L. Awange
Erik W. Grafarend

Preface to the second edition

This work is in essence the second edition of the 2005 book by Awange and Grafarend "Solving Algebraic Computational Problems in Geodesy and Geoinformatics". This edition represents a major expansion of the original work in several ways.

Realizing the great size of some realistic *geodetic* and *geoinformatic* problems that cannot be solved by pure *symbolic algebraic* methods, combinations of the *symbolic and numeric techniques*, so called *hybrid techniques* have been introduced. As a result, new chapters have been incorporated to cover such numeric methods. These are: *Linear Homotpy* in Chapter 6, and the *Extended Newton-Raphson* in Chapter 8, with each Chapter accompanied by new numerical examples. Other chapters dealing with the Basics of polynomial theory, LPS-GPS orientations and vertical deflections, as well as GNSS meteorology in environmental monitoring, have been refined. We also point out that Computer algebra system (Chapter 16) of the first edition has been omitted in the present book due to the rapid changing of computational algorithms.

In the mean time, since the date of the first edition, some earlier methods have been improved. Therefore, chapters like Procrustes Solution and Datum Transformation Problems have been expanded and the associated improvements in the symbolic-numeric methods are demonstrated for the case of affine transformation with 9 parameters.

In order to emphasize the theoretical background of the methods and their practical applications to geodetic and geoinformatic problems, as well as to compare and qualify them for different applications, the book has been split into two parts. Part I covers the theoretical concepts of the algebraic symbolic and numeric methods and as such, readers already familiar with these can straight away move to the applications covered in part II of the book. Indeed, part II provides in-depth practical applications in geodesy and geoinformatics.

Perhaps the most considerable extension from a *theoretical* as well as from a *practical* point of view is the electronic supplement to the book. This extra electronic material contains 20 chapters and about 50 problems solved with different *symbolic*, *numeric* and *hybrid* techniques using one of the leading Computer Algebraic Systems CAS's *Mathematica*. The notebooks provide the possibility of carrying out real-time computations with different data or models. In addition, some Mathematica modules representing algorithms discussed in the book are supplied to make it easier for the reader to solve his/her own real geodetic/geoinformatic problems. These modules are open source, therefore they can be easily modified by users to suit their own special purposes. The effectiveness of the different methods are compared and qualified for different problems and some practical recipes given for the choice of the appropriate method. The actual evaluation of the codes as parallel computation on multi-core/porcessor machines is also demonstrated. For users not familiar with the Mathematica system, the pdf versions of the notebooks are also provided.

Last, but not least, the company of the authors has also been extended, demonstrating that nowadays the cooperation of peoples from different scientific fields is indispensable when writing such a comprehensive book.

Overall, in this second edition, we have tried to bring together the *basic theories* and their geodetic/geoinformatic applications as well as the practical realization of these algorithms. In addition, the extensive references listing should help interested readers to immerse themselves in the different topics more deeply.

We have attempted to correct the various errors that were inadvertently left in the first edition, however readers are encouraged to contact us about errors or omissions in the current edition.

Many thanks go to *Prof. B. Buchberger*, the "*father*" of *Groebner basis method*, for his positive comments on our Groebner basis solutions and for agreeing to write a foreword for our book, *Prof. R. Lewis* for explaining his EDF method (Early Discovery Factor) to compute Dixon resultant as well as for carrying out some computations with his algebraic computer code Fermat. We are also grateful to *Dr. D. Lichtblau* for helping us to learn the proper and efficient use of Mathematica, especially the Groebner basis algorithm as well as to writing some appreciating words for the back cover of the book. Special thanks to Dr. *K. Flemming* and Dr. *C. Hirt* of Curtin University for sparing time to proof read this edition and for providing valuable comments. Further thanks to *K. Flemming* for preparing Fig. 17.4.

J.L. Awange wishes to express his utmost sincere thanks to *Prof. B. Heck* of Department of Physical Geodesy (Karlsruhe Institute of Technology, Germany) for hosting him during the period of the Alexander von Humboldt Fellowship (2008-2011). In particular, your ideas, suggestions and motivation on Chapter 6 enriched the book. *J.L. Awange* is further grateful to *Prof. B. Veenendaal* (HoD, Spatial Sciences, Curtin University of Technology, Australia) and *Prof. F. N. Onyango* (Vice Chancellor, Maseno University, Kenya) for the support and motivation that enabled the preparation of this edition. Last, but not least, *J.L. Awange's* stay at Curtin University of Technology is supported by *Curtin Research Fellowship*, while his stay at Karlsruhe Institute of Technology is supported by *Alexander von Humboldt's Ludwig Leichhardt's Memorial Fellowship*: The author is very grateful for the financial support. This is a TIGeR publication No 220.

Perth (Australia), Stuttgart (Germany), and Budapest (Hungary), October 2009

Joseph L. Awange
Erik W. Grafarend
Béla Paláncz
Piroska Zaletnyik.

Foreword

I compliment the authors on this book because it brings together mathematical methods for the solution of multi-variable polynomial equations that are hardly covered side by side in any ordinary mathematical book: The book explains both *algebraic* ("*exact*") and *numerical* ("*approximate*") methods. It also points to the recent combination of algebraic and numerical methods ("*hybrid*" methods), which is currently one of the most promising directions in the area of computer mathematics. The reason why this book manages to bring the algebraic and the numerical aspect together is because it is strictly goal-oriented towards the solution of fundamental problems in the areas of geodesy and geoinformatics - e.g. the positioning problem - and the solution of application problems does not allow purism in methodology but, rather, has to embrace different approaches with different benefits in different circumstances.

Personally, it is very fulfilling for me to see that my *Groebner bases* methodology, mainly by the work of the authors, finds now also useful applications in the areas of geodesy and geoinformatics. Since the book compares, in the applications, *Groebner bases* and *resultants* as the two main algebraic approaches, it also gives a lot of new motivation for further mathematical research in the relationship between these two approaches, which is still not well understood.

All good wishes for the further success of this book in the community of both *geoinformatics* and *computer mathematics*!

Austria, September 2009
Prof. Dr.phil. Dr.h.c.mult. Bruno Buchberger
Professor of Computer Mathematics, and Head of Softwarepark Hagenberg.

Contents

1 **Introduction** ... 1
- 1-1 Motivation ... 1
- 1-2 Modern challenges ... 1
- 1-3 Facing the challenges ... 2
- 1-4 Concluding remarks ... 5

Part I Algebraic symbolic and numeric methods

2 **Basics of ring theory** ... 9
- 2-1 Some applications to geodesy and geoinformatics ... 9
- 2-2 Numbers from operational perspective ... 9
- 2-3 Number rings ... 13
- 2-4 Concluding remarks ... 16

3 **Basics of polynomial theory** ... 17
- 3-1 Polynomial equations ... 17
- 3-2 Polynomial rings ... 18
  - 3-21 Polynomial objects as rings ... 18
  - 3-22 Operations "addition" and "multiplication" ... 20
- 3-3 Factoring polynomials ... 21
- 3-4 Polynomial roots ... 21
- 3-5 Minimal polynomials ... 22
- 3-6 Univariate polynomials with real coefficients ... 23
  - 3-61 Quadratic polynomials ... 23
  - 3-62 Cubic polynomials ... 25
  - 3-63 Quartic polynomials ... 26
- 3-7 Methods for investigating roots ... 27
  - 3-71 Logarithmic and contour plots on complex plane ... 27
  - 3-72 Isograph simulator ... 29
  - 3-73 Application of inverse series ... 29
- 3-8 Computation of zeros of polynomial systems ... 30
- 3-9 Concluding remarks ... 32

4 **Groebner basis** ... 33
- 4-1 The origin ... 33
- 4-2 Basics of Groebner basis ... 34
- 4-3 Buchberger algorithm ... 40
  - 4-31 Mathematica computation of Groebner basis ... 45
  - 4-32 Maple computation of Groebner basis ... 46
- 4-4 Concluding remarks ... 47

5 Polynomial resultants 49
5-1 Resultants: An alternative to Groebner basis 49
5-2 Sylvester resultants 49
5-3 Multipolynomial resultants 51
5-31 F. Macaulay formulation: 52
5-32 B. Sturmfels' formulation 54
5-33 The Dixon resultant 55
5-331 Basic concepts 55
5-332 Formulation of the Dixon resultant 56
5-333 Dixon's generalization of the Cayley-Bézout method 57
5-334 Improved Dixon resultant - Kapur-Saxena-Yang method 58
5-335 Heuristic methods to accelerate the Dixon resultant 59
5-336 Early discovery of factors: the EDF method 60
5-4 Concluding remarks 61

6 Linear homotpy 63
6-1 Introductory remarks 63
6-2 Background to homotopy 63
6-3 Definition and basic concepts 64
6-4 Solving nonlinear equations via homotopy 65
6-41 Tracing homotopy path as initial value problem 69
6-42 Types of linear homotopy 71
6-421 Fixed point homotopy 71
6-422 Newton homotopy 71
6-423 Start system for polynomial systems 72
6-5 Concluding remarks 76

7 Solutions of Overdetermined Systems 79
7-1 Estimating geodetic and geoinformatics unknowns 79
7-2 Algebraic LEast Square Solution (ALESS) 80
7-21 Transforming overdetermined systems to determined 80
7-22 Solving the determined system 82
7-3 Gauss-Jacobi combinatorial algorithm 84
7-31 Combinatorial approach: the origin 84
7-32 Linear and nonlinear Gauss-Markov models 87
7-33 Gauss-Jacobi combinatorial formulation 88
7-331 Combinatorial solution of nonlinear Gauss-Markov model 92
7-332 Construction of minimal combinatorial subsets 92
7-333 Optimization of combinatorial solutions 93
7-334 The Gauss-Jacobi combinatorial algorithm 97
7-4 Concluding remarks 98

8 Extended Newton-Raphson method 101
8-1 Introductory remarks 101
8-2 The standard Newton-Raphson approach 102
8-3 Examples of limitations of the standard approach 102
8-31 Overdetermined polynomial systems 102
8-32 Overdetermined non-polynomial system 104
8-33 Determined polynomial system 105
8-34 Underdetermined polynomial system 106
8-4 Extending the Newton-Raphson approach using pseudoinverse 107
8-5 Applications of the Extended Newton-Raphson method 108
8-6 Concluding remarks 110

9 Procrustes solution 111
9-1 Motivation 111
9-2 Procrustes: Origin and applications 112
9-21 Procrustes and the magic bed 112
9-22 Multidimensional scaling 114
9-23 Applications of Procrustes in medicine 114
9-231 Gene recognition 114
9-232 Identification of malaria parasites 115
9-3 Partial procrustes solution 116
9-31 Conventional formulation 116
9-32 Partial derivative formulation 118
9-4 The General Procrustes solution 119
9-5 Extended general Procrustes solution 130
9-51 Mild anisotropy in scaling 130
9-52 Strong anisotropy in scalling 131
9-6 Weighted Procrustes transformation 133
9-7 Concluding remarks 135

Part II Applications to geodesy and geoinformatics

10 LPS-GNSS orientations and vertical deflections 139
10-1 Introductory remarks 139
10-2 Positioning systems 140
10-3 Global positioning system (GPS) 141
10-4 Local positioning systems (LPS) 142
10-41 Local datum choice in an LPS 3-D network 142
10-42 Relationship between global and local level reference frames 144
10-43 Observation equations 145
10-5 Three-dimensional orientation problem 146
10-51 Procrustes solution of the orientation problem 147
10-52 Determination of vertical deflection 149
10-6 Example: Test network Stuttgart Central 150
10-7 Concluding remarks 153

11 Cartesian to ellipsoidal mapping 155
11-1 Introductory remarks 155
11-2 Mapping topographical points onto reference ellipsoid 155
11-3 Mapping geometry 158
11-4 Minimum distance mapping 160
11-41 Grafarend-Lohse's mapping of $\mathbb{T}^2 \longrightarrow \mathbb{E}^2_{a,a,b}$ 163
11-42 Groebner basis' mapping of $\mathbb{T}^2 \longrightarrow \mathbb{E}^2_{a,a,b}$ 164
11-43 Extended Newton-Raphson's mapping of $\mathbb{T}^2 \longrightarrow \mathbb{E}^2_{a,a,b}$ 166
11-5 Concluding remarks 171

12 Positioning by ranging 173
12-1 Applications of distances 173
12-2 Ranging by global navigation satellite system (GNSS) 174
12-21 The pseudo-ranging four-points problem 174
12-211 Sturmfels' approach 177
12-212 Groebner basis approach 179
12-22 Ranging to more than four GPS satellites 181
12-221 Extended Newton-Raphson solution 185
12-222 Homotopy solution of GPS N-point problem 187
12-223 Least squares versus Gauss-Jacobi combinatorial . . . 189
12-3 Ranging by local positioning systems (LPS) 191
12-31 Planar ranging 192
12-311 Conventional approach 192
12-312 Sylvester resultants approach 193
12-313 Reduced Groebner basis approach 194
12-314 Planar ranging to more than two known stations . . . 198
12-315 ALESS solution of overdetermined planar ranging problem 200
12-32 Three-dimensional ranging 203
12-321 Closed form three-dimensional ranging 203
12-322 Conventional approaches 204
12-323 Solution by elimination approach-2 205
12-324 Groebner basis approach 206
12-325 Polynomial resultants approach 207
12-326 N-point three-dimensional ranging 211
12-327 ALESS solution 212
12-328 Extended Newton-Raphson's solution 215
12-4 Concluding remarks 216

13 Positioning by resection methods 217
13-1 Resection problem and its importance 217
13-2 Geodetic resection 220
13-21 Planar resection 220
13-211 Conventional analytical solution 220
13-212 Groebner basis approach 222

13-213 Sturmfels' resultant approach ... 223
13-22 Three-dimensional resection ... 226
13-221 Exact solution ... 226
13-222 Solution of Grunert's distance equations ... 226
13-223 Groebner basis solution of Grunert's equations ... 228
13-224 Polynomial resultants' solution of Grunert's distance equations ... 230
13-225 Linear homotopy solution ... 233
13-226 Grafarend-Lohse-Schaffrin approach ... 237
13-227 3d-resection to more than three known stations ... 240
13-3 Photogrammetric resection ... 244
13-31 Grafarend-Shan Möbius photogrammetric resection ... 245
13-32 Algebraic photogrammetric resection ... 246
13-4 Concluding remarks ... 248

14 Positioning by intersection methods ... 249
14-1 Intersection problem and its importance ... 249
14-2 Geodetic intersection ... 250
14-21 Planar intersection ... 250
14-211 Conventional solution ... 250
14-212 reduced Groebner basis solution ... 251
14-22 Three-dimensional intersection ... 254
14-221 Closed form solution ... 254
14-222 Conventional solution ... 255
14-223 Reduced Groebner basis solution ... 255
14-224 Sturmfels' resultants solution ... 256
14-225 Intersection to more than three known stations ... 259
14-3 Photogrammetric intersection ... 261
14-306 Grafarend-Shan Möbius approach ... 262
14-307 Commutative algebraic approaches ... 263
14-4 Concluding remarks ... 263

15 GNSS environmental monitoring ... 265
15-1 Satellite environmental monitoring ... 265
15-2 GNSS remote sensing ... 269
15-21 Space borne GNSS meteorology ... 269
15-22 Ground based GNSS meteorology ... 271
15-3 Refraction (bending) angles ... 273
15-31 Transformation of trigonometric equations to algebraic ... 274
15-32 Algebraic determination of bending angles ... 276
15-321 Application of Groebner basis ... 276
15-322 Sylvester resultants solution ... 277
15-4 Algebraic analysis of some CHAMP data ... 278
15-5 Concluding remarks ... 286

16 Algebraic diagnosis of outliers 289
16-1 Outliers in observation samples 289
16-2 Algebraic diagnosis of outliers 290
16-21 Outlier diagnosis in planar ranging 292
16-22 Diagnosis of multipath error in GNSS positioning 296
16-3 Concluding remarks 300

17 Datum transformation problems 303
17-1 The 7-parameter datum transformation and its importance 303
17-2 Algebraic solution of the 7-parameter transformation problem 305
17-21 Groebner basis transformation 305
17-22 Dixon resultant solution 309
17-23 Gauss-Jacobi combinatorial transformation 313
17-3 The 9-parameter (affine) datum transformation 318
17-4 Algebraic solution of the 9-parameter transformation 320
17-41 The 3-point affine transformation problem 320
17-411 Simplifications for the symbolic solution 320
17-412 Symbolic solution with Dixon resultant 321
17-413 Symbolic solution with reduced Groebner basis 323
17-42 The N-points problem 325
17-421 ALESS approach to overdetermined cases 327
17-422 Homotopy solution of the ALESS-determined model 329
17-43 Procrustes solution 333
17-5 Concluding remark 337

Appendix 339
Appendix A-1: Definitions 339
Appendix A-2: C. F. Gauss combinatorial approach 340
Appendix A-3: Linear homotopy 343
Appendix A-4: Determined system of the 9-parameter transformation N point problem 344

References 349

Index 371

1 Introduction

1-1 Motivation

A potential answer to modern challenges faced by geodesists and geoinformatics (see, e.g., Sect. 1-3), lies in the application of *algebraic* computational techniques. The present book provides an in-depth look at algebraic computational methods and combines them with special *local* and *global numerical methods* like the *Extended Newton-Raphson* and the *Homotopy continuation method* to provide smooth and efficient solutions to real life-size problems often encountered in geodesy and geoinformatics, but which cannot be adequately solved by algebraic methods alone.

Algebra has been widely applied in fields such as robotics for kinematic modelling, in engineering for offset surface construction, in computer science for automated theorem proving, and in Computer Aided Design (CAD). The most well-known application of algebra in geodesy could perhaps be the use of Legendre polynomials in spherical harmonic expansion studies. More recent applications of algebra in geodesy are shown in the works of Biagi and Sanso [77], Awange [14], Awange and Grafarend [41], and Lannes and Durand [259], the latter proposing a new approach to differential GPS based on algebraic graph theory.

The present book is divided into two parts. Part I focuses on the algebraic and numerical methods and presents powerful tools for solving algebraic computational problems inherent in geodesy and geoinformatics. The algebraic methods are presented with numerous examples of their applicability in practice. Part I can therefore be skipped by readers with an advanced knowledge in algebraic methods, and who are more interested in the applications of the methods which are presented in part II.

1-2 Modern challenges

In daily geodetic and geoinformatic operations, *nonlinear equations* are encountered in many situations, thus necessitating the need for developing efficient and reliable *computational tools*. Advances in computer technology have also propelled the development of precise and accurate measuring devices capable of collecting large amount of data. Such advances and improvements have brought new challenges to practitioners in fields of geosciences and engineering, which include:

- Handling in an efficient and manageable way the *nonlinear systems of equations* that relate observations to unknowns. These nonlinear systems of equations whose exact (algebraic) solutions have mostly been difficult to solve, e.g., the transformation problem presented in Chap. 17 have been a thorn in the side of users. In cases where the number of observations n and the number of unknowns m are equal, i.e., $n = m$, the unknown parameters may be obtained by solving explicitly (in a closed form) nonlinear systems of equations.

J.L. Awange et al., *Algebraic Geodesy and Geoinformatics*, 2nd ed.,
DOI 10.1007/978-3-642-12124-1_1, © Springer-Verlag Berlin Heidelberg 2010

Because of the difficulty in practise of obtaining reliable closed form procedures, approximate numerical methods have been adopted. Such procedures depend on some *approximate* starting values, *linearization* and *iterations.* In some cases, the numerical methods used are unstable or the iterations fail to converge, depending on the initial "guess" [334, pp. 340–342]. The other shortcoming of approximate numerical procedures has been pointed out by Cox et al. [118, pp. 28–32], who in their book have shown that systems of equations with exact solutions become vulnerable to small errors introduced during the process of establishing the roots. In the case of extending the partial solution to the complete solutions of the system, errors may accumulate and thus become very large (ill-conditioned problems). If the partial solution was derived by iterative procedures, then the errors incurred during the root-finding may blow up during the extension of the partial solution to the complete solution (back substitution). There exists therefore a strong need for unified procedures that can be applied in general to offer exact solutions to nonlinear systems of equations.

- Managing large amounts of data. In GPS meteorology for example, more than 1000 satellite occultations are obtained on a daily basis, from which the bending angles of the signals are to be computed. In practice, the nonlinear system of equations for bending angles is often solved using numerical approaches such as Newton's method iteratively. An explicit formula could, however, be derived from the nonlinear system of equations, as presented in Chap. 15.
- Obtaining a unified closed form solution (e.g., Awange et al. [39]) for different problems. For a particular problem, several procedures are often put forward in an attempt to offer exact solutions. The GPS pseudo-range problem for example, has attracted several exact solution procedures, as outlined in the works of [53, 181, 242, 243, 272, 363]. It is desirable in such a case to have a unified solution approach which can easily be applied to all problems in general.
- Controlling approximate numerical algorithms that are widely used (see, e.g., Awange [36]).
- Obtaining computational procedures that are time saving.
- Having computational procedures that do not peg their operations on approximate starting values, linearization or iterations.
- Taking advantage of the large storage capacity and fast speed of modern computers to solve problems which have hitherto evaded solution.
- Prove the validity of theorems and formulae that are in use, which were derived based on a trial and error basis.
- Perform rigorous analysis of the nonlinearity effects on most models that are in operation, but assume or ignore nonlinearity.

1-3 Facing the challenges

These challenges and many others had existed before, and earlier researchers had acknowledged the fact and realized the need for addressing them through the development of explicit solutions. Merritt [301] had, for example, listed the advantages of explicit solutions as;

1. the provision of satisfaction to the users (photogrammetrists and mathematicians) of the methods,
2. the provision of data tools for checking the iterative methods,
3. the desire by geodesists whose task of control network densification does not favour iterative procedures,
4. the provision of solace and,
5. the requirement of explicit solutions rather than iterative by some applications.

Even though such advantages had been noted, their actual realization was out of reach as the equations involved were large and required more than a pen and paper to solve. Besides, another drawback was that these exact solutions were like a rare jewel. The reason for this was partly because the methods required extensive computations and partly because the resulting symbolic expressions were too large and required computers with large storage capacity. Until recently, computers that were available could hardly handle large computations due to the lack of sufficiently fast Central Processing Unit (CPU), shortage of Random Access Memory (RAM) and limited hard disk storage capacity. The other setback was that some of the methods, especially those from algebraic fields, were formulated based on theoretical concepts that were hard to realize or comprehend without the help of computers. For a long time therefore, these setbacks hampered progress in the development of explicit procedures. The painstaking efforts to obtain exact solutions discouraged practitioners to the extent that the use of numerical approaches were the order of the day. Most of these numerical approaches had no independent methods for validation, while other problems evaded numerical solutions and required closed form solutions.

In some applications, explicit formulae rather than numerical solutions are desirable. In such cases, explicit procedures are usually employed. The resulting explicit formulae often consist of *univariate polynomials* relating the unknown parameters (unknown variables) to the known variables (observations). By inserting numeric values into these explicit formulae, solutions can immediately be computed for the unknown variables. In order to gain a deeper understand of this discussion, let us consider a case where students have been asked to integrate the function $f(x) = x^5$ with respect to x. In this case, the power of x, i.e., 5 is definite and the integration can easily be performed. Assume now that for a specific purpose, the power of x can be varied, taking on different values say $n = 1, 2, 3, \ldots$. In such a case, it is not prudent to integrate x raised to each power, but to seek a general explicit formula by integrating

$$\int x^n \mathrm{d}x, \tag{1.1}$$

to give

$$\frac{x^{n+1}}{n+1}. \tag{1.2}$$

One thereafter needs only to insert a given value of n in (1.2) to obtain a solution. In practice, several problems require explicit formulae as they are performed repeatedly.

Besides the requirement of exact solutions by some applications, there also exists the problem of exact solutions of overdetermined systems (i.e., where more observations than unknown exist). In reality, field measurements often result in more data being collected than is required to determine the unknown parameters, with exact solutions to the overdetermined problems being just one of the challenges faced. In some applications, such as the 7-parameter datum transformation discussed in Chap. 17, where coordinates have to be transformed from local coordinate systems to the global coordinate system and vice versa, the handling of stochasticities of these systems still poses a serious challenge to users. Approximate numerical procedures which are applied in practice do not offer a tangible solution to this problem. Other than the stochasticity issues, numerical methods employed to solve the 7-parameter datum transformation problem require some initial starting values, linearization and iterations as already mentioned. In Photogrammetry, where the rotation angles are very small, the initial starting values are often set to zero. This, unfortunately, may not be the case for other applications in geosciences. In Chap. 9 we present powerful analytical and algebraic techniques developed from the fields of multidimensional scaling and abstract algebra to solve the problem. In particular, the *Procrustes algorithm*, which enjoys wide use in fields such as medicine and sociology, is straightforward and easy to program. The advantages of these techniques are; the non-requirements of the conditions that underpin approximate numerical solutions, and their capability to take into consideration weights of the systems involved.

The solution of unknown parameters from nonlinear observational equations are only meaningful if the observations themselves are pure and uncontaminated by gross errors (outliers). This raises the issue of outlier detection and management. Traditionally, statistical procedures have been put forward for detecting outliers in observational data sample. Once detected, the observations that are considered to be outliers are isolated and the remaining pure observations used to estimate the unknown parameters. Huber [233] and Hampel et al [205], however, point out the dangers that exist in such an approach, namely false rejection of otherwise good observations, and the false retention of contaminated observations. To circumvent these dangers, robust estimation procedures were proposed in 1964 by the father of robust statistics, P. J. Huber [231] to manage outliers without necessarily rejecting outlying observations. Since then, as we shall see in Chap. 16, several contributions to outlier management using robust techniques have been put forward. Chapter 16 deviates from the statistical approaches to present an algebraic outlier diagnosis tool that enjoys the advantages already discussed.

On the instrumentation front, there has been a tremendous improvement in computer technology. Today's laptops are made with large storage capacity with high memory, thus enabling faster computations. The wide spread availability of multicore processors as well as some Computer Algebraic System (CAS) such as *Mathematica* and *MATLAB* offer the possibility for parallel computing without any special knowledge. Problems can now be solved using algebraic methods that would have been impossible to solve by hand.

1-4 Concluding remarks

This book covers both algebraic ("exact") and numerical ("approximate") methods, therefore presents modern and efficient techniques for solving geodetic and geoinformatics algebraic problems, with the aim of meeting the challenges addressed above. Examples are illustrated using Mathematica software package to demonstrate the computer algebra techniques of Groebner basis and resultants. Global and local numerical solution methods e.g., extended Newton-Raphson and homotopy, are also presented. An accompanying CD-ROM is included which contains Mathematica notebooks with computational illustrations and application packages to solve real-life problems in geodesy and geoinformatics, such as resection, intersection, and orientation problems.

Part I

Algebraic symbolic and numeric methods

2 Basics of ring theory

2-1 Some applications to geodesy and geoinformatics

This chapter presents the concepts of *ring theory* from a geodetic and geoinformatics perspective. The presentation is such that the mathematical formulations are augmented with examples from the two fields. Ring theory forms the basis upon which polynomial rings operate. As we shall see later, exact solution of nonlinear systems of equations are pinned to the operations on polynomial rings. In Chap. 3, polynomials will be discussed in detail. In order to understand the concept of polynomial rings, one needs first to be familiar with the basics of ring theory. This chapter is therefore a preparation for the understanding of the polynomial rings presented in Chap. 3. Ring of numbers which is presented in Sect. 2-2 plays a significant role in daily operations. They permit operations addition, subtraction, multiplication and division of numbers. For those engaged in data collection, ring of numbers play the following role;

- they specify the number of sets of observations to be collected,
- they specify the number of observations or measurements per set,
- they enable manipulation of these measurements to determine the unknown parameters.

We start by presenting *ring of numbers*. Elementary introduction of the sets of *natural numbers*, *integers*, *rational numbers*, *real numbers*, *complex numbers* and *quaternions* are first given before defining the ring. We strive to be as simple as possible so as to make the concepts clear to readers with less or no knowledge of rings.

2-2 Numbers from operational perspective

When undertaking operations such as measurements of angles, distances, gravity, photo coordinates, digitizing of points etc., numbers are often used. Measured values are normally assigned numbers. A measured distance for example can be assigned a value of 100 m to indicate the length. Numbers, e.g., $1, 2, ...$, also find use as;

- counters to indicate the frequency of taking measurements,
- counters indicating the number of points to be observed or,
- passwords to;
  - the processing hardware (e.g. computers),
  - softwares (such as those of Geographical Information Systems (GIS) packages) and,
  - accessing pin numbers in the bank!

J.L. Awange et al., *Algebraic Geodesy and Geoinformatics*, 2nd ed.,
DOI 10.1007/978-3-642-12124-1_2, © Springer-Verlag Berlin Heidelberg 2010

In all these cases, one operates on a set of *natural numbers*

$$\mathbb{N} = \{0, 1, 2, ...\}, \tag{2.1}$$

with 0 added. The number 0 was invented by the Babylonians in the third century B.C.E, re-invented by the Mayans in the fourth century C.E and in India in the fifth century [235, p. 69]. The set $\mathbb{N}$ in (2.1) is closed under;

- addition, in which case the sum of two numbers is also a natural number (e.g., $3+6=9$) and,
- multiplication, in which case the product of two numbers is a natural number (e.g., $3 \times 6 = 18$).

Subtraction, i.e., the difference of two natural numbers is however not necessarily a natural number (e.g. $3-6=-3$). To circumvent the failure of the natural numbers to be closed under subtraction, negative numbers were introduced and added in front of natural numbers. For a natural number n for example, $-n$ is written. This expanded set

$$\mathbb{Z} = \{-2, -1, 0, 1, 2, ...\}, \tag{2.2}$$

is the set of *integers*. The letter $\mathbb{Z}$ is adopted from the first letter of the German word for integers "**Z**ahl". The set $\mathbb{Z}$ is said to have:

- an "additive identity" number 0 which when added to any integer n preserves the "identity" of n, e.g., $0+13=13$,
- "additive inverse" $-n$ which when added to an integer n results in an identity 0, e.g., $-13+13=0$. The number -13 is an additive inverse of 13.

The set $\mathbb{Z}$ with the properties "addition" and "additive inverse" enables one to manipulate numbers by being able to add and subtract. This is particularly helpful when handling measured values. It allows for instance the solution of equations of type $y+m=0$, where m is an integer. In-order to allow them to divide numbers as is the case with distance ratio observations, "multiplicative identity" and "inverse" have to be specified as:

- "multiplicative identity" is the integer 1 which when multiplied with any integer n preserves the "identity" of n, e.g., $1 \times 13 = 13$,
- "multiplicative inverse" is an integer m such that its multiplication with an integer n results in an identity 1, e.g., $m \times n = 1$.

For a non-zero integer n, therefore, a multiplicative inverse $\frac{1}{n}$ has to be specified. The multiplicative inverse of 5 for example is $\frac{1}{5}$. This leads to an expanded set comprising of both integers and their multiplicative inverses as

$$\mathbb{Q} = \{-2, -\frac{1}{2}, -1, 0, 1, 2, \frac{1}{2}, ...\}, \tag{2.3}$$

where a new number has been created for each number except $-1, 0, 1$. Except for 0, which is a special case, the set $\mathbb{Q}$ is closed under "additive" and "multiplicative inverses" but not "addition" and "multiplication". This is circumvented by incorporating all products of integers and multiplicative inverses $m \times \frac{1}{n} = \frac{m}{n}$, which are ratios of integers resulting into a set of *rational numbers* $\mathbb{Q}$. $\mathbb{Q}$ is the first letter of Quotient and is closed since:

- For every rational number, there exist an additive inverse which is also a rational number, e.g., $-\frac{1}{13} + \frac{1}{13} = 0$.
- Every rational number except 0 has a multiplicative inverse which is also a rational number, e.g., $13 \times \frac{1}{13} = 1$.
- The set of rational numbers is closed under addition and multiplication, e.g., $\frac{1}{3} + \frac{1}{3} = \frac{2}{3}$ and $\frac{1}{3} \times \frac{1}{3} = \frac{1}{9}$.

The set $\mathbb{Q}$ is suitable as it permits addition, subtraction, multiplication and division. It therefore enables the solution of equations of the form $ny - m = 0$, $\{m, n\}$ being arbitrary integers, with $n \neq 0$. This set is however not large enough as it leaves out the square root of numbers and thus cannot measure the Pythagorean length. In geodesy, as well as geoinformatics, the computation of distances from station coordinates by Pythagoras demands the use of square root of numbers. The set of quotient $\mathbb{Q}$ is thus enlarged to the set of real numbers $\mathbb{R}$, where the positive real numbers are the ones required to measure distances as shall be seen in Chaps. 12, 13 and 14. Negative real numbers are included to provide additive inverses. The set $\mathbb{R}$ also possesses multiplicative inverses. This set enables the solution of equations of the form $y^2 - 3 = 0 \Rightarrow y = \pm\sqrt{3}$, which is neither integer nor rational. The set $\mathbb{R}$ is however not large enough to provide a solution to an equation of the form $y^2 + 1 = 0$. It therefore gives way to the set $\mathbb{C}$ of complex numbers, where $\mathrm{i}^2 = -1$.

The set $\mathbb{C}$ can be expanded further into a set $\mathbb{H}$ of *quaternions* which was discovered by W. R. Hamilton on the 16th of October 1843, having worked on the problem for 13 years (see Note 2.1 on p. 12). Even as he discovered the quaternions, it occurred to him that indeed Euler had known of the existence of the four square identity in 1748 and that quaternion multiplication had been used by Rodrigues in 1840 to compute the product of rotations in $\mathbb{R}^3$ [371]. Indeed as we shall see in Chap. 7, Gauss knew of quaternions even before Hamilton, but unfortunately, he never published his work. In geodesy and geoinformatics, quaternions have been used to solve the three-dimensional resection problem by [187]. They have also found use in the solution of the similarity transformation problem discussed in Chap. 17 as evidenced in the works of [362, 382, 383, 434]. Quaternion is defined as the matrix

$$\begin{bmatrix} (a + d\mathrm{i}) & (b + c\mathrm{i}) \\ (-b + c\mathrm{i}) & (a - d\mathrm{i}) \end{bmatrix} \mid \{a, b, c, d\} \in \mathbb{R}, \tag{2.4}$$

which is expressed in terms of unit matrices $\mathbf{1}, \mathbf{i}, \mathbf{j}, \mathbf{k}$ as

$$\left[\begin{array}{l}\begin{bmatrix} a+di & b+ci \\ -b+ci & a-di \end{bmatrix} = a\begin{bmatrix} 1 & 0 \\ 0 & 1 \end{bmatrix} + b\begin{bmatrix} 0 & 1 \\ -1 & 0 \end{bmatrix} + c\begin{bmatrix} 0 & i \\ i & 0 \end{bmatrix} + d\begin{bmatrix} i & 0 \\ 0 & -i \end{bmatrix} \\ \qquad\qquad a\mathbf{1} + b\mathbf{i} + c\mathbf{j} + d\mathbf{k}, \end{array}\right. \tag{2.5}$$

where $\mathbf{1}, \mathbf{i}, \mathbf{j}, \mathbf{k}$ are quaternions of norm 1 that satisfy

$$\left[\begin{array}{l} \mathbf{i}^2 = \mathbf{j}^2 = \mathbf{k}^2 = -\mathbf{1} \\ \quad \mathbf{ij} = \mathbf{k} = -\mathbf{ji} \\ \quad \mathbf{jk} = \mathbf{i} = -\mathbf{kj} \\ \quad \mathbf{ki} = \mathbf{j} = -\mathbf{ik}. \end{array}\right. \tag{2.6}$$

The norm of the quaternions is the determinant of the matrix (2.4) and gives

$$det \begin{bmatrix} (a+di) & (b+ci) \\ (-b+ci) & (a-di) \end{bmatrix} = a^2 + b^2 + c^2 + d^2, \tag{2.7}$$

which is a four square identity. This matrix definition is due to Cayley, while Hamilton wrote the rule $\mathbf{i}^2 = \mathbf{j}^2 = \mathbf{k}^2 = \mathbf{ijk} = -\mathbf{1}$ that define quaternion multiplication from which he derived the four square identity [371, p. 156].

We complete this section by defining algebraic integers as

Definition 2.1 (Algebraic). *A number $n \in \mathbb{C}$ is algebraic if*

$$a_n\alpha^n + a_{n-1}\alpha^{n-1} + ... + a_1\alpha + a_0 = 0, \tag{2.8}$$

and it takes on the degree n if it satisfies no such equation of lower degree and $a_0, a_1, ..., a_n \in \mathbb{Z}$.

We shall see in Chap. 3 that Definition (2.1) satisfies the definition of a univariate polynomial.

Note 2.1 (Hamilton's Letter). How can one dream about such a "quaternion algebra" $\mathbb{H}$? *W.* R. Hamilton (16th October 1843) invented *q*uaternion numbers as outlined in a letter (1865) to his son *A.* H. Hamilton for the following reason:

> "If I may be allowed to speak of *myself* in connection with the subject, I might do so in a way which would bring *you* in, by referring to an *antequaternionic* time, when you were a mere *child*, but had caught from me the conception of a vector, as represented by a *triplet*; and indeed I happen to be able to put the finger of memory upon the year and month –October, 1843– when having recently returned from visits to Cork and Parsonstown, connected with a meeting of the British Association, the desire to discover the laws of the multiplication referred to regained with me a certain strength and earnestness, which had for years been dormant, but was then on the point of being gratified, and was occasionally talked of with you. Every morning in the early part of the above cited month, on my coming down to breakfast, your (then) little brother William Edwin, and yourself, used to ask me, "well, Papa, can you *multiply* triplets"? Whereto I was

always obliged to reply, with a sad shake of the head: "No, I can only *add* and *subtract* them."

But on the 16th day of the same month – which happened to be a Monday, and a Council day of the Royal Irish Academy – I was walking in to attend and preside, and your mother was walking with me, along the Royal Canal, to which she had perhaps driven; and although she talked with me now and then, yet an *under-current* of thought was going on in my mind, which gave at last a *result*, whereof it is not too much to say that I felt *at once* the importance. An *electric* circuit seemed to *close*; and a spark flashed forth. The herald (as I *foresaw, immediately*) of many long years to come of definitely directed thought and work, by *myself* if spared, and at all events on the part of *others*, if should even be allowed to live long enough distinctly to communicate the discovery. Nor could I resist the impulse – unphilosophical as it may have been – to cut with a knife on a stone of Brougham Bridge, as we passed it, the fundamental formula with the symbols, i, j, k; namely

$$\mathbf{i}^2 = \mathbf{j}^2 = \mathbf{k}^2 = \mathbf{ijk} = \mathbf{-1},$$

which contains the *solution of the problem*, but of course, as an inscription, has long since mouldered away. A more durable notice remains, however, on the Council Books of the Academy for that day (October 16th, 1843), which records the fact, that I then asked for and obtained base to read a paper on *quaternion*, at the *First General Meeting* of the Session: which reading took place accordingly, on Monday the 13th of the November following."

2-3 Number rings

In everyday lives of geodesists and geoinformatists, rings are used albeit without being noticed: A silent tool without which perhaps they might find the going tough. In the preceding section, the sets of integers $\mathbb{Z}$, rational numbers $\mathbb{Q}$, real numbers $\mathbb{R}$ and complex numbers $\mathbb{C}$ were introduced as being closed under addition and multiplication. Loosely speaking, a system of numbers that is closed under addition and multiplication is a ring. A more precise definition of a ring based on linear algebra will be given later.

It suffices at this point to think of the sets $\mathbb{Z}$, $\mathbb{Q}$, $\mathbb{R}$ and $\mathbb{C}$, upon which we manipulate numbers, as being a collection of numbers that can be added, multiplied, have additive identity 0 and multiplicative identity 1. In addition, every number in these sets has an additive inverse thus forming a ring. Measurements of distances, angles, directions, photo coordinates, gravity etc., comprise the set $\mathbb{R}$ of real numbers. This set as we saw earlier is closed under addition and multiplication. Its elements were seen to possess additive and multiplicative identities, and also additive inverses, thus qualifying to be a ring.

In algebra books, one often encounters the term *field* which seems somewhat confusing with the term *ring*. In the brief outline of the number ring above, whereas the sets $\mathbb{Z}$, $\mathbb{Q}$, $\mathbb{R}$ and $\mathbb{C}$ qualified as rings, the set $\mathbb{N}$ of natural numbers failed as it lacked additive inverse. The sets $\mathbb{Q}$, $\mathbb{R}$ and $\mathbb{C}$ also have an additional property that every number $n \neq 0$ in the ring has a multiplicative inverse. A ring in which every $n \neq 0$ has a multiplicative inverse is called a field. The set $\mathbb{Z}$ therefore is not a field as it does not have multiplicative inverse. In this book, the terms ring and field will be used interchangeably to refer to the sets $\mathbb{Q}$, $\mathbb{R}$ and $\mathbb{C}$ which qualify both as rings and as fields.

A curious reader will note that the term number ring was selected as the heading for this section and used in the discussion. This is because we have several other types of rings that do not use numbers as objects. In our examples, we used numbers to clarify closeness under addition and multiplication. We will see later in Chap. 3 that polynomials, which are objects and not numbers, also qualify as rings. For daily measurements and manipulation of observations, number rings and polynomial rings suffices. Other forms of rings such as fruit rings, modular arithmetic rings and congruence rings are elaborately presented in algebra books such as [235] and [298]. In-order to give a precise definition of a ring, we begin by considering the definition of *linear algebra*. Detailed treatment of linear algebra is presented in [63, 64, 303, 385].

Definition 2.2 (Linear algebra). *Algebra can be defined as a set S of elements and a finite set M of operations. In linear algebra the elements of the set S are vectors over the field $\mathbb{R}$ of real numbers, while the set M is basically made up of two elements of internal relation namely "additive" and "multiplicative". An additional definition of the external relation expounds on the term linear algebra as follows: A linear algebra over the field of real numbers $\mathbb{R}$ consists of a set R of objects, two internal relation elements (either "additive" or "multiplicative") and one external relation as follows:*

$$
\begin{array}{l}
(opera)_1 =: \alpha : R \times R \rightarrow R \\
(opera)_2 =: \beta : \mathbb{R} \times R \rightarrow R \, or \, R \times \mathbb{R} \rightarrow R \\
(opera)_3 =: \gamma : R \times R \rightarrow R.
\end{array}
$$

The three cases are outlined as follows:
** With respect to the internal relation α ("join"), R as a linear space in a vector space over $\mathbb{R}$, an Abelian group written "additively" or "multiplicatively":*

$$\mathbf{a}, \mathbf{b}, \mathbf{c} \in R$$

	Axiom	"Additively" written Abelian group	"Multiplicatively" written Abelian group
		$\alpha(\mathbf{a},\mathbf{b}) =: \mathbf{a}+\mathbf{b}$	$\alpha(\mathbf{a},\mathbf{b}) =: \mathbf{a}\circ\mathbf{b}$
1	Associativity	$G1+ : (\mathbf{a}+\mathbf{b})+\mathbf{c} = = \mathbf{a}+(\mathbf{b}+\mathbf{c})$ (additive assoc.)	$G1\circ : (\mathbf{a}\circ\mathbf{b})\circ\mathbf{c} = = \mathbf{a}\circ(\mathbf{b}\circ\mathbf{c})$ (multiplicative assoc.)
2	Identity	$G2+ : \mathbf{a}+\mathbf{0} = \mathbf{a}$ (additive identity, neutral element)	$G2\circ : \mathbf{a}\circ\mathbf{1} = \mathbf{a}$ (multiplicative identity neutral element)
3	Inverse	$G3+ : \mathbf{a}+(-\mathbf{a}) = \mathbf{0}$ (additive inverse)	$G3\circ : \mathbf{a}\circ\mathbf{a}^{-1} = \mathbf{1}$ (multiplicative inverse)
4	Commutativity	$G4+ : \mathbf{a}+\mathbf{b} = \mathbf{b}+\mathbf{a}$ (additive commutativity, Abelian axiom)	$G4\circ : \mathbf{a}\circ\mathbf{b} = \mathbf{b}\circ\mathbf{a}$ (multiplicative comm., Abelian axiom)

with the triplet of axioms $\{G1+, G2+, G3+\}$ or $\{G1\circ, G2\circ, G3\circ\}$ constituting the set of *group axioms* and $\{G4+, G4\circ\}$ the *Abelian axioms*. Examples of groups include:

1. The group of integer $\mathbb{Z}$ under addition.
2. The group of non-zero rational number $\mathbb{Q}$ under multiplication.
3. The set of rotation about the origin in the Euclidean plane under the operation of composite function.

* With respect to the external relation β the following compatibility conditions are satisfied

$$\mathbf{a}, \mathbf{b} \in R,\ t, u \in \mathbb{R}$$
$$\beta(t, \mathbf{a}) =: t \times \mathbf{a}$$

1 distr. $D1+ : t \times (\mathbf{a} + \mathbf{b}) = (\mathbf{a} + \mathbf{b}) \times t =$ $= t \times \mathbf{a} + t \times \mathbf{b} = \mathbf{a} \times t + \mathbf{b} \times t$	$D1\circ : t \times (\mathbf{a} \circ \mathbf{b}) = (\mathbf{a} \circ \mathbf{b}) \times t$ $= (t \times \mathbf{a}) \circ \mathbf{b} = \mathbf{a} \circ (\mathbf{b} \times t)$
1st additive distributivity	1st multiplicative distributivity
2 distr. $D2+ : (t + u) \times \mathbf{a} = \mathbf{a} \times (t + u) =$ $= t \times \mathbf{a} + u \times \mathbf{a} = \mathbf{a} \times t + \mathbf{a} \times u$	$D2\circ : (t \circ u) \times \mathbf{a} = \mathbf{a} \times (t \circ u)$ $= t \circ (u \times \mathbf{a}) = (\mathbf{a} \times t) \circ u$
2nd additive distributivity	2nd multiplicative distributivity

$$D3 : 1 \times \mathbf{a} = \mathbf{a} \times 1 = \mathbf{a}\,(left\ and\ right\ identity)$$

* With respect to the *internal relation* γ (*"meet"*) the following conditions are satisfied

$$\mathbf{a}, \mathbf{b}, \mathbf{c} \in R,\ t \in \mathbb{R}$$
$$\gamma(\mathbf{a}, \mathbf{b}) =: \mathbf{a} * \mathbf{b}$$

Axiom			Comments
1	Ass.	$G1* : (\mathbf{a} * \mathbf{b}) * \mathbf{c} = \mathbf{a} * (\mathbf{b} * \mathbf{c})$	Associativity w.r.t internal multiplication
1	dist.	$D1 * +; \mathbf{a} * (\mathbf{b} + \mathbf{c}) = \mathbf{a} * \mathbf{b} + \mathbf{a} * \mathbf{c}$ $(\mathbf{a} + \mathbf{b}) * \mathbf{c} = \mathbf{a} * \mathbf{c} + \mathbf{b} * \mathbf{c}$	Left and Right additive dist. w.r.t internal multiplication
1	dist.	$D1 * \circ; \mathbf{a} * (\mathbf{b} \circ \mathbf{c}) = (\mathbf{a} * \mathbf{b}) \circ \mathbf{c}$ $(\mathbf{a} \circ \mathbf{b}) * \mathbf{c} = \mathbf{a} \circ (\mathbf{b} * \mathbf{c})$	left and right multiplicative dist. w.r.t internal multiplication
2	dist.	$D2 * \times; t \times (\mathbf{a} * \mathbf{b}) = (t \times \mathbf{a}) * \mathbf{b}$ $(\mathbf{a} * \mathbf{b}) \times t = \mathbf{a} * (\mathbf{b} \times t)$	left and right dist. of internal and external multiplication

Definition 2.3 (Ring). *A sub-algebra is called a ring with identity if the following two conditions encompassing (seven conditions) hold:*
(a) The set R is an Abelian group with respect to addition, i.e., four conditions $\{G1+, G2+, G3+, G4+\}$ of Abelian group hold.

(b) The set R is a semi-group with respect to multiplication; that is, $\{G1*, G2*\}$ holds. In other words, the set R comprises a monoid (i.e. a set with two operations, associativity and identity with respect to multiplication $(*)$). The last condition is the left and right additive distributivity with respect to internal multiplication $\{D1*+\}$ which connects the Abelian group and the monoid. *In total the four conditions forming the Abelian group (a) and the three forming the semi-group in (b) add up to form seven conditions enclosed in a ring in Fig. 2.1.*

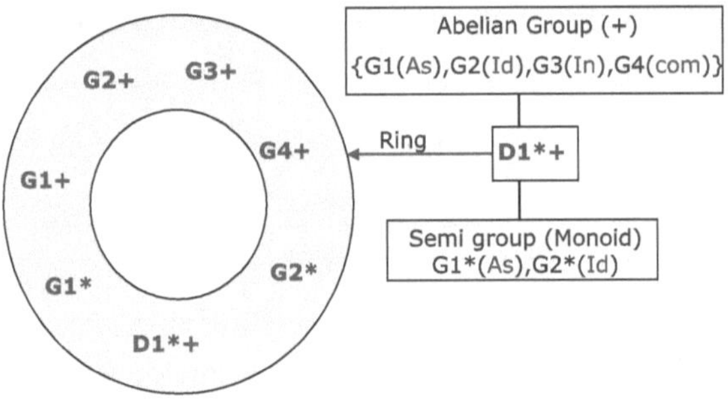

Figure 2.1. Ring

Condition $G2*$ makes R a *"ring with identity"*, while the inclusion of $G3*$ makes the ring be known as the *"division ring"* if every non-zero element of the ring has a multiplicative inverse. The ring becomes a *"commutative ring"* If it has the commutative multiplicative $G4*$. Examples of *rings* include:

- Field k of real numbers $\mathbb{R}$, complex numbers $\mathbb{C}$ and rational numbers $\mathbb{Q}$. In particular, a *ring* becomes a field if every non zero element of the *ring* has a multiplicative inverse as already discussed.
- Integers $\mathbb{Z}$.
- The set $\mathbb{H}$ of quaternions that are non commutative.
- Polynomial function P in n variables over a ring R expressed as $P = R[x_1, \ldots, x_n]$.

For the solution of algebraic computational problems in geodesy and geoinformatics, it suffices to consider a ring as being commutative and to include identity element.

2-4 Concluding remarks

The concept of numbers and ring of numbers have been presented from a geodetic and geoinformatics perspective. In the next chapter, the number ring will provide the framework for discussing polynomial rings, the main algebraic tool that permits the solution of nonlinear systems of equations. The basics of ring algebra discussed provides fundamentals required to understand the materials that will be presented in latter chapters. For more detailed coverage of rings, we refer to [258].

3 Basics of polynomial theory

3-1 Polynomial equations

In geodesy and geoinformatics, most observations are related to unknowns parameters through equations of algebraic (polynomial) type. In cases where the observations are not of polynomial type, as exemplified by the GPS meteorology problem of Chap. 15, they are converted via *Theorem* 3.1 on p. 19 into polynomials. The unknown parameters are then be obtained by solving the resulting polynomial equations. Such solutions are only possible through application of operations addition and multiplication on polynomials which form elements of polynomial rings. This chapter discusses polynomials and the properties that characterize them. Starting from the definitions of *monomials*, basic polynomial aspects that are relevant for daily operations are presented. A *monomial* is defined as

Definition 3.1 (Monomial). *A monomial is a multivariate product of the form* $x_1^{\alpha_1}x_2^{\alpha_2}\ldots x_n^{\alpha_n}$, $(\alpha_1, \ldots, \alpha_n) \in \mathbb{Z}_+^n$ *in the variables* $x_1, \ldots, x_n$.

In Definition 3.1 above, the set $\mathbb{Z}_+^n$ comprises positive elements of the set of integers (2.2) that we saw in Chap. 2, p. 10.

Example 3.1 (Monomial). Consider the system of equations for solving distances in the three-dimensional resection problem given as (see, e.g., (13.44) on p. 230)

$$\begin{bmatrix} x_1^2 + 2a_{12}x_1x_2 + x_2^2 + a_o = 0 \\ x_2^2 + 2b_{23}x_2x_3 + x_3^2 + b_o = 0 \\ x_3^2 + 2c_{31}x_3x_1 + x_1^2 + c_o = 0 \\ where \;\; x_1 \in \mathbb{R}^+,\, x_2 \in \mathbb{R}^+, x_3 \in \mathbb{R}^+. \end{bmatrix}$$

The variables $\{x_1, x_2, x_3\}$ are unknowns while the other terms are known constants. The products of variables $\{x_1^2, x_1x_2, x_2^2, x_2x_3, x_3^2, x_3x_1\}$ are monomials in $\{x_1, x_2, x_3\}$.

Summation of monomials form polynomials defined as

Definition 3.2 (Polynomial). *A polynomial* $f \in k[x_1, \ldots, x_n]$ *in variables* $x_1, \ldots, x_n$ *with coefficients in the field k is a finite linear combination of monomials with pairwise different terms expressed as*

$$f = \sum_{\alpha} a_\alpha x^\alpha, \quad a_\alpha \in k, \quad x^\alpha = (x^{\alpha_1}, \ldots, x^{\alpha_n}),\; \alpha = (\alpha_1, \ldots, \alpha_n), \tag{3.1}$$

where a_α *are coefficients in the field k, e.g.,* $\mathbb{R}$ *or* $\mathbb{C}$ *and* x^α *the monomials.*

Example 3.2 (Polynomials). Equations

$$\begin{bmatrix} x_1^2 + 2a_{12}x_1x_2 + x_2^2 + a_o = 0 \\ x_2^2 + 2b_{23}x_2x_3 + x_3^2 + b_o = 0 \\ x_3^2 + 2c_{31}x_3x_1 + x_1^2 + c_o = 0, \end{bmatrix}$$

J.L. Awange et al., *Algebraic Geodesy and Geoinformatics*, 2nd ed.,
DOI 10.1007/978-3-642-12124-1_3,

in *Example* 3.1 are *multivariate polynomials.* The first expression is a multivariate polynomial in two variables $\{x_1, x_2\}$ and a linear combination of monomials $\{x_1^2, x_1x_2, x_2^2\}$. The second expression is a *multivariate polynomial* in two variables $\{x_2, x_3\}$ and a linear combination of the monomials $\{x_2^2, x_2x_3, x_3^2\}$, while the third expression is a *multivariate polynomial* in two variables $\{x_3, x_1\}$ and a linear combination of the monomials $\{x_3^2, x_3x_1, x_1^2\}$.

In *Example* 3.2, the coefficients of the polynomials are elements of the set $\mathbb{Z}$. In general, the coefficients can take on any sets $\mathbb{Q}$, $\mathbb{R}$, $\mathbb{C}$ of number rings or other rings such as modular arithmetic rings. These coefficients can be added, subtracted, multiplied or divided, and as such play a key role in determining the solutions of polynomial equations. The definition of the set to which the coefficients belong determines whether a polynomial equation is solvable or not. Consider the following example:

Example 3.3. Given an equation $9w^2 - 1 = 0$ with the coefficients in the integral domain, obtain the integer solutions. Since the coefficient $9 \in \mathbb{Z}$, the equation does not have a solution. If instead the coefficient $9 \in \mathbb{Q}$, then the solution $w = \pm\frac{1}{3}$ exist.

From *Definition* 2.1 of algebraic, polynomials become algebraic once (3.1) is equated to 0. The fundamental problem of algebra can thus be stated as the solution of equations of form (3.1) equated to 0.

3-2 Polynomial rings

In Sect. 2-3 of Chap. 2, the theory of rings was introduced with respect to numbers. Apart from the number rings, polynomials are objects that also satisfy ring axioms leading to *"polynomial rings"* upon which operations *"addition"* and *"multiplication"* are implemented.

3-21 Polynomial objects as rings

Polynomial rings are defined as

Definition 3.3 (Polynomial ring). *Consider a ring R say of real numbers $\mathbb{R}$. Given a variable $x \notin R$, a univariate polynomial $f(x)$ is formed (see Definition 3.2 on p. 17) by assigning coefficients $a_i \in R$ to the variable and obtaining summation over finite number of distinct integers. Thus*

$$f(x) = \sum_{\alpha} c_\alpha x^\alpha, c_\alpha \in R, \alpha \geq 0$$

is said to be a univariate polynomial over R . If two polynomials are given such that $f_1(x) = \sum_i c_i x^i$ and $f_2(x) = \sum_j d_j x^j$, then two binary operations "addition" *and "multiplication" can be defined on these polynomials such that:*

(a) Addition: $f_1(x) + f_2(x) = \sum_k e_k x^k, e_k = c_k + d_k, e_k \in R$

(b) Multiplication: $f_1(x).f_2(x) = \sum_k g_k x^k, g_k = \sum_{i+j=k} c_i d_j, g_k \in R.$

A collection of polynomials with these "additive" and "multiplicative" rules form a commutative ring with zero element and identity 1. A univariate polynomial $f(x)$ obtained by assigning elements c_i belonging to the ring R to the variable x is called a polynomial ring and is expressed as $f(x) = R[x]$. In general the entire collection of all polynomials in $x_1, \ldots, x_n$, with coefficients in the field k that satisfy the definition of a ring above are called a polynomial rings.

Designated P, polynomial rings are represented by n unknown variables x_i over k expressed as $P := k[x_1, \ldots, x_n]$. Its elements are polynomials known as *univariate* when $n = 1$ and *multivariate* otherwise. The distinction between a polynomial ring and a polynomial is that the latter is the sum of a finite set of monomials (see e.g., Definition 3.1 on p. 17) and is an element of the former.

Example 3.4. Equations

$$\begin{bmatrix} x_1^2 + 2a_{12}x_1x_2 + x_2^2 + a_o = 0 \\ x_2^2 + 2b_{23}x_2x_3 + x_3^2 + b_o = 0 \\ x_3^2 + 2c_{31}x_3x_1 + x_1^2 + c_o = 0 \end{bmatrix}$$

of *Example* 3.1 are said to be polynomials in three variables $[x_1, x_2, x_3]$ forming elements of the polynomial ring P over the field of real numbers $\mathbb{R}$ expressed as $P := \mathbb{R}[x_1, x_2, x_3]$.

Polynomials that we use in solving unknown parameters in various problems, as we shall see later, form elements of polynomial rings. Polynomial rings provide means and tools upon which to manipulate the polynomial equations. They can either be added, subtracted, multiplied or divided. These operations on polynomial rings form the basis of solving systems of equations algebraically as will be made clear in the chapters ahead. Next, we state the theorem that enables the solution of nonlinear systems of equations in geodesy and geoinformatics.

Theorem 3.1. *Given n algebraic (polynomial) observational equations, where n is the dimension of the observation space $\mathbb{Y}$ of order l in m unknown variables , and m is the dimension of the parameter space $\mathbb{X}$, the application of least squares solution (LESS) to the algebraic observation equations gives $(2l-1)$ as the order of the set of nonlinear algebraic normal equations. There exists m normal equations of the polynomial order $(2l-1)$ to be solved.*

:*proof*:

Given nonlinear algebraic equations $f_i \in k\{\xi_1, \ldots, \xi_m\}$ expressed as

$$\left[\begin{array}{l} f_1 \in k\{\xi_1, \ldots, \xi_m\} \\ f_2 \in k\{\xi_1, \ldots, \xi_m\} \\ \quad\cdot \\ \quad\cdot \\ \quad\cdot \\ f_n \in k\{\xi_1, \ldots, \xi_m\}, \end{array}\right. \tag{3.2}$$

with the order considered as l, we write the objective function to be minimized as

$$\|f\|^2 = f_1^2 + \ldots. + f_n^2 \mid \forall f_i \in k\{\xi_1, \ldots, \xi_m\}, \tag{3.3}$$

and obtain the partial derivatives (first derivatives of 3.3) with respect to the unknown variables $\{\xi_1, \ldots, \xi_m\}$. The order of (3.3) which is l^2 then reduces to $(2l-1)$ upon differentiating the objective function with respect to the variables $\xi_1, \ldots, \xi_m$. Thus resulting in m normal equations of the polynomial order $(2l-1)$. ♣

Example 3.5 (Pseudo-ranging problem). For pseudo-ranging or distance equations, the order of the polynomials in the algebraic observational equations is $l = 2$. If we take the "pseudo-ranges squared" or "distances squared", a necessary procedure in-order to make the observation equations "algebraic" or "polynomial", and implement least squares solution (LESS), the objective function which is of order $l = 4$ reduces by one to order $l = 3$ upon differentiating once. The normal equations are of order $l = 3$ as expected.

The significance of Theorem 3.1 is that all observational equations of interest are successfully converted to *"algebraic"* or *"polynomial"* equations. This implies that problems requiring exact algebraic solutions must first have their equations converted into algebraic. This will be made clear in Chap. 15 where trigonometric nonlinear system on equations are first converted into algebraic.

3-22 Operations "addition" and "multiplication"

Definition 3.3 implies that a polynomial ring qualifies as a ring based on the applications of operations "*addition*" and "multiplication" on its coefficients. In this case, the axioms that follow the Abelian group with respect to "addition" and the semi group with respect to "*multiplication*" readily follow. Of importance in manipulating polynomial rings using operations "*addition*" and "*multiplication*" is the concept of division of polynomials defined as

Definition 3.4 (Polynomial division). *Consider the* polynomial ring $k[x]$ *whose elements are polynomials* $f(x)$ *and* $g(x)$. *There exists unique polynomials* $p(x)$ *and* $r(x)$ *also elements of polynomial ring* $k[x]$ *such that*

$$\boxed{f(x) = g(x)p(x) + r(x)},$$

with either $r(x) = 0$ *or degree of* $r(x)$ *is less than the degree of* $g(x)$.

For univariate polynomials, as in *Definition* 3.4, the Euclidean algorithm employs operations "addition" and "multiplication" to factor polynomials in-order to reduce them to satisfy the definition of division algorithm.

3-3 Factoring polynomials

In-order to understand the factorization of polynomials, it is essential to revisit some of the properties of prime numbers of integers. This is due to the fact that polynomials behave much like integers. Whereas for integers, any integer $n > 1$ is either prime (i.e., can only be factored by 1 and n itself) or a product of prime numbers, a polynomial $f(x) \in k[x]$ is either irreducible in $k[x]$ or factors as a product of irreducible polynomials in the field $k[x]$. The polynomial $f(x)$ has to be of positive degree. Factorization of polynomials play an important role as it enables solution of polynomial roots as will be seen in the next section. Indeed, the Groebner basis algorithm presented in Chap. 4 makes use of the factorization of polynomials.

3-4 Polynomial roots

More often than not, the most encountered interaction with polynomials is perhaps the solution of its roots. Finding the roots of polynomials is essential for most computations that we undertake in practice. As an example, consider a simple planar ranging case where distances have been measured from two known stations to an unknown station (see e.g, Fig. 4.1 on p. 34). In such a case, the measured distances are normally related to the coordinates of the unknown station by multivariate polynomial equations. If for instance a station P_1, whose coordinates are $\{x_1, y_1\}$ is occupied, the distance s_1 can be measured to an unknown station P_0. The coordinates $\{x_0, y_0\}$ of this unknown station are desired and have to be determined from distance measurements. The relationship between the measured distance and the coordinates is given by

$$s_1 = \sqrt{(x_1 - x_0)^2 + (y_1 - y_0)^2}. \tag{3.4}$$

Applying *Theorem* 3.1, a necessary step to convert (3.4) into polynomial, (3.4) is squared to give a multivariate quadratic polynomial

$$s_1^2 = (x_1 - x_0)^2 + (y_1 - y_0)^2. \tag{3.5}$$

Equation (3.5) has two unknowns thus necessitating a second distance measurement to be taken. Measuring this second distance s_2 from station P_2, whose coordinates $\{x_2, y_2\}$ are known, to the unknown station P_0 leads to a second multivariate quadratic polynomial equation

$$s_2^2 = (x_2 - x_0)^2 + (y_2 - y_0)^2. \tag{3.6}$$

The intersection of the two equations (3.5) and (3.6) results in two quadratic equations $ax_0^2 + bx_0 + c = 0$ and $dy_0^2 + ey_0 + f = 0$ whose roots give the desired coordinates x_0, y_0 of the unknown station P_0. In Sect. 4-1, we will expound further on the derivation of these multivariate quadratic polynomial equations.

In Sect. 3-6, we will discuss the types of polynomials with real coefficients. Suffice to mention at this point that polynomials, as defined in *Definition* 3.2 with the coefficients in the field k, has a solution ξ such that on replacing the variable x^{α}, one obtains

$$\boxed{a_n\xi^n + a_{n-1}\xi^{n-1} + ... + a_1\xi + a_0 = 0}. \tag{3.7}$$

From high school algebra, we learnt that if ξ is a solution of a polynomial $f(x)$, also called the root of $f(x)$, then $(x - \xi)$ divides the polynomial $f(x)$. This fact enables the solution of the remaining roots of the polynomial as we already know. The division of $f(x)$ by $(x - \xi)$ obeys the division rule discussed in Sect. 3-22. In a case where $f(x) = 0$ has many solutions (i.e., multiple roots $\xi_1, \xi_2, ..., \xi_m$), then $(x - \xi_1), (x - \xi_2), ..., (x - \xi_m)$ all divide $f(x)$ in the field k.

In general, a polynomial of degree n will have n roots that are either real or complex. If one is operating in the real domain, i.e., the polynomial coefficients are real, the complex roots normally results in a pair of conjugate roots. Polynomial coefficients play a significant role in the determination of the roots. A slight change in the coefficients would significantly alter the solutions. For ill-conditioned polynomials, such a change in the coefficients can lead to disastrous results. Methods of determining polynomial roots have been elaborately presented by [334]. We should point out that for polynomials of degree n in the field of real numbers $\mathbb{R}$ however, the *radical solutions* exist only for polynomials up to degree 4. Above this, *Niels Henrick Abel* (1802-1829) proved through his *impossibility theorem* that the roots are insolvable, while *Evariste Galois* (1811-1832) gave a more concrete proof that for every integer n greater than 4, there can not be a formula for the roots of a general n^th degree polynomial in terms of coefficients.

3-5 Minimal polynomials

In Sect. 2-3, we presented the number rings concept and extended the sets from that of natural numbers $\mathbb{N}$ to the complex number $\mathbb{C}$ in-order to cater for expanded operations. For polynomials, roots may fail to exist in one set say $\mathbb{Q}$ but exist in another set $\mathbb{R}$ as we saw in Sect. 2-2. The polynomial $y^2 - 12 = 0$, for example, has no roots in $\mathbb{Q}[y]$ but the roots ± 12 exist in $\mathbb{R}$. The expansion of the set from $\mathbb{Q}$ to $\mathbb{R}$ is also called *field extension* of k. It may occur however that in the polynomial ring $k[x]$, the solution ξ satisfy not only the polynomial $p(x)$ but also another polynomial $h(x)$, where $p(x)$ and $h(x)$ are both elements of $k[x]$. In case several polynomials in $k[x]$ have ξ as a root, and the polynomials are multiples of a polynomial of least degree that also contains ξ as root, this polynomial of least degree is termed the *minimal polynomial.*

As an example, consider two polynomials $-1-2x^2+x^4$ and $x-x^2+2x^3-2x^4-x^5+x^6$ with a similar root $\sqrt{1+\sqrt{2}}$ illustrated in Figs 3.1 and 3.2 respectively. The polynomial $-1-2x^2+x^4$ having the lowest degree is the minimum polynomial.

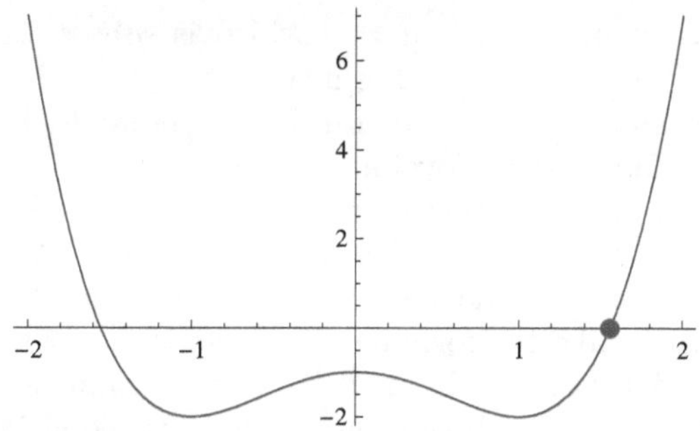

Figure 3.1. The minimal polynomial $-1 - 2x^2 + x^4$ for the root a $\sqrt{1+\sqrt{2}}$

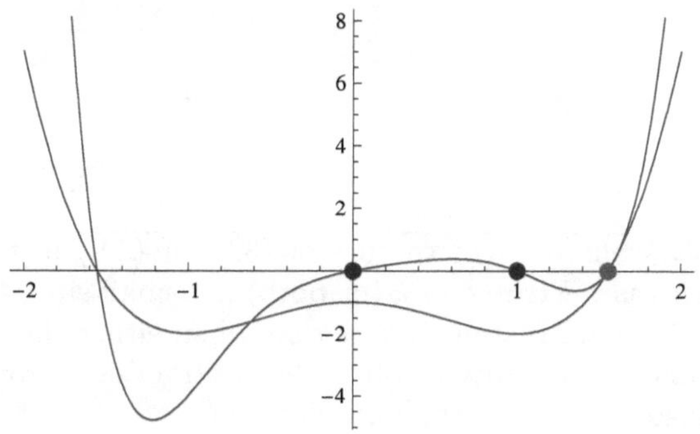

Figure 3.2. A polynomial $x - x^2 + 2x^3 - 2x^4 - x^5 + x^6$ with the same root $\sqrt{1+\sqrt{2}}$, but not a minimal polynomial for this root

In dealing with Groebner basis in Chap. 4 for example, it will be seen that several polynomials in the field $k[x]$ contain the same root ξ. This property will be used to reduce several multivariate polynomials to univariate polynomials whose solutions fulfill the multivariate polynomials.

3-6 Univariate polynomials with real coefficients

As we shall see later, the solution of a system of polynomial equations can be reduced to the solution of a univariate polynomial. Therefore it is useful to study the solution of polynomials with a single variable. In this section we revisit the various types of univariate polynomials with the coefficients in the field $\mathbb{R}$ of reals, which we often use to manipulate measurements. We recapture the basic high school mathematics of inferring the roots of polynomials from the coefficients.

3-61 Quadratic polynomials

In Sect. 3-4 we introduced the quadratic equations and demonstrated their association with distance measurements. In general, the simplest polynomial is the linear polynomial $cx + d = 0$ which is solved for x by simply multiplying both sides of the equation by the inverse of c, provided that $c \neq 0$ holds. The solution thus becomes $x = -c^{-1}d$. Linear polynomials, i.e., polynomials of degree 1 find use in manipulating levelling and gravimetric observations. In these cases, they are manipulated in vector space through the solution of linear algebraic equation $\mathbf{Ax} + \mathbf{y} = \mathbf{0}$ to give the solution $\mathbf{x} = (\mathbf{A'A^{-1}})\mathbf{A'y}$, provided that $\mathbf{A'A}$ is regular.

Polynomials of degree 2 are known as quadratic polynomials. For univariate cases, they take the form $ax^2 + bx + c = 0$. For simple cases such as $x^2 + 2x = 0$, the solution can be obtained by factorization, e.g., $x(x+2)$ leading to $x = 0$ or $x = -2$. The general solution of quadratic equations of the form $ax^2 + bx + c = 0$ with real coefficients $\{a, b, c\}$ is given by the quadratic formulae

$$x = \frac{-b \pm \sqrt{b^2 - 4ac}}{2a}, \tag{3.8}$$

or

$$x = \frac{2c}{-b \pm \sqrt{b^2 - 4ac}}. \tag{3.9}$$

Press [334] discourages the use of (3.8) or (3.9) in the determination of the two roots for cases where a or c (or both) are small since this leads to inaccurate solutions. The main reason cited is that when either the coefficient a or c (or both) is small, one of the roots involves the subtraction b from a very nearly equal value. They instead propose the formular

$$\boxed{q = -\frac{1}{2}[b + sgn(b)\sqrt{b^2 - 4ac}]}, \tag{3.10}$$

with $Sign(x)$ giving -1, 0 or 1 depending on whether x is negative, zero, or positive. The two roots are then given by

$$x_1 = \frac{q}{a}, \quad x_2 = \frac{c}{q}. \tag{3.11}$$

In computer algebra software of Matlab and Mathematica, the roots of a quadratic polynomial are obtained via

- Matlab: $x = roots([a\ b\ c])$, where $[a\ b\ c]$ is a vector containing the coefficients in the field $\mathbb{R}$ of reals. The quadratic equation can also be solved using the solve command, e.g., $solve(`ax^2 + bx + c = 0, x')$, where x indicates the variable to be solved.
- Mathematica: $x = Root[f, k]$, where f is the quadratic equation and k the kth root. The quadratic equation can also be solved using the solve command, e.g., $Solve[ax^2 + bx + c == 0, x]$.

In general, every quadratic polynomial has exactly two real or two complex roots. From the coefficients, if $b^2 - 4ac > 0$, the roots are real but if $b^2 - 4ac < 0$ the roots are a pair of non real complex numbers. The case where $b^2 - 4ac = 0$ gives real and identical roots and is also known as the bifurcation point upon which the roots change sign.

3-62 Cubic polynomials

These are polynomials of degree 3 and take the form $ax^3 + bx^2 + cx + d = 0$.

Like quadratic polynomials, simple cases can also be solved via factorization e.g., $x^3 - 2x = 0$ is factored as $x(x^2 - 2)$ to give the solutions $x = 0, x = -\sqrt{2}$ or $x = +\sqrt{2}$. Another approach would be to reduce the cubic polynomial such that the polynomials of degree 2 are eliminated to give a simplified version of the form $y^3 + ey + f = 0$ known as a reduced cubic polynomial. The simplified version can then be solved for the roots via Cardano's formula as

$$y = \sqrt[3]{-\frac{f}{2} + \sqrt{T}} + \sqrt[3]{-\frac{f}{2} - \sqrt{T}}, \tag{3.12}$$

where $T = (\frac{e}{3})^3 + (\frac{f}{2})^2$. Once one real root say ξ_1 has been obtained, the polynomial $y^3 + ey + f = 0$ is divided by $(y - \xi_1)$ and the resulting quadratic polynomial solved for the remaining roots. An alternative approach is presented by [334] who proceed as follows: Let $\{a, b, c\}$ be the real coefficients of a cubic polynomial. Compute

$$\left[\begin{array}{l} K \equiv \dfrac{a^2 - 3b}{g} \\[2ex] L \equiv \dfrac{2a^3 - gab + 27c}{54}. \end{array}\right. \tag{3.13}$$

If K and L are real, and $L < K$, then the cubic polynomial has three real roots computed by

$$\left[\begin{array}{l} x_1 = -2\sqrt{K}cos(\dfrac{\Theta}{3}) - \dfrac{a}{3} \\[2ex] x_2 = -2\sqrt{K}cos(\dfrac{\Theta + 2\pi}{3}) - \dfrac{a}{3} \\[2ex] x_3 = -2\sqrt{K}cos(\dfrac{\Theta - 2\pi}{3}) - \dfrac{a}{3}, \end{array}\right. \tag{3.14}$$

where[1]

$$\Theta = cos^{-1}(\frac{L}{\sqrt{K^3}}).$$

Using computer algebra software of Matlab and Mathematica the roots of a cubic polynomial are obtained via

- Matlab: $x = roots([a\ b\ c\ d])$, where $[a\ b\ c\ d]$ is a vector containing the coefficients in the field $\mathbb{R}$ of reals. The quadratic equation can also be solved using the solve command, e.g., $solve(`ax^3 + bx^2 + cx + d = 0, x')$.

[1] The origin of the equation is traced by the authors to chapter VI of François Viète's treatise "De emendatione" Published in 1615

- Mathematica: $x = Root[f, k]$, where f is the cubic equation, and k, the kth root. The quadratic equation can also be solved using the solve command, e.g., $Solve[`ax^3 + bx^2 + cx + d == 0, x']$. Considering a $= 1$, the solution is

$$x = \begin{cases} -(b/3) - \frac{(2^{(1/3)}(-b^2+3c))}{(3(-2b^3+9bc-27d+3\sqrt{3}\sqrt{-b^2c^2+4c^3+4b^3d-18bcd+27d^2})^{(1/3)})} + \\ + \frac{(-2b^3+9bc-27d+3\sqrt{3}\sqrt{-b^2c^2+4c^3+4b^3d-18bcd+27d^2})^{(1/3)}}{(3\times 2^{(1/3)})}, \end{cases}$$

$$x = \begin{cases} -(b/3) + \frac{((1+i\sqrt{3})(-b^2+3c))}{(3\times 2^{(2/3)}(-2b^3+9bc-27d+3\sqrt{3}\sqrt{-b^2c^2+4c^3+4b^3d-18bcd+27d^2})^{(1/3)})} - \\ - \frac{((1-i\sqrt{3})(-2b^3+9bc-27d+3\sqrt{3}\sqrt{-b^2c^2+4c^3+4b^3d-18bcd+27d^2})^{(1/3)})}{(6\times 2^{(1/3)})}, \end{cases}$$

$$x = \begin{cases} -(b/3) + \frac{((1-i\sqrt{3})(-b^2+3c))}{(3\times 2^{(2/3)}(-2b^3+9bc-27d+3\sqrt{3}\sqrt{-b^2c^2+4c^3+4b^3d-18bcd+27d^2})^{(1/3)})} - \\ - \frac{((1+i\sqrt{3})(-2b^3+9bc-27d+3\sqrt{3}\sqrt{-b^2c^2+4c^3+4b^3d-18bcd+27d^2})^{(1/3)})}{(6\times 2^{(1/3)})} \end{cases}$$

In general, if ξ_1, ξ_2, ξ_3 are the roots of a cubic polynomial, the discriminant D can de defined as

$$D = (\xi_1 - \xi_2)^2(\xi_1 - \xi_3)^2(\xi_2 - \xi_3)^2, \tag{3.15}$$

and computed from the coefficients a, b, c, d to infer on the nature of the roots. Considering $a = 1$, [235, p. 156, Exercise 10.17] gives the formula of the discriminant D from the coefficients b, c, d as

$$D = 18bcd - 4b^3d + b^2c^2 - 4c^3 - 27d^2. \tag{3.16}$$

If $D > 0$ then the roots of the cubic polynomial are real and distinct. If $D < 0$, then one of the roots is real and the remaining two are non real complex conjugate. In a case where $D = 0$, multiple roots all which are real are given. In case the coefficients b, c, d are all positive, then all the three roots will be negative, while if b, d are negative and c positive, all the roots will be positive.

3-63 Quartic polynomials

Quartic polynomials are those of degree 4. In a case where one root ξ_1 exist for a polynomial $p(x) = 0$, the division algorithm can be applied to obtain the factor $(x - \xi_1)f(x)$. Here, $f(x)$ is a cubic polynomial that can be solved as discussed in Sect. 3-62 to give at least one real root. The quartic polynomial $ax^4 + bx^3 + cx^2 + dx + e = 0$ therefore has at least two real roots. The following conditions may apply for a quartic polynomial:

- $p(x)$ has four real roots.
- $p(x)$ has two real roots and two complex conjugate roots.
- $p(x)$ has no real roots.

The solution of a quartic polynomial proceeds via substitution approach in-order to reduce it. Considering a case where $a = 1$, the quartic polynomial $x^4+bx^3+cx^2+dx+e = 0$ is reduced by substituting $x = z+a$, with $a \in \mathbb{R}$, to $Z^4+CZ^2+EZ+F = 0$ which is solved for $g(Z) = 0$. The solutions of $g(Z) = 0$ satisfies those of $p(x) = 0$ (see Sect. 3-5). $Z^4 + CZ^2 + EZ + F = 0$ is called the reduced quartic polynomial which can be solved as discussed by [235, pp. 159-166].

Solution of the roots of quartic polynomials using Computer algebra software of Matlab and Mathematica is as follows:

- Matlab: $x = roots\left([a\ b\ c\ d\ e]\right)$, where $[a\ b\ c\ d\ e]$ is a vector containing the coefficients in the field $\mathbb{R}$ of reals. The quadratic equation can also be solved using the solve command, e.g., $solve('ax^4 + bx^3 + cx^2 + dx + e = 0, x')$
- Mathematica: $x = Root[f, k]$, where f is the quartic equation and k the kth root. The quadratic equation can also be solved using the solve command, e.g., $Solve[ax^4 + bx^3 + cx^2 + dx + e == 0, x]$. In the accompanying CD, one of the Mathematica solution when $a = 1$ is presented.

In general, if $\xi_1, \xi_2, \xi_3, \xi_4$ are the roots of a quartic polynomial, the discriminant D can de defined as

$$D = (\xi_1 - \xi_2)^2(\xi_1 - \xi_3)^2(\xi_1 - \xi_4)^2(\xi_2 - \xi_3)^2(\xi_2 - \xi_4)^2(\xi_3 - \xi_4)^2, \tag{3.17}$$

and computed from the coefficients b, c, d, e to infer on the nature of the roots. Considering $a = 1$, [235, p. 171] gives the formula of the discriminant D from the coefficients b, c, d, e as

$$D = \left[\begin{array}{c} 18bcd^3 + 18b^3cde - 80bc^2de - 6b^2d^2e + 144cd^2e \\ +144b^2ce^2 - 128c^2e^2 - 192bde^2 + b^2c^2d^2 - 4b^3d^3 - 4c^3d^2 \\ -4b^2c^3e + 16c^4e - 27d^4 - 27b^4e^2 + 256e^3. \end{array}\right. \tag{3.18}$$

If $D > 0$ then all the roots of the quartic polynomial are real and distinct or all the four roots are pairs of non real complex conjugates. If $D < 0$, then two roots are real and distinct while the other two are complex conjugates. For a case where $D = 0$, at least two of the roots coincide.

3-7 Methods for investigating roots

3-71 Logarithmic and contour plots on complex plane

Here we introduce two graphical and an algebraic - numeric method to find roots of polynomials. Let us consider the following polynomial

$$f = -2829 + 65593z - 228035z^2 + 536375z^3 - 295200z^4 + 222000z^5,$$

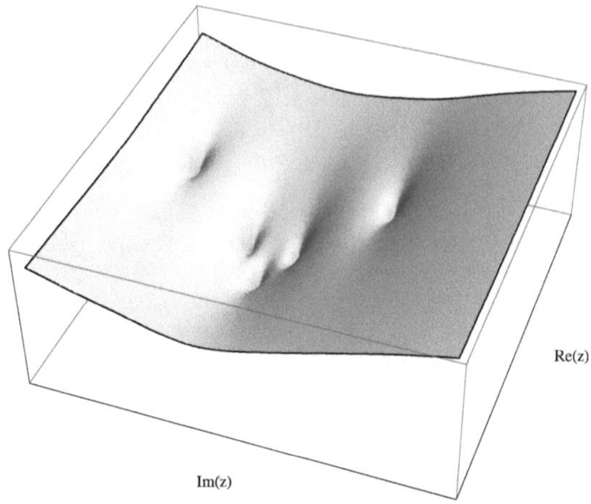

Figure 3.3. The surface of the function $log(abs(f(z)) + 1)$ on the complex plane

Plotting the logarithmic of the absolute value of $f(z)+1$ on the complex plane leads to Fig. 3.3. The additive constant is used to avoid singularity at $abs(f(z)) = 0$. The "holes" on the surface show the approximate positions of the roots. To get better approximation, we can display $Re(f(z)) = 0$ and $Im(f(z)) = 0$ contours on the complex plane in Fig. 3.4. The cross points of these contours represent locations of the roots.

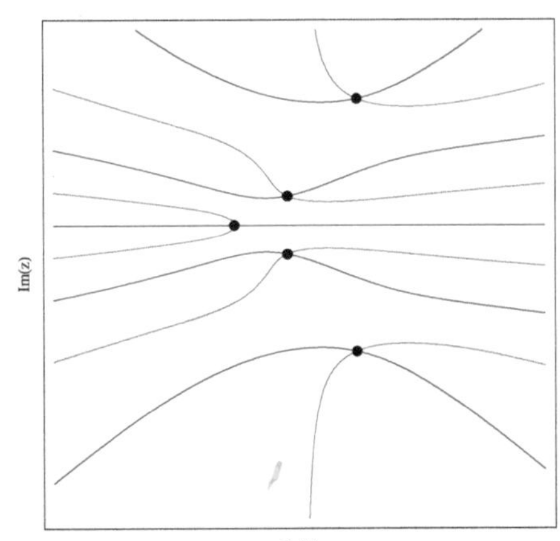

Figure 3.4. The contour plot of $f(z)$ on the complex plane. Along the lines the real and imaginary parts are zero

3-72 Isograph simulator

Let us consider the exponential form of the complex numbers $z = r^{exp(i\phi)}$, and display $f(z)$ as function of ϕ at constant r. Changing r values, find the proper r value at which the curve of $f(z)$ crosses the origin of the coordinate system of the complex plane (0, 0). The function IsoGraphSimulator in Mathematica simulates this process. The numerical values of the roots are computed via NSolve and are as presented in Fig. 3.5.

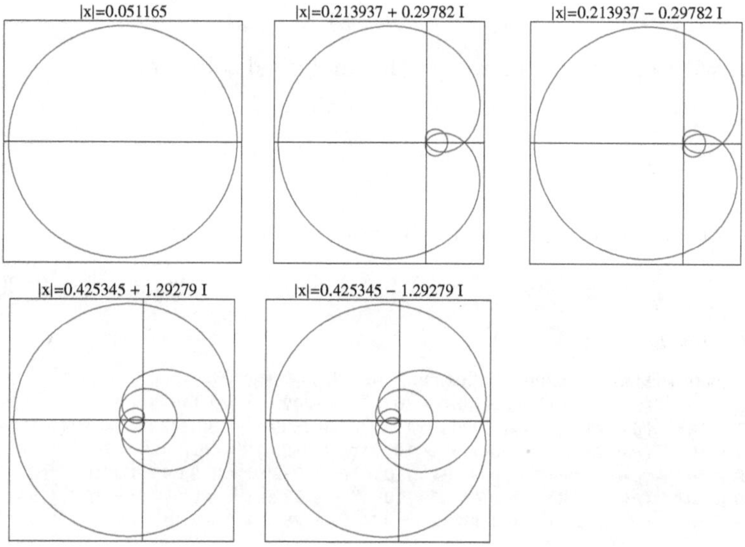

Figure 3.5. The contour plot of $f(z)$ on the complex plane. Along the lines the real and imaginary parts are zero

3-73 Application of inverse series

We consider the $x = r(y)$, as the inverse mapping of $y = f(x)$. Then the roots of $f(x)$ can be computed as $r(0)$, while $f(r(0)) = 0$. In order to illustrate the idea let us consider the power series expansion for following second order polynomial, about the point $x = 0$ to order four,

$$y = f(x) = c + bx + ax^2$$

The inverse series provides the approximation of the inverse function $r(y)$,

$$r(y) = \frac{(y-c)}{b} - \frac{(a(y-c)^2)}{b^3} + \frac{(2a^2(y-c)^3)}{b^5} - \frac{(5a^3(y-c)^4)}{b^7}$$

The $r(0)$ gives the root of the polynomial $f(x)$ as

$$r(0) = -\frac{c}{b} - \frac{ac^2}{b^3} - \frac{2a^2c^3}{b^5} - \frac{5a^3c^4}{b^7}.$$

One may recognize that the $(k+1)^t h$ term can be expressed as

$$-\frac{a^k b^{-1-2k} c^{1+k} Binomial[2k,k]}{1+k},$$

indeed, for example, considering the first four terms, we get

$$-\frac{c}{b} - \frac{ac^2}{b^3} - \frac{2a^2c^3}{b^5} - \frac{5a^3c^4}{b^7}.$$

However, with infinity terms we get the analytical solution,

$$\frac{b\left(-1+\sqrt{\frac{b^2-4ac}{b^2}}\right)}{2a}$$

Now we can apply this method to the polynomial considered as an example,

$$f(x) = -2829 + 65593z - 228035z^2 + 536375z^3 - 295200z^4 + 222000z^5$$

whose inverse is

$$\begin{aligned}
&r(y) = 0.0000152455(y+2829.) + 8.08033\times10^{-10}(y+2829.)^2 + \\
&5.66776\times10^{-14}(y+2829.)^3 + 3.91367\times10^{-18}(y+2829.)^4 + 2.14685\times10^{-22}(y+2829.)^5 + \\
&2.00147\times10^{-27}(y+2829.)^6 - 1.89902\times10^{-30}(y+2829.)^7 - 3.98433\times10^{-34}(y+2829.)^8 - \\
&5.76614\times10^{-38}(y+2829.)^9 - 6.85252\times10^{-42}(y+2829.)^{10} - 6.79426\times10^{-46}(y+2829.)^{11} - \\
&5.16349\times10^{-50}(y+2829.)^{12} - 1.66894\times10^{-54}(y+2829.)^{13} + 3.69854\times10^{-58}(y+2829.)^{14} + \\
&1.05096\times10^{-61}(y+2829.)^{15} + 1.77627\times10^{-65}(y+2829.)^{16} + 2.38232\times10^{-69}(y+2829.)^{17} + \\
&2.64939\times10^{-73}(y+2829.)^{18} + 2.32786\times10^{-77}(y+2829.)^{19} + 1.18834\times10^{-81}(y+2829.)^{20}
\end{aligned}$$

One of the real root is

$$r(0) = 0.051165$$

Using this root, we can reduce the order of our polynomial to a quartic. For further details, see Chap. 2 of the accompanying CD.

3-8 Computation of zeros of polynomial systems

In one variable case, the polynomial,

$$p(x) = a_d x^d + a_{d-1}x^{(d-1)} + \ldots + a_2x^2 + a_1x + a_0$$

has d roots, counting multiplicities, in the field $\mathbb{C}$ of complex numbers according to the fundamental theorem of algebra. However in multivariable case, the situation is more complicated. To keep the discussion easy and transparent, we consider a system of two polynomials. Let us consider the following system,

$$\begin{aligned}
G &= a_1 + a_2x + a_3xy + a_4y \\
H &= b_1 + b_2x^2y + b_3xy^2
\end{aligned}$$

Theorem 3.2 (Bézout's Theorem). *Consider two polynomial equations in two unknowns:* $g(x, y) = h(x, y) = 0$. *If this system has only finitely many zeros* $(x, y) \in \mathbb{C}^2$, *then the number of zeros is at most* $deg(g) * deg(h)$. *Here* $deg(g)$ *and* $deg(h)$ *are the total degree of g(x,y) and h(x,y), respectively, see Sect. 3-1*

These two polynomials have precisely four distinct zeros $(x, y) \in \mathbb{C}^2$ for generic choices of coefficients a_i and b_j. It means that a certain polynomial in the coefficients a_i, b_j, called the discriminant, should be non- zero. The discriminant can be computed via the function Resultant, which is zero if and only if the polynomials have a common root, see details in Chapter 5.

Bézout' s theorem would predict $2 * 3 = 6$ common complex zeros for our system. Indeed, in projective geometry we would expect the cubic curve $(G = 0)$ and the quadratic curve $(H = 0)$ to intersect in six points. However, the system has actually 4 common roots. To understand why is four and not six let us consider convex polygons associated with our system. A *polytope* is a subset of $\mathbb{R}^n$ which is the convex hull of a finite set of points. A 2-dimensional polytope is called *polygon*. Consider a polynomial $f(x, y)$,

$$f(x, y) = a_1 x^{u_1} y^{v_1} + a_2 x^{u_2} y^{v_2} + \ldots + a_m x^{u_m} y^{v_m}$$

This is a polygon in $\mathbb{R}^2$ having at most m vertices. In general, every polynomial in n unknowns can be represented by a *Newton polytope* in $\mathbb{R}^n$. Let us consider the Newton polygons of our systems. The Newton polygon of the polynomial G is quadrangel with points (0,0), (1,0),(1,1) and (0,1), while that of the polynomial H is a triangle (0,0), (2,1), (1,2), see Fig. 3.6.

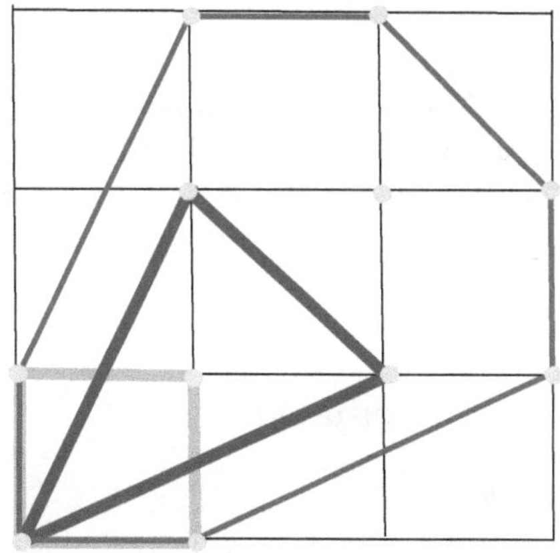

Figure 3.6. Mixed area of Newton-polygons of the two polynomials

The *Minkowski* sum of the two polygons, P and Q in the plane is

$$P + Q = p + q : p \in P, q \in Q$$

The Newton polygon of the product of two polynomials is the Minkowski sum of the Newton polyon of these two polynomials, namely

$$New(G * H) = New(G) + New(H)$$

The corresponding points can be seen on Fig. 3.6. So the Minkowski sum of the polygons mirrors the algebraic operation of multiplying polynomials. If P and Q ar any two polynomials then their mixed area can be defined as it follows.

$$\mathcal{M}(P, Q) = area(P + Q) - area(P) - area(Q)$$

In our case,

$$\mathcal{M}(New(G), New(H)) = \frac{13}{2} - 1 - \frac{3}{2} = 4$$

This number coincides with the number of common zeros of G and H. This is not an accident, but is an instance of the general theorem of *Bernstein.*

Theorem 3.3 (Bernstein's Theorem). *If G and H are two generic bivariate polynomials, then the number of non-zero solutions of* $G(x, y) = H(x, y) = 0 \in \mathbb{C}^2$ *equals the mixed area* $\mathcal{M}(New(G), New(H))$.

Elimination theory deals with the problem of eliminating one or more variables from a system of polynomial equations, thus reducing the problem to a smaller problem in fewer variables. For instance, if we wish to solve,

$$U = a_0 + a_1x + a_2x^2; V = b_0 + b_1x + b_2x^2$$

for $U = V = 0$ with $a_2 \neq 0$ and $b_2 \neq 0$, we can eliminate the variable x to get a polynomial of degree 4 which is the resultant. This resultant vanishes if and only if the given quadratic polynomials have common complex root x. This resultant method can be extended for case $n > 2$ and various determinants formulae are known for multivariate resultant, see Chap. 5. Other approaches include Groebner basis and homotopy which are discussed in the Chaps. 4 and 7 respectively.

3-9 Concluding remarks

What we have presented is just a nutshell of the topic "polynomials". Several books, e.g., [55, 298, 335, 436] are dedicated specifically to it.

4 Groebner basis

"There are no good, general methods for solving systems of more than one nonlinear equation. Furthermore, it is not hard to see why (very likely) there never will be any good, general methods:..." W. H. Press et al.

4-1 The origin

This chapter presents you the reader with one of the most powerful computer algebra tools, besides the polynomial resultants (discussed in the next chapter), for solving nonlinear systems of equations which you may encounter. The basic tools that you will require to develop your own algorithms for solving problems requiring closed form (exact) solutions are presented. This powerful tool is the "Gröbner basis" written in English as Groebner basis. It was first suggested by *B. Buchberger* in 1965, a PhD student of *Wolfgang Groebner* (1899 - 1980). *Groebner*, already in 1949, had suggested a method for finding a linearly independent basis of the vector space of the residue class ring of the polynomial ring modulo a polynomial ideal. In studying termination of this method, *Buchberger* came up both with the notion of *Groebner bases* (certain generating sets for *polynomial ideals*) and with an always terminating algorithm for computing them. In 1964, H. Hironaka (1931-) had independently introduced an analogous notion for the domain of *power series* in connection with his work on resolution of singularities in algebraic geometry and named it *standard basis* [262, p. 187]. However, he did not give any method for computing these bases. *B. Buchberger* decided to honour his thesis supervisor *W. Groebner* by naming the standard basis for *Ideals* in polynomial rings $k[x_1, \ldots, x_n]$ as *Groebner basis* [95].

In this book, as in modern books, we will adopt the term Groebner basis and present the subject in the simplest form that can easily be understood from geodetic as well as geoinformatics perspective.

As a recipe, consider that most problems in nature, here in geodesy, geoinformatics, machine vision, robotics, surveying etc., can be modelled by *nonlinear systems of equations*. Let us consider a simple case of planar distance measurements in Fig. 4.1. Equations relating these measured distances to the coordinates of an unknown station were already presented in Sect. 3-4. In that section, we did relate the measured distances $\{s_i,\, i = 1, 2\}$ to the coordinates of the unknown station by (3.5) and (3.6). We stated that the intersection of these two equations lead to univariate polynomials whose solution give the desired position of an unknown station. We did not however give any explanation on how the univariate polynomials are derived from the set of multivariate quadratic polynomials (3.5) and (3.6). The derivation of the univariate polynomials from systems of nonlinear equations form one of the major tasks of Groebner basis. Let us denote the distance $\{s_i,\, i = 1, 2\}$ by $\{d_i,\, i = 1, 2\}$ and re-write (3.5) and (3.6) respectively as

$$d_1^2 = (x_1 - x_0)^2 + (y_1 - y_0)^2 \tag{4.1}$$

and

J.L. Awange et al., *Algebraic Geodesy and Geoinformatics*, 2nd ed.,
DOI 10.1007/978-3-642-12124-1_4, © Springer-Verlag Berlin Heidelberg 2010

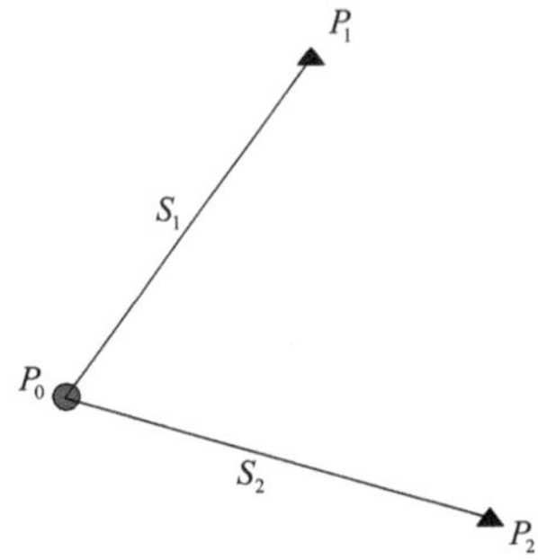

Figure 4.1. Planar distance observations

$$d_2^2 = (x_2 - x_0)^2 + (y_2 - y_0)^2. \tag{4.2}$$

The task confronting us now is to obtain from these two nonlinear equations the coordinates $\{x_0, y_0\}$ of the unknown station P_0. In case (4.1) and (4.2) were linear, the solution for $\{x_0, y_0\}$ would have been much easier. One could simply solve them using either matrix inversion, graphically, Gauss-Jordan or Gauss elimination techniques. Unfortunately they are nonlinear and can not be solved using the procedures above. Groebner basis and polynomial resultant approaches are algebraic techniques that are proposed to offer solutions to nonlinear systems of equations such as (4.1) and (4.2).

4-2 Basics of Groebner basis

Groebner basis, in some sense, is a generalization to the multivariate case of the notion of greatest common divisors for systems of univariate polynomials. Its direct application is the elimination of variables in nonlinear systems of equations. Let us start by the problem of finding the greatest common divisors in Example 4.1:

Example 4.1 (Greatest common divisors (gcd)). Given the numbers 12, 20, and 18, find their greatest common divisor. We proceed by writing the factors as

$$\left.\begin{array}{r} 12 = 2^2.3^1.5^0 \\ 20 = 2^2.3^0.5^1 \\ 20 = 2^1.3^2.5^0 \end{array}\right] \rightarrow 2^1.3^0.5^0 = 2, \tag{4.3}$$

leading to 2 as the greatest common divisor of 12, 20 and 18. Next, let us consider the case of univariate polynomials $f_1, f_2 \in k[x]$ in (4.4).

$$\left.\begin{array}{l} f_1 = 3x^4 - 3x^3 + 8x^2 + 2x - 5 \\ f_2 = 5x^4 - 4x^2 - 9x + 21 \end{array}\right] \rightarrow \textit{Euclidean algorithm} = f \in k[x]. \tag{4.4}$$

Equation (4.4) employs the Euclidean algorithm which obtains one univariate polynomial as the gcd of the two univariate polynomials f_1 and f_2. If on the other hand expressions in (4.4) were not univariate but multivariate, e.g., $g_1, g_2 \in k[x, y]$ as in equation (4.5), then one applies the Buchberger algorithm which is discussed in Sect. 4-3.

$$\left.\begin{array}{l} g_1 = xy + x - y - 1 \\ g_2 = xy - x - y + 1 \end{array}\right] \rightarrow \mathit{Buchberger\ algorithm} = \mathit{Groebner\ basis}. \tag{4.5}$$

Groebner basis therefore, is the greatest common divisors of a multivariate system of polynomial equations $\{g_1, g_2\}$.

Groebner basis as stated earlier is useful for eliminating variables in nonlinear systems of equations. Gauss elimination technique on the other hand is applicable for linear cases as shown in Example 4.2.

Example 4.2 (Gauss elimination technique). Solve the linear system of equations

$$\left[\begin{array}{c} -x + y + 2z = 2 \\ 3x - y + z = 6 \\ -x + 3y + 4z = 4. \end{array}\right. \tag{4.6}$$

The first step is to eliminate x in the second and third expressions of (4.6). This is achieved by multiplying the first expression by 3 and adding to the second expression to give the second expression of (4.7). The third expression of (4.7) is obtained by subtracting the first expression from the third expression in (4.6).

$$\left[\begin{array}{l} -x + y + 2z = 2 \\ \;\;2y + 7z = 12 \\ \;\;2y + 2z = 2. \end{array}\right. \tag{4.7}$$

The second step is to eliminate y in the second and third expressions of (4.7). This is achieved by subtracting the second expression from the third expression in (4.7) to give (4.8).

$$\left[\begin{array}{l} -x + y + 2z = 2 \\ \;\;2y + 7z = 12 \\ \;\;-5z = -10. \end{array}\right. \tag{4.8}$$

The solution of $z = 2$ in (4.8) can now be substituted back into the second equation $2y + 7z = 12$ to give the value of $y = -1$, which together with the value of $z = 2$ are substituted into the first equation to give the value of $x = 1$ to complete the Gauss elimination technique.

In many applications however, equations relating unknown variables to the measured (observed) quantities are normally nonlinear and often consist of many variables (multivariate). In such cases, the Gauss elimination technique for the univariate polynomial equations employed in *Example* 4.2 gives way to Groebner basis as illustrated in *Examples* 4.3 and 4.4. In general, the Groebner basis algorithm reduces a system of multivariate polynomial equations. This is done by employing operations "*addition*" and "*multiplication*" on a polynomial ring (see Sect. 3-22) to give more simplified expressions. Given a system of polynomial equations which are to be solved explicitly for unknowns, e.g., (4.1) and (4.2), Groebner basis algorithm is applied to reduce the set of polynomials into another set (e.g., from a system $F(x, y, z)$ to another system $G(x, y, Z)$) of polynomials with suitable properties that allow solution. If $F(x, y, z)$ is a set of nonlinear system of polynomial equations, Groebner basis eliminates variables in a manner similar to Gauss elimination technique for linear cases to reduce it to $G(x, y, z)$. With Lexicographic ordering of the monomials (see *Definition* A.2 in Appendix A-1 on p. 339), one expression in $G(x, y, z)$ always turns out to be a univariate polynomial. Its roots are easily obtained using algebraic software of Matlab, Mathematica or Maple, and can be substituted in the other elements of the set $G(x, y, z)$ to obtain a complete solution which also satisfy the original set $F(x, y, z)$. *Examples* 4.3 and 4.4 elaborate on the application of Groebner basis.

Example 4.3 (Groebner basis computation). Let us consider a simple example from [98]. Consider a set $F(x, y) = \{f_1, f_2\}$ to have as its elements

$$\begin{bmatrix} f_1 = xy - 2y \\ f_2 = 2y^2 - x^2, \end{bmatrix} \tag{4.9}$$

where $\{f_1, f_2\} \in I$ are the generators of the Ideal I (see definition of Ideal on p. 37). We now seek a simplified set of generators of this Ideal using Buchberger algorithm. By employing operations "addition" and "multiplication", the Groebner basis algorithm (also called Buchberger algorithm) reduces the system of nonlinear equations (4.9) into another set G of F as

$$G := \{-2x^2 + x^3, -2y + xy, -x^2 + 2y^2\}. \tag{4.10}$$

In Mathematica software, using the lexicographic order $x > y$, i.e., x comes before y, the Groebner basis could simply be computed by entering the command

$$GroebnerBasis[F, \{x, y\}]. \tag{4.11}$$

The set G in (4.10) contains one univariate polynomial $-2x^2 + x^3$, which can easily be solved using roots command in Matlab for solutions $\{x = 0, x = 0, x = 2\}$ and substituted in any of the remaining elements of the set G to solve for y. The solutions of G, i.e., the roots $\{x = 0, x = 0, x = 2\}$) and those of y satisfy polynomials in F. This can be easily tested by substituting these solutions into (4.9) to give 0.

Let us consider as a second example an optimization problem.

Example 4.4 (Minimum and maximization problem). Find the minimum and maximum of $f(x, y, z) = x^3 + 2xyz - z^2$, such that $g(x, y, z) = x^2 + y^2 + z^2 - 1$. First, we obtain the partial derivatives of $f - L\,g = 0$ with respect to $\{x, y, z, L\}$, where L is the lagrangean multiplier as

$$\frac{\partial f}{\partial\{x, y, z, L\}} := F = \begin{bmatrix} 3x^2 + 2yz - 2xL = 0 \\ 2xz - 2yL = 0 \\ 2xy - 2z - 2zL = 0 \\ x^2 + y^2 + z^2 - 1 = 0. \end{bmatrix} \tag{4.12}$$

Groebner basis is invoked in Mathematica by

$$GroebnerBasis[\{F\}, \{x, y, L, z\}],$$

which leads to

$$G = \begin{bmatrix} L - 1.5x - 1.5yz - 43.7z^6 - 62.2z^4 - 17.5z^2 = 0 \\ x^2 + y^2 + z^2 - 1 = 0 \\ y^2z - 1.8z^5 + 2.8z^3 - z = 0 \\ z^7 - 1.5z^5 + 0.6z^3 - 0.04z = 0. \end{bmatrix} \tag{4.13}$$

The solution of z in (4.13) can then be substituted into the third equation $y^2z - 1.8z^5 + 2.8z^3 - z = 0$ to give the value of y. The obtained values of z and y are then substituted into the second equation to give the value of x, and thus complete the Groebner basis solution. Later in the chapter, we will introduce the *reduced Groebner basis* which can be used to obtain directly the last expression of (4.13), i.e., the univariate polynomial in z.

The theory behind the operation of Groebner basis is however not so simple. In the remainder of this chapter, we will try to present in a simplified form the algorithm behind the computation of Groebner bases. In Chap. 3, we learnt that polynomials are elements of a ring and that they satisfy the ring axioms of addition and subtraction. The computation of Groebner basis is achieved by the capability to manipulate the polynomials to generate *Ideals* defined as

Definition 4.1 (Ideal). *An Ideal is generated by a family of generators as consisting of the set of linear combinations of these generators with polynomial coefficients. Let $f_1, \ldots, f_s$ and $c_1, \ldots, c_s$ be polynomials in $k[x_1, \ldots, x_n]$, then*

$$< f_1, \ldots, f_s > = \sum_{i=1}^{s} c_i f_i. \tag{4.14}$$

In (4.14), $< f_1, \ldots, f_s >$ is an Ideal and if a subset $I \subset k[x_1, \ldots, x_n]$ is an Ideal, it must satisfy the following conditions [117, p. 29];

- $0 \in I$,
- *If $f, g \in I$, then $f + g \in I$ (i.e., I is an additive subgroup of the additive group of the field k),*
- *If $f \in I$ and $c \in k[x_1, \ldots, x_n]$, then $cf \in I$ (i.e., I is closed under multiplication ring element).*

Example 4.5 (Ideal). Equations (4.1) and (4.2) are expressed algebraically as

$$\begin{bmatrix} f_1 := (x_1 - x_0)^2 + (y_1 - y_0)^2 - d_1^2 \\ f_2 := (x_2 - x_0)^2 + (y_2 - y_0)^2 - d_2^2, \end{bmatrix} \tag{4.15}$$

where polynomials $\{f_1, f_2\}$ belong to the polynomial ring $\mathbb{R}[x_0, y_0]$. If the polynomials

$$\begin{bmatrix} c_1 := 4x_0 + 6 \\ c_2 := x_0 + y_0 \end{bmatrix} \tag{4.16}$$

also belong to the same polynomial ring $\mathbb{R}[x_0, y_0]$, an Ideal is generated by a linear combination

$$I := \begin{bmatrix} < f_1, f_2 >= c_1 f_1 + c_2 f_2 \\ \\ = (4x_0 + 6) f_1 + (x_0 + y_0) f_2. \end{bmatrix} \tag{4.17}$$

In this case, $\{f_1, f_2\}$ are said to be generators of the *Ideal* I.

Definition (4.1) of an *Ideal* can be presented in terms of polynomial equations $f_1, \ldots, f_s \in k[x_1, \ldots, x_n]$. This is done by expressing the system of polynomial equations as

$$\begin{bmatrix} f_1 = 0 \\ f_2 = 0 \\ . \\ . \\ f_s = 0, \end{bmatrix} \tag{4.18}$$

and using them to derive others by multiplying each individual equation f_i by another polynomial $c_i \in k[x_1, \ldots, x_n]$ and summing to get $c_1 f_1 + c_2 f_2 + \ldots + c_s f_s = 0$ (cf., 4.14). The *Ideal* $< f_1, \ldots, f_s >$ thus consists of a system of equations $f_1 = f_2 = \ldots = f_s = 0$, thus indicating that if $f_1, \ldots, f_s \in k[x_1, \ldots, x_n]$, then $< f_1, \ldots, f_s >$ is an *Ideal* generated by $f_1, \ldots, f_s$, i.e., being the *basis* of the *Ideal* I.

In this case, a collection of these *nonlinear algebraic equations* forming *Ideals* are referred to as the set of polynomials generating the *Ideal* and forms the elements of this *Ideal*. Perhaps a curious reader may begin to wonder why the term *Ideal* is used. To quench this curiosity we refer to [310, p. 220] and quote from [63, p. 59] who wrote:

> "On the origin of the term *Ideal*, the concept is attributed to *Dedekind* who introduced it as a set theoretical version of *Kummer's "Ideal number"* to circumvent the failure of unique factorization in certain natural extension of the domain $\mathbb{Z}$. The relevance of *Ideal* in the theory of *polynomial rings* was highlighted by *Hilbert Basis Theorem*. The systematic development of *Ideal* theory; in more general rings is largely due to *E. Noether*. In the older literature, the term "module" is sometimes used for *"Ideal"* (cf., [281]). The term "ring" seems to be due to *D. Hilbert*; *Kronecker* used the term "order" for ring".

Example 4.6 (Ideal). Consider example (4.3) with polynomials in $\mathbb{R}[x, y]$. The *Ideal* $I =< xy - 2y, 2y^2 - x^2 >$.

The generators of an Ideal can be computed using the *division algorithm* defined as (cf., Definition 3.4 of polynomial division on p. 20)

Definition 4.2 (Division algorithm). *Fix a monomial order of polynomials say $x > y$ for polynomials $F = (h_1, \ldots, h_s)$. Then every $f \in k[x, y]$ can be written in the form $f = a_1 h_1 + a_2 h_2 + \ldots + a_s h_s + r$, where $a_i, r \in k[x, y]$ and either $r = 0$ or a linear combination with coefficients in k of monomials, none of which is divisible by any of $LT(f_1), \ldots, LT(f_s)$ (see Definition A.5 on p. 340 for leading term LT).*

Example 4.7 (Division algorithm in a univariate case). Divide the polynomial $f = x^3 + 2x^2 + x + 5$ by $h = x^2 - 2$. We proceed as follows:

$$\left[\begin{array}{r} x+2 \\ x^2-2 \mid x^3+2x^2+x+5 \\ \underline{x^3-2x} \\ 2x^2+3x+5 \\ \underline{2x^2-4} \\ 3x+1, \end{array}\right. \tag{4.19}$$

implying
$x^3 + 2x^2 + x + 5 = (x + 2)(x^2 - 2) + (3x + 1)$, with $a = (x + 2)$ and $r = (3x + 1)$.

The *division algorithm* given in definition (4.2) fits well to the case of *univariate polynomials* as the remainder r can uniquely be determined. For *multivariate polynomials*, the remainder may not be uniquely determined as this depends on the order of the divisors. The division of the polynomial F by $\{f_1, f_2\}$ where f_1 comes before f_2 may not necessarily give the same remainder as the division of F by $\{f_2, f_1\}$ in whose case the order has been changed. This problem is overcome if we pass over to Groebner basis where the existence of every *Ideal* is assured by the Hilbert Basis Theorem [117, pp. 47–61]. The Hilbert Basis Theorem assures that every Ideal $I \subset k[x_1, \ldots, x_n]$ has a finite generating set, that is $I = < g_1, \ldots, g_s >$ for some $\{g_1, \ldots, g_s\} \in I$. The finite generating set G in *Hilbert Basis Theorem* is what is known as a *basis*. Suppose every non-zero polynomial is written in decreasing order of its monomials:

$$\sum_{i=1}^{n} d_i x_i, \quad d_i \neq 0, \; x_i > x_{i+1}, \tag{4.20}$$

if we let the system of generators of the *Ideal* be in a set G, a polynomial f is reduced with respect to G if no leading monomial of an element of G (LM (G)) divides the leading monomial of f (LM(f)). The polynomial f is said to be *completely reduced* with respect to G if no monomials of f is divisible by the leading monomial of an element of G [125, pp. 96–97].

The *basis* G, which completely reduces the polynomial f and uniquely determines the remainder r is also known as the Groebner basis and is defined as follows:

Definition 4.3 (Groebner basis). *A system of generators G of an Ideal I is called a Groebner basis (with respect to the order $<$) if every reduction of $f \in I$ to* *a reduced polynomial (with respect to G) always gives zero as a remainder. This definition is a special case of a more general definition given as: Fix a monomial order and let* $G = \{g_1, \ldots, g_t\} \subset k[x_1, \ldots, x_n]$. *Given* $f \in k[x_1, \ldots, x_n]$, *then* f *reduces to zero Modulo G, written as*

$$f \rightarrow_G 0, \tag{4.21}$$

if f *can be written in the form (cf., 4.18 on p. 38)*

$$f = a_1 g_1 + \ldots + a_t g_t \tag{4.22}$$

such that whenever $a_i g_i \neq 0$, *we have multideg(f)≥ multideg(*$a_i g_i$ *) (see Definition A.5 on p. 340 for leading term LT, LM and Multideg).*

Following the *Definition* 4.3, the reader can revisit *Examples* (4.3) and (4.4) which present the Groebner basis G of the original system F of equations.

Groebner basis has become a household name in algebraic manipulations and finds application in fields such as mathematics and engineering for solving partial differential equations e.g., [273, p. 432]. It has found use as a tool for discovering and proving theorems to solving systems of polynomial equations as elaborated in publications by [99]. Groebner basis also give a solution to the *Ideal* membership problem. By reducing a given polynomial f with respect to the Groebner basis G, f is said to be a member of the *Ideal* if zero remainder is obtained. This implies that if $G = \{g_1, \ldots, g_s\}$ is a Groebner basis of an *Ideal* $I \subset k[x_1, \ldots, x_n]$ and $f \in k[x_1, \ldots, x_n]$ a polynomial, $f \in I$ if and only if the remainder on division of f by G is zero. Groebner bases can also be used to show the equivalence of polynomial equations. Two sets of polynomial equations will generate the same *Ideal* if and only if their Groebner bases are equal with respect to any term ordering, e.g., the solutions of (4.10) satisfy those of (4.9). This property is important in that the solutions of the Groebner basis will satisfy the original system formed by the generating set of nonlinear equations. It implies that a system of polynomial equations $f_1(x_1, \ldots, x_n) = 0, \ldots, f_s(x_1, \ldots, x_n) = 0$ will have the same solutions with a system arising from any Groebner basis of $f_1, \ldots, f_s$ with respect to any term ordering. This is the main property of Groebner basis that is used to solve systems of polynomial equations as will be explained in the next section.

4-3 Buchberger algorithm

The *B. Buchberger algorithm* is the algorithm that computes Groebner bases from given systems of polynomial equations by cancelling the *leading terms* of these polynomials. With the *lexicographic* ordering chosen, one of the elements of the resulting Groebner basis is often a univariate polynomial whose roots can be obtained using Matlab's *"roots"* command. Given polynomials $g_1, \ldots, g_s \in I$, the algorithm seeks to derive the Groebner basis of this *Ideal.* Systems of equations $g_1 = 0, \ldots, g_s = 0$ to be solved in practice are normally formed by these same polynomials which here generate the *Ideal.* The algorithm computes the Groebner basis by making use of pairs of polynomials from the original polynomials $g_1, \ldots, g_s \in I$ and computes the subtraction polynomial known as the $S-polynomial$ defined [117, p. 81] as:

Definition 4.4 (S–polynomial[1]). *Let $f, g \in k[x_1, \ldots x_n]$ be two non-zero polynomials. If multideg $(f) = \alpha$ and multideg $(g) = \beta$, then let $\gamma = \gamma_1, \ldots, \gamma_n$, where $\gamma_i = \max\{\alpha_i, \beta_i\}$ for each i. x^γ is called the Least Common Multiple (LCM) of LM(f) and LM(g) expressed as $x^\gamma = LCM\{LM(f), LM(g)\}$. The $S-$polynomial of f and g is given as*

$$\boxed{S(f,g) = \frac{x^\gamma}{LT(f)} f - \frac{x^\gamma}{LT(g)} g.} \tag{4.23}$$

Expression (4.23) gives S as a linear combination of the monomials

$$\frac{x^\gamma}{LT(f)}, \frac{x^\gamma}{LT(g)},$$

with polynomial coefficients f and g and thus belongs to the *Ideal* generated by f and g (e.g., Definition (4.1) for Ideal on p. 37).

Example 4.8 (S–Polynomial). Consider two polynomials in variables $\{x, y, z\}$ as

$$\begin{bmatrix} g_1 = x^2 + 2a_{12}xy + y^2 + a_{oo} \\ g_2 = y^2 + 2b_{23}yz - +z^2 + b_{oo}. \end{bmatrix} \tag{4.24}$$

with the *lexicographic ordering* defined as $x > y > z$, the $S-$ polynomial $S(g_1, g_2)$ is computed as

$$\begin{bmatrix} LM(g_1) = x^2,\ LM(g_2) = y^2, LT(g_1) = x^2,\ LT(g_2) = y^2 \\ LCM(LM(g_1), LM(g_2)) = x^2y^2 \\ \\ S = \frac{x^2y^2}{x^2}(x^2 + 2a_{12}xy + y^2 + a_{oo}) - \frac{x^2y^2}{y^2}(y^2 + 2b_{23}yz + x_3^2 + b_{oo}) \\ \\ = y^2x^2 + 2a_{12}xy^3 + y^4 + a_{oo}y^2 - x^2y^2 - 2b_{23}x^2yx_3 - x^2x_3^2 - b_{oo}x^2) \\ \\ = -b_{oo}x^2 - 2b_{23}x^2yx_3 - x^2x_3^2 + 2a_{12}xy^3 + y^4 + a_{oo}y^2 \end{bmatrix} \tag{4.25}$$

[1]For the terms appearing in this definition, refer to Appendix A-1, Definition A.5 on p. 340

Theorem 4.1 (Groebner basis in terms of $S-$ polynomial). *A basis G is Groebner basis if and only if for every pair of polynomials f and g of G, $S(f,g)$ reduces to zero with respect to G.*

Theorem 4.1 (and the notion of S-polynomial) together with its proof is exclusively due to Buchberger [95, 96] and is the essence of his PhD thesis, namely, that the theorem shows that, given a polynomial set F, it is possible to decide algorithmically whether or not F is a Groebner basis! It is therefore on the basis of this *Theorem* that the *Buchberger algorithm* for constructing *Groebner bases hinges*!

The implication of *Definition* (4.1) is the following: Given two polynomials $f, g \in G$ such that $LCM\{LM(f), LM(g)\} = LM(f).LM(g)$, the leading monomials of f and g are relatively prime leading to $S(f,g) \rightarrow_G 0$. The concept of prime integer is documented in [234, pp. 1–17].

Example 4.9 (Computation of Groebner basis from the $S-$polynomials). By completing the example given by [125, pp. 101–102], we illustrate how the Buchberger algorithm works. Let us consider the Ideal generated by the polynomial equations

$$\begin{bmatrix} g_1 = x^3yz - xz^2 \\ g_2 = xy^2z - xyz \\ g_3 = x^2y^2 - z, \end{bmatrix} \tag{4.26}$$

with the lexicographic ordering $x > y > z$ adopted. The S–polynomials to be formed are $S(g_1, g_2)$, $S(g_2, g_3)$ and $S(g_1, g_3)$. We consider first $S(g_2, g_3)$ and show that the result is used to suppress g_1. Consequently any pair $S(g_1, g_i)$ (e.g., $S(g_1, g_2)$ and $S(g_1, g_3)$) containing g_1 will not be considered. With $LT(g_2) = xy^2z$ and $LT(g_3) = x^2y^2$ the $LCM(g_2, g_3) = x^2y^2z$. The S–polynomials is then computed as

$$\begin{bmatrix} S(g_2, g_3) = \dfrac{x^2y^2z}{xy^2z}g_2 - \dfrac{x^2y^2z}{x^2y^2}g_3 \\ = (x^2y^2z - x^2yz) - (x^2y^2z - z^2) \\ = -x^2yz + z^2. \end{bmatrix} \tag{4.27}$$

One immediately notes that the leading term of the resulting polynomial LT$(S(g_2, g_3))$ is not divisible by any of the leading terms of the elements of G. The remainder upon the division of $S(g_2, g_3)$ by the polynomials in G is not zero (i.e., when reduced with respect to G). The set G therefore is *not* a Groebner basis. The resulting polynomial is denoted g_4, and its negative (to make calculations more reliable) added to the initial set of G leading to

$$\begin{bmatrix} g_1 = x^3yz - xz^2 \\ g_2 = xy^2z - xyz \\ g_3 = x^2y^2 - z \\ g_4 = x^2yz - z^2. \end{bmatrix} \tag{4.28}$$

The S–polynomials to be formed are now $S(g_1, g_2)$, $S(g_1, g_3)$, $S(g_1, g_4)$, $S(g_2, g_4)$ and $S(g_3, g_4)$. In the set of G, one can write $g_1 = xg_4$ leading, without any change, to the suppression of g_1 leaving only $S(g_2, g_4)$ and $S(g_3, g_4)$ to be considered. Then

$$\left[\begin{array}{l} S(g_2, g_4) = xg_2 - yg_4 \\ \quad = -x^2yz + yz^2, \end{array}\right. \tag{4.29}$$

is reduced by adding g_4 to give $g_5 = yz^2 - z^2$, a non zero value. The set G, which is still *not* a Groebner basis now becomes

$$\left[\begin{array}{l} g_2 = xy^2z - xyz, \\ g_3 = x^2y^2 - z, \\ g_4 = x^2yz - z^2, \\ g_5 = yz^2 - z^2. \end{array}\right. \tag{4.30}$$

The S–polynomials to be considered are now $S(g_3, g_4)$, $S(g_2, g_5)$, $S(g_3, g_5)$ and $S(g_4, g_5)$. We have

$$\left[\begin{array}{l} S(g_3, g_4) = zg_3 - yg_4 \\ \quad = yz^2 - z^2, \end{array}\right. \tag{4.31}$$

which upon subtraction from g_5 reduces to zero. Further,

$$\left[\begin{array}{l} S(g_2, g_5) = zg_2 - xyg_5 \\ \quad = -xyz^2 + xyz^2 \\ \quad = 0 \end{array}\right. \tag{4.32}$$

and

$$\left[\begin{array}{l} S(g_4, g_5) = zg_4 - x^2yg_5 \\ \quad = x^2z^2 - z^3, \end{array}\right. \tag{4.33}$$

which is added to G as g_6 giving

$$\left[\begin{array}{l} g_2 = xy^2z - xyz, \\ g_3 = x^2y^2 - z, \\ g_4 = x^2yz - z^2, \\ g_5 = yz^2 - z^2, \\ g_6 = x^2y^2 - z^3. \end{array}\right. \tag{4.34}$$

The S polynomials to be formed next are $S(g_3, g_5)$, $S(g_2, g_6)$, $S(g_3, g_6)$, $S(g_4, g_6)$ and $S(g_5, g_6)$. We now complete the example by illustrating that all these S – *polynomials* reduce to zero as follows:

$$\boxed{\left[\begin{array}{l} S(g_3, g_5) = z^2g_3 - x^2yg_5 = x^2yz^2 - z^3 - zg_4 = 0 \\ S(g_2, g_6) = xzg_2 - y^2g_6 = -x^2y^2z^2 + y^2z^3 + y^2g_4 = 0 \\ S(g_3, g_6) = z^2g_3 - y^2g_6 = y^2z^3 - z^3 - (yz - z)g_5 = 0 \\ S(g_4, g_6) = zg_4 - yg_6 = yz^3 - z^3 - zg_5 = 0 \\ S(g_5, g_6) = x^2g_5 - yg_6 = -x^2z^2 + yz^3 + g_6 - zg_5 = 0, \end{array}\right.} \tag{4.35}$$

comprising the Groebner basis of the original set in (4.26).

The importance of S–polynomials is that they lead to the cancellation of the leading terms of the polynomial pairs involved. In so doing, polynomial variables are systematically eliminated according to the ordering chosen. For example if the lexicographic ordering $x > y > z$ is chosen, x will be eliminated first, followed by y and the final expression may consist only of the variable z. Cox et al [118, p. 15] has indicated the advantage of lexicographic ordering as being the ability to produce Groebner basis with systematic elimination of variables. *Graded lexicographic ordering* (see *Definition* A.3 of Appendix A-1 on p. 339), on the other hand has the advantage of minimizing the amount of computational space needed to produce the Groebner basis.

Buchberger algorithm is therefore a *generalization* of the *Gauss elimination procedure* for linear systems of equations as shown in *Examples* 4.2, 4.3 and 4.4. If we now put our system of polynomial equations to be solved in a set G, S–pair combinations can be formed from the set of G as illustrated in *Examples* 4.1 and 4.9. The *theorem*, known as the *Buchberger's S*–pair *polynomial criterion*, gives the criterion for deciding whether a given basis is a Groebner basis or not. It suffices to compute all the S–polynomials and check whether they reduce to zero. Should one of the polynomials not reduce to zero, then the basis fails to be a Groebner basis. Since the reduction is a linear combination of the elements of G, it can be added to the set G without changing the *Ideal* generated. Buchberger [97] gives an *optimization criterion* that reduces the number of the S–polynomials already considered in the algorithm. The criterion states that if there is an element h of G such that the leading monomial of h, i.e., LM(h, divides the LCM($f, g \in G$), and if $S(f,h)$, $S(h,g)$ have already been considered, then there is no need of considering $S(f,g)$ as this reduces to zero.

The essential observation in using Groebner bases to solve systems of polynomial equations is that the variety (simultaneous solution of systems of polynomial equations) does not depend on the original system of the polynomials $F := \{f_1, \ldots, f_s\}$, but instead on the *Ideal* I generated by F. This therefore means that the variety $V = V(I)$. One makes use of the special generating set (Groebner basis) instead of the actual system F. Since the Ideal is generated by F, the solutions obtained by solving the affine variety of this Ideal satisfies the original system F of equations as already stated. Buchberger [96] proved that $V(I)$ is void, and thus giving a test as to whether a system of polynomial F can be solved. The solution can be obtained if and only if the computed Groebner basis of Ideal I has 1 as its element. Buchberger [96] further gives the criterion for deciding if $V(I)$ is finite. If the system has been proved to be solvable and finite then [415, theorem 8.4.4, p. 192] gives a theorem for deciding whether the system has finitely or infinitely many solutions. The Theorem states that if G is a Groebner basis, then a solvable system of polynomial equations has finitely many solutions if and only if for every x_i, $1 \leq i \leq n$, there is a polynomial $g_i \in G$ such that $LM(g_i)$ is a pure power of x_i. The process of addition of the remainder after the reduction by the S–polynomials, and thus expanding the generating set is shown by [96], [118, p. 88] and [125, p. 101] to terminate.

The Buchberger algorithm thus makes use of the subtraction polynomials known as the S–polynomials in Definition (4.4) to eliminate the leading terms of a pair of polynomials. In so doing, and if lexicographic ordering is chosen, the process ends up with one of the computed S–polynomials being a univariate polynomial which can be solved and substituted back in the other S–polynomials using the *extension theorem* [118, pp. 25–26] to obtain the other variables.

4-31 Mathematica computation of Groebner basis

Groebner basis can be computed using algebraic softwares of Mathematica (e.g.,version 7 onwards). In Mathematica, Groebner basis command is executed by writing

$$In[1] := GroebnerBasis[\{polynomials\}, \{variables\}], \tag{4.36}$$

where In[1]:= is the Mathematica prompt which computes the Groebner basis for the Ideal generated by the polynomials with respect to the *monomial order* specified by *monomial order options.*

Example 4.10 (Mathematica computation of Groebner basis). In Example (4.3) on p. 36, the systems of polynomial equations were given as

$$\begin{bmatrix} f_1 = xy - 2y \\ f_2 = 2y^2 - x^2. \end{bmatrix}$$

Groebner basis of this system would be computed by

$$In[1] := GroebnerBasis[\{f_1, f_2\}, \{x, y\}], \tag{4.37}$$

leading to the same values as in (4.10).

With this approach, however, one obtains too many elements of Groebner basis which may not be relevant to the task at hand. In a case where the solution of a specific variable is desired, one can avoid computing the undesired variables, and alleviate the need for back-substitution by simply computing the *reduced Groebner basis*. In this case (4.36) modifies to

$$In[1] := GroebnerBasis[\{polynomials\}, \{variables\}, \{elims\}], \tag{4.38}$$

where *elims* is for elimination order. Whereas the term reduced Groebner basis is widely used in many Groebner basis literature, we point out that within Mathematica software, the concept of a reduced basis has a different technical meaning. In this book, as in its predecessor, and to keep with the tradition, we maintain the use of the term reduced Groebner basis.

Example 4.11 (Mathematica computation of reduced Groebner basis). In Example 4.10, one would compute the reduced Groebner basis using (4.38) as

$$In[1] := GroebnerBasis[\{f_1, f_2\}, \{x, y\}, \{y\}], \tag{4.39}$$

which will return only $-2x^2 + x^3$. Note that this form is correct only when the *retained* and *eliminated* variables are disjoint. If they are not, there is absolutely no guarantee as to which category a variable will be put in! As an example, consider the solution of three polynomials $x^2 + y^2 + z^2 - 1$, $xy - z + 2$, and $z^2 - 2x + 3y$. In the approach presented in (4.39), the solution would be

$$In[1] := GroebnerBasis[\{x^2+y^2+z^2-1, xy-z+2, z^2-2x+3y\}, \{x, y, z\}, \{x, y\}], \tag{4.40}$$

leading to

$$\{1024 - 832z - 215z^2 + 156z^3 - 25z^4 + 24z^5 + 13z^6 + z^8\}.$$

Lichtblau (Priv. Comm.) however suggests that the retained variable, in this case (z) and the eliminated variables (x, y) be separated, i.e.,

$$In[1] := GroebnerBasis[\{x^2 + y^2 + z^2 - 1, xy - z + 2, z^2 - 2x + 3y\}, \{z\}, \{x, y\}]. \tag{4.41}$$

The results of (4.41) are the same as those of (4.40), but with the advantage of a vastly better speed if one uses an ordering better suited for the task at hand, e.g., elimination of variables in this example. For the problems that are solved in our books, the condition of the retained and eliminated variables being disjoint is true, and hence the approach in (4.40) has been adopted.

The univariate polynomial $-2x^2 + x^3$ is then solved for x using the roots command in Matlab (see e.g., [207, p. 146]) by

$$roots(\,[1 \;\; -2 \;\; 0 \;\; 0]\,). \tag{4.42}$$

The values of the row vector in (4.42) are the coefficients of the cubic polynomial $x^3 - 2x^2 + 0x + 0 = 0$ obtained from (4.39) (see Sect. 3-62 for solutions of cubic polynomials).

The values of y from Example 4.3 can equally be computed from (4.39) by replacing y in the option part with x and thus removing the need for back substitution. We leave it for the reader to compute the values of y from Example 4.3 and also those of z in Example 4.4 using reduced Groebner basis (4.38) as an exercise. The reader should confirm that the solution of y leads to $y^3 - 2y$ with the roots $y = 0$ or $y = \pm 1.4142$. From experience, we recommend the use of reduced Groebner basis for applications in geodesy and geoinformatics. This will; fasten the computations, save on computer space, and alleviates the need for back-substitution.

4-32 Maple computation of Groebner basis

In Maple Version 10 the command is accessed by typing > *with (Grobner);* Basis (WL, T) is for Groebner, *gbasis* (WL, T) is for reduced Groebner basis. WL list or set of polynomials, T monomial order description, e.g. 'plex'(variables) - pure lexicographic order and > is the Maple prompt and the semicolon ends the

Maple command). Once the Groebner basis package has been loaded, the execution command then becomes > *gbasis (polynomials, variables)* which computes the Groebner basis for the *ideal* generated by the *polynomials* with respect to the *monomial ordering* specified by *variables* in the executable command. We should point out that at the time of writing this book, August 2009, version 13 of Maple has been released.

4-4 Concluding remarks

Using the Groebner basis, most systems of nonlinear equations that are encountered in geodesy and geoinformatics can be solved. All that is required of the user is to write algorithms that can easily be run in Mathematica or Maple using the steps discussed. In latter chapters, we will demonstrate how algorithms using Groebner basis can be written for various tasks. Application of the technique in geodesy can be found in the works of [14, 15, 17, 24, 27, 31, 33]. Several publications exist on the subject, e.g., [63, 64, 99, 115, 117, 118, 125, 262, 331, 374, 392, 415]. For readers who may be interested in exploring the subject further, these literature and similar others are worth reading. The Gröebner bases approach presented in this chapter adds to the treasures of methods that are useful for solving nonlinear algebraic systems of equations in geodesy, geoinformatics, machine vision, robotics and surveying.

Finally, we begun the chapter by a quote from [334]. We think that indeed, systems of more than one nonlinear equations are solvable, and the answer lies in commutative algebra!

5 Polynomial resultants

5-1 Resultants: An alternative to Groebner basis

Besides Groebner basis approach discussed in Chap. 4, the other powerful algebraic tools for solving nonlinear systems of equations are the polynomial resultants approaches. While Groebner basis may require large storage capacity during its computations, polynomial resultants approaches presented herein offers remedy to users who may not be lucky to have computers with large storage capacities. This chapter presents polynomial resultants approaches starting from the resultants of two polynomials, known as the "*Sylvester resultants*", to the resultants of more than two polynomials in several variables known as "*multipolynomial resultants*". In normal matrix operations in linear algebra, one is often faced with the task of computing determinants. Their applications to least squares approach are well known.

For polynomial resultants approaches discussed herein, the ability to compute determinants of matrices is the essential requirement. We will look at how they are formed and applied to solve nonlinear systems of equations. Indeed [358] had already used the resultant technique to the $R^2 \rightarrow R^2$ mapping of gravitation lens. Such mapping describes the light rays which run from a deflector plane (lens) to an observer. For simple lenses such as point masses in galactic fields, [358] observed the global mapping to be an algebraic expression whose inversion led to the problem of solving a polynomial in two variables. Further use of polynomial resultants in geodesy is exemplified in the works of [277, pp. 72–76] and [22, 28, 34].

5-2 Sylvester resultants

Sylvester resultants approach is useful for solving explicitly nonlinear systems of equations with two polynomials in two variables. Problems in this category could be those of two dimensional nature such as planar ranging, planar resection etc., as shall be seen in subsequent chapters. Polynomial resultants approach is based on homogeneous polynomials defined as

Definition 5.1 (Homogeneous polynomial). *If monomials of a polynomial p with non zero coefficients have the same total degree, the polynomial p is said to be homogeneous.*

Example 5.1 (Homogeneous polynomial equation). A homogeneous polynomial equation of total degree 2 is $s = x^2 + y^2 + z^2 + xy + xz + yz$, since the monomials $\{x^2, y^2, z^2, xy, xz, yz\}$ all have the sum of their powers (total degree) being 2.

To set the ball rolling, let us examine next the resultant of two univariate polynomials $s, t \in k[x]$ of *positive degree* as

$$\begin{bmatrix} s = k_0 x^i + \ldots . + k_i,\ k_0 \neq 0,\ i > 0 \\ t = l_0 x^j + \ldots . + l_j,\ l_0 \neq 0,\ j > 0. \end{bmatrix} \tag{5.1}$$

J.L. Awange et al., *Algebraic Geodesy and Geoinformatics*, 2nd ed.,
DOI 10.1007/978-3-642-12124-1_5,

The resultant of s and t, denoted Res(s, t), is the $(i+j)\times(i+j)$ determinant

$$Res\,(s,\,t) = det \begin{bmatrix} k_0 & k_1 & k_2 & . & . & . & k_i & 0 & 0 & 0 & 0 & 0 \\ 0 & k_0 & k_1 & k_2 & . & . & . & k_i & 0 & 0 & 0 & 0 \\ 0 & 0 & k_0 & k_1 & k_2 & . & . & . & k_i & 0 & 0 & 0 \\ 0 & 0 & 0 & k_0 & k_1 & k_2 & . & . & . & k_i & 0 & 0 \\ 0 & 0 & 0 & 0 & k_0 & k_1 & k_2 & . & . & . & k_i & 0 \\ 0 & 0 & 0 & 0 & 0 & k_0 & k_1 & k_2 & . & . & . & k_i \\ l_0 & l_1 & l_2 & . & . & . & l_j & 0 & 0 & 0 & 0 & 0 \\ 0 & l_0 & l_1 & l_2 & . & . & . & l_j & 0 & 0 & 0 & 0 \\ 0 & 0 & l_0 & l_1 & l_2 & . & . & . & l_j & 0 & 0 & 0 \\ 0 & 0 & 0 & l_0 & l_1 & l_2 & . & . & . & l_j & 0 & 0 \\ 0 & 0 & 0 & 0 & l_0 & l_1 & l_2 & . & . & . & l_j & 0 \\ 0 & 0 & 0 & 0 & 0 & l_0 & l_1 & l_2 & . & . & . & l_j \end{bmatrix}, \tag{5.2}$$

where the coefficients of the first polynomial s in (5.1) occupy j rows, while those of the second polynomial t occupy i rows. The empty spaces are occupied by zeros as shown above such that a square matrix is obtained. This resultant is known as the *Sylvester resultant* and has the following properties [118, §3.5] and [375];

1. Res(s, t) is a polynomial in $k_0, \ldots, k_i, l_0, \ldots, l_j$ with integer coefficients.
2. Res(s, t) $= 0$ if and only if $s(x)$ and $t(x)$ have a common factor in $k[x]$.
3. There exist a polynomial $q,\ r \in k[x]$ such that $qs + rt = Res(s,\,t)$.

Sylvester resultants can be used to solve systems of polynomial equations in two variables as shown in Example (5.2). It should also be pointed out that Mathematica has a built in function for computing Sylvester resultant written simply as

$$Resultant[2x^3 + x^2 - 3x + 6, -5x^2 + 2x - 13, x]$$

where equations are placed in the brackets with the last element of the bracket indicating the variable to be eliminated. This illustration gives the resultant as -17908.

Example 5.2 (Sylvester resultants solution of systems of nonlinear equations). Consider the system of equations given in [375, p. 72] as

$$\begin{bmatrix} p := xy - 1 = 0 \\ q := x^2 + y^2 - 4 = 0. \end{bmatrix} \tag{5.3}$$

In-order to eliminate one variable e.g., x, the variable y is *hidden*, i.e., the variable say y is considered as a constant (polynomial of degree zero). We then have the *Sylvester resultant* from (5.2) as

$$Res\,(s,\,t,x) = det \begin{bmatrix} y & -1 & 0 \\ 0 & y & -1 \\ 1 & 0 & y^2-4 \end{bmatrix} = y^4 - 4y^2 + 1, \tag{5.4}$$

which can be readily solved for the variable y and substituted back in any of the equations in (5.3) to obtain the values of the variable x. Alternatively, the procedure can be applied to derive x directly. Hiding x, one obtains with (5.2)

$$Res\,(s,\,t,y) = det \begin{bmatrix} x & -1 & 0 \\ 0 & x & -1 \\ 1 & 0 & x^2-4 \end{bmatrix} = x^4 - 4x^2 + 1. \tag{5.5}$$

The roots of the univariate polynomials (5.4) and (5.5) are then obtained using the Matlab's root command as

$$\{x, y\} = roots([\,1\;0\;-4\;0\;1]\;) = \pm 1.9319\,or\, \pm 0.5176. \tag{5.6}$$

In (5.6), the row vector $[\,1\;0\;-4\;0\;1]$ are the coefficients of the quartic polynomials in either (5.4) or (5.5). Zeros are the coefficients of the variables $\{x^3, y^3\}$ and $\{x, y\}$. The solutions in (5.6) satisfy the polynomials in (5.4) and (5.5). They also satisfy the original nonlinear system of equations (5.3). In (5.4) and (5.5), the determinant can readily be obtained from MATLAB software by typing $det(\mathbf{A})$, where $\mathbf{A}$ is the matrix whose determinant is desired.

For two polynomials in two variables, the construction of resultants is relatively simpler and algorithms for the execution are incorporated in computer algebra systems. Resultants of more than 2 polynomials of multiple variables are however complicated. For their construction, we turn to the *multipolynomial resultants.*

5-3 Multipolynomial resultants

Whereas the resultant of two polynomials in two variables is well known and algorithms for computing it well incorporated into computer algebra packages such as Maple, *multipolynomial resultants*, i.e., the resultant of more than two polynomials still remain an active area of research. This section therefore extends on the use of *Sylvester resultants* to resultants of more than two polynomials of multiple variables, known as *multipolynomial resultants.*

The need for multipolynomial resultants method in geodesy and geoinformatics is due to the fact that many problems encountered require the solution of more than two polynomials of multiple variables. This is true since we are living in a three-dimensional world. We shall therefore understand the term multipolynomial resultants to mean resultants of more than two polynomials. We treat it as a tool besides Groebner bases, and perhaps more powerful to eliminate variables in systems of polynomial equations. In defining it, [288] writes:

> "Elimination theory, a branch of classical algebraic geometry, deals with conditions for common solutions of a system of polynomial equations. Its main result is the construction of a single resultant polynomial of n homogeneous polynomial equations in n unknowns, such that the vanishing of the resultant is a *necessary* and *sufficient* condition for the given system

to have a non-trivial solution. We refer to this resultant as the multipolynomial resultant and use it in the algorithm presented in the paper".

In the formation of the design matrix whose determinants are needed, several approaches can be used as discussed in [284, 285, 286, 287, 288] who applies the eigenvalue-eigenvector approach, [100] who uses characteristic polynomial approach, and [373, 375] who proposes a more compact approach for solving the resultants of a ternary quadric using the Jacobian determinant approach. In this book, two approaches are presented; first the approach based on F. Macaulay [280] formulation (the pioneer of resultants approach) and then a more modern approach based on B. Sturmfels' [375] formulation.

5-31 F. Macaulay formulation:

With n polynomials, the construction of the matrix whose entries are the coefficients of the polynomials $f_1, \ldots, f_n$ can be done in five steps as follows:

Step 1: The given polynomials $f_1 = 0, \ldots, f_n = 0$ are considered to be homogeneous equations in the variables $x_1, \ldots, x_n$ and if not, they are homogenized. Let the degree of the polynomial f_i be d_i. The first step involves the determination of the *critical degree* given by [49] as

$$d = 1 + \sum (d_i - 1). \tag{5.7}$$

Step 2: Once the critical degree has been established, the given monomials of the polynomial equations are multiplied with each other to generate a set X. The elements of this set consists of monomials whose total degree equals the critical degree. Thus if we are given polynomial equations $f_1 = 0, \ldots, f_n = 0$, each monomial of f_1 is multiplied by those of $f_2, \ldots, f_n$, those of f_2 are multiplied by those of $f_3, \ldots, f_n$ until those of f_{n-1} are multiplied by those of f_n. The set X of monomials generated in this form is

$$X^d = \{x^d \mid d = \alpha_1 + \alpha_2 + \ \ldots \ + \alpha_n\}, \tag{5.8}$$

with the variable $x^d = x_1^{\alpha_1} \ldots x_n^{\alpha_n}$.

Step 3: The set X containing monomials each of total degree d is now partitioned according to the following criteria [100, p. 54]

$$\left[\begin{array}{ll} X_1^d = & \{x^\alpha \in X^d \mid \alpha_1 \geq d_1\} \\ X_2^d = & \{x^\alpha \in X^d \mid \alpha_2 \geq d_2 \ and \ \alpha_1 < d_1\} \\ \cdot \quad \cdot & \cdot \\ \cdot \quad \cdot & \cdot \\ \cdot \quad \cdot & \cdot \\ X_n^d = & \{x^\alpha \in X^d \mid \alpha_n \geq d_n \ and \ \alpha_i < d_i, \ for \, i = 1, \ldots, n-1\}. \end{array}\right. \tag{5.9}$$

The resulting sets of X_i^d are disjoint and every element of X^d is contained in exactly one of them.

Step 4: From the resulting subsets $X_i^d \subset X^d$, a set of polynomials F_i which are homogeneous in n variables are defined as

$$F_i = \frac{X_i^d}{x_i^{d_i}} f_i. \tag{5.10}$$

From (5.10), a *square matrix* **A** is now formed with the row elements being the *coefficients* of the monomials of the polynomials $F_i \mid_{i=1,\ldots,n}$ and the columns corresponding to the N monomials of the set X^d. The formed square matrix **A** is of the order

$$\binom{d+n-1}{d} \times \binom{d+n-1}{d},$$

and is such that for a given polynomial F_i in (5.10), the row of the square matrix **A** is made up of the symbolic coefficients of each polynomial. The square matrix **A** has a special property that the non trivial solution of the homogeneous equations F_i which also form the solution of the original equations f_i are in its null space. This implies that the matrix must be singular or its determinant, $det(\mathbf{A})$, must be zero. For the determinant to vanish, therefore, the original equations f_i and their homogenized counterparts F_i must have the same non trivial solutions.

Step 5: After computing the determinant of the square matrix **A** above, [280] suggests the computation of *extraneous factor* in-order to obtain the resultant. Cox et al. [118, Proposition 4.6, p. 99] explains the *extraneous factors* to be integer polynomials in the coefficients of $\bar{F}_0, \ldots, \bar{F}_{n-1}$, where $\bar{F}_i = F_i(x_0, \ldots, x_{n-1}, 0)$. It is related to the determinant via

$$determinant = Res(F_1, \ldots, F_n).Ext, \tag{5.11}$$

with the determinant computed as in step 4, $Res(F_1, \ldots, F_n)$ being the multipolynomial resultant and Ext the extraneous factor. This expression was established as early as 1902 by F. Macaulay [280] and this procedure of *resultant* formulation thus named after him. Macaulay [280] determines the extraneous factor from the sub-matrix of the $N \times N$ square matrix **A** and calls it a factor of minor obtained by deleting rows and columns of the $N \times N$ matrix **A**. A monomial x^α of total degree d is said to be reduced if $x_i^{d_i}$ divides x^α for exactly one i. The extraneous factor is obtained by computing the determinant of the sub-matrix of the coefficient matrix **A**, obtained by deleting rows and columns corresponding to reduced monomials x^α.

For our purpose, it suffices to solve for the unknown variable hidden in the coefficients of the polynomials f_i by obtaining the determinant of the $N \times N$ square matrix **A** and equating it to zero neglecting the extraneous factor. This is because the extraneous factor is an integer polynomial and as such not related to the variable in the determinant of **A**. The existence of the non-trivial solutions provides the *necessary* and *sufficient* conditions for the vanishing of the determinant.

5-32 B. Sturmfels' formulation

Given three homogeneous equations of degree 2 as

$$\begin{array}{l} F_1 := a_{11}x^2 + a_{12}y^2 + a_{13}z^2 + a_{14}xy + a_{15}xz + a_{16}yz = 0 \\ F_2 := a_{21}x^2 + a_{22}y^2 + a_{23}z^2 + a_{24}xy + a_{25}xz + a_{26}yz = 0 \\ F_3 := a_{31}x^2 + a_{32}y^2 + a_{33}z^2 + a_{34}xy + a_{35}xz + a_{36}yz = 0, \end{array} \tag{5.12}$$

the Jacobian determinant is computed by

$$J = det \begin{bmatrix} \frac{\partial F_1}{\partial x} & \frac{\partial F_1}{\partial y} & \frac{\partial F_1}{\partial z} \\ \frac{\partial F_2}{\partial x} & \frac{\partial F_2}{\partial y} & \frac{\partial F_2}{\partial z} \\ \frac{\partial F_3}{\partial x} & \frac{\partial F_3}{\partial y} & \frac{\partial F_3}{\partial z} \end{bmatrix}, \tag{5.13}$$

resulting in a cubic polynomial in the coefficients $\{x,\ y,\ z\}$. Since the determinant polynomial J in (5.13) is a cubic polynomial, its partial derivatives will be quadratic polynomials in variables $\{x, y, z\}$ and are written in the form

$$\begin{array}{l} \frac{\partial J}{\partial x} := b_{11}x^2 + b_{12}y^2 + b_{13}z^2 + b_{14}xy + b_{15}xz + b_{16}yz = 0 \\ \frac{\partial J}{\partial y} := b_{21}x^2 + b_{22}y^2 + b_{23}z^2 + b_{24}xy + b_{25}xz + b_{26}yz = 0 \\ \frac{\partial J}{\partial z} := b_{31}x^2 + b_{32}y^2 + b_{33}z^2 + b_{34}xy + b_{35}xz + b_{36}yz = 0. \end{array} \tag{5.14}$$

The coefficients b_{ij} in (5.14) are cubic polynomials in a_{ij} of (5.12). The final step in computing the resultant of the initial system (5.12) involves the computation of the determinant of a 6×6 matrix given by

$$Res_{222}(F_1, F_2, F_3) = det \begin{bmatrix} a_{11} & a_{12} & a_{13} & a_{14} & a_{15} & a_{16} \\ a_{21} & a_{22} & a_{23} & a_{24} & a_{25} & a_{26} \\ a_{31} & a_{32} & a_{33} & a_{34} & a_{35} & a_{36} \\ b_{11} & b_{12} & b_{13} & b_{14} & b_{15} & b_{16} \\ b_{21} & b_{22} & b_{23} & b_{24} & b_{25} & b_{26} \\ b_{31} & b_{32} & b_{33} & b_{34} & b_{35} & b_{36} \end{bmatrix}. \tag{5.15}$$

The resultant (5.15) vanishes if and only if (5.12) have a common solution $\{x,\ y,\ z\}$, where $\{x,\ y,\ z\}$ are complex numbers or real numbers not all equal zero. The subscripts on the left-hand-side of (5.15) indicate the degree of the polynomials in (5.12) whose determinants are sought.

5-33 The Dixon resultant

5-331 Basic concepts

Consider a system of polynomials

$$\mathcal{F} = \{f_0, f_1, ..., f_d\} \tag{5.16}$$

where

$$f_i = \sum_{\alpha \in \mathcal{A}_i} c_{i,\alpha} \boldsymbol{x}^{\alpha} \tag{5.17}$$

and

$$\boldsymbol{x}^{\alpha} = x_1^{\alpha_1} x_2^{\alpha_2} ... x_d^{\alpha_d} \tag{5.18}$$

for each $i = 0,...,d$, and with the polynomial coefficient $c_{i,\alpha}$, which are also sometimes referred to as the parameters of the system. In general, we consider a collection of polynomials in $x_1,...,x_d$ with coefficients in the field of an arbitrary ring $\mathcal{A}_i \subset \mathbf{N}^d$ [113]. Elimination theory tells us how to construct a single polynomial of d homogeneous polynomial equations in d unknowns, such that its vanishing is a necessary and sufficient condition for a given system to have a non-trivial solution. This single polynomial is called a resultant polynomial [288].

One way to compute the resultant of a given polynomial system is to construct a matrix with the property that whenever the polynomial system has a solution, such a matrix has a deficient rank, thereby implying that the determinant of any maximal minor is a multiple of the resultant. A simple way to construct a resultant matrix is to use the dialytic method [76], i.e. multiply each polynomial with a finite set of monomials, and rewrite the resulting system in matrix form. We call such matrix the dialytic matrix.

This alone, however, does not guarantee that a matrix so constructed is a resultant matrix. Note that such matrices are usually quite sparse: matrix entries are either zero or coefficients of the polynomials in the original system. Good examples of resultant dialytic matrices are Sylvester [377] for the univariate case, and Macaulay [281], as well as the Newton sparse matrices of Canny and Emiris [102] for the multivariate case, where they all differ only in the selection of multiplier monomial sets.

In contrast to dialytic matrices, the *Dixon matrix* is dense since its entries are determinants of the coefficients of the polynomials in the original system. It has the advantage of being an order of magnitude smaller in comparison to a dialytic matrix, which is important as the computation of the symbolic determinant of a matrix is sensitive to its size. The *Dixon matrix* is constructed through the computation of the Dixon polynomial, which is expressed in matrix form.

Below, the generalized multivariate Dixon formulation for simultaneously eliminating many variables from a polynomial system and computing its resultant is briefly reviewed [114].

Let $\pi_i(\boldsymbol{x}^{\alpha}) = \sigma_1^{\alpha_1}...\sigma_i^{\alpha_i} x_{i+1}^{\alpha_{i+1}}...x_d^{\alpha_d}$ where $i \in \{0, 1, ..., d\}$ and the $\sigma_i' s$ are new variables; $\pi_0(\boldsymbol{x}^{\alpha}) = \boldsymbol{x}^{\alpha}$. Now π_i is extended to polynomials in a natural way as

$$\pi_i\left(f_j\left(x_1,...,x_d\right)\right)=f_j\left(\sigma_1,...\sigma_i,x_{i+1},...,x_d\right) \tag{5.19}$$

Given a polynomial system $\mathcal{F}=\{f_0,\ f_1,...f_d\}$, the *Dixon polynomial* is defined as

$$\theta\left(f_0,f_1,...f_d\right)=\delta\left(x_1,\ldots,x_d,\sigma_1,\ldots,\sigma_d\right)=\prod_{i=1}^{d}\frac{1}{\sigma_i-x_i}\mathrm{Det}\begin{pmatrix}\pi_0\left(f_0\right) & \pi_0\left(f_1\right) & \ldots & \pi_0\left(f_d\right)\\ \pi_1\left(f_0\right) & \pi_1\left(f_1\right) & \ldots & \pi_1\left(f_d\right)\\ \cdot & \cdot & \cdots & \cdot\\ \cdot & \cdot & \cdots & \cdot\\ \cdot & \cdot & \cdots & \cdot\\ \pi_d\left(f_0\right) & \pi_d\left(f_1\right) & \ldots & \pi_d\left(f_d\right)\end{pmatrix}. \tag{5.20}$$

The order in which the original variables in $\mathbf{x}$ are replaced by new variables in $\boldsymbol{\sigma}$ is significant in the sense that the *Dixon polynomial* computed using two distinct variable orderings may differ.

A *Dixon polynomial* $\delta(\mathbf{x},\ \boldsymbol{\sigma})$ can be written in bilinear form as [114],

$$\delta\left(x_1,...,x_d,\sigma_1,...\sigma_d\right)=\boldsymbol{\Xi\Theta X}^T, \tag{5.21}$$

where $\boldsymbol{\Xi}=(\sigma^{\beta_1},...,\sigma^{\beta_k})$ and $\mathbf{X}=(x^{\alpha_1},...,x^{\alpha_s})$ are row vectors. The $k\times s$ matrix $\boldsymbol{\Theta}$ is called the Dixon matrix.

5-332 Formulation of the Dixon resultant

Cayley's Formulation of Bézout's Method

Let us recall here the Cayley's [104] formulation of *Bezout's method* [309] for solving two polynomial equations in the univariate case. However, this method is actually due to Euler (e.g., Salmon [352]).

Let us consider two univariate polynomials $f(x)$ and $g(x)$, and let

$$deg=max(degree(f),degree(g)) \tag{5.22}$$

and σ be an auxiliary variable. The quantity

$$\delta(x,\sigma)=\frac{1}{x-\sigma}\det\begin{pmatrix}f(x) & g(x)\\ f(\sigma) & g(\sigma)\end{pmatrix}=\frac{f(x)g(\sigma)-f(\sigma)g(x)}{x-\sigma} \tag{5.23}$$

is a symmetric polynomial in x and σ of *deg*-1 which is called the Dixon polynomial of f and g. Every common zero of f and g is a zero of $\delta(x,\sigma)$ for all values of σ.

Example 5.3. Let us consider two univariate polynomials with parameter π,

$$\begin{aligned}f(x)&=(x-1)(x+3)(x-4)=12-11x-2x^2+x^3\\ g(x)&=(x-\pi)(x+4)=-4\pi+4x-\pi x+x^2.\end{aligned} \tag{5.24}$$

The *Dixon polynomial* is then expressed as

$$\delta = \frac{1}{x-\sigma} \det \begin{pmatrix} f(x) & g(x) \\ f(\sigma) & g(\sigma) \end{pmatrix}, \tag{5.25}$$

which on substituting Eq.(5.24) gives

$$\delta = \frac{1}{x-\sigma} \det \begin{pmatrix} 12 - 11x - 2x^2 + x^3 & -4\pi + 4x - \pi x + x^2 \\ 12 - 11\sigma - 2\sigma^2 + \sigma^3 & -4\pi + 4\sigma - \pi\sigma + \sigma^2 \end{pmatrix}, \tag{5.26}$$

and evaluating the determinant and simplifying leads to $-48 + 56\pi - 12x + 8\pi x - 4\pi x^2 - 12\sigma + 8\pi\sigma + 3x\sigma - 2\pi x\sigma + 4x^2\sigma - \pi x^2\sigma - 4\pi\sigma^2 + 4x\sigma^2 - \pi x\sigma^2 + x^2\sigma^2$.

Hence, at a common zero, each coefficient of σ^i in δ vanishes, leading to

$$\begin{array}{r} \sigma^2 : -4\pi + 4x - \pi x + x^2 = 0 \\ \sigma^1 : -12 + 8\pi + 3x - 2\pi x + 4x^2 - \pi x^2 = 0 \\ \sigma^0 : -48 + 56\pi - 12x + 8\pi x - 4\pi x^2 = 0. \end{array} \tag{5.27}$$

It is a homogeneous system in variables x^0, x^1 and x^2.

$$\begin{pmatrix} -4\pi & 4-\pi & 1 \\ -12+8\pi & 3-2\pi & 4-\pi \\ -48+56\pi & -12+8\pi & -4\pi \end{pmatrix} \begin{pmatrix} x^0 \\ x^1 \\ x^2 \end{pmatrix} = \begin{pmatrix} 0 \\ 0 \\ 0 \end{pmatrix} \tag{5.28}$$

This system has non-trivial solutions if and only if its determinant D is zero. D is called the *Dixon resultant* of f and g. The matrix of the system $\mathbf{M}$ is the Dixon matrix,

$$\mathbf{M} = \begin{pmatrix} -4\pi & 4-\pi & 1 \\ -12+8\pi & 3-2\pi & 4-\pi \\ -48+56\pi & -12+8\pi & -4\pi \end{pmatrix} \tag{5.29}$$

and its determinant is

$$D = -480 + 440\pi + 80\pi^2 - 40\pi^3 \tag{5.30}$$

The polynomials have common zeros if D vanishes. Indeed, solving the equation $D = 0$ we get
$\pi_1 = -3, \pi_2 = 1$and $\pi_3 = 4$

5-333 Dixon's generalization of the Cayley-Bézout method

Dixon [128] generalized *Cayley's approach* to *Bezout's method* to systems of three polynomials equations in three unknowns. Let

$$\begin{array}{r} f(x,y,z) = 0 \\ g(x,y,z) = 0 \\ h(x,y,z) = 0. \end{array} \tag{5.31}$$

Now, the *Dixon polynomial* is defined by

$$\delta(x,y,z,\sigma,\xi) = \frac{1}{(x-\sigma)(y-\xi)} \det \begin{pmatrix} f(x,y,z) & g(x,y,z) & h(x,y,z) \\ f(\sigma,y,z) & g(\sigma,y,z) & h(\sigma,y,z) \\ f(\sigma,\xi,z) & g(\sigma,\xi,z) & h(\sigma,\xi,z) \end{pmatrix} \tag{5.32}$$

Example 5.4. Let us consider the three-variate polynomials

$$\begin{aligned} f &= x^2 + y^2 - 1 \\ g &= x^2 + z^2 - 1 \\ h &= y^2 + z^2 - 1. \end{aligned} \tag{5.33}$$

The *Dixon polynomial* is

$$\delta(x, y, z, \sigma, \xi) = \frac{1}{(x - \sigma)(y - \xi)} det \begin{pmatrix} x^2 + y^2 - 1 & x^2 + z^2 - 1 & y^2 + z^2 - 1 \\ \sigma^2 + y^2 - 1 & \sigma^2 + z^2 - 1 & y^2 + z^2 - 1 \\ \sigma^2 + \xi^2 - 1 & \sigma^2 + z^2 - 1 & \xi^2 + z^2 - 1 \end{pmatrix}. \tag{5.34}$$

Equation (5.34) leads to $yz - 2x^2yz + y\xi - 2x^2y\xi + z\sigma - 2x^2z\sigma + \xi\sigma - 2x^2\xi\sigma$
Now we eliminate the variables y and z. Considering that

$$\delta(x, y, z, \sigma, \xi) = \sigma\xi\left(1 - 2x^2\right) + \xi y\left(1 - 2x^2\right) + \sigma z\left(1 - 2x^2\right) + yz\left(1 - 2x^2\right), \tag{5.35}$$

and the system of equations to be

$$\begin{array}{rcccc} . & y^0z^0 & y^1z^0 & y^0z^1 & y^1z^1 \\ \sigma^0\xi^0: & 0 & 0 & 0 & 1 - 2x^2 \\ \sigma^1\xi^0: & 0 & 0 & 1 - 2x^2 & 0 \\ \sigma^0\xi^1: & 0 & 1 - 2x^2 & 0 & 0 \\ \sigma^1\xi^1: & 1 - 2x^2 & 0 & 0 & 0 \end{array}, \tag{5.36}$$

The Dixon matrix is then given as

$$\mathbf{M} = \begin{pmatrix} 0 & 0 & 0 & 1 - 2x^2 \\ 0 & 0 & 1 - 2x^2 & 0 \\ 0 & 1 - 2x^2 & 0 & 0 \\ 1 - 2x^2 & 0 & 0 & 0 \end{pmatrix}, \tag{5.37}$$

and its determinant, the Dixon resultant, as $D = 1 - 8x^2 + 24x^4 - 32x^6 + 16x^8$.

Dixon (1908) proved that for the three polynomials of degree two, the vanishing of D is a necessary condition for the existence of a common zero. Furthermore, D is not identically zero, namely, it is not zero for every value of x.

5-334 Improved Dixon resultant - Kapur-Saxena-Yang method

Dixon's method and proofs easily generalize to a system of *n+1* generic n degree polynomials in n unknowns (for more details, see, e.g., [239]). Recall that a polynomial is generic if all its coefficients are independent parameters, unrelated to each other. A polynomial in n variables is n-degree if all powers to the maximum of each variable appear in it. Dixon's method only applies to generic n degree polynomials. If this condition fails, then one can face the following problems:

(a) The Dixon matrix, **M**, may be singular;
(b) After removal of the rows and columns containing zeros, the vanishing of the determinant of the Dixon matrix, D, may not give a necessary condition for the existence of a common zero;
(c) After removal of the rows and columns containing zeros, the Dixon matrix may not even be square. Hence, its determinant cannot be defined.

Kapur et al. [239] addressed all three problems successfully, provided a certain precondition holds. Namely, assuming that the column that corresponds to the monomial $1 = x_1^0 x_2^0 \ldots$, the *Dixon matrix* is not a linear combination of the remaining ones. If this precondition is true, then $D = 0$ is a necessary condition of common zeros. This theorem yields a simple algorithm for obtaining the necessary condition $D = 0$, which is called the *Kapur-Saxena-Yang-Dixon* resultant [308]. The algorithm involves the following:

(1) Compute the Dixon matrix **M**. If the precondition holds, continue;
(2) Reduce the **M** rows without scaling to row echelon form, **M**'.
(3) Compute the product D of the pivots of **M**'.

This algorithm was implemented into *Mathematica* by Nakos and Williams [309, 308]. We employ this to illustrate how the last example could be solved. In *Mathematica 5.2* [386], the Dixon resultant prompt is called by typing

$$<< ResultantDixon.$$

In solving Eq.(5.34), we first compute the Dixon polynomial using

$$DixonPolynomial[\{f, g, h\}, \{y, z\}, \{\sigma, \xi\}],$$

followed by the Dixon matrix using

$$\mathbf{M} = DixonMatrix[\{f, g, h\}, \{y, z\}, \{\sigma, \xi\}]$$

which leads to

$$\begin{pmatrix} 0 & 0 & 0 & 1-2x^2 \\ 0 & 0 & 1-2x^2 & 0 \\ 0 & 1-2x^2 & 0 & 0 \\ 1-2x^2 & 0 & 0 & 0 \end{pmatrix}. \tag{5.38}$$

The Dixon resultant is then computed via $DixonResultant[\{f, g, h\}, \{y, z\}, \{\sigma, \xi\}]$ which gives $1 - 8x^2 + 24x^4 - 32x^6 + 16x^8$. Since all these functions are in-built, one can proceed to compute the Dixon resultant directly without going through Dixon polynomial and matrix.

5-335 Heuristic methods to accelerate the Dixon resultant

The basic idea of the Dixon method is to construct a square matrix M whose determinant D is a multiple of the resultant. Usually M is not unique, it is obtained as a maximal minor, in a larger matrix we shall call M^+, and there are usually many maximal minors - any one of which will do. The entries in M are polynomials in parameters. The factors of D that are not the resultant are called the *spurious factors*, and their product is sometimes referred to as the *spurious factor*.

The naive way to proceed is to compute D, factor it, and separate the spurious factor from the actual resultant. But there are problems. On one the hand, the determinant may be so large as for it to be impractical or even impossible to compute, even though the resultant is relatively small, the spurious factor is huge. On the other hand, the determinant may be so large that factoring it is impractical.

Lewis developed three heuristic methods to overcome these problems [267, 268]. The first may be used on any polynomial system. It uses known factors of D to compute other factors. The second also may be used on any polynomial system and it discovers factors of D so that the complete determinant is never produced. The third applies only when the resultant appears as a factor of D in certain exponential patterns. These methods were discovered by experimentation and may apply to other resultant formulations, such as the Macaulay.

5-336 Early discovery of factors: the EDF method

This method is to exploit the observed fact that D has many factors. In other words, we try to turn the existence of spurious factors to our advantage. By elementary row and column manipulations (Gaussian elimination) we discover probable factors of D and extract them from $M_0 \equiv M$. This produces a smaller matrix M_1, still with polynomial entries, and a list of discovered numerators and denominators. Here is a very simple example.

Example 5.5. Given initially

$$M_0 = \begin{pmatrix} 9 & 2 \\ 4 & 4 \end{pmatrix} \qquad \text{numerators : ---} \qquad \text{denominators : ---}$$

We factor a 2 out of the second column, then a 2 from the second row. Thus

$$M_0 = \begin{pmatrix} 9 & 1 \\ 2 & 1 \end{pmatrix} \qquad \text{numerators : } 2, 2 \qquad \text{denominators : ---}$$

Note that $9 \times 4 - 2 \times 4 = 2 \times 2 \times (9 \times 1 - 2 \times 1)$.
We change the second row by subtracting 2/9 of the first

$$M_0 = \begin{pmatrix} 9 & 1 \\ 0 & 7/9 \end{pmatrix} \qquad \text{numerators : } 2, 2 \qquad \text{denominators : ---}$$

We pull out the denominator 9 from the second row, and factor out 9 from the first column:

$$M_0 = \begin{pmatrix} 1 & 1 \\ 0 & 7 \end{pmatrix} \qquad \text{numerators : } 2, 2, 9 \quad \text{denominators : } 9$$

Note that $9 \times 7/9 - 1 \times 0 = \frac{2\times 2\times 9}{9} \times (1 \times 7 - 1 \times 0)$.

We "clean up" or consolidate by dividing out the common factor of 9 from the numerator and denominator lists; any one that occurs may be erased and the list compacted since the first column is canonically simple. We have hence finished one step of the algorithm, and have produced a smaller M_1

$$M_1 = (7) \qquad \text{numerators} : 2, 2 \qquad \text{denominators} : \_\_\_$$

The algorithm terminates by pulling out the 7:

$$\text{numerators} : 2, 2, 7 \quad \text{denominators} : \_\_\_$$

Realize, however, that the $det(M_0) = det \begin{pmatrix} 9 & 2 \\ 4 & 4 \end{pmatrix} = 2 * 2 * 7 = 28$

As expected (since the original matrix contained all integers) the denominator list is empty. The product of all the entries in the numerator list is the determinant, but we never needed to deal with any number larger than 9.

The accelerated Dixon resultant by the Early Discovery Factors (Dixon - EDF) algorithm, was suggested and implemented in the computer algebra system *Fermat* by Lewis [267, 269].

The Dixon resultant is a very attractive tool for solving system of multivariate polynomial geodetic equations (see [319]). Comparing it to other multipolynomial resultants like Strumfels's method, it has the advantages of (i) its small size and high density of the Dixon matrix, (ii) faster computational speed and (iii) being robust.

5-4 Concluding remarks

With modern computers, the polynomial resultant approaches discussed can easily be used to develop algorithms for solving systems of nonlinear equations. Compared to Groebner basis, these have the advantage of not computing extra parameters, thus requiring less computer memory. Its shortcoming, however, lies in the formation of the design matrix which become more complicated and cumbersome as the number of polynomials and variables increases. Unless Groebner basis fails, we recommend it for solving geodetic and geoinformatics nonlinear systems of equations. On the other hand, the polynomial resultants approach comes in handy when the computer's space is limited. With modern computer storage capacity though, most problems requiring algebraic solutions in the fields mentioned above can easily be handled by Groebner basis without fear of a computer breakdown. Publications on the subject include: [22, 28, 34, 49, 100, 101, 103, 118, 128, 148, 149, 195, 254, 279, 280, 281, 282, 284, 285, 286, 287, 288, 352, 289, 290, 291, 292, 305, 373, 375, 401, 404].

Besides Groebner bases and polynomial resultants techniques, there exists another approach for eliminating variables developed by W. WU [418] using the ideas proposed by [347]. This approach is based on Ritt's characteristic set construction and was successfully applied to automated geometric theorem by Wu. This algorithm is referred as the *Ritt-Wu's algorithm* [292].

6 Linear homotpy

6-1 Introductory remarks

A fundamental task in geodesy is the solving of systems of equations. Many geodetic problems are represented as systems of multivariate polynomials. A common problem in solving such systems is improper initial starting values for iterative methods, leading to the convergence to solutions with no physical meaning, or convergence that requires *global method*. Although symbolic methods such as Groebner bases or resultants have been shown to be very efficient, i.e., providing solutions for determined systems such as 3-point problem of 3D affine transformation, the symbolic algebra can be very time consuming, even with special Computer Algebra Systems (CAS). This Chapter proposes the *Linear Homotopy* method that can be implemented easily in high level computer languages like C++ and Fortran, which are faster than CAS by at least two orders of magnitude. Using *Mathematica*, the power of Homotopy is demonstrated by solving three nonlinear geodetic problems: resection, GPS positioning and affine transformation. The method enlarging the domain of convergence is found to be efficient, less sensitive to rounding errors, and has a lower complexity compared to other local methods like Newton-Raphson.

The Chapter is organized as follows: In Section 6-2, a background to linear homotopy is presented. Section 6-3 presents the definition and basic concepts of homotopy required to understand the solution of nonlinear equations. For simplicity, a quadratic and a third degree polynomial equations are used to illustrate the approach. Although the suggested technique can be applied to square systems, overdetermined problems can also be solved after transforming the N-point problem into a square one in a least squares sense. In part II of the book, it will be demonstrated how the technique can be used to solve three nonlinear geodetic problems.

6-2 Background to homotopy

Solving nonlinear systems, especially algebraic polynomial systems, is a fundamental problem that occurs frequently in various fields of science and engineering, like robotics, computational chemistry, computer vision, computational geometry, and signal processing. Using numerical algorithms to solve polynomial systems with tools originating from algebraic geometry is the main activity in so-called *Numerical Algebraic Geometry*, see e.g., [366, 367]. This is a new and developing field on the crossroads of algebraic geometry, numerical analysis, computer science and engineering.

The homotopy continuation method is a global numerical method for solving nonlinear systems in general, and also polynomial systems, see [193]. This method is used to locate all geometrically isolated roots as well as decompose positive dimensional solution sets into irreducible components [394]. Homotopy continuation

J.L. Awange et al., *Algebraic Geodesy and Geoinformatics*, 2nd ed.,
DOI 10.1007/978-3-642-12124-1_6, © Springer-Verlag Berlin Heidelberg 2010

has been established as a reliable and efficient method for solving polynomial systems over the last two decades, originating from the works of [129, 146]. Some computer codes have been developed for homotopy methods, for example Bertini [59], HOM4PS [263], HOMPACK [402], PHCpack [393], and PHoM [200]. Other codes are written in *Maple* (see e.g., [270]) and *Matlab* [127]. Recently, codes for fixed point and Newton homotopy were developed in *Mathematica* see [80, 81].

In geodesy, several algebraic procedures have been put forward for solving nonlinear systems of equations (see, e.g., [41, 320]). The procedures suggested in these studies, such as Groebner bases and resultant approaches, are, however, normally restricted by the size of the nonlinear systems of equations involved. In most cases, the symbolic computations are time consuming. These symbolic computations were necessitated by the failure to obtain suitable starting values for numerical iterative procedures. In situations where large systems of equations are to be solved (e.g., affine transformation), and where symbolic methods are insufficient, there exists the need to investigate the suitability of other alternatives. One such alternative is *linear homotopy*.

In this study linear homotopy continuation methods are introduced (see [249]). The definitions and basic ideas are considered and illustrated. The general algorithms of the linear homotopy method are considered: iterated solution of homotopy equations via Newton-Raphson method; and as an initial value problem of a system of ordinary differential equations. The efficiency of these methods are illustrated here by solving polynomial equation systems in geodesy, namely the solution of 3D resection, GPS navigation and 3D affine transformation problems. Our computations were carried out with Mathematica and Fermat computer algebra systems on a HP workstation xw 4100 with XP operation system, 3 GHz P4 Intel processor and 1 GB RAM. The details can be found in MathSource (Wolfram Research Inc.) in [320], while the mathematical background of the algorithms is discussed in this chapter, and in the literature, e.g. [367]. The application of the implemented Mathematica functions is illustrated in the *Appendix A-3.*

6-3 Definition and basic concepts

The continuous deformation of an object to another object is known as homotopy. Let us consider two univariate polynomials p_1 and q_1 of the same degree,

$$p_1(x) = x^2 - 3 \tag{6.1}$$

$$q_1(x) = -x^2 - x + 1. \tag{6.2}$$

The linear convex function, H, expressed by the variables x and λ, gives the homotopy function defined as

$$\boxed{H(x, \lambda) = (1 - \lambda)p_1(x) + \lambda q_1(x)}, \tag{6.3}$$

where $\lambda \in [0, 1]$.

In geometric terms, the homotopy H provides us with a continuous deformation from p_1, which is obtained for $\lambda = 0$ by $H(x, 0)$, to q_1, which is obtained for $\lambda = 1$ by $H(x, 1)$. The homotopy function for different λ values, $\lambda = 0, 0.1, \ldots, 1$ can be seen on Fig. 6.1. It is called linear homotopy because H is a linear function of the variable λ. [1]

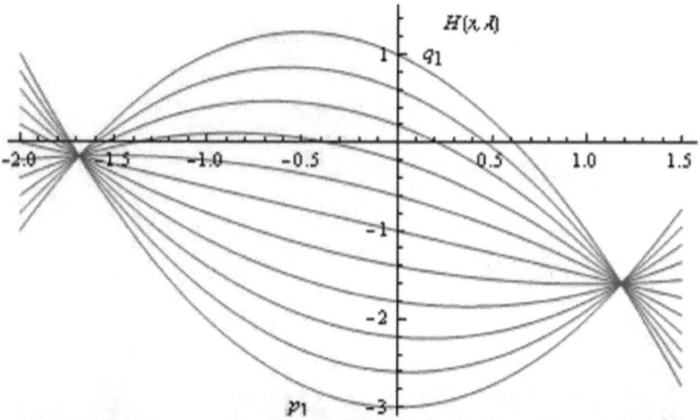

Figure 6.1. The homotopy H for the polynomials p_1 and q_1 (Eqns. 6.1 to 6.3) for different values of λ. Note $\lambda = 0\,(p_1)$ and $\lambda = 1\,(q_1)$.

The polynomial p_1 is called the start system and the polynomial q_1 is called the target system.

6-4 Solving nonlinear equations via homotopy

The homotopy continuation method deforms continuously the known roots of the start system into the roots of the target system. Now, let us look at how homotopy can be used to solve a simple polynomial equation. Considering a polynomial of degree two (i.e., a quadratic equation),

$$q(x) = x^2 + 8x - 9, \tag{6.4}$$

whose roots ($q(x) = 0$) are desired. Instead of this polynomial, we consider a simpler one by deleting the middle term. The roots of this simpler polynomial can be found easily by inspection,

$$p(x) = x^2 - 9. \tag{6.5}$$

---

[1] *Remark:* In some cases when the starting values are not complex, it is reasonable to use the "gamma trick" (see [367]), introducing a random complex constant gamma in the homotopy function: $H(x, \lambda) = (1 - \lambda)p(x) + \lambda\gamma q(x)$, because some singular solutions may lead to problems for Newton's method.

This polynomial also has two roots and will be considered as the start system for the target system, equation (6.4). The linear homotopy can then be defined as follows (see Fig. 6.2)

$$\begin{aligned} H(x,\lambda) &= (1-\lambda)p(x) + \lambda q(x) \\ &= (1-\lambda)\left(x^2-9\right) + \lambda\left(x^2+8x-9\right) = x^2+8x\lambda-9. \end{aligned} \tag{6.6}$$

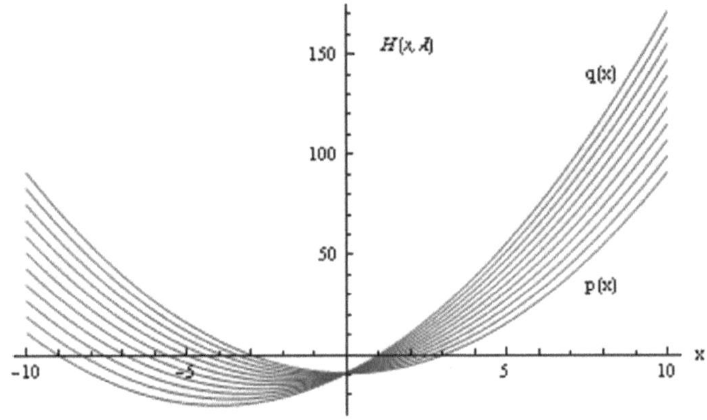

Figure 6.2. The homotopy H (Eqn. 6.6) for the polynomials $p(x)$ and $q(x)$ (Eqns. 6.4 and 6.5, respectively)

The homotopy continuation method deforms continuously the known roots of the start system p(x) = 0, into the roots of the target system q(x) = 0 (see Fig. 6.3 and Fig. 6.5). Let us solve the equation $H(x,\lambda) = 0$ for different values of λ as a parameter. Considering $x_0 = 3$, one of the solutions of $p(x) = 0$, as an initial guess value, and solving $H(x,\lambda_1) = 0$, where $\lambda_1 = 0.2$, we employ the Newton-Raphson method repeatedly, with $\lambda_i = 0.2, 0.4, 0.6, 0.8, 1.0$ (see *Table* 6.1 and Fig. 6.3) by

$$x_i = NewtonRaphson\left(H\left(x,\lambda_i\right), \{x, x_{i-1}\}\right), \tag{6.7}$$

where x_{i-1} is the initial value for the Newton-Raphson method.

The homotopy path is the function $x = x(\lambda)$, where in every point $H(x,\lambda) = 0$, see Fig. 6.4. The points in Fig. 6.4. represent the discrete points of the homotopy path, where $H(x_i,\lambda_i) = 0$ for every $i = 0,\ldots,5$. The second root of Eq(6.4) can be computed similarly by starting from $x_0 = -3$. We now obtain the following values, shown in *Table* 6.2 and Fig. 6.5, with the points of the homotopy path shown in Fig. 6.6.

Table 6.1. The roots of the deforming polynomials for the case of $x_0 = 3$ (Eqn. 6.7)

i	λ_i	x_i
0	0.0	3.00000
1	0.2	2.30483
2	0.4	1.80000
3	0.6	1.44187
4	0.8	1.18634
5	1.0	1.00000

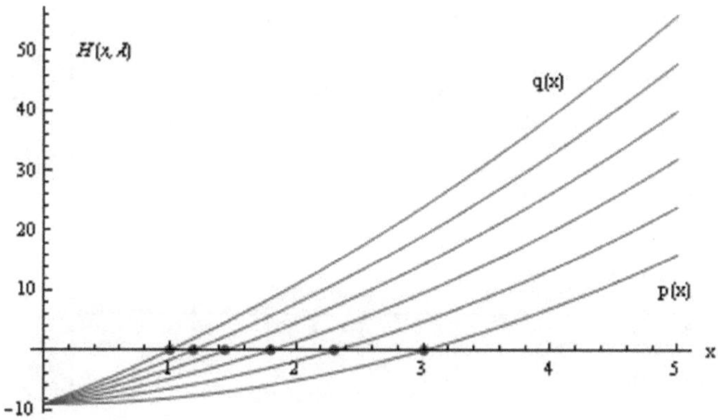

Figure 6.3. Deformation of a root of the polynomial $p(x)$ into a root of the polynomial $q(x)$ for the case of $x_0 = 3$

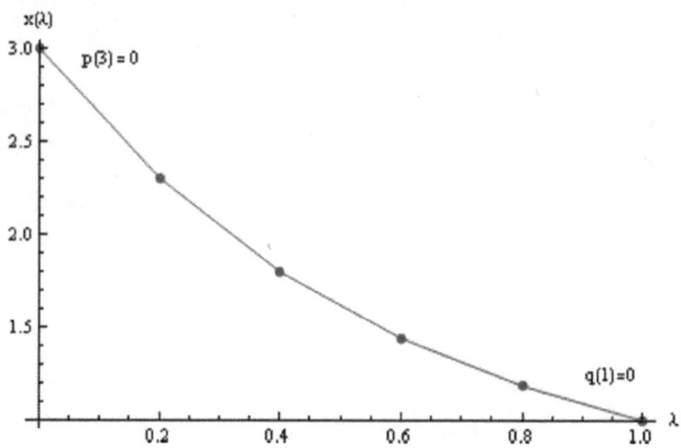

Figure 6.4. The path of homotopy transforming the root of $p(x)$ into the root of $q(x)$ for the case of $x_0 = 3$

Table 6.2. The roots of the deforming polynomials for the case of $x_0 = -3$

i	λ_i	x_i
0	0.0	-3.00000
1	0.2	-3.90483
2	0.4	-5.00000
3	0.6	-6.24187
4	0.8	-7.58634
5	1.0	-9.00000

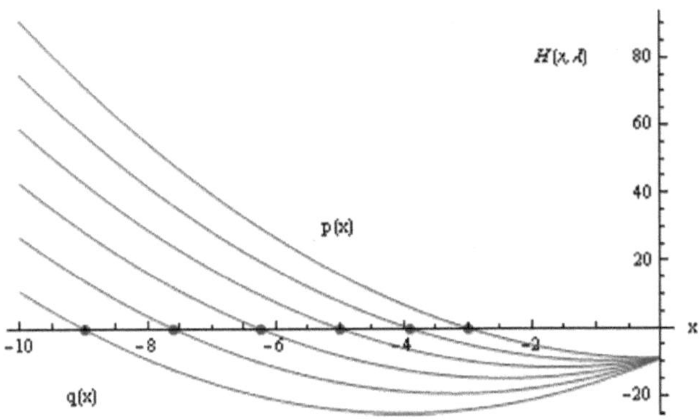

Figure 6.5. Deformation of a root of the polynomial $p(x)$ into a root of the polynomial $q(x)$ for the case of $x_0 = -3$

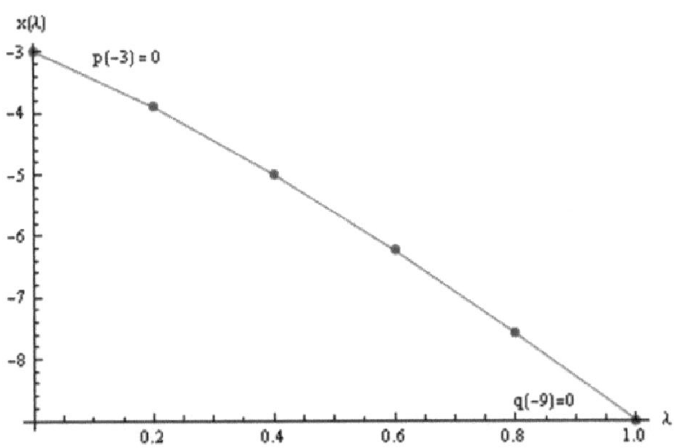

Figure 6.6. The path of homotopy transforming the root of $p(x)$ into the root of $q(x)$ for the case of $x_0 = -3$

6-41 Tracing homotopy path as initial value problem

Comparing the homotopy solution with the traditional Newton-Raphson solution, if $\Delta\lambda$ is small enough, the convergence may be ensured in every step. One can consider this root-tracing problem as an initial value problem of an ordinary differential equation. Considering $H(x,\lambda)=0$ for every $\lambda \in [0,1]$, therefore

$$\mathrm{d}H(x,\lambda) = \frac{\partial H}{\partial x}\mathrm{d}x + \frac{\partial H}{\partial \lambda}\mathrm{d}\lambda \equiv 0 \qquad \lambda \in [0,1]. \tag{6.8}$$

Then the initial value problem is

$$H_x \frac{\mathrm{d}x(\lambda)}{\mathrm{d}\lambda} + H_\lambda = 0 \tag{6.9}$$

with

$$x(0) = x_0. \tag{6.10}$$

Here, H_x is the Jacobian of H with respect to x_i, $i = 1, \ldots, n$, for the case of n nonlinear equations with n variables. In our single variable case, see Eq(6.6),

$$\frac{\partial H}{\partial x} = 2x + 8\lambda \tag{6.11}$$

and

$$\frac{\partial H}{\partial \lambda} = 8x. \tag{6.12}$$

Considering that $x = x(\lambda)$, we obtain

$$\frac{\mathrm{d}x(\lambda)}{\mathrm{d}\lambda} = -\frac{H_\lambda}{H_x} = -\frac{8x(\lambda)}{2x(\lambda)+8\lambda} = -\frac{4x(\lambda)}{x(\lambda)+4\lambda} \tag{6.13}$$

with the initial value

$$x(0) = 3. \tag{6.14}$$

The initial value problem given by Eqns. (6.13) and (6.14) can be solved by any numerical integration technique. The most simple is the Euler method and the most popular is the Runge-Kutta method and its numerous variations, see [106]. Here we employed the built-in solver of *Mathematica*, **NDSolve**, see [208, 320]). The numerical solution of this initial value problem can be seen in Fig. 6.7. for the case of $x_0 = 3$.

This is the continuous path of the homotopy function, Eq(6.6), where in every point $H(x,\lambda) \equiv 0$. Then one of the roots of the target system, Eq(6.4) is $x(1) = 1$. Considering $x_0 = -3$ as the initial value, we get the other root, $x(1) = -9$. In the multivariate case we have to compute the inverse of the Jacobian, namely the initial value problem which is expressed as

$$\frac{\mathrm{d}x(\lambda)}{\mathrm{d}\lambda} = -H_x^{-1}H_\lambda \tag{6.15}$$

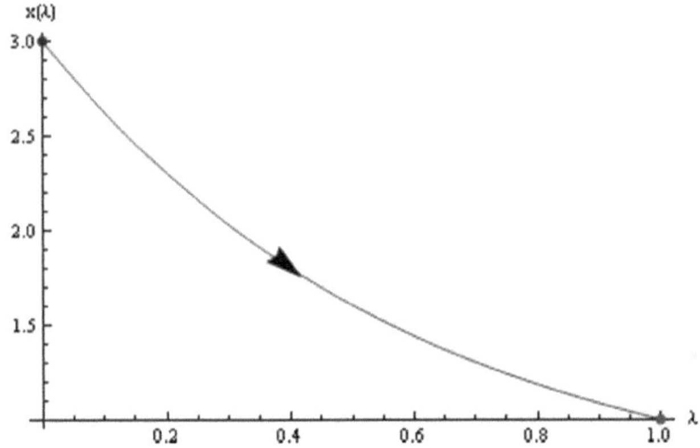

Figure 6.7. Solution of homotopy equation as an the initial value problem

with

$$x(0) = x_0. \tag{6.16}$$

Here H_x^{-1} is the inverse of the Jacobian H_x with respect to x_i, $i = 1, \ldots, n$, in the case of n nonlinear equations with n variables. To carry out the numerical integration, any numerical method can be used, for example the *Runge-Kutta method*. The inverse can be computed in a symbolic way for a limited size like $n = 6$ or 8. For larger systems, a special method is employed, which is more effective than any other standard numerical computation of the inverse of a matrix, see [147]. We consider a new parameter in equation (6.8), namely the parameter t, then

$$\mathrm{d}H(x(t), \lambda(t)) = \frac{\partial H}{\partial x}\mathrm{d}x(t) + \frac{\partial H}{\partial \lambda}\mathrm{d}\lambda(t) \equiv 0 \qquad \lambda(t) \in [0, 1]. \tag{6.17}$$

This means that with $s(t) = \{x(t), \lambda(t)\}$ we obtain

$$H_s(s(t))\frac{\mathrm{d}s}{\mathrm{d}t} = 0, \tag{6.18}$$

where H_s is the Jacobian of the homotopy function with respect to s having $n+1$ dimensions. Then the ith derivative function can be expressed as

$$\frac{\mathrm{d}s_i}{\mathrm{d}t} = (-1)^{i+1} \det\left(H_s^{(1)}, \ldots, H_s^{(i-1)}, H_s^{(i+1)}, \ldots, H_s^{(n+1)}\right), \tag{6.19}$$

where $H_s^{(i)}$ is the ith column of the Jacobian matrix $H_s(s(t))$, see [147]. However, in that case one needs to check the upper bound of the integration parameter, since integration should be carried out up to t^*, where $\lambda(t^*) = 1$, see e.g., [80]. Therefore, it is reasonable to eliminate the last derivative of vector s, namely

$$\frac{\mathrm{d}s_{n+1}}{\mathrm{d}t} = \frac{\mathrm{d}\lambda}{\mathrm{d}t}. \tag{6.20}$$

Considering that

$$\frac{\mathrm{d}s_i}{\mathrm{d}t} = \frac{\mathrm{d}x_i}{\mathrm{d}t} = \frac{\mathrm{d}x_i}{\mathrm{d}\lambda}\frac{\mathrm{d}\lambda}{\mathrm{d}t} = \frac{\mathrm{d}x_i}{\mathrm{d}\lambda}\frac{\mathrm{d}s_{n+1}}{\mathrm{d}t}, \tag{6.21}$$

and

$$\frac{\mathrm{d}x_i}{\mathrm{d}\lambda} = \frac{\mathrm{d}x_i}{\mathrm{d}t} / \frac{\mathrm{d}\lambda}{\mathrm{d}t} = \frac{\mathrm{d}s_i}{\mathrm{d}t} / \frac{\mathrm{d}s_{n+1}}{\mathrm{d}t}, \tag{6.22}$$

for $i = 1, 2, \ldots, n$. The integration now can be carried out with the independent variable λ on [0, 1].

Sometimes the integration can fail because of a singularity in the Jacobi matrix of $H(x, \lambda)$. In order to avoid this, we consider a modified complex homotopy function,

$$H(x, \lambda) = \gamma(1 - \lambda)p_1(x) + \lambda q_1(x) \tag{6.23}$$

where γ is a random complex number. For almost all choices of a *complex* constant γ, all solution paths defined by the homotopy above are regular, i.e.: for all $\lambda \in [0, 1]$, the Jacobian matrix of $H(x, \lambda)$ is regular and no path diverges.

6-42 Types of linear homotopy

As we have seen, the start system can be constructed intuitively, reducing the original system (target system) to a more simple system (start system), whose roots can be easily computed. In order to obtain all roots of the target system, the start system should have as many roots as the target system[2]. The start system can be constructed in many ways, however there are two typical types of the start systems which are usually employed.

6-421 Fixed point homotopy

The start system can be considered as

$$p(x) = x - x_0 \tag{6.24}$$

where x_0 is a guess value for the root of the target system $q(x) = 0$. In this case, the homotopy function is

$$H(x, \lambda) = (1 - \lambda)(x - x_0) + \lambda q(x). \tag{6.25}$$

6-422 Newton homotopy

Another construction for the start system is to consider $p(x)$ as

$$p(x) = q(x) - q(x_0). \tag{6.26}$$

In this case the homotopy function is

[2] *Remark:* However, there are certain theorems that place smoothness restrictions on the input functions, under which all solutions of the target system are found. See [9]

$$H(x,\lambda) = (1-\lambda)(q(x) - q(x_0)) + \lambda q(x) \tag{6.27}$$

or

$$H(x,\lambda) = q(x) - (1-\lambda)q(x_0). \tag{6.28}$$

Although the Newton homotopy is one of the easiest homotopies to use, one does not have guarantees that all solutions will be found, see [109]. There are other methods to construct start systems for linear homotopy. In the following section, we shall see how one can define a start system for polynomial systems automatically.

6-423 Start system for polynomial systems

A fundamental question is, how can we find the proper start system, which will provide all solutions of the target system? This problem can be solved if the nonlinear system is specially a system of polynomial equations.

Let us consider the case where we are looking for the homotopy solution of $f(x) = 0$, where $f(x)$ is a polynomial system, $f(x) : \mathbb{R}^n \to \mathbb{R}^n$. To get all solutions, one should find a proper polynomial system as the start system, $g(x) = 0$, where $g(x) : \mathbb{R}^n \to \mathbb{R}^n$ with known and easily computable solutions. An appropriate start system can be generated in the following way, see [215].

Let $f_i(x_1,\ldots,x_n)$, $i = 1,\ldots,n$ be a system of n polynomials. We are interested in the common zeros of the system, namely $f = (f_1(x),\ldots,f_n(x)) = 0$. Let d_j denote the degree of the jth polynomial - that is the degree of the highest order monomial in the equation. Then such a starting system is,

$$g_j(x) = e^{\boldsymbol{i}\phi_j}\left(x_j^{d_j} - \left(e^{\boldsymbol{i}\theta_j}\right)^{d_j}\right) = 0, \qquad j = 1,\ldots,n \tag{6.29}$$

where ϕ_j and θ_j are random real numbers in the interval $[0,\ 2\ \pi]$ and here $\boldsymbol{i}$ is the imaginary value. If no random complex numbers are introduced for the initial values, this may lead to singularity in the Jacobian of the Newton's method. The equation above has the obvious particular solution $x_j = e^{\boldsymbol{i}\theta_j}$ and the complete set of the starting solutions for $j = 1,\ldots,n$ is given by

$$e^{\boldsymbol{i}\left(\theta_j + \frac{2\pi k}{d_j}\right)}, \qquad k = 0,1,\ldots,d_j - 1. \tag{6.30}$$

Bezout's theorem, see [249], states that the number of isolated roots of such a system is bounded by the total degree of the system, $d = d_1 d_2 \cdots d_n$. We illustrate the discussion above by means of *Example* 6.1. However, we should mention, that there is a tighter upper bound for the number of the solutions introduced as the BKK-bound (due to contributions by Bernstein, Khovanskii and Kushnirenko) and using "mixed volumes", see [74, 75]. For dense systems, i.e., if all monomials have non-zero coefficients, this quantity gives us back the Bezout bound, whereas in many other cases it might be much smaller.

Example 6.1. Let us consider the following system

$$f_1(x, y) = x^2 + y^2 - 1 \tag{6.31}$$

$$f_2(x, y) = x^3 + y^3 - 1. \tag{6.32}$$

With the degrees of the polynomials being $d_1 = 2$ and $d_2 = 3$, this system has six pairs of roots (i.e., $d_1 d_2 = 2 \times 3 = 6$) listed in *Table* 6.3 resulting from the direct solution.

Table 6.3. Roots of system Eqs. (6.31)-(6.32)

x	y
0	1
0	1
1	0
1	0
$-1 - \frac{i}{\sqrt{2}}$	$-1 + \frac{i}{\sqrt{2}}$
$-1 + \frac{i}{\sqrt{2}}$	$-1 - \frac{i}{\sqrt{2}}$

Now we compute a start system and its solutions by considering equations (6.29) and (6.30) with $n = 2$ and $x_1 = x$, $and\, x_2 = y$, the resulting start system consisting of two equations. Employing Eq (6.29), the first equation is

$$g_1(x, y) = e^{i\phi_1}(x^{d_1} - \left(e^{i\theta_1}\right)^{d_1}) \tag{6.33}$$

or in its trigonometric form

$$g_1(x, y) = (\cos\phi_1 + i\sin\phi_1)\left[x^2 - (\cos\theta_1 + i\sin\theta_1)^2\right]. \tag{6.34}$$

which can be simplified as

$$g_1(x, y) = \left(x^2 - \cos 2\theta_1 - i\sin 2\theta_1\right)(\cos\phi_1 + i\sin\phi_1) \tag{6.35}$$

Generating two real random numbers, θ_1 and ϕ_1 in the interval $[0,\ 2\ \pi]$, we may get

$$g_1(x, y) = (0.673116 + 0.739537i)\left(-0.933825 + 0.35773i + x^2\right). \tag{6.36}$$

The second equation can be generated in the same way with randomly generated real numbers, θ_2 and ϕ_2.

$$g_2(x, y) = (-0.821746 - 0.569853i)\left(-0.957532 - 0.288325i + y^3\right). \tag{6.37}$$

The first equation has the following roots, considering Eq (6.30) with $d_1 = 2$

$$x_1 = e^{i\left(\theta_1 + \frac{2\pi 0}{2}\right)} = 0.983317 - 0.1819i, \tag{6.38}$$

and

$$x_2 = e^{i\left(\theta_1 + \frac{2\pi 1}{2}\right)} = -0.983317 + 0.1819i. \tag{6.39}$$

Similarly, the second equation has three roots, $d_2 = 3$

$$y_1 = e^{i\left(\theta_2 + \frac{2\pi 0}{3}\right)} = -0.413328 - 0.910582i \tag{6.40}$$

$$y_2 = e^{i\left(\theta_2 + \frac{2\pi 1}{3}\right)} = 0.995251 + 0.0973382i \tag{6.41}$$

$$y_3 = e^{i\left(\theta_2 + \frac{2\pi 2}{3}\right)} = -0.581923 + 0.813244i. \tag{6.42}$$

All combinations of these roots $\{x_i, y_j\}$ are the roots of the start system,

$$\{x_1, y_1\} = \{0.983317 - 0.1819i, -0.413328 - 0.910582i\} \tag{6.43}$$

$$\{x_1, y_2\} = \{0.983317 - 0.1819i, 0.995251 + 0.0973382i\} \tag{6.44}$$

$$\{x_1, y_3\} = \{0.983317 - 0.1819i, -0.581923 + 0.813244i\} \tag{6.45}$$

$$\{x_2, y_1\} = \{-0.983317 + 0.1819i, -0.413328 - 0.910582i\} \tag{6.46}$$

$$\{x_2, y_2\} = \{-0.983317 + 0.1819i, 0.995251 + 0.0973382i\} \tag{6.47}$$

$$\{x_2, y_3\} = \{-0.983317 + 0.1819i, -0.581923 + 0.813244i\} \tag{6.48}$$

The homotopy function is given by

$$H(x, y, \lambda) = (1 - \lambda)\begin{pmatrix} g_1(x, y) \\ g_2(x, y) \end{pmatrix} + \lambda \begin{pmatrix} f_1(x, y) \\ f_2(x, y) \end{pmatrix}, \tag{6.49}$$

and the corresponding differential equation system

$$\frac{d}{d\lambda}\begin{pmatrix} x(\lambda) \\ y(\lambda) \end{pmatrix} = -H_{x,y}^{-1} H_\lambda \tag{6.50}$$

where

$$H_{x,y} = \begin{pmatrix} \lambda \frac{\partial f_1}{\partial x} + (1-\lambda)\frac{\partial g_1}{\partial x} & \lambda \frac{\partial f_1}{\partial y} + (1-\lambda)\frac{\partial g_1}{\partial y} \\ \lambda \frac{\partial f_2}{\partial x} + (1-\lambda)\frac{\partial g_2}{\partial x} & \lambda \frac{\partial f_2}{\partial y} + (1-\lambda)\frac{\partial g_2}{\partial y} \end{pmatrix} \tag{6.51}$$

and

$$H_\lambda = \begin{pmatrix} f_1(x, y) - g_1(x, y) \\ f_2(x, y) - g_2(x, y) \end{pmatrix}. \tag{6.52}$$

Hence, we must solve this system with six initial values. These initial values - the solutions of the start system - will provide the start points of the six homotopy paths. The end points of these paths are the six desired solutions of the original system $f_1 = 0$ and $f_2 = 0$. In this case, the trajectories of the solution of the system of differential equations, Eq (6.50) will be complex. The curves in Fig. 6.8. shows the homotopy paths belonging to the six initial values given by, Eq (6.43) - Eq (6.48). The curves show the paths on the complex plane, where axis *Re* stands for the real and axis *Im* stands for the imaginary parts of a complex number. The parameter value λ is increasing along the path from the solution of the start system, $\lambda = 0$ up to the solution of the target system, $\lambda = 1$, the end point of the curve. Along the paths the homotopy function is $H = 0$ for all λ. The end points of the homotopy paths are listed in *Table* 6.4

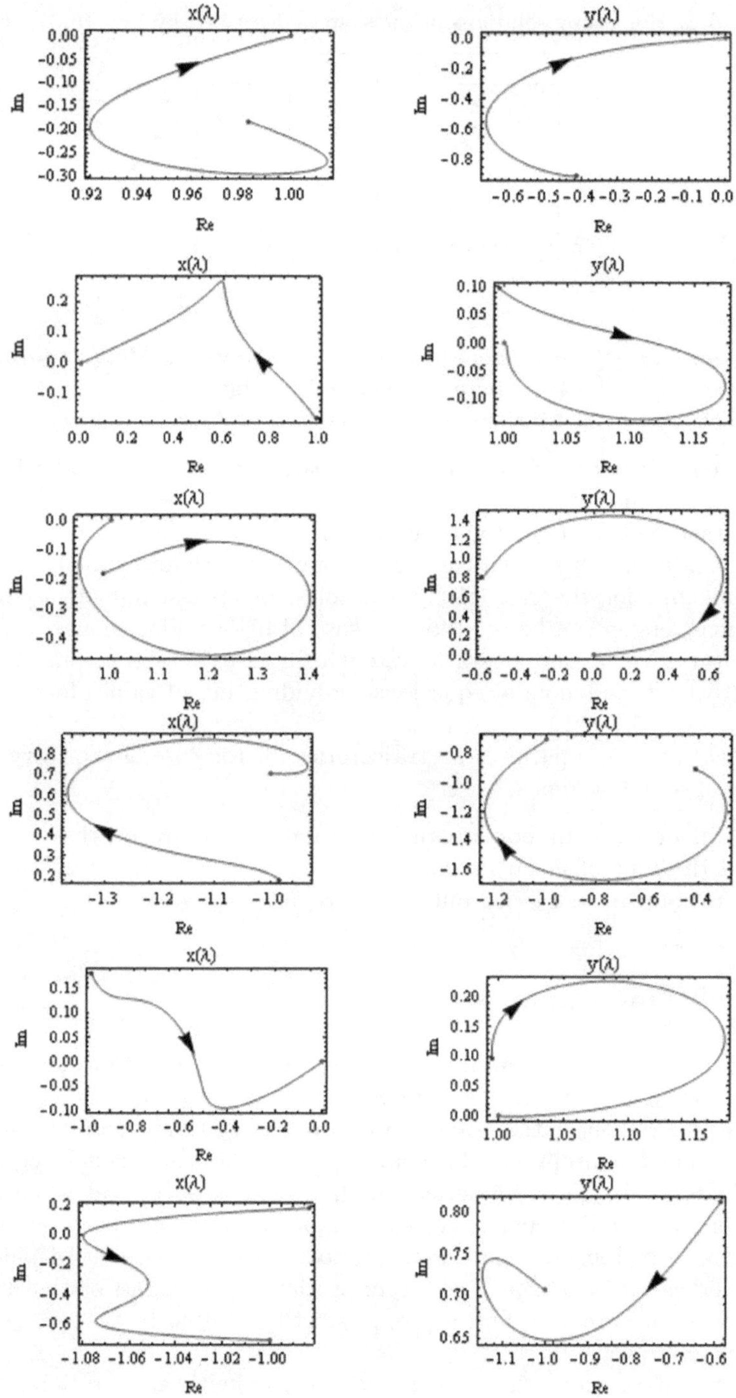

Figure 6.8. Homotopy paths starting from the six starting values expressed by (6.43)-(6.48)

Table 6.4. Homotopy solutions of the system described by Eqs. (6.31)-(6.32)

x	y
0.	1.
0.	1.
1.	0.
1.	0.
$-1. - 0.707107i$	$-1. + 0.707107i$
$-1. + 0.707107i$	$-1. - 0.707107i$

Computations can be easily achieved using the CAS system *Mathematica*, as illustrated in the *Appendix A-3* where this example is solved.

Some features of our Mathematica implementation are as follows:

1. Direct computation of the homotopy paths using the standard Newton-Raphson method.
2. Computation of the homotopy paths using numerical integration. An implicit differential equation system can be transformed into an explicit one by computer algebra for the case of 6-8 variables, or special numerical techniques without inversion can be used for the case of higher dimensions.
3. Computation of a proper start system using Bezout's theorem for systems of multivariate polynomial equations, providing initial values for all paths of roots.
4. Visualization of the paths of all trajectories for the case of arbitrary numbers of variables and systems.

Generally, all of these functions can be parameterized freely, that is the start system and the type of linear homotopy can be defined by the user. In addition, any computation can be carried out to any degree of precision.

6-5 Concluding remarks

As demonstrated in these examples, the linear homotopy method proves to be a powerful solution tool in solving nonlinear geodetic problems, especially if it is difficult to find proper initial values to ensure the convergence of local solution methods. Linear homotopy is robust and enlarges the convergency region of the local methods. This global numerical method can be successful when symbolic computation based on Groebner basis or Dixon resultant fail because of the size and complexity of the problem. However, to reduce the number of paths to be traced as indicated by *Bezout's* theorem as the upper bound of the number of solutions, it is important to find a proper starting system to ensure fewer initial value problems to solve.

This method provides the geodesy community with an additional powerful mathematical tool that is useful, not only in root finding, but also in solving complex problems that can be transformed into systems of polynomial equations. We have also shown that it offers faster computations and in some cases solves complex problems where existing methods such as Groebner basis or even local numerical methods, such as Newton-type methods, fail.

7 Solutions of Overdetermined Systems

"Pauca des Matura" –a few but ripe – C. F. Gauss

7-1 Estimating geodetic and geoinformatics unknowns

In geodesy and geoinformatics, field observations are normally collected with the aim of estimating parameters. Very frequently, one has to handle *overdetermined systems* of nonlinear equations. In such cases, there exist *more equations* than *unknowns*, therefore "the solution" of the system can be interpreted only in *least squares* sense.

In geodynamics for example, GPS and gravity measurements are undertaken with the aim of determining crustal deformation. With improvement in instrumentation, more observations are often collected than the unknowns. Let us consider a simple case of measuring structural deformation. For deformable surfaces, such as mining areas, or structures (e.g., bridges), several observable points are normally marked on the surface of the body. These points would then be observed from a network of points set up on a non-deformable stable surface. Measurements taken are distances, angles or directions which are normally more than the unknown positions of the points marked on the deformable surface leading to *redundant observations*.

Procedures that are often used to estimate the unknowns from the measured values will depend on the nature of the equations relating the observations to the unknowns. If these equations are linear, then the task is much simpler. In such cases, any procedure that can invert the normal equation matrix such as *least squares*, *linear Gauss-Markov model* etc., would suffice. Least squares problems can be *linear* or *nonlinear*. The *linear least squares* problem has a closed form (exact) solution while the *nonlinear* problem does not. They first have to be *linearized* and the unknown parameters estimated by *iterative* refinements; at each iteration the system is approximated by a *linear* one.

Procedures for estimating parameters in linear models have been documented in [244]. Press et al. [334] present algorithms for solving linear systems of equations. If the equations relating the observations to the unknowns are nonlinear as already stated, they have first to be linearized and the unknown parameters estimated iteratively using numerical methods. The operations of these numerical methods require some approximate starting values. At each iteration step, the preceding estimated values of the unknowns are improved. The iteration steps are repeated until the difference between two consecutive estimates of the unknowns satisfies a specified threshold. Procedures for solving nonlinear problems such as the Steepest-descent, Newton's, Newton-Rapson and Gauss-Newton's have been discussed in [334, 379]. In particular, [379] recommends the Gauss-Newton's method as it exploits the structure of the objective function (sum of squares) that is to be minimized. In [381], the manifestation of the nonlinearity of a function during the various stages of adjustment is considered. While extending

J.L. Awange et al., *Algebraic Geodesy and Geoinformatics*, 2nd ed.,
DOI 10.1007/978-3-642-12124-1_7,

the work of [252] on *nonlinear adjustment* with respect to geometric interpretation, [178, 179] have presented the *necessary* and *sufficient* conditions for least squares adjustment of *nonlinear Gauss-Markov model*, and provided the geometrical interpretation of these conditions. Another geometrical approach include the work of [83], while non geometrically treatment of nonlinear problems have been presented by [61, 255, 297, 332, 350, 354].

This Chapter presents different approaches for solving the problem. Two procedures; the **A**lgebraic **LE**ast **S**quare **S**olution (ALESS) discussed in Sect. 7-2 and the Gauss-Jacobi combinatorial approach presented in Sect. 7-3. For the ALESS approach, the original problem is transformed into a *minimization problem* constructing the *objective function* symbolically. The overdetermined system is then converted into a determined one by defining the *objective function* as the sum of the square of residuals of the equations. The necessary condition for the minimum is set such that each *partial derivatives* of the objective function should be zero. In this case the determined model will consist of as many equations as many parameters that were in the original overdetermined model.

7-2 Algebraic LEast Square Solution (ALESS)

7-21 Transforming overdetermined systems to determined

Let Δ be the objective function to be minimized

$$\Delta(p_1, p_2, ..., p_n) = \sum f_i^2, \tag{7.1}$$

where n is the number of the unknown parameters - $p_j \quad (j = 1 \ldots n)$ - and $f_i \quad (i = 1 \ldots m)$ are the observational equations. The objective function should be minimized according to the necessary condition of the minimum,

$$\frac{\partial \Delta}{\partial p_1} = 0, \frac{\partial \Delta}{\partial p_2} = 0, \ldots, \frac{\partial \Delta}{\partial p_n} = 0. \tag{7.2}$$

Now the system consists of as many equations as the number of the unknown parameters. The solution of the original, overdetermined system (in least square sense) will also be the solution of this "*square determined system*".

Let us suppose, that our nonlinear system is a system of multivariate polynomial equations, then the following theorem can be considered:

Theorem 7.1. *Given m algebraic (polynomial) observational equations, where m is the dimension of the observation space Y of order l in n unknown variables, and n is the dimension of the parameter space X. Then there exists n normal equations of the polynomial order $(2l-1)$ to be solved with algebraic methods.*

This solution will be the algebraic least square solution (ALESS) of the overdetermined system.

Proof

Let us consider the following system,

$$e_1(x,y) = x^2 + y - 3$$
$$e_2(x,y) = x + \frac{1}{8}y^2 - 1$$
$$e_3(x,y) = x - y$$

here $n = 2$, $m = 3$ and $l = 2$. The objective function to be minimized is

$$\Delta = e_1^2 + e_2^2 + e_3^2 = 10 - 2x - 4x^2 + x^4 - 6y - 2xy + 2x^2y + \frac{7}{4}y^2 + \frac{1}{4}xy^2 + \frac{1}{64}y^4.$$

The order of the objective function is $2l = 4$. The overdetermined system has one solution from the point of view of least square sense, the minimum of this objective function.

Now, We need to find the solutions of the original system by solving the determined problem. Considering the necessary condition for the minimum as

$$f_1 = \frac{\partial \Delta}{\partial x} = -2 - 8x + 4x^3 - 2y + 4xy + \frac{1}{4}y^2 = 0$$
$$f_2 = \frac{\partial \Delta}{\partial y} = -6 - 2x + 2x^2 + \frac{7}{2}y + \frac{1}{2}xy + \frac{1}{16}y^3 = 0,$$

one obtains the determined system whose order is $2l - 1 = 3$. Solving this square system leads to 6 complex solutions and 3 real ones. The real solutions are

$$x_1 = -1.371 \qquad y_1 = -0.177989$$
$$x_2 = 1.24747 \qquad y_2 = 1.27393$$
$$x_3 = -2.69566 \qquad y_3 = -4.24638,$$

which upon being substituted to the objective function Δ results into the residuals (r_i) as

$$r_1 = 8.71186$$
$$r_2 = 0.232361$$
$$r_3 = 4.48362.$$

From these residuals, the admissible real solutions is the one which provides the least value of the residual (r_i), i.e., the second solution. In this case, the solution in least squares sense is

$$x_2 = 1.24747 \qquad y_2 = 1.27393.$$

♣

7-22 Solving the determined system

It is possible to find the solutions of the determined square system using local or global methods. Local methods such as the extended Newton-Raphson (see Chap. 9) or homotopy (see Chap. 7)) can be used if a good initial values are known. Usually, these initial values can be calculated from the solution of a minimal subset (see Palancz et al [322, 323, 430]). Using global methods, one should find all of the *real solutions* of the determined system representing the original overdetermined system. This is then followed by selecting solutions that provide the *least value* of the *objective function.*

Let us examine two methods of finding all of the roots of the previous polynomial system. The two main types of the algebraic methods that are at our disposal, which we have encountered in the previous chapters are:

- symbolic solutions using computer algebra as *resultants* or *Groebner basis*
- global numerical methods like *linear homotopy*

As an illustration, considering our problem, we solve the polynomial system via reduced Groebner basis using *Mathematica* as

$$GroebnerBasis[\{f_1, f_2\}, \{x, y\}, \{y\}]$$
$$GroebnerBasis[\{f_1, f_2\}, \{x, y\}, \{x\}],$$

which leads to

$$-318 + 256x - 231x^2 - 292x^3 + 166x^4 + 186x^5 + 44x^6 - 56x^7 + 8x^9 = 0,$$
$$-24576 - 110592y + 147456y^2 - 39168y^3 + 2016y^5 - 48y^6 + 104y^7 + y^9 = 0.$$

Solving the above univariate polynomials provide the real solution for variable x as

$$x_1 = -2.69566 \quad x_2 = -1.371 \quad x_3 = 1.24747,$$

and those of the variable y as

$$y_1 = -4.24638 \quad y_2 = -0.177989 \quad y_3 = 1.27393,$$

which are the same solutions we obtained in the previous section. The solutions giving minimum residuals are

$$x = 1.24747 \qquad y = 1.27393.$$

In general, for exact solution of determined systems, *NSolve*, the built-in function of Mathematica, which utilize numerical Groebner basis seems to be a good choice. However, if the system has many roots without physical meaning and one does not need to compute all of the roots, the linear homotopy of fixed point or *FindRoot* built-in Mathematica can be an appropriate method. Whereas the statement above refers to Mathematica, it is essential for users to know that other algebraic packages, e.g., Matlab and Maple also have similar capabilities as Mathematica.

To demonstrate the global numerical method, let us employ the *linear homotopy*. As our example is a polynomial system, we can use the automatically generated start systems with random complex numbers (see Sect. 6-423). The target system is given by

$$f_1(x,y) = -2 - 8x + 4x^3 - 2y + 4xy + \frac{1}{4}y^2 = 0$$
$$f_2(x,y) = -6 - 2x + 2x^2 + \frac{7}{2}y + \frac{1}{2}xy + \frac{1}{16}y^3 = 0.$$

with x, y variables. The degrees of the polynomials are $d_1 = 3$ and $d_2 = 3$. The start system is given by

$$g_1(x,y) = (-0.814932 + 0.579556\boldsymbol{i})(-0.550639 + 0.834743\boldsymbol{i} + x^3)$$
$$g_2(x,y) = (0.858366 - 0.513038\boldsymbol{i})(-0.77 - 0.638044\boldsymbol{i} + y^3)$$

and its initial values, the solutions of the start system are presented in *Table* 7.1.

Table 7.1. Initial values of the homotopy function

x	y
$-0.193155 + 0.981168i$	$-0.288775 - 0.957397i$
$-0.193155 + 0.981168i$	$0.973518 + 0.228612i$
$-0.193155 + 0.981168i$	$-0.684743 + 0.728785i$
$-0.753139 - 0.657861i$	$-0.288775 - 0.957397i$
$-0.753139 - 0.657861i$	$0.973518 + 0.228612i$
$-0.753139 - 0.657861i$	$-0.684743 + 0.728785i$
$0.946294 - 0.323307i$	$-0.288775 - 0.957397i$
$0.946294 - 0.323307i$	$0.973518 + 0.228612i$
$0.946294 - 0.323307i$	$-0.684743 + 0.728785i$

The number of paths is 9. Employing the direct path tracing technique we get as solutions in *Table* 7.2.

Table 7.2. End points of the homotopy paths

x	y
-1.371	-0.177989
$-0.987112 + 1.25442i$	$2.21036 + 1.61614i$
-2.69566	-4.24638
$0.33941 + 0.679091i$	$-0.606933 - 7.81887i$
$-0.987112 - 1.25442i$	$2.21036 - 1.61614i$
$0.33941 - 0.679091i$	$-0.606933 + 7.81887i$
$2.0573 + 1.28006i$	$-0.0282082 - 7.43985i$
1.24747	1.27393
$2.0573 - 1.28006i$	$-0.0282082 + 7.43985i$

The real solutions are the same as we have seen before. The paths of the good solution can be seen in Fig. 7.1.

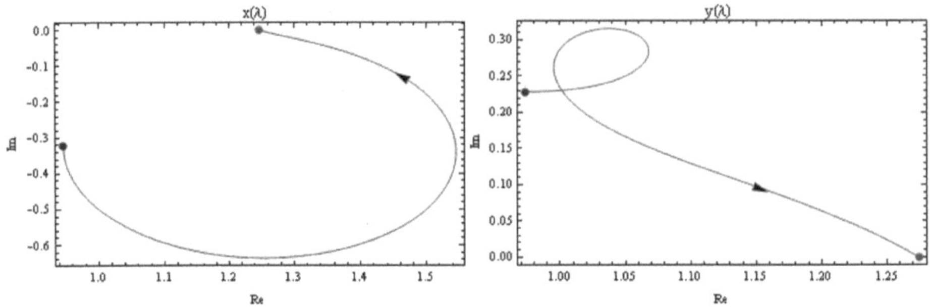

Figure 7.1. The paths of the good solution

7-3 Gauss-Jacobi combinatorial algorithm

In this section a combinatorial method is presented. The *advantage* of this method is that solving a sub-problems in symbolic form, the numerical solution of all of the combinatorial subproblems can be speedily computed, and using proper weighting technique, the solutions can be easily achieved. However, the *disadvantage* of the method is that for a vastly overdetermined problem, combinatorial explosion results. To avoid this, one of the appropriate approaches is to solve a few sub-problems in a closed form and choose the initial solutions that are in the correct vicinity of the desired solutions and use them as starting values for a *least square optimization*. In Mathematica for example, choosing initial solutions that are in the correct vicinity would involve calling *FindMinimum*, (Lichtblau, Private Communication) and in Lichtblau[269]. Alternatively, another local method, e.g., the Extended Newton-Raphson method discussed in Chap. 8 can also be employed starting with the weighted sub-problem solutions.

7-31 Combinatorial approach: the origin

Presented in this chapter is an alternative approach to traditional iterative numerical procedures for solving overdetermined problems, i.e., where more observations than unknown exist. This approach, which we call the *Gauss-Jacobi combinatorial* has the following advantages:

1. From the start, the objective is known.
2. It does not require linearization.
3. The need for iteration does not exist.
4. The variance-covariance matrices of all parameters are considered.
5. It can be exploited for outlier diagnosis.

The combinatorial approach traces its roots to the work of *C. F. Gauss* which was published posthumously (see Appendix A-2). Whereas the procedures presented in Chaps. 4 and 5 solve nonlinear systems of equations where the number of observations n and unknowns m are equal, i.e., $n = m$, Gauss-Jacobi combinatorial solves the case where $n > m$. In Fig. 4.1 on p. 34 for example, two distance measurements from known stations P_1 and P_2 were used to determine the position of unknown station P_0. Let us assume that instead of the two known stations, a third distance was measured from point P_3 as depicted in Fig. 7.2. In such a

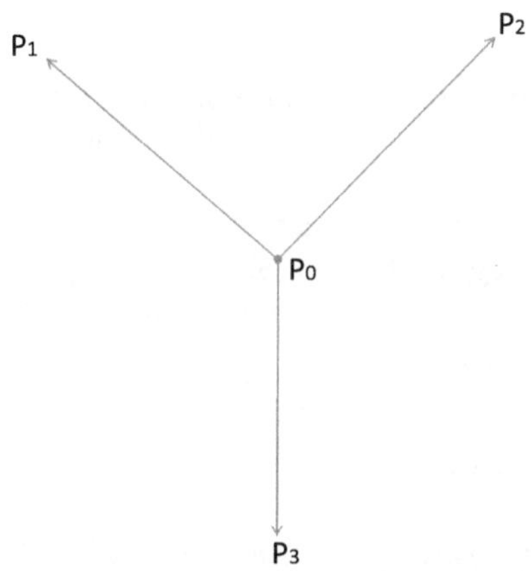

Figure 7.2. Planar distance observations

case, there exist three possibilities (combinations) for determining the position of the unknown station P_0. Recall that for Fig. 4.1 on p. 34, two nonlinear distance equations were written (e.g., 4.1 and 4.2). For Fig. 7.2, systems of distance equations could be written for combinations $\{P_1P_0P_2\}$, $\{P_1P_0P_3\}$ and $\{P_2P_0P_3\}$. For combination $\{P_1P_0P_2\}$ for example, one writes

$$d_1^2 = (x_1 - x_0)^2 + (y_1 - y_0)^2 \tag{7.3}$$

and

$$d_2^2 = (x_2 - x_0)^2 + (y_2 - y_0)^2. \tag{7.4}$$

Equations (7.3) and (7.4) lead to solutions $\{x_0, y_0\}_{1,2}$ as position of the unknown station P_0, where the subscripts indicate the combinations used. Combination $\{P_1P_0P_3\}$ gives

$$d_1^2 = (x_1 - x_0)^2 + (y_1 - y_0)^2 \tag{7.5}$$

and

$$d_3^2 = (x_3 - x_0)^2 + (y_3 - y_0)^2, \tag{7.6}$$

leading to solutions $\{x_0, y_0\}_{1,3}$ as the position of the unknown station P_0. The last combination $\{P_2P_0P_3\}$ has

$$d_2^2 = (x_2 - x_0)^2 + (y_2 - y_0)^2 \tag{7.7}$$

and

$$d_3^2 = (x_3 - x_0)^2 + (y_3 - y_0)^2, \tag{7.8}$$

as its system of equations leading to solutions $\{x_0, y_0\}_{2,3}$. The solutions $\{x_0, y_0\}_{1,2}$, $\{x_0, y_0\}_{1,3}$ and $\{x_0, y_0\}_{2,3}$ from these combinations are however not the same due to unavoidable effects of random errors. It is in attempting to harmonize these solutions to give the correct position of point P_0 that *C. F. Gauss* proposed the combinatorial approach. He believed that plotting these three combinatorial solutions resulted in an error figure with the shape of a triangle. He suggested the use of weighted arithmetic mean to obtain the final position of point P_0. In this regard the weights were obtained from the products of squared distances $\overline{P_0P_1}$, $\overline{P_0P_2}$ and $\overline{P_0P_3}$ (from unknown station to known stations) and the square of the perpendicular distances from the sides of the error triangle to the unknown station. According to [310, pp. 272–273], the motto in Gauss seal read *"pauca des matura"* meaning *few but ripe*. This belief led him not to publish most of his important contributions. For instance, [310, pp. 272–273] writes

> "Although not all his results were recorded in the diary (many were set down only in letters to friends), several entries would have each given fame to their author if published. *Gauss* knew about the quaternions before *Hamilton...*".

Unfortunately, the combinatorial method, like many of his works, was later to be published after his death (see e.g., Appendix A-2). Several years later, the method was independently developed by *C. G. I. Jacobi* [236] who used the square of the determinants as the weights in determining the unknown parameters from the arithmetic mean. *Werkmeister* [409] later established the relationship between the area of the error figure formed from the combinatorial solutions and the standard error of the determined point. In this book, the term combinatorial is adopted since the algorithm uses combinations to get all the finite solutions from which the optimum value is obtained. The optimum value is obtained by minimizing the sum of square of errors of pseudo-observations formed from the combinatorial solutions. For combinatorial optimization techniques, we refer to [140]. We will refer to this *combinatorial approach* as the *Gauss-Jacobi combinatorial algorithm* in appreciation of the work done by both *C. F. Gauss* and *C. G. I. Jacobi*.

In the approaches of *C. F. Gauss* and later *C. G. I. Jacobi*, one difficulty however remained unsolved. This was the question of how the various nonlinear systems of equations, e.g., (7.3 and 7.4), (7.5 and 7.6) or (7.7 and 7.8) could be solved explicitly! The only option they had was to linearize these equations, which in essence was a negation of what they were trying to avoid in the first place. Had they been aware of algebraic techniques that we saw in Chaps. 4 and 5, they could have succeeded in providing a complete algebraic solution to the overdetermined problem. In this chapter, we will complete what was started by these two gentlemen and provide a complete algebraic algorithm which we name in their honour. This algorithm is designed to provide a solution to the *nonlinear Gauss-Markov model*. First we define both the linear and nonlinear Gauss-Markov model and then formulate the *Gauss-Jacobi combinatorial algorithm* in Sect. 7-33.

7-32 Linear and nonlinear Gauss-Markov models

Linear and *nonlinear Gauss-Markov models* are commonly used for parameter estimation. Koch [244] presents various models for estimating parameters in linear models, while [180] divide the models into non-stochastic, stochastic and mixed models. We limit ourselves in this book to the simple or special Gauss Markov model with full rank. For readers who want extensive coverage of parameter estimation models, we refer to the books of [180, 244]. The use of the Gauss-Jacobi combinatorial approach proposed as an alternative solution to the nonlinear Gauss-Markov model will require only the special linear Gauss-Markov model during optimization. We start by defining the *linear Gauss-Markov model* as follows:

Definition 7.1 (Special linear Gauss-Markov model). *Given a real $n \times 1$ random vector $\mathbf{y} \in \mathbb{R}^n$ of observations, a real $m \times 1$ vector $\boldsymbol{\xi} \in \mathbb{R}^m$ of unknown fixed parameters over a real $n \times m$ coefficient matrix $\mathbf{A} \in \mathbb{R}^{n \times m}$, a real $n \times n$ positive definite dispersion matrix $\boldsymbol{\Sigma}$, the functional model*

$$\mathbf{A}\boldsymbol{\xi} = E\{\mathbf{y}\}, E\{\mathbf{y}\} \in R(\mathbf{A}), rk\mathbf{A} = m, \boldsymbol{\Sigma} = D\{\mathbf{y}\}, rk\boldsymbol{\Sigma} = n \tag{7.9}$$

is called special linear Gauss-Markov model with full rank.

The unknown vector $\boldsymbol{\xi}$ of fixed parameters in the *special linear Gauss-Markov model* (7.9) is normally estimated by **B***est* **L***inear* **U***niformly* **U***nbiased* **E***stimation* **BLUUE**, defined in [180, p. 93] as

Definition 7.2 (Best Linear Uniformly Unbiased Estimation BLUUE). *An $m \times 1$ vector $\hat{\boldsymbol{\xi}} = \mathbf{Ly} + \boldsymbol{\kappa}$ is $V - BLUUE$ for $\boldsymbol{\xi}$ (**B**est **L**inear **U**niformly **U**nbiased **E**stimation) respectively the $(V - Norm)$ in (7.9) when on one hand it is uniformly unbiased in the sense of*

$$E\{\hat{\boldsymbol{\xi}}\} = E\{\mathbf{Ly} + \boldsymbol{\kappa}\} = \boldsymbol{\xi} \, for \, all \, \boldsymbol{\xi} \in \mathbb{R}^m, \tag{7.10}$$

and on the other hand in comparison to all other linear uniformly unbiased estimators give the minimum variance and therefore the minimum mean estimation error in the sense of

$$\begin{aligned} trD\{\hat{\boldsymbol{\xi}}\} &= E\{(\hat{\boldsymbol{\xi}} - \boldsymbol{\xi})'(\hat{\boldsymbol{\xi}} - \boldsymbol{\xi})\} = \\ &= \sigma^2 \mathbf{L\Sigma L} = \|\mathbf{L}\|_V^2 = \min_L, \end{aligned} \tag{7.11}$$

where $\mathbf{L}$ is a real $m \times n$ matrix and $\boldsymbol{\kappa}$ an $m \times 1$ vector.

Using (7.11) to estimate the unknown fixed parameters' vector $\boldsymbol{\xi}$ in (7.9) leads to

$$\hat{\boldsymbol{\xi}} = (\mathbf{A}'\boldsymbol{\Sigma}^{-1}\mathbf{A})^{-1}\mathbf{A}'\boldsymbol{\Sigma}^{-1}\mathbf{y}, \tag{7.12}$$

with its regular dispersion matrix

$$D\{\hat{\boldsymbol{\xi}}\} = (\mathbf{A}'\boldsymbol{\Sigma}^{-1}\mathbf{A})^{-1}. \tag{7.13}$$

Equations (7.12) and (7.13) are the two main equations that are applied during the combinatorial optimization. The dispersion matrix (variance-covariance matrix) $\boldsymbol{\Sigma}$ is *unknown* and is obtained by means of estimators of type MINQUE, BIQUUE or BIQE as in [160, 338, 339, 340, 341, 342, 353]. In Definition 7.1, we used the term 'special'. This implies the case where the matrix $\mathbf{A}$ has full rank and $\mathbf{A}'\boldsymbol{\Sigma}^{-1}\mathbf{A}$ is invertible, i.e., regular. In the event that $\mathbf{A}'\boldsymbol{\Sigma}^{-1}\mathbf{A}$ is not regular (i.e., $\mathbf{A}$ has a rank deficiency), the rank deficiency can be overcome by procedures such as those presented by [180, pp. 107–165], [244, pp. 181–197] and [94, 176, 177, 299, 302, 329] among others.

Definition 7.3 (Nonlinear Gauss-Markov model). *The model*

$$E\{\mathbf{y}\} = \mathbf{y} - \mathbf{e} = \mathbf{A}(\boldsymbol{\xi}), D\{\mathbf{y}\} = \boldsymbol{\Sigma}, \tag{7.14}$$

with a real $n \times 1$ random vector $\mathbf{y} \in \mathbb{R}^n$ of observations, a real $m \times 1$ vector $\boldsymbol{\xi} \in \mathbb{R}^m$ of unknown fixed parameters, $n \times 1$ vector $\mathbf{e}$ of random errors (with zero mean and dispersion matrix $\boldsymbol{\Sigma}$), $\mathbf{A}$ being an injective function from an open domain into $n-$dimensional space $\mathbb{R}^n (m < n)$ and E the "expectation" operator is said to be a nonlinear Gauss-Markov model.

While the solution of the linear Gauss-Markov model by **B***est* **L***inear* **U***niformly* **U***nbiased* **E***stimator* (BLUUE) is straight forward, the solution of the nonlinear Gauss-Markov model is not straight forward owing to the *nonlinearity* of the *injective function* (or map function) $\mathbf{A}$ that maps $\mathbb{R}^m$ to $\mathbb{R}^n$. The difference between the linear and nonlinear Gauss-Markov models therefore lies on the injective function $\mathbf{A}$. For the linear Gauss-Markov model, the injective function $\mathbf{A}$ is linear and thus satisfies the algebraic axiom discussed in Chap. 2, i.e.,

$$\mathbf{A}(\alpha\xi_1 + \beta\xi_2) = \alpha\mathbf{A}(\xi_1) + \beta\mathbf{A}(\xi_2), \alpha, \beta \in \mathbb{R}, \xi_1, \xi_2 \in \mathbb{R}^m. \tag{7.15}$$

The m–dimensional manifold traced by $\mathbf{A}(.)$ for varying values of $\boldsymbol{\xi}$ is flat. For the nonlinear Gauss-Markov model on the other hand, $\mathbf{A}(.)$ is a *nonlinear* vector function that maps $\mathbb{R}^m$ to $\mathbb{R}^n$ tracing an m–dimensional manifold that is curved. The immediate problem that presents itself is that of obtaining a global minimum. Procedures that are useful for determining global minimum and maximum can be found in [334, pp. 387–448].

In geodesy and geoinformatics, many nonlinear functions are normally assumed to be moderately nonlinear thus permitting linearization by Taylor series expansion and then applying the linear model (Definition 7.1, Eqs. 7.12 and 7.13) to estimate the unknown fixed parameters and their dispersions [244, pp. 155–156]. Whereas this may often hold, the effect of nonlinearity of these models may still be significant on the estimated parameters. In such cases, the Gauss-Jacobi combinatorial algorithm presented in Sect. 7-33 can be used as we will demonstrate in the chapters ahead.

7-33 Gauss-Jacobi combinatorial formulation

The *C. F. Gauss* and *C. G. I Jacobi* [236] *combinatorial Lemma* is stated as follows:

Lemma 7.1 (Gauss-Jacobi combinatorial). *Given n algebraic observation equations in m unknowns, i.e.,*

$$\begin{aligned} a_1x + b_1y - y_1 &= 0 \\ a_2x + b_2y - y_2 &= 0 \\ a_3x + b_3y - y_3 &= 0, \\ &\dots . \end{aligned} \tag{7.16}$$

for the determination of the unknowns x and y, there exist no set of solutions $\{x, y\}_{i,j}$ *from any combinatorial pair in (7.16) that satisfy the entire system of equations. This is because the solutions obtained from each combinatorial pair of equations differ from the others due to the unavoidable random measuring errors. If the solutions from the pair of the combinatorial equations are designated* $x_{1,2}, x_{2,3}, \dots$ *and* $y_{1,2}, y_{2,3}, \dots$ *with the subscripts indicating the combinatorial pairs, then the combined solutions are the sum of the weighted arithmetic mean*

$$\boxed{x = \frac{\pi_{1,2}x_{1,2} + \pi_{2,3}x_{2,3} + \dots .}{\pi_{1,2} + \pi_{2,3} + \dots .}, \; y = \frac{\pi_{1,2}y_{1,2} + \pi_{2,3}y_{2,3} + \dots .}{\pi_{1,2} + \pi_{2,3} + \dots .}}, \tag{7.17}$$

with $\{\pi_{1,2}, \pi_{2,3}, \dots\}$ *being the weights of the combinatorial solutions given by the square of the determinants as*

$$\begin{aligned} \pi_{1,2} &= (a_1b_2 - a_2b_1)^2 \\ \pi_{2,3} &= (a_2b_3 - a_3b_2)^2 \\ &\dots . \end{aligned} \tag{7.18}$$

The results are identical to those of least squares solution.

The proof of *Lemma* 7.1 is given in [230] and [405, pp. 46–47]. For nonlinear cases however, the results of the combinatorial optimization may not coincide with those of least squares as will be seen in the coming chapters. This could be attributed to the remaining traces of nonlinearity following linearization of the nonlinear equations in the least squares approach or the generation of weight matrix by the combinatorial approach. We will later see that the combinatorial approach permits linearization only for generation of the weight matrix during optimization process.

Levelling is one of the fundamental tasks carried out in engineering, geodynamics, geodesy and geoinformatics for the purpose of determining heights of stations. In carrying out levelling, one starts from a point whose height is known and measures height differences along a levelling route to a closing point whose height is also known. In case where the starting point is also the closing point, one talks of loop levelling. The heights of the known stations are with respect to the mean sea level as a reference. In Example 7.1, we use loop levelling network to illustrate Lemma 7.1 of the Gauss-Jacobi combinatorial approach.

Example 7.1 (Levelling network). Consider a levelling network with four-points in Fig. 7.3 below.

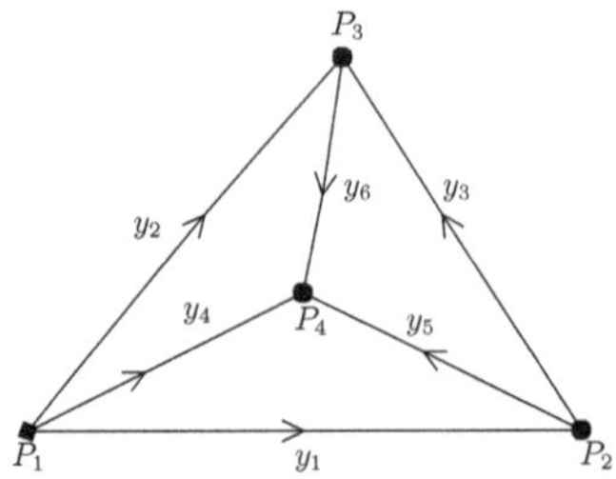

Figure 7.3. Levelling network

Let the known height of point P_1 be given as h_1. The heights h_2 and h_3 of points P_2 and P_3 respectively are unknown. The task at hand is to carry out loop levelling from point P_1 to determine these unknown heights. Given three stations with two of them being unknowns, there exist

$$\binom{3}{2} = \frac{3!}{2!(3-2)!} = 3$$

number of combinatorial routes that can be used to obtain the heights of points P_2 and P_3. If station P_4 is set out for convenience along the loop, the levelling routes are $\{P_1 - P_2 - P_4 - P_1\}$, $\{P_2 - P_3 - P_4 - P_2\}$, and $\{P_3 - P_1 - P_4 - P_3\}$. These combinatorials sum up to the outer loop $P_1 - P_2 - P_3 - P_1$. The observation equations formed by the height difference measurements are written as

$$\begin{aligned} x_2 - h_1 &= y_1 \\ x_3 - h_1 &= y_2 \\ x_3 - x_2 &= y_3 \\ x_4 - h_1 &= y_4 \\ x_4 - x_2 &= y_5 \\ x_4 - x_3 &= y_6, \end{aligned} \tag{7.19}$$

which can be expressed in the form of the special linear Gauss-Markov model (7.9) on p. 87 as

$$E\left\{\begin{bmatrix} y_1 + h_1 \\ y_2 + h_1 \\ y_3 \\ y_4 + h_1 \\ y_5 \\ y_6 \end{bmatrix}\right\} = \begin{bmatrix} 1 & 0 & 0 \\ 0 & 1 & 0 \\ -1 & 1 & 0 \\ 0 & 0 & 1 \\ -1 & 0 & 1 \\ 0 & -1 & 1 \end{bmatrix} \begin{bmatrix} x_2 \\ x_3 \\ x_4 \end{bmatrix}, \tag{7.20}$$

where $y_1, y_2, \ldots, y_6$ are the observed height differences and x_2, x_3, x_4 the unknown heights of points P_2, P_3, P_4 respectively. Let the dispersion matrix $D\{\mathbf{y}\} = \mathbf{\Sigma}$ be chosen such that the correlation matrix is unit (i.e., $\mathbf{\Sigma} = \mathbf{I}_3 = \mathbf{\Sigma}^{-1}$ *positive definite*, $rk\mathbf{\Sigma}^{-1} = 3 = n$), the decomposition matrix $\mathbf{Y}$ and the normal equation matrix $\mathbf{A}'\mathbf{\Sigma}^{-1}\mathbf{A}$ are given respectively by

$$\mathbf{Y} = \begin{bmatrix} y_1 + h_1 & 0 & 0 \\ 0 & 0 & y_2 + h_1 \\ 0 & y_3 & 0 \\ -(y_4 + h_1) & 0 & y_2 + h_1 \\ y_5 & -y_5 & 0 \\ 0 & y_6 & -y_6 \end{bmatrix}, \quad \mathbf{A}'\mathbf{\Sigma}^{-1}\mathbf{A} = \begin{bmatrix} 3 & -1 & -1 \\ -1 & 3 & -1 \\ -1 & -1 & 3 \end{bmatrix}. \tag{7.21}$$

The columns of $\mathbf{Y}$ correspond to the vectors of observations $\mathbf{y}_1$, $\mathbf{y}_2$ and $\mathbf{y}_3$ formed from the combinatorial levelling routes. We compute the heights of points P_2 and P_3 using (7.12) for each combinatorial levelling routes as follows:

- Combinatorials route(1):=$P_1 - P_2 - P_4 - P_1$. Equations (7.21) and (7.12) leads to the partial solutions

$$\hat{\xi}_{route(1)} = \frac{1}{2}\begin{bmatrix} y_1 + \dfrac{h_1}{2} - \dfrac{y_5}{2} - \dfrac{y_4}{2} \\ \dfrac{y_1}{2} - \dfrac{y_4}{2} \\ \dfrac{y_1}{2} - \dfrac{h_1}{2} + \dfrac{y_5}{2} - y_4 \end{bmatrix} \tag{7.22}$$

- Combinatorials route(2):=$P_2 - P_3 - P_4 - P_2$ gives

$$\hat{\xi}_{route(2)} = \frac{1}{2}\begin{bmatrix} \dfrac{y_5}{2} - \dfrac{y_3}{2} \\ \dfrac{y_3}{2} - \dfrac{y_6}{2} \\ \dfrac{y_6}{2} - \dfrac{y_5}{2} \end{bmatrix}. \tag{7.23}$$

- Combinatorials route(3):=$P_3 - P_1 - P_4 - P_3$ gives

$$\hat{\xi}_{route(3)} = \frac{1}{2}\begin{bmatrix} \dfrac{y_4}{2} - \dfrac{y_2}{2} \\ \dfrac{y_4}{2} + \dfrac{y_6}{2} - \dfrac{h_1}{2} - y_2 \\ \dfrac{h_1}{2} - \dfrac{y_2}{2} - \dfrac{y_6}{2} + y_4 \end{bmatrix}. \tag{7.24}$$

The heights of the stations x_2, x_3, x_4 are then given by the summation of the combinatorial solutions

$$\begin{bmatrix} x_2 \\ x_3 \\ x_4 \end{bmatrix} = \hat{\xi}_l = \hat{\xi}_{route(1)} + \hat{\xi}_{route(2)} + \hat{\xi}_{route(3)} = \frac{1}{2} \begin{bmatrix} y_1 + \frac{h_1}{2} - \frac{y_3}{2} - \frac{y_2}{2} \\ \frac{y_1}{2} + \frac{y_3}{2} - \frac{h_1}{2} - y_2 \\ \frac{y_1}{2} - \frac{y_2}{2} \end{bmatrix}. \quad (7.25)$$

If one avoids the combinatorial routes and carries out levelling along the outer route $P_1 - P_2 - P_3 - P_1$, the heights could be obtained directly using (7.12) as

$$\begin{bmatrix} x_2 \\ x_3 \\ x_4 \end{bmatrix} = \hat{\xi}_l = (\mathbf{A}'\mathbf{\Sigma}^{-1}\mathbf{A})^{-1}\mathbf{A}'\mathbf{\Sigma}^{-1} \begin{bmatrix} y_1 + h_1 \\ -(y_2 + h_1) \\ y_3 \\ 0 \\ 0 \\ 0 \end{bmatrix} = \frac{1}{2} \begin{bmatrix} y_1 + \frac{h_1}{2} - \frac{y_3}{2} - \frac{y_2}{2} \\ \frac{y_1}{2} + \frac{y_3}{2} - \frac{h_1}{2} - y_2 \\ \frac{y_1}{2} - \frac{y_2}{2} \end{bmatrix}, \quad (7.26)$$

In which case the results are identical to (7.25). For linear cases therefore, the results of Gauss-Jacobi combinatorial algorithm gives solution (7.25) which is identical to that of least squares approach in (7.26), thus validating the postulations of Lemma 7.1.

♣

7-331 Combinatorial solution of nonlinear Gauss-Markov model

The Gauss-Jacobi combinatorial Lemma 7.1 on p. 89 and the levelling example were based on a linear case. In case of nonlinear systems of equations, such as (7.3 and 7.4), (7.5 and 7.6) or (7.7 and 7.8), the nonlinear Gauss-Markov model (7.14) is solved in two steps:

- Step 1: Combinatorial minimal subsets of observations are constructed and rigorously solved by means of either Groebner basis or polynomial resultants.
- Step 2: The combinatorial solution points obtained from step 1, which are now linear, are reduced to their final adjusted values by means of **B**est **L**inear **U**niformly **U**nbiased **E**stimator (BLUUE). The dispersion matrix of the real valued random vector of pseudo-observations from *Step 1* are generated via the *nonlinear error propagation law* also known as the *nonlinear variance-covariance propagation*.

7-332 Construction of minimal combinatorial subsets

Since $n > m$ we construct minimal combinatorial subsets comprising m equations solvable in closed form using either Groebner basis or polynomial resultants. We begin by giving the following elementary definitions:

Definition 7.4 (Permutation). *Let us consider that a set S with elements $\{i, j, k\} \in S$ is given, the arrangement resulting from placing $\{i, j, k\} \in S$ in some sequence is known as permutation. If we choose any of the elements say i first, then each of the remaining elements j, k can be put in the second position, while the third position is occupied by the unused letter either j or k. For the set S, the following permutations can be made:*

$$\begin{bmatrix} ijk\ ikj\ jik \\ jki\ kij\ kji. \end{bmatrix} \tag{7.27}$$

From (7.27) there exist three ways of filling the first position, two ways of filling the second position and one way of filling the third position. Thus the number of permutations is given by $3 \times 2 \times 1 = 6$. *In general, for n different elements, the number of permutation is equal to* $n \times \ldots \times 3 \times 2 \times 1 = n!$

Definition 7.5 (Combination). *If for n elements only m elements are used for permutation, then we have a combination of the mth order. If we follow the definition above, then the first position can be filled in n ways, the second in $\{n-1\}$ ways and the mth in $\{n-(m-1)\}$ ways. In (7.27), the combinations are identical and contain the same elements in different sequences. If the arrangement is to be neglected, then we have for n elements, a combination of mth order being given by*

$$C_k = \binom{n}{m} = \frac{n!}{m!(n-m)!} = \frac{n(n-1)\ldots(n-m+1)}{m \times \ \ldots \ \times 3 \times 2 \times 1}. \tag{7.28}$$

Given n nonlinear equations to be solved, we first form C_k minimal combinatorial subsets each consisting of m elements (where m is the number of the unknown elements). Each minimal combinatorial subset C_k is then solved using either of the algebraic procedures discussed in Chaps. 4 and 5.

Example 7.2 (Combinatorial). In Fig.7.2 for example, $n = 3$ and $m = 2$, which with (7.28) leads to three combinations given by (7.3 and 7.4), (7.5 and 7.6) and (7.7 and 7.8). Groebner basis or polynomial resultants approach is then applied to each combinatorial pair to give the combinatorial solutions $\{x_0, y_0\}_{1,2}$, $\{x_0, y_0\}_{1,3}$ and $\{x_0, y_0\}_{2,3}$.

7-333 Optimization of combinatorial solutions

Once the *combinatorial minimal subsets* have been solved using either Groebner basis or polynomial resultants, the resulting sets of solutions are considered as pseudo-observations. For each combinatorial, the obtained minimal subset solutions are used to generate the dispersion matrix via the nonlinear error propagation law/variance-covariance propagation e.g., [180, pp. 469–471] as follows:

From the nonlinear observation equations that have been converted into its algebraic (polynomial) via Theorem 3.1 on p. 19, the *combinatorial minimal subsets* consist of polynomials $f_1, \ldots, f_m \in k[x_1, \ldots, x_m]$, with $\{x_1, \ldots, x_m\}$ being the unknown variables (fixed parameters) to be determined. The variables $\{y_1, \ldots, y_n\}$

are the known values comprising the pseudo-observations obtained following closed form solutions of the minimum combinatorial subsets. We write the polynomials as

$$\left[\begin{array}{l} f_1 := g(x_1, \ldots, x_m, y_1, \ldots, y_n) = 0 \\ f_2 := g(x_1, \ldots, x_m, y_1, \ldots, y_n) = 0 \\ \qquad\qquad\qquad \cdot \\ \qquad\qquad\qquad \cdot \\ \qquad\qquad\qquad \cdot \\ f_m := g(x_1, \ldots, x_m, y_1, \ldots, y_n) = 0, \end{array}\right. \tag{7.29}$$

which are expressed in matrix form as

$$\mathbf{f} := \mathbf{g}(\mathbf{x}, \mathbf{y}) = \mathbf{0}. \tag{7.30}$$

In (7.30) the unknown variables $\{x_1, \ldots, x_m\}$ are placed in a vector $\mathbf{x}$ and the known variables $\{y_1, \ldots, y_n\}$ in $\mathbf{y}$. Error propagation is then performed from pseudo-observations $\{y_1, \ldots, y_n\}$ to parameters $\{x_1, \ldots, x_m\}$ which are to be explicitly determined. They are characterized by the *first moments*, the expectations $E\{\mathbf{x}\} = \boldsymbol{\mu}_x$ and $E\{\mathbf{y}\} = \boldsymbol{\mu}_y$, as well as the *second moments*, the variance-covariance matrices/dispersion matrices $D\{\mathbf{x}\} = \boldsymbol{\Sigma}_x$ and $D\{\mathbf{y}\} = \boldsymbol{\Sigma}_y$. From [180, pp. 470–471], we have up to nonlinear terms

$$D\{\mathbf{x}\} = \mathbf{J}_x^{-1}\mathbf{J}_y\boldsymbol{\Sigma}_y\mathbf{J}_y'(\mathbf{J}_x^{-1})', \tag{7.31}$$

with $\mathbf{J}_x, \mathbf{J}_y$ being the partial derivatives of (7.30) with respect to $\mathbf{x}$,$\mathbf{y}$ respectively at the Taylor points $(\boldsymbol{\mu}_x, \boldsymbol{\mu}_y)$. The approximate values of unknown parameters $\{x_1, \ldots, x_m\} \in \mathbf{x}$ appearing in the Jacobi matrices $\mathbf{J}_x, \mathbf{J}_y$ are obtained either from Groebner basis or polynomial resultants solution of the nonlinear system of equations (7.29).

Given $\mathbf{J}_i = \mathbf{J}_{x_i}^{-1}\mathbf{J}_{y_i}$ from the ith combination and $\mathbf{J}_j = \mathbf{J}_{x_j}^{-1}\mathbf{J}_{y_j}$ from the jth combination, the correlation between the ith and j^{th} combinations is given by

$$\boldsymbol{\Sigma}_{ij} = \mathbf{J}_j\boldsymbol{\Sigma}_{y_j y_i}\mathbf{J}_i^{'}. \tag{7.32}$$

The sub-matrices variance-covariance matrix for the individual combinatorials $\boldsymbol{\Sigma}_1, \boldsymbol{\Sigma}_2, \boldsymbol{\Sigma}_3, \ldots, \boldsymbol{\Sigma}_k$ (where k is the number of combinations) obtained via (7.31) and the correlations between combinatorials obtained from (7.32) form the variance-covariance/dispersion matrix

$$\boldsymbol{\Sigma} = \begin{bmatrix} \boldsymbol{\Sigma}_1 & \boldsymbol{\Sigma}_{12} & . & \ldots & \boldsymbol{\Sigma}_{1k} \\ \boldsymbol{\Sigma}_{21} & \boldsymbol{\Sigma}_2 & . & \ldots & \boldsymbol{\Sigma}_{2k} \\ . & & \boldsymbol{\Sigma}_3 & & \\ . & & & . & \\ . & & & . & \\ \boldsymbol{\Sigma}_{k1} & . & . & . & \boldsymbol{\Sigma}_k \end{bmatrix} \tag{7.33}$$

for the entire k combinations. This will be made clear by Example 7.4. The obtained dispersion matrix $\boldsymbol{\Sigma}$ is then used in the *linear Gauss-Markov model* (7.12) to obtain the estimates $\hat{\boldsymbol{\xi}}$ of the unknown parameters $\boldsymbol{\xi}$. The combinatorial solutions are considered as pseudo-observations and placed in the vector $\mathbf{y}$ of observations, while the design matrix $\mathbf{A}$ comprises of integer values 1 which are the coefficients of the unknowns as in (7.36). The procedure thus optimizes the combinatorial solutions by the use of BLUUE. Consider the following example.

Example 7.3. From Fig. 7.2 on p. 85, three possible combinations each containing two nonlinear equations necessary for solving the two unknowns are given and solved as discussed in Example 7.2 on p. 93. Let the combinatorial solutions $\{x_0, y_0\}_{1,2}$, $\{x_0, y_0\}_{1,3}$ and $\{x_0, y_0\}_{2,3}$ be given in the vectors $\mathbf{z}_I(y_1, y_2)$, $\mathbf{z}_{II}(y_1, y_3)$ and $\mathbf{z}_{III}(y_2, y_3)$ respectively. If the solutions are placed in a vector $\mathbf{z}_J = [\mathbf{z}_I \; \mathbf{z}_{II} \; \mathbf{z}_{III}]^{'}$, the adjustment model is then defined as

$$E\{\mathbf{z}_J\} = \mathbf{I}_{6\times3}\boldsymbol{\xi}_{3\times1}, D\{\mathbf{z}_J\} \text{ from variance/covariance propagation.} \tag{7.34}$$

Let

$$\boldsymbol{\xi}^n = \mathbf{L}\mathbf{z}_J \text{ subject to } \mathbf{z}_J := \begin{bmatrix} \mathbf{z}_I \\ \mathbf{z}_{II} \\ \mathbf{z}_{III} \end{bmatrix} \in \mathbb{R}^{6\times1}, \tag{7.35}$$

such that the postulations $trD\{\boldsymbol{\xi}^n\} = min$, i.e., "*best*," and $E\{\boldsymbol{\xi}^n\} = \boldsymbol{\xi}$ for all $\boldsymbol{\xi}^n \in \mathbb{R}^m$ i.e., "*uniformly unbiased*" holds. We then have from (7.33), (7.34) and (7.35) the result

$$\hat{\boldsymbol{\xi}} = (\mathbf{I}^{'}_{3\times6}\boldsymbol{\Sigma}_{\mathbf{z}_J}\mathbf{I}_{6\times3})\mathbf{I}^{'}_{3\times6}\boldsymbol{\Sigma}^{-1}_{\mathbf{z}_J}\mathbf{z}_J \tag{7.36}$$

$$\hat{\mathbf{L}} = arg\{trD\{\boldsymbol{\xi}^n\} = tr\,\mathbf{L}\boldsymbol{\Sigma}_y\mathbf{L}^{'} = min \mid UUE\}$$

The dispersion matrix $D\{\hat{\boldsymbol{\xi}}\}$ of the estimates $\hat{\boldsymbol{\xi}}$ is obtained via (7.13). The shift from arithmetic weighted mean to the use of *linear Gauss Markov model* is necessitated as we do not readily have the weights of the minimal combinatorial subsets but instead have their dispersion matrices obtained via *error propagation/variance-covariance propagation.* If the equivalence *Theorem* of [180, pp. 339–341] is applied, an adjustment using linear Gauss Markov model instead of weighted arithmetic mean in Lemma 7.1 is permissible.

Example 7.4 (Error propagation for planar ranging problem). For the unknown station $P_0(X_0, Y_0) \in \mathbb{E}^2$ of the planar ranging problem in Fig. 7.2 on p. 85, let distances S_1 and S_2 be measured to two known stations $P_1(X_1, Y_1) \in \mathbb{E}^2$ and $P_2(X_2, Y_2) \in \mathbb{E}^2$ respectively. The distance equations are expressed as

$$\begin{bmatrix} S_1^2 = (X_1 - X_0)^2 + (Y_1 - Y_0)^2 \\ S_2^2 = (X_2 - X_0)^2 + (Y_2 - Y_0)^2, \end{bmatrix} \tag{7.37}$$

which are written algebraically as

$$\begin{bmatrix} f_1 := (X_1 - X_0)^2 + (Y_1 - Y_0)^2 - S_1^2 = 0 \\ f_2 := (X_2 - X_0)^2 + (Y_2 - Y_0)^2 - S_2^2 = 0. \end{bmatrix} \tag{7.38}$$

On taking total differential of (7.38), we have

$$\left[\begin{array}{l} \mathrm{d}f_1 := 2(X_1-X_0)\mathrm{d}X_1 - 2(X_1-X_0)\mathrm{d}X + 2(Y_1-Y_0)\mathrm{d}Y_1- \\ \qquad -2(Y_1-Y_0)\mathrm{d}Y - 2S_1\mathrm{d}S_1 = 0 \\ \\ \mathrm{d}f_2 := 2(X_2-X_0)\mathrm{d}X_2 - 2(X_2-X_0)\mathrm{d}X + 2(Y_2-Y_0)\mathrm{d}Y_2- \\ \qquad -2(Y_2-Y_0)\mathrm{d}Y - 2S_2\mathrm{d}S_2 = 0. \end{array}\right. \tag{7.39}$$

Arranging (7.39) with the unknown terms $\{X_0, Y_0\} = \{x_1, x_2\} \in \mathbf{x}$ on the left-hand-side and the known terms

$$\{X_1, Y_1, X_2, Y_2, S_1, S_2\} = \{y_1, y_2, y_3, y_4, y_5, y_6\} \in \mathbf{y},$$

on the right-hand-side leads to

$$\mathbf{J}_x \begin{bmatrix} \mathrm{d}X_0 \\ \mathrm{d}Y_0 \end{bmatrix} = \mathbf{J}_y \begin{bmatrix} \mathrm{d}S_1 \\ \mathrm{d}X_1 \\ \mathrm{d}Y_1 \\ \mathrm{d}S_2 \\ \mathrm{d}X_2 \\ \mathrm{d}Y_2 \end{bmatrix}, \tag{7.40}$$

with

$$\mathbf{J}_x = \begin{bmatrix} \frac{\partial f_1}{\partial X_0} & \frac{\partial f_1}{\partial Y_0} \\ \\ \frac{\partial f_2}{\partial X_0} & \frac{\partial f_2}{\partial Y_0} \end{bmatrix} = \begin{bmatrix} -2(X_1-X_0) & -2(Y_1-Y_0) \\ -2(X_2-X_0) & -2(Y_2-Y_0) \end{bmatrix}, \tag{7.41}$$

and

$$\left[\begin{array}{l} \mathbf{J}_y = \begin{bmatrix} \frac{\partial f_1}{\partial S_1} & \frac{\partial f_1}{\partial X_1} & \frac{\partial f_1}{\partial Y_1} & 0 & 0 & 0 \\ \\ 0 & 0 & & \frac{\partial f_2}{\partial S_2} & \frac{\partial f_2}{\partial X_2} & \frac{\partial f_2}{\partial Y_2} \end{bmatrix} = \\ \\ = \begin{bmatrix} 2S_1 & -2(X_1-X_0) & -2(Y_1-Y_0) & 0 & 0 & 0 \\ 0 & 0 & 0 & 2S_2 & -2(X_2-X_0) & -2(Y_2-Y_0) \end{bmatrix}. \end{array}\right. \tag{7.42}$$

If we consider that

$$D\{\mathbf{x}\} = \mathbf{\Sigma}_x = \begin{bmatrix} \sigma^2_{X_0} & \sigma_{X_0Y_0} \\ \sigma_{Y_0X_0} & \sigma^2_{Y_0} \end{bmatrix}$$

$$D\{\mathbf{y}\} = \mathbf{\Sigma}_y = \begin{bmatrix} \sigma^2_{S_1} & \sigma_{S_1X_1} & \sigma_{S_1Y_1} & \sigma_{S_1X_2} & \sigma_{S_1S_2} & \sigma_{S_1Y_2} \\ \sigma_{X_1S_1} & \sigma^2_{X_1} & \sigma_{X_1Y_1} & \sigma_{X_1S_2} & \sigma_{X_1X_2} & \sigma_{X_1Y_2} \\ \sigma_{Y_1S_1} & \sigma_{Y_1X_1} & \sigma^2_{Y_1} & \sigma_{Y_1S_2} & \sigma_{Y_1X_2} & \sigma_{Y_1Y_2} \\ \sigma_{S_2S_1} & \sigma_{S_2X_1} & \sigma_{S_2Y_1} & \sigma^2_{S_2} & \sigma_{S_2X_2} & \sigma_{S_2Y_2} \\ \sigma_{X_2S_1} & \sigma_{X_2X_1} & \sigma_{X_2Y_1} & \sigma_{X_2S_2} & \sigma^2_{X_2} & \sigma_{X_2Y_2} \\ \sigma_{Y_2S_1} & \sigma_{Y_2X_1} & \sigma_{Y_2Y_1} & \sigma_{Y_2S_2} & \sigma_{Y_2X_2} & \sigma^2_{Y_2} \end{bmatrix}, \tag{7.43}$$

we obtain with (7.40), (7.41) and (7.42) the dispersion (7.31) of the unknown variables $\{X_0, Y_0\} = \{x_1, x_2\} \in \mathbf{x}$.

7-334 The Gauss-Jacobi combinatorial algorithm

The Gauss-Jacobi combinatorial program operates in *three* phases. In the *first* phase, one forms *minimal combinations* of the nonlinear equations using (7.28) on p. 93. Using either Groebner basis or polynomial resultants, the desired combinatorial solutions are obtained. The combinatorial results form pseudo-observations, which are within the solution space of the desired values. This first phase in essence *projects a nonlinear* case into a *linear* case.

Once the first phase is successfully carried out with the solutions of the various subsets forming pseudo-observations, the nonlinear variance-covariance/error propagation is carried out in the *second phase* to obtain the *weight matrix*. This requires that the stochasticity of the initial observational sample be known in-order to propagate them to the pseudo-observations.

The *final phase* entails the adjustment step, which is performed to obtain the *barycentric values.* Since the pseudo-observations are linearly independent, the special linear Gauss-Markov model (see Definition 7.1 on p. 87) is employed.

Stepwise, the *Gauss-Jacobi combinatorial algorithm* proceeds as follows:

- **Step 1**: Given an overdetermined system with n observations in m unknowns, using (7.28), form minimal combinations from the n observations that comprise m equations in m unknowns.
- **Step 2**: Solve each set of m equations from step 1 above for the m unknowns using either Groebner basis or polynomial resultant algebraic techniques.
- **Step 3**: Perform the nonlinear error/variance-covariance propagation to obtain the variance-covariance matrix of the combinatorial solutions obtained in Step 2.
- **Step 4**: Using the pseudo-observations of step 2, and the variance-covariance matrix from step 3, adjust the pseudo-observations via the special linear Gauss-Markov model to obtain the adjusted position of the unknown station.

Figure 7.4 summarizes the operations of the Gauss-Jacobi combinatorial algorithm which employs Groebner basis or polynomial resultants as computing engines to solve nonlinear systems of equations (e.g., Fig.7.5).

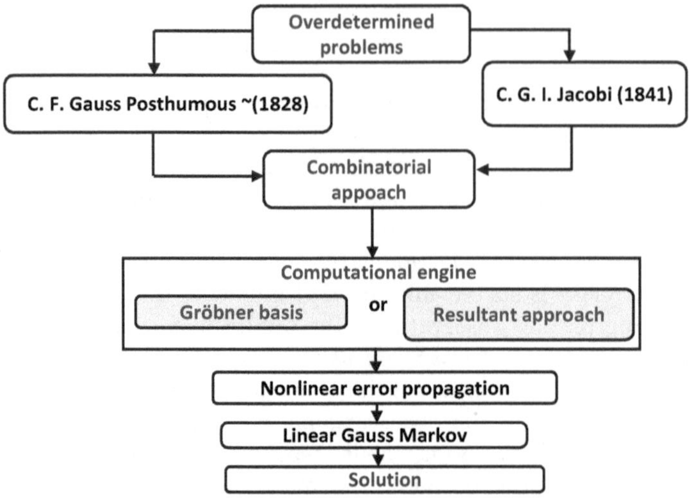

Figure 7.4. Gauss-Jacobi combinatorial algorithm

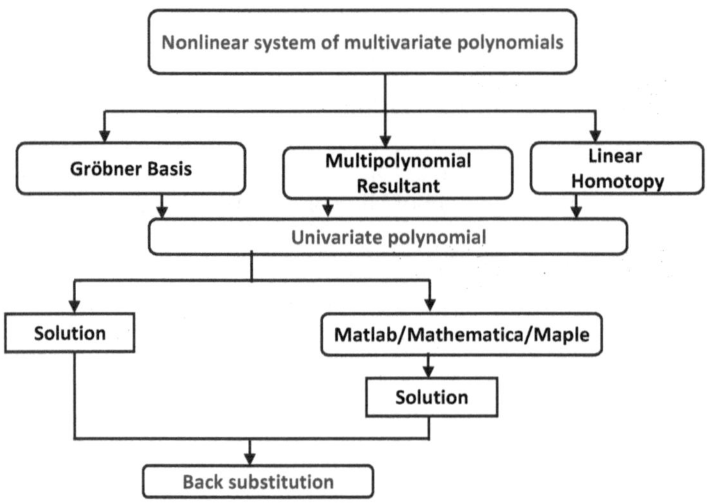

Figure 7.5. Combinatorial computing engine

7-4 Concluding remarks

In Chaps. 4 and 5, Groebner basis and polynomial resultants algorithms were presented for solving in exact form the nonlinear systems of equations. It was demonstrated in this chapter how they play a leading role in overcoming the major difficulty that was faced by *C. F. Gauss* and *C. G. I. Jacobi.* The key to success is to use these algebraic techniques as the computing engine of the Gauss-Jacobi combinatorial algorithm. In so doing, an alternative procedure to linearized and iterative numerical procedures that peg their operations on approximate starting values was presented. Such algebraic technique for solving overdetermined problems requires neither approximate starting values nor linearization (except for the generation of the weight matrix). With modern computing technology, the combinatorial formation and computational time for geodetic or geoinformatics' algebraic computational problems is immaterial. In the chapters ahead, the power of this technique will be demonstrated. Further materials on the topic are presented in [199, 201].

ALESS method has provided a different approach for solving overdetermined systems by transforming them into *square systems*. Using this method, the number of equations equals the number of the unknown variables, independently from the number of equation of the original system. Consequently, overdetermined system can be solved efficiently with this method (unlike *Gauss-Jacobi combinatorial algorithm* where the combinatorial explosion impedes the solution of big overdetermined systems). The solution of the transformed square system is possible also with local and with global methods. Fig. 7.6 gives a summary of the algebraic algorithms and show when each procedure can be applied.

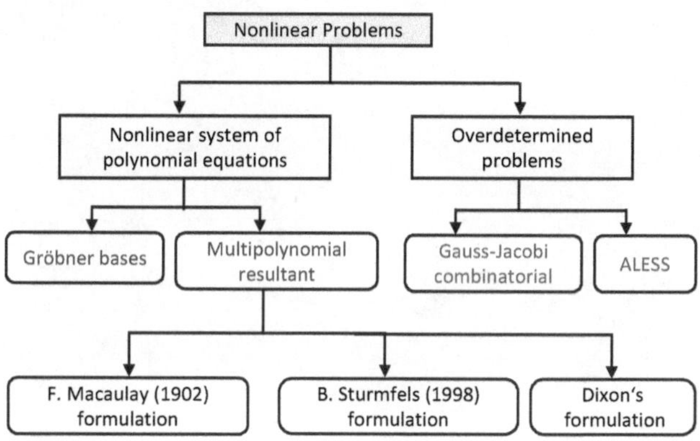

Figure 7.6. Algebraic solution approach. Here, the nonlinear system of equations implies determined nonlinear system of equations.

8 Extended Newton-Raphson method

8-1 Introductory remarks

In Chap. 7, we have seen that overdetermined nonlinear systems are common in geodetic and geoinformatic applications, that is there are frequently more measurements than it is necessary to determine unknown variables, consequently the number of the variables n is less then the number of the equations m. Mathematically, a solution for such systems can exist in a least square sense. There are many techniques to handle such problems, e.g.,:

- *Direct minimization* of the *residual* of the system, namely the minimization of the sum of the *least square* of the errors of the equations as the objective. This can be done by using *local methods*, like gradient type methods, or by employing *global methods*, like genetic algorithms.
- *Gauss-Jacobi combinatorial* solution. Having more independent equations, m, than variables, n, so $m > n$, the solution - in a *least-squares* sense - can be achieved by solving the
$$\left\{ \begin{matrix} m \\ n \end{matrix} \right\}$$
combinatorial square subsets $(n \times n)$ of a set of m equations, and then weighting these solutions properly. The square systems can be solved again via *local methods*, like *Newton-type methods* or by applying computer algebra (*resultants*, *Groebner basis*) or global numerical methods, like *linear homotopy* presented in Chap. 6.
- Considering the *necessary condition* of the minimum of the least square error, the overdetermined system can be transformed into a square one via computer algebra (see ALESS in Sect. 7-2). Then, the square system can be solved again by *local* or *global methods.* It goes without saying that this technique works for non-polynomial cases as well.
- For the special type of *overdetermined systems* arising mostly from datum transformation problems, the so called *Procrustes algorithm* can be used. There exist different types of them, *partial*, *general* and *extended Procrustes* algorithms. These methods are global and practically they need only a few or no iterations.

In this chapter, a special numerical method is introduced, which can solve overdetermined or underdetermined nonlinear systems directly. In addition, it is robust enough to also handle determined systems when the Jacobian is *ill-conditioned*. Our problem is to solve a set of nonlinear equations

$$\boxed{f(x) = 0} \tag{8.1}$$

where $f{:}\mathfrak{R}^m \to \mathfrak{R}^n$, namely $x \in \mathfrak{R}^m$ and $f \in \mathfrak{R}^n$.

If $n = m$ and the Jacobi matrix has a full rank everywhere, in other words the system of equations is regular, and if in addition, the initial value of the iteration

J.L. Awange et al., *Algebraic Geodesy and Geoinformatics*, 2nd ed.,
DOI 10.1007/978-3-642-12124-1_8, © Springer-Verlag Berlin Heidelberg 2010

is close enough to the solution, the *Newton-Raphson* method ensures quadratic convergence. If one of these conditions fails, for example the system is *over or under determined*, or if the Jacobi matrix is *singular*, one can use the *Extended Newton-Raphson method*, see e.g. [68, 69, 137, 214, 336]. In addition isolated multiple roots may cause low convergency. To restore quadratic convergence, a deflation method can be used, see *Zhao* [435]. Let us illustrate these problems with the following examples.

8-2 The standard Newton-Raphson approach

Consider the set of nonlinear equations,

$$f_i\left(x_1, x_2, \ldots, x_n\right) = 0, i = 1, \ldots, n$$

which can be written in compact form as

$$f(x) = 0$$

We wish to find the set of x_i satisfying this system, therefore let us to expand it in a *Taylor series* about the x^k iterate, where the superscript k denotes the iterative number

$$f_i\left(x^{k+1}\right) = f_i\left(x^k\right) + \sum_{j=1}^{n} \left.\frac{\partial f_i}{\partial x_j}\right|_{x^k} \left(x_j^{k+1} - x_j^k\right) + \ldots$$

which define the Jacobian matrix,

$$J_{i,j}^k = \left.\frac{\partial f_i}{\partial x_j}\right|_{x^k}$$

and the set $f_i\left(x^{k+1}\right) = 0$. Then,

$$J^k\left(x^{k+1} - x^k\right) + f\left(x^k\right) = 0$$

or the x^{k+1} iteration can be expressed as

$$x^{k+1} = x^k - \left(J^k\right)^{-1} f\left(x^k\right)$$

Quadratic convergence means that the number of the correct significant digits is doubled at each iteration.

8-3 Examples of limitations of the standard approach

8-31 Overdetermined polynomial systems

Let us consider a simple monomial system (see *Sommese and Wampler* [367]

$$\begin{aligned} f_1 &= x^2 = 0 \\ f_2 &= xy = 0 \\ f_3 &= y^2 = 0 \end{aligned} \tag{8.2}$$

This system is a "*monomial ideal*" and trivial for computer algebra. Its Groebner basis is

$$G = \{y^2, xy, x^2\} \tag{8.3}$$

whose solutions are,

$$S = \{\{y = 0, y = 0\}, \{x = 0, y = 0\}, \{x = 0, x = 0\}\}. \tag{8.4}$$

We can see that the origin is an isolated singular root with multiplicity of 3. Global polynomial solvers using numerical Groebner basis can also solve this system. However, the standard Newton-Raphson method fails, since the system is overdetermined, although we can transform the overdetermined system into a determined one in a least squares sense. The objective function is the sum of the square of the residium of the equations,

$$\mathrm{W} = x^4 + x^2y^2 + y^4. \tag{8.5}$$

Considering the necessary conditions for the minimum, we obtain the following system,

$$\begin{aligned} g_1 &= 4x^3 + 2xy^2 = 0 \\ g_2 &= 2x^2y + 4y^3 = 0 \end{aligned} \tag{8.6}$$

Then, using the Newton-Raphson method, with initial values (x_0= 1, y_0= 1), the result is,

$$x = 1.78642 \times 10^{-8},\ y = 1.78642 \times 10^{-8},$$

The convergence is slow and the accuracy of the solution is poor (even if the initial guess are changed) because of the existence of multiple roots, see e.g. *Chapra* and *Canale* [106]. Fig. 8.1 shows the steps of the iterations.

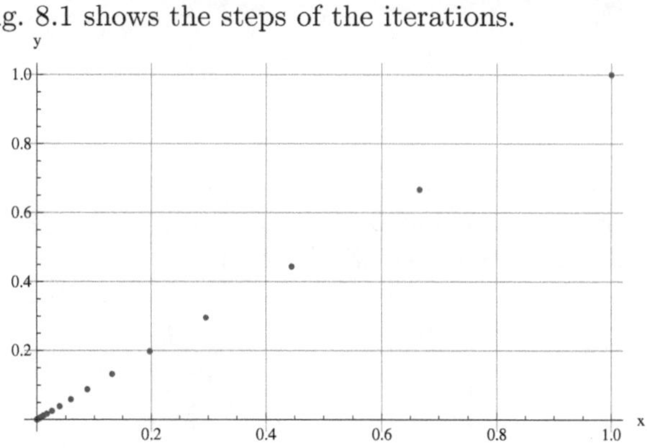

Figure 8.1. Slow convergence for the case of multiple roots (i.e., a plot of x versus y solutions starting with $x = 1, y = 1$).

The norm of the error of the procedure after 45 iteration steps is 2.52639×10^{-8}. Even global minimization with genetic algorithm will give a bad approximation,

$$x = 1.82247 \times 10^{-8},\ y = 5.29274 \times 10^{-9}$$

The reason for the slow convergence, as well as for the poor accuracy, is the increasing multiplicity of the roots (x = 0, y = 0) from 3 up to 9, after transforming the overdetermined system (f_1, f_2, f_3) into a square one $(g_1,\ g_2)$.

8-32 Overdetermined non-polynomial system

A usual test problem for parameter estimation procedures is the *Bard* [56] problem. The prototype equation of the system of these nonlinear equations is,

$$f_i = p_1 + \frac{b_i}{c_i p_2 + d_i p_3} - a_i = 0 \tag{8.7}$$

where the numerical values of the coefficients are presented in *Table 8.1*.

Table 8.1 (Coefficients of the Bard equations):

i	a_i	b_i	c_i	d_i
1	0.14	1	15	1
2	0.18	2	14	2
3	0.22	3	13	3
4	0.25	4	12	4
5	0.29	5	11	5
6	0.32	6	10	6
7	0.35	7	9	7
8	0.39	8	8	8
9	0.37	9	7	7
10	0.5	10	6	6
11	0.73	11	5	5
12	0.96	12	4	4
13	1.34	13	3	3
14	2.1	14	2	2
15	4.39	15	1	1

We have 15 equations and 3 unknown parameters, (p_1, p_2, p_3). The system is overdetermined and not a polynomial one. In this case, the global polynomial solver, as well as the Newton-Raphson method, fail. The minimization of the sum of square of errors of the equations,

$$W(p_1, p_2, p_3) = \sum_{i=1}^{15} f_i^2 \tag{8.8}$$

can be done directly via global minimization using, e.g., genetic algorithm leading to,

$$p_1 = 0.0679969, p_2 = 0.889355, p_3 = 2.57247.$$

Alternatively, the overdetermined problem can be transformed into a determined one, like in the case of polynomials (see ALESS). The necessary conditions of the minimum.

$$g_j\left(p_1, p_2, p_3\right) = \frac{\partial W}{\partial p_j} = 0, j = 1, 2, 3 \tag{8.9}$$

To solve this system (g_1, g_2, g_3), the *Newton-Raphson method* can be successful, with initial values $(p_{10}, p_{20}, p_{30}) = (1, 1, 1)$ leading to,

$$p_1 = 0.0679969,\ p_2 = 0.889355,\ p_3 = 2.57247$$

However, the method may fail, when the starting values are far from the solution and the Jacobian becomes singular, which is the case when $(p_{10}, p_{20}, p_{30}) =$ (-1, 1, 1). A global polynomial solver provides a solution, but besides the correct solution, other real solutions also arise (see *Table 8.2*).

Table 8.2 Solution of the square system (g_1, g_2, g_3):

p_1	p_2	p_3
$0.170869 + 0.0606932i$	$-5.99409 + 7.168i$	$9.59168 - 7.07904i$
$0.170869 - 0.0606932i$	$-5.99409 - 7.168i$	$9.59168 + 7.07904i$
0.22783	-4.44817	8.44473
$0.147587 + 0.0939871i$	$-3.42647 + 2.4683i$	$6.98725 - 2.33614i$
$0.147587 - 0.0939871i$	$-3.42647 - 2.4683i$	$6.98725 + 2.33614i$
0.216599	-2.61935	6.70349
$0.114883 + 0.112266i$	$-1.99681 + 1.21785i$	$5.51067 - 1.06593i$
$0.114883 - 0.112266i$	$-1.99681 - 1.21785i$	$5.51067 + 1.06593i$
0.181359	-1.55169	5.58774
$0.077954 + 0.114988i$	$-1.15986 + 0.720556i$	$4.62469 - 0.570547i$
$0.077954 - 0.114988i$	$-1.15986 - 0.720556i$	$4.62469 + 0.570547i$
0.120499	-0.852832	4.69769
$0.0413009 + 0.102133i$	$-0.621754 + 0.449866i$	$4.04119 - 0.321571i$
$0.0413009 - 0.102133i$	$-0.621754 - 0.449866i$	$4.04119 + 0.321571i$
0.0450109	-0.386462	3.96801
$0.0119518 + 0.0689036i$	$-0.227926 + 0.275464i$	$3.61455 - 0.192522i$
$0.0119518 - 0.0689036i$	$-0.227926 - 0.275464i$	$3.61455 + 0.192522i$
0.0679969	0.889355	2.57247

However, only one positive real solution appears, see last row in *Table 8.2.* This is because the determined system has more solutions than the original overdetermined system. In addition, the computation time will also increase.

8-33 Determined polynomial system

Now let us consider a determined system of polynomial equations, which is a somewhat modified version of the example of *Ojika* [314],

$$\begin{aligned} f_1(x, y) &= x^2 + y - 3 \\ f_2(x, y) &= x + \tfrac{y^2}{8} - 1 \end{aligned} \tag{8.10}$$

First, let us try to solve the problem using the *Newton-Raphson method*, starting with (x_0= -1, y_0=-1). The result is

$$x = 1.56598,\ y = 1.27716$$

which is not correct (compare with Fig. 8.3). The norm of error is 1.06057. The situation is the same with (x_0= 1, y_0= -1) which gives

$$x = 1.27083,\ y = 1.57379.$$

Again, this is not a solution. The norm of error is 0.061036, so the Newton-Raphson method fails.

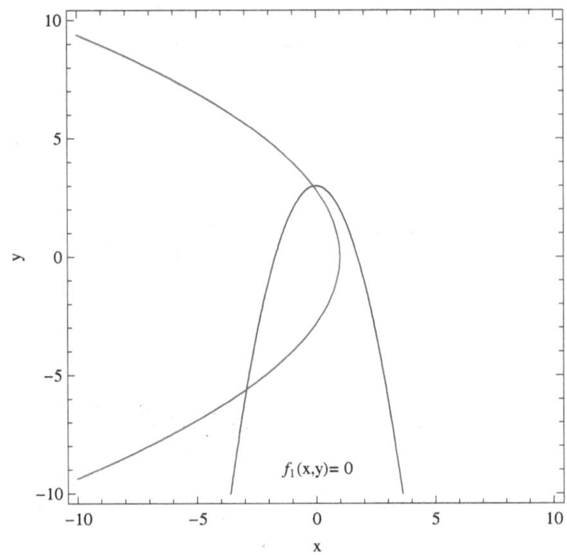

Figure 8.2. Real roots of the system of the two polynomials of Eqn. 8.10

8-34 Underdetermined polynomial system

Let us consider the following underdetermined system, see *Quoc-Nam Tran* [337]

$$\begin{aligned} f_1 &= (x-u)^2 + (y-v)^2 - 1 \\ f_2 &= 2v(x-u) + 3u^2(y-v) \\ &\ (3wu^2-1)(2wv-1) \end{aligned} \tag{8.11}$$

It goes without saying that the system has infinite roots. To make the solution unique, one may consider the solution with the minimal norm. This constraint leads to the following minimization problem,

$$W(x,y,u,v,w) = \sqrt{x^2+y^2+u^2+v^2+w^2} \rightarrow \min_{x,y,u,v,w} W \tag{8.12}$$

under the constrains,

$$f_i = 0, i = 1,2,3 \tag{8.13}$$

It is clear that the Newton-Raphson method can not be used. Employing genetic algorithm, we get,

$x = 0.187018$, $y = -1.08919 \times 10^{-6}$, $u = 0.81298$, $v = 0.000124852$, $w = 0.504332$

with a norm of 0.975.

8-4 Extending the Newton-Raphson approach using pseudoinverse

Now, in order to avoid these difficulties an extension of the Newton - Raphson method can be introduced, see [337]. This method uses the pseudoinverse of the Jacobian, instead of its inverse. The computation of the pseudoinverse is based on the singular value decomposition technique. Every A matrix $m \times n$, $m \geq n$ can be decomposed as

$$A = \mathbf{U\Sigma V}^T \tag{8.14}$$

where $(.)^T$ denotes the transposed matrix, $\mathbf{U}$ an $m \times n$ matrix, and $\mathbf{V}$ an $n \times n$ matrix satisfying

$$\mathbf{U}^T\mathbf{U} = \mathbf{V}^T\mathbf{V} = \mathbf{V}\mathbf{V}^T = I_n \tag{8.15}$$

and $\mathbf{\Sigma} =< \sigma_1, ..., \sigma_n >$ a diagonal matrix. These σ_i's, $\sigma_1 \geq \sigma_2 \geq, ..., \sigma_n \geq 0$ are the square root of the non negative eigenvalues $\mathbf{A}^T\mathbf{A}$ and are called the *singular values* of matrix $\mathbf{A}$. As it is known from linear algebra e.g. *Bernstein* [73], singular value decomposition (SVD) is a technique used to compute the *pseudoinverse* for a singular or ill-conditioned matrix of linear systems. In addition, this method provides a *least square solution for overdetermined systems* and *minimal norm solutions for the case of undetermined systems.*

The pseudoinverse of a matrix $\mathbf{A}$ of $m \times n$ is a matrix $\mathbf{A}^+$ of $n \times m$ satisfying

$$\mathbf{AA^+A} = \mathbf{A}, \mathbf{A^+AA^+} = \mathbf{A}^+, \ (\mathbf{A^+A})^* = \mathbf{A^+A}, (\mathbf{AA^+})^* = \mathbf{AA}^+ \tag{8.16}$$

where $(.)^*$ denotes the conjugate transpose of the matrix. There always exists a unique $\mathbf{A}^+$ which can be computed using SVD: (a) If $m \geq n$ and $\mathbf{A} = \mathbf{U\Sigma V}^T$ then

$$\mathbf{A}^+ = \mathbf{V\Sigma}^{-1}\mathbf{U}^T \tag{8.17}$$

where $\Sigma^{-1} =< 1/\sigma_1, ..., 1/\sigma_n >$. b) If $m < n$ then compute the $\left(\mathbf{A}^T\right)^+$, pseudoinverse of $\mathbf{A}^T$ and then

$$\mathbf{A}^+ = \left(\left(\mathbf{A}^T\right)^+\right)^T \tag{8.18}$$

The idea of using pseudoinverse in order to generalize the Newton-Raphson method is not new. It means that in the iteration formula (8.19), the pseudoinverse of the Jacobian matrix will be employed instead,

$$x_{i+1} = x_i - J^+(x_i) f(x_i) \tag{8.19}$$

The pseudoinverse can be computed in a symbolic as well as a numeric form. For example, considering the first example, see *Sommese and Wampler's* [367] equations in Sect. 16-21, the *Jacobi matrix* is

$$J(x,y) = \begin{pmatrix} 2x & 0 \\ y & x \\ 0 & 2y \end{pmatrix} \tag{8.20}$$

and its pseudoinverse

$$J^+(x,y) = \begin{pmatrix} \frac{x^3+4xy^2}{2(x^4+4x^2y^2+y^4)} & \frac{y^3}{x^4+4x^2y^2+y^4} & -\frac{xy^2}{2(x^4+4x^2y^2+y^4)} \\ -\frac{x^2y}{2(x^4+4x^2y^2+y^4)} & \frac{x^3}{x^4+4x^2y^2+y^4} & \frac{4x^2y+y^3}{2(x^4+4x^2y^2+y^4)} \end{pmatrix} \tag{8.21}$$

The Jacobian matrix at the point ($x_i = 1$, $y_i = 1$) is,

$$J(1,1) = \begin{pmatrix} 2 & 0 \\ 1 & 1 \\ 0 & 2 \end{pmatrix} \tag{8.22}$$

and the pseudoinverse is

$$J^+(1,1) = \begin{pmatrix} \frac{5}{12} & \frac{1}{6} & -\frac{1}{12} \\ -\frac{1}{12} & \frac{1}{6} & \frac{5}{12} \end{pmatrix} \tag{8.23}$$

The new values (x_{i+1}, y_{i+1}) in the next iteration step are

$$\begin{pmatrix} 1 \\ 1 \end{pmatrix} - J^+(1,1) \begin{pmatrix} f_1(1,1) \\ f_2(1,1) \\ f_3(1,1) \end{pmatrix} = \begin{pmatrix} \frac{1}{2} \\ \frac{1}{2} \end{pmatrix} \tag{8.24}$$

8-5 Applications of the Extended Newton-Raphson method

Example 8.1 (Overdetermined polynomial system). Let us recall the *Sommese and Wampler's* [367] equations solved in Sect. 8-31. Using the Extended Newton-Raphson method with initial values ($x_0 = 1$, $y_0 = 1$), we get the solution (after 40 iterations) as

$$x = 1.8189894035458573 \times 10^{-12},\ y = 1.81899 \times 10^{-12}.$$

The convergence is shown in Fig. 8.3 and the norm of error is $2.5724394843074972 \times 10^{-12}$. We therefore achieved a considerably more accurate result with approximately the same number of iteration steps as that of Sect. 8-31. This means faster convergency.

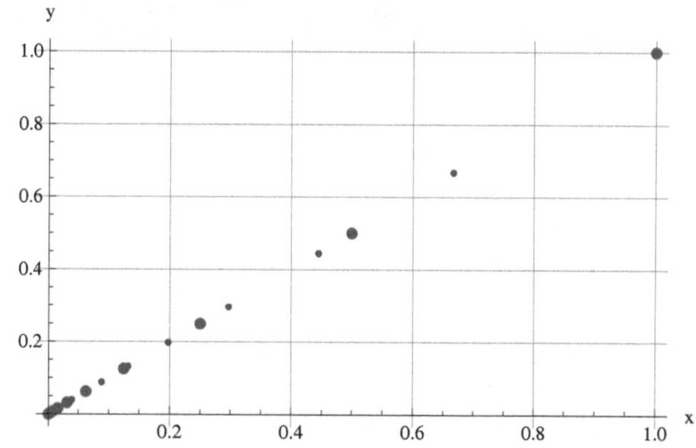

Figure 8.3. Convergence of the Extended Newton-Raphson method (big points) while that of the standard Newton-Raphson (small points)

Example 8.2 (Overdetermined non-polynomial system). Extended Newton-Raphson method can solve directly the original overdetermined system of Sect. 8-32 with both initial conditions more quicker than the standard methods (see Chap. 7) when solving the transformed square one system of equations. In addition, no transformation of the overdetermined model into a square one is necessary.

Example 8.3 (Determined system with singular Jacobian). The Extended Newton Raphson method converges from all of the different initial values for the case of the *Ojika* [314] problem that we attempted to solve in Sect. 8-33 using the standard Newton-Raphson method (i.e., obtaining the roots of the system of the two polynomials of Eqn. 8.10). Figs.8.4- 8.6 illustrate the fast convergence obtained when Extended Newton-Raphson is obtained as opposed to the standard Newton-Raphson (see Fig.).

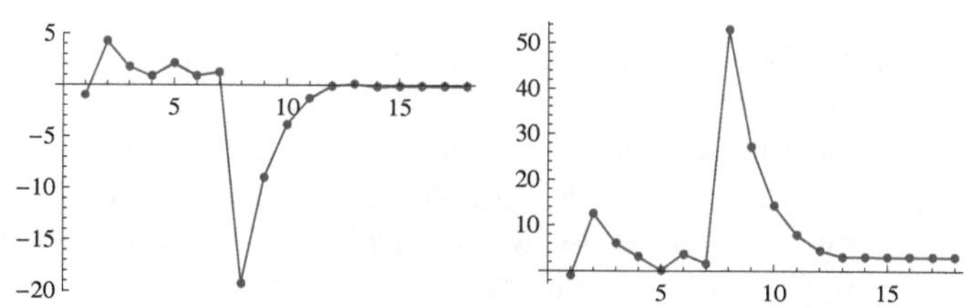

Figure 8.4. Convergence when starting values are (-1, -1)

Example 8.4 (Underdetermined system). The result produced by the Extended Newton-Raphson method depends on the initial condition, which is natural in

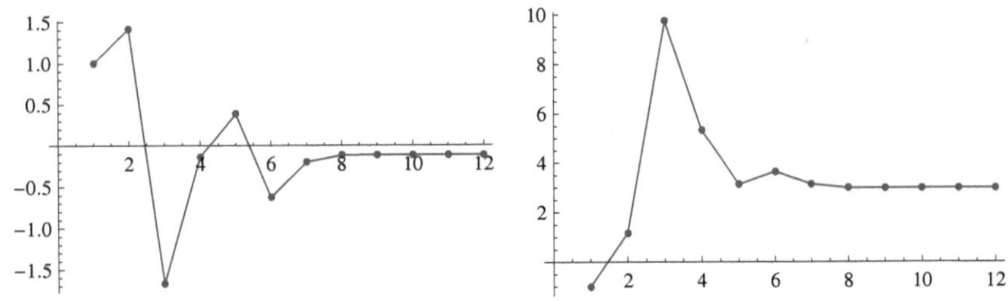

Figure 8.5. Convergency when starting values are (1, -1)

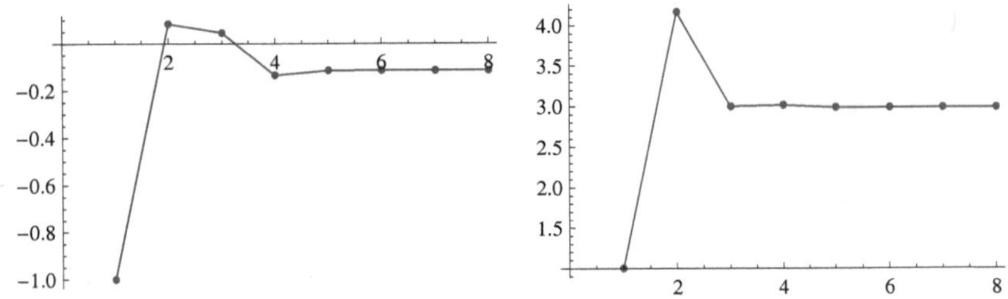

Figure 8.6. Convergency when starting values are (-1, 1)

the case of a local method, but no divergence takes place. In order to get the solution with the minimal norm, we computed the solutions with 500 initial values generated randomly in the interval [- 0.5, 0.5]. The computation took 1.422 sec and the norm of the solution was 0.977. While using global constrained minimization via genetic algorithm, the computation needed 2.390 sec and the norm of the solution was 0.975.

8-6 Concluding remarks

The examples demonstrate the efficiency and effectiveness of the suggested method, especially for cases of *overdetermined systems*. Although the *Extended Newton-Raphson* procedure is a local method, the solution of a square subset of the overdetermined system can provide a good initial guess for the computation. As a side effect, the method can also solve determined ill- conditioned problems frequently arising from geometrically ill-posed configurations e.g. in GPS positioning [41] or in solving datum transformation problems, see [430]. In addition, the implementation of this method is quite easy, since the computation of inverse of the Jacobian matrix could be simply replaced by the pseudoinverse in any Newton - Raphson procedure.

9 Procrustes solution

"It seems very strange that up to now Procrustes analysis has not been widely applied in geodetic literature. With this technique linearization problems of non linear equations system and iterative procedures of computation could be avoided, in general, with significant time saving and less analytical difficulties" F. Crosilla.

9-1 Motivation

This chapter presents the minimization approach known as *"Procrustes"* which falls within the multidimensional scaling techniques discussed in Sect. 9-22. Procrustes analysis is the technique of matching one configuration into another in-order to produce a measure of match. In adjustment terms, the partial Procrustes problem is formulated as the least squares problem of transforming a given matrix $\mathbf{A}$ into another matrix $\mathbf{B}$ by an orthogonal transformation matrix $\mathbf{T}$ such that the sum of squares of the residual matrix $\mathbf{E} = \mathbf{A} - \mathbf{BT}$ is minimum. This technique has been widely applied in shape and factor analysis. It has also been used for multidimensional rotation and also in scaling of different matrix configurations. In geodesy and geoinformatics, data analysis often require *scaling*, *rotation* and *translation* operations of different matrix configurations. Photogrammetrists, for example, have to determine the orientation of the camera during aerial photogrammetry and transform photo coordinates into ground coordinates. This is achieved by employing scaling, translation and rotation operations. These operations are also applicable to remote sensing and Geographical Information System (GIS) where map coordinates have to be transformed to those of the digitizing table. In case of robotics, the orientation of the robotic arm has to be determined, while for machine and computer visions, the orientation of the Charge-Coupled Device (CCD) cameras has to be established. In practice, positioning with satellites, particularly the Global Navigation Satellite Systems (GNSS) such us GPS and GLONASS has been on rise. The anticipated GALILEO satellites will further increase the use of satellites in positioning. This has necessitated the transformation of coordinates from the Global Positioning System (WGS 84) into local geodetic systems and vice versa.

A classical problem in geodesy and geoinformatics that would benefit from this technique is transformation, and in particular the 7–parameter datum transformation problem. The traditional approach of solving this problem, for instance, has been to linearize the nonlinear equations and then apply least squares method iteratively. With the proposed Procrustes approach, all that is required of the user is to insert the coordinates of one system (e.g., local coordinate system) in say, the matrix $\mathbf{A}$, and those of the other system (e.g., GPS in WGS-84) into the matrix $\mathbf{B}$. Using Procrustes analysis technique presented in this chapter, and later in Chap. 17, the desired scale, rotation and translation parameters can be obtained directly.

J.L. Awange et al., *Algebraic Geodesy and Geoinformatics*, 2nd ed.,
DOI 10.1007/978-3-642-12124-1_9, © Springer-Verlag Berlin Heidelberg 2010

Although long applied in other fields such as; sociology, to map crime versus cities, and also in medicine as we will see in Sect. 9-23, Procrustes method is relatively new to the fields of geodesy and geoinformatics. Its first entry into geodesy can be traced back to the work of [120, 121] where the method was used in the creation of the criterion matrix used for deformation analysis. Further applications include the works of [16, 168, 169] who applies it to compute the three-dimension orientation parameters, deflection of the vertical, and 7-parameter datum transformation.

Recent application of the approach in geoinformatics can be found in the works of [123] who employs it to solve the photogrammetric block adjustment by independent models, [65] who applies it for size and shape three-dimensional object reconstructions, and [66] who uses the technique to update cadastral maps. At the beginning of the Chapter, we quoted F. Crosilla [122], the *father* of Procrustes in geodesy and geoinformatics. He wonders why such an amazing technique has not been widely applied in geodesy.

Procrustes method is a very effective method for determining the Helmert's datum transformation parameters since it requires neither *initial starting values* nor *iteration*. Due to this attractive attribute, the *ABC-Procrustes* algorithm extended it to solve the 3D affine transformation problem where *scale factors* are different in the 3 principal directions X, Y, Z. In this study, it is illustrated that such direct extension is restricted to cases of *mild anisotropy* in scaling. For strong anisotropy, however, the procedure fails. The PZ-method is proposed as an extension of the ABC algorithm in case of strong anisotropy. The procedures are applied to determine transformation parameters for; (i) transforming Australian Geodetic Datum (AGD 84) to Geocentric Datum Australia (GDA 94), i.e., mild anisotropy and (ii) the Hungarian datum (strong anisotropy). The results indicate that the PZ-algorithm leads to a global minimization as opposed to the *ABC-algorithm* albeit with slightly longer computational time. However, the ABC-method is found to be useful for computing proper initial values for PZ-method thereby increasing its efficiency.

In this chapter, the *partial*, also called *simple* Procrustes algorithm which is sufficient for solving only the rotation elements is presented. It will be demonstrated how the approach solves the three-dimensional orientation and the vertical deflection (direction of local gravity vector) problems. In Chap. 17, the *general Procrustes algorithm* will be presented and used to solve the 7-parameter similarity transformation problem which is often encountered in practice.

9-2 Procrustes: Origin and applications

9-21 Procrustes and the magic bed

The origin of the name, and perhaps the concept is traced back to Greece. Somewhere in Attica in Greece lived a robber whose name was Procrustes. His house was so well positioned besides the road such that he was frequented by visitors who had to spend the night. In his house, Procrustes also known as Damastes kept a special bed: So special was the bed such that the visitors were required to fit in it. Unfortunately for Procrustes, neither were all his visitors of the same height nor length of the magic bed. All the same, the visitors were somehow forced in some "magic" way to fit into the magic bed. This was not done by adjusting the bed, but to the contrary its occupants! Procrustes devised ways to fit his quests onto his bed. Guests who were shorter for the bed were stretched by hammering or racking their bodies to fit the bed, while those who were longer had their legs chopped off! In both cases, the victims died. As fate would have it, Procrustes was himself adjusted to fit his own bed by Theseus, a young Attic hero whose mission was to eliminate robbers. The Encyclopedia of Greek Mythology writes[1]:

> "Procrustes (proh-KRUS-teez). A host who adjusted his guests to their bed. Procrustes, whose name means "he who stretches", was arguably the most interesting of Theseus's challenges on the way to becoming a hero. He kept a house by the side of the road where he offered hospitality to passing strangers, who were invited in for a pleasant meal and a night's rest in his very special bed (see Fig. 9.1[1]). Procrustes described it as having the unique property that its length exactly matched whomsoever lay down upon it. What Procrustes didn't volunteer was the method by which this "one-size-fits-all" was achieved, namely as soon as the guest lay down Procrustes went to work upon him, stretching him on the rack if he was too short for the bed and chopping off his legs if he was too long. Theseus turned the tables on Procrustes, fatally adjusting him to fit his own bed."

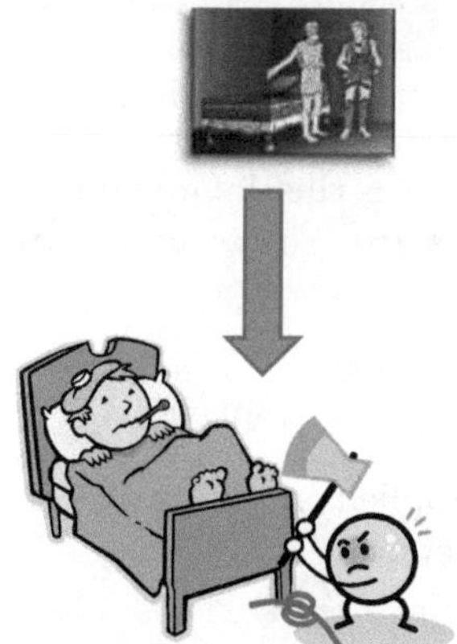

Figure 9.1. Procrustes and his "magical" bed

This magic bed of Procrustes has become a saying for arbitrarily - and perhaps ruthlessly - *forcing someone or something to fit into an unnatural scheme or pattern.*

[1] http://www.mythweb.com/encyc/gallery/procrustes_c.html ©Mythweb.com

9-22 Multidimensional scaling

Multidimensional scaling (MDS) is a method that represents measurements of similarity (or dissimilarity) among pairs of objects such as distances between points of low-dimensional multidimensional space. Let us consider for example that data consists of intelligence tests and that one desires to see the correlation between the tests. MDS can be used to represent these data in a plane such that the correlation can be studied. The more closer the points are (i.e., the shorter the distances between the points), the more correlated they are. MDS thus gives an advantage of graphical visualization of hidden adherent properties between objects. MDS has been described by [87] as:

- An approach for representing similarity and dissimilarity data as exemplified by distances of low dimensional space. This is done in-order to make this data accessible for visual inspection and exploration.
- An approach for testing if and how certain criteria by which one distinguishes among different objects of interest are mirrored in a corresponding empirical differences of this object (i.e., correlated).
- A data analytic approach that allows one to discover the three-dimensions that underlie judgements of dissimilarity and similarity.
- A psychological model that explains judgements of dissimilarity in terms of a rule that mimics a particular type of distance function.

Procrustes approach therefore is a procedure that is applied to realize the goals of MDS. In other words, it is a tool of MDS concerned with the fitting of one configuration into another as close as possible.

9-23 Applications of Procrustes in medicine

As a motivational urge to embrace this long overdue powerful tool, this section presents briefly two areas where Procrustes procedure has found practical application. These are:

- Procrustes application software for gene recognition [150].
- Identification of malarial parasites [130].

The technique has also been applied in various fields ranging from biology, psychology, to structural analysis etc.

9-231 Gene recognition

Gene recognition started as a statistical analysis and splicing sites. The statistical procedures however could not deal with other types of genes such as eukaryotic (i.e., a single-celled or multicellular organism whose cells contain a distinct membrane-bound nucleus). To solve this problem, researchers in the field developed PROCRUSTES software, which uses similarity-based approach to gene recognition [150]. This was achieved using spliced alignment technique. The software is reported by Human Genome News[2] to be able to identify with remarkable accuracy human version of genes that are in other forms of life. The human genes are broken into smaller segments known as exons. Searching for exons is analogous to following a magazine article that appears in, say, pp. 5, 23, 84, and 93, with almost identical advertisement and other articles appearing between. The software is applied to construct all these pages that contain the required article and automatically combine them into a best fitting set. The technique is said to work best when a "target protein" from the nonhuman sample guides the search, thus ensuring an accuracy that approaches 100%. In this technique, if a genomic sequence and a set of candidate exons are given, the algorithm explores all possible exon assemblies and finds a chain with the best fit to relate target protein. Instead of trying to identify the correct exons by statistical means (which predicts the correlation between the predicted and the actual gene to 70%, with just 40-50% exons predicted correctly), PROCRUSTES considers all possible chain with the maximum global similarity to the target protein. The procedure predicts a correlation of about 99% between the predicted and the actual gene [150]. The tool is useful in pinpointing elusive human version of cancer-causing gene!

9-232 Identification of malaria parasites

Dryden [130] applies Procrustes to identify proteins by comparing the electrophoretic gel images (Fig. 9.2[3]). The gels are obtained from strains of parasite which carry malaria. The procedure uses Procrustes matching and affine shape registration to match the gels. It applies some biological material to the left corner of the two images of gels A and B in Fig. 9.2. The material is then separated down the gel according to molecular weight (with the highest on top) and across the gel according to isoelectric point (with the highest on the right of the gel). Gel image is then used to identify strains of parasites using pattern of spots marked by (+). Dark spots appearing on the gels indicate the composition of protein and are marked by some expert in both gels A and B. Ten spots are marked in each gel and then classified as either invariant or variant spots.

The invariant spots are considered to be present for all parasites. The arrangement of the variant spots is of particular interest as it helps in the identification of malarial parasite. The field problem sighted by [130] however is that gels are prone to deformation such as *translation*, *scaling*, *rotation*, *affine transformation* and *smooth-linear bending*. Gel images therefore need to be registered by matching each image using a set of transformation to alleviate the deformations above [130]. This is achieved through the use of Procrustes analysis.

---

[2] July-September 1996; 8:(1)

[3] ©Chapman and Hall Press

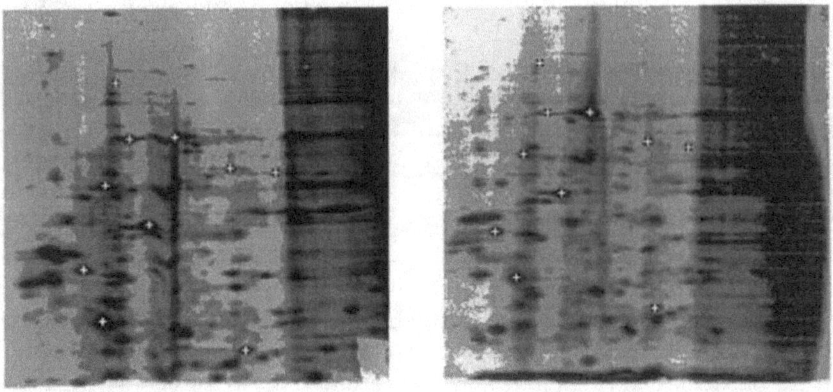

Figure 9.2. The electrophoretic gels from gel A and gel B. Ten invariant spots have been marked by (+) in white above ©Chapman and Hall Press

9-3 Partial procrustes solution

9-31 Conventional formulation

Procrustes being a technique of matching one configuration into another and producing a measure of match, seeks the *isotropic dilatation* and the rigid *translation*, *reflection* and *rotation* needed to best match one configuration to another [116, p. 92]. In this chapter, the term **partial** shall be used to mean ***optimal rotation*** in-order to avoid confusion since the term is used differently by different authors. For example, [197, 198] considers a case where the configuration matrix has several unknown elements in the minimization of the Frobenius norm as the partial Procrustes problem. Dryden [130] on the other hand uses the term partial Procrustes to refer to minimization of the Frobenius norm only over the translation and rotation. The general Procrustes solution is used as the minimization over the full set of similarity transformation as shall be seen in Chap. 17. In the solution of partial Procrustes problem, we refer to Table 9.1 for some matrix properties which will be of use.

The Procrustes problem is concerned with fitting a configuration **B** into **A** as close as possible. The simplest Procrustes case is one in which both configurations have the same dimensionality and the same number of points, which can be brought into a $1-1$ correspondence by substantive considerations [87, p. 339]. Let us consider the case where both **A** and **B** are of the same dimension. The partial Procrustes problem is then formulated as

$$\mathbf{A} = \mathbf{B}\mathbf{T} \tag{9.1}$$

The rotation matrix **T** in (9.1) is then solved by measuring the distances between corresponding points in both configurations, square these values, and add them to obtain the sum of squares $\|\mathbf{A}-\mathbf{BT}\|^2$ which is then minimized. One proceeds via Frobenius norm as follows:

$$\begin{array}{l} min \, \|\mathbf{X}-\mathbf{YT}\| := \sqrt{tr(\mathbf{X}' - \mathbf{T}'\mathbf{Y}')(\mathbf{X}-\mathbf{YT})} \\ \mathbf{T}'\mathbf{T} = \mathbf{I}. \end{array} \tag{9.2}$$

In-order to obtain $\mathbf{T}$ in (9.2), the following properties of a matrix in Table (9.1) are essential.

Table 9.1. Matrix properties for procrustes analysis

(a)	$\mathrm{tr}\,\mathbf{A} = \sum_{i=1}^{n} a_{ii}$	Definition of trace function
(b)	$\mathrm{tr}\,\mathbf{A} = \mathrm{tr}\,\mathbf{A}'$	Invariant under transpose
(c)	$\mathrm{tr}\,\mathbf{ABC} = \mathrm{tr}\,\mathbf{C}\,\mathbf{AB} = \mathrm{tr}\,\mathbf{BCA}$	Invariant under 'cyclic' permutation
(d)	$\mathrm{tr}\,\mathbf{A}'\mathbf{B} = \mathrm{tr}\,(\mathbf{A}'\mathbf{B})' = \mathrm{tr}\,\mathbf{B}'\mathbf{A} = \mathrm{tr}\,\mathbf{AB}'$	Combining properties (b) and (c)
(e)	$\mathrm{tr}\,(\mathbf{A}+\mathbf{B}) = \mathrm{tr}\,\mathbf{A} + \mathrm{tr}\mathbf{B}$	Summation rule

Using (9.2) and the properties of Table 9.1, one writes

$$\left[\begin{array}{l} \|\mathbf{A}-\mathbf{BT}\|^2 := tr(\mathbf{A}' - \mathbf{T}'\mathbf{B}')(\mathbf{A}-\mathbf{BT}) \\ \mathbf{T}'\mathbf{T} = \mathbf{I} \\ = tr(\mathbf{A}'\mathbf{A} - 2\mathbf{A}'\mathbf{BT} + \mathbf{T}'\mathbf{B}'\mathbf{BT}) \\ = tr\mathbf{A}'\mathbf{A} - 2tr\mathbf{A}'\mathbf{BT} + tr\mathbf{B}'\mathbf{B} \\ tr\mathbf{T}'\mathbf{B}'\mathbf{BT} = tr\mathbf{TT}'\mathbf{B}'\mathbf{B} = tr\mathbf{B}'\mathbf{B}. \end{array}\right. \tag{9.3}$$

The simplification $tr\mathbf{T}'\mathbf{B}'\mathbf{BT} = tr\mathbf{B}'\mathbf{B}$ in (9.3) is obtained by using the property of invariance of the trace function under cyclic permutation (i.e., property (c) in Table 9.1). Since $\mathrm{tr}(\mathbf{A}'\mathbf{A})$ and $\mathrm{tr}(\mathbf{B}'\mathbf{B})$ are not dependent on $\mathbf{T}$, we note from (9.3) that

$$\begin{array}{l} \|\mathbf{A}-\mathbf{BT}\|^2 = \min \ \Leftrightarrow tr(\mathbf{A}'\mathbf{BT}) = \max \\ \mathbf{T}'\mathbf{T} = \mathbf{TT}' = \mathbf{I}_k. \end{array} \tag{9.4}$$

If $\mathbf{U\Sigma V}'$ is the singular value decomposition of $\mathbf{A}'\mathbf{B}$ and $\mathbf{C} = \mathbf{A}'\mathbf{B}$, then we have

$$\left[\begin{array}{l} \mathbf{A}'\mathbf{B} = \mathbf{U\Sigma V}' \\ \text{if}\, \mathbf{C} = \mathbf{U\Sigma V}', \ \ \mathbf{U}, \mathbf{V}' \in \mathrm{SO}(3) \\ \mathbf{\Sigma} = \mathrm{Diag}(\sigma_1, \ . \ . \ . \ , \ \sigma_k)\text{then} \\ \mathrm{tr}(\mathbf{CT}) \le \ \sum_{i=1}^{k} \sigma_k, \\ \text{with} \\ k = 3. \end{array}\right. \tag{9.5}$$

The proof for (9.5) is given by [294, p. 34] as follows: Substituting for $\mathbf{C}$ from its singular value decomposition and with the property (c) in Table 9.1, one writes

$$\left[\begin{array}{l}\mathrm{tr}(\mathbf{CT}) = \mathrm{tr}(\mathbf{U\Sigma V}'\mathbf{T}) = \mathrm{tr}(\mathbf{\Sigma V}'\mathbf{TU})\\ \text{taking}\\ \mathbf{R} = (ij)1 \le i,j \le k = \mathbf{V}'\mathbf{TU}\ \text{orthogonal and}\ |r_{ii}| \le 1\\ \text{then}\\ \mathrm{tr}(\mathbf{\Sigma V}'\mathbf{TU}) = \sum_{i=1}^{k} \sigma_i r_{ii} \le \sum_{i=1}^{k} \sigma_i.\end{array}\right. \tag{9.6}$$

From (9.5) and (9.6), one notes that

$$\mathrm{tr}(\mathbf{A}'\mathbf{BT}) = \max \Leftrightarrow \mathrm{tr}(\mathbf{A}'\mathbf{BT}) \le \sum_{i=1}^{k} \gamma_i, \tag{9.7}$$

subject to the singular value decomposition

$$\mathbf{A}'\mathbf{B} = \mathbf{U\Sigma V}',\ \mathbf{U},\mathbf{V} \in SO(3)\ \text{and orthogonal}. \tag{9.8}$$

Finally, the maximum value is obtained as

$$\max(\mathrm{tr}\mathbf{A}'\mathbf{BT}) = \sum_{i=1}^{k} \gamma_i \ \Leftrightarrow \mathbf{T} = \mathbf{VU}'. \tag{9.9}$$

Thus the solution of the rotation matrix by Procrustes method is

$$\mathbf{T} = \mathbf{VU}'. \tag{9.10}$$

9-32 Partial derivative formulation

This approach is attributed to *P. H. Schonemann* [355] as well as [356]. Proceeding from the Frobenius norm in (9.2) and using (9.1) leads to

$$d_1 = tr\,\mathbf{A}'\mathbf{A} - 2tr\,\mathbf{A}'\mathbf{BT}, +\mathbf{T}'\mathbf{B}'\mathbf{BT}, \tag{9.11}$$

while the condition that $\mathbf{T}'\mathbf{T} = \mathbf{I}$ leads to

$$d_2 = \mathbf{\Lambda}(\mathbf{T}'\mathbf{T} - \mathbf{I}). \tag{9.12}$$

where $\mathbf{\Lambda}$ is the m x m unknown matrix of Lagrange multipliers. Equations (9.11) and (9.12) are added to give

$$d = d_1 + d_2. \tag{9.13}$$

The derivative of (9.13) are obtained with respect to $\mathbf{T}$ as

$$\left[\begin{array}{l}\frac{\partial d}{\partial \mathbf{T}} = \frac{\partial d_1}{\partial \mathbf{T}} + \frac{\partial d_2}{\partial \mathbf{T}}\\ = \frac{\partial\left(tr\,\mathbf{A}'\mathbf{A} - 2tr\,\mathbf{A}'\mathbf{BT} + tr\,\mathbf{T}'\mathbf{B}'\mathbf{BT}\right)}{\partial \mathbf{T}} + \frac{\partial\left(\mathbf{\Lambda T}'\mathbf{T} - \mathbf{\Lambda I}\right)}{\partial \mathbf{T}}\\ = -2\mathbf{B}'\mathbf{A} + \mathbf{B}'\mathbf{BT} + \mathbf{B}'\mathbf{BT} + \mathbf{T\Lambda} + \mathbf{T\Lambda}'\\ = \left(\mathbf{B}'\mathbf{B} + \mathbf{B}'\mathbf{B}\right)\mathbf{T} - 2\mathbf{B}'\mathbf{A} + \mathbf{T}\left(\mathbf{\Lambda} + \mathbf{\Lambda}'\right).\end{array}\right. \tag{9.14}$$

From (9.14), let

$$\mathbf{B}'\mathbf{B} = \mathbf{B}^*, \ \mathbf{B}'\mathbf{A} = \mathbf{C} \text{ and } \left(\mathbf{\Lambda} + \mathbf{\Lambda}'\right) = 2\mathbf{\Lambda}^*. \tag{9.15}$$

For an extremum value of d, we set $\frac{\partial d}{\partial T} = 0$ such that

$$\begin{bmatrix} 2\mathbf{C} = 2\mathbf{B}^*\mathbf{T} + 2\mathbf{T}\mathbf{\Lambda}^* \\ \mathbf{C} = \mathbf{B}^*\mathbf{T} + \mathbf{T}\mathbf{\Lambda}^*, \end{bmatrix} \tag{9.16}$$

leading to both $\mathbf{B}^*$ and $\mathbf{\Lambda}^*$ being symmetric. Hence

$$\mathbf{\Lambda}^* = \mathbf{T}'\mathbf{C} - \mathbf{T}'\mathbf{B}'\mathbf{T}. \tag{9.17}$$

But $\mathbf{B}^* \rightarrow$ symmetric and thus $\mathbf{T}'\mathbf{B}^*\mathbf{T}$ is also symmetric. $\mathbf{T}'\mathbf{C}$ is therefore symmetric or

$$\begin{bmatrix} \mathbf{T}'\mathbf{C} = \mathbf{C}'\mathbf{T} \\ \text{from the side condition} \\ \mathbf{T}'\mathbf{T} = \mathbf{T}\mathbf{T}' = \mathbf{I}_3 \\ \text{we have that} \\ \mathbf{C} = \mathbf{T}\mathbf{C}'\mathbf{T}. \end{bmatrix} \tag{9.18}$$

From (9.8), we had $\mathbf{C} = \mathbf{A}'\mathbf{B} = \mathbf{U}\mathbf{\Sigma}\mathbf{V}'$ by SVD. In the present case we note that $\mathbf{C} = \mathbf{B}'\mathbf{A}$, thus $\mathbf{C} = \mathbf{B}'\mathbf{A} = \mathbf{V}\mathbf{\Sigma}\mathbf{U}'$. From (9.18) we have

$$\begin{bmatrix} \text{with } \mathbf{U}'\mathbf{U} = \mathbf{U}\mathbf{U}' = \mathbf{V}'\mathbf{V} = \mathbf{V}\mathbf{V}' = \mathbf{I}_3 \\ \mathbf{C} = \mathbf{T}\mathbf{C}'\mathbf{T} \\ \mathbf{V}\mathbf{\Sigma}\mathbf{U}' = \mathbf{T}\mathbf{U}\mathbf{\Sigma}\mathbf{V}'\mathbf{T} \\ \mathbf{V} = \mathbf{T}\mathbf{U} \\ \text{or } \mathbf{T} = \mathbf{V}\mathbf{U}', \end{bmatrix} \tag{9.19}$$

which is identical to (9.10).

9-4 The General Procrustes solution

In Sect. 9-3, we presented the partial Procrustes algorithm and referred to it as "partial" because it was applied to solve only the rotation component of the datum transformation problem. The analysis of the parameterized conformal group $\mathbb{C}_7(3)$ (7-parameter similarity transformation) as presented in Sect. 17-1, however, requires the estimation of the *dilatation* unknown (scale factor), unknown vector of *translation* and the *unknown matrix of rotation*. These unknowns are determined from a given matrix data set of Cartesian coordinates as pseudo-observations. In addition to the unknown rotation matrix which was determined in Sect. 9-3, therefore, one has to determine the *scale* and *translation* elements. The *partial*

Procrustes algorithm gives way to the *general Procrustes algorithm*. The transpose which was indicated by $\{'\}$ in Sect. 9-3 will be denoted by $\{*\}$ in this section. In Sect. 17-1, the 7-parameter datum transformation problem will be formulated such that the solution of (17.1) lead to the desired seven parameters.

The unknown parameters for the 7-parameter transformation problem are a scalar-valued *scale factor* $x_1 \in \mathbb{R}$, a vector-valued *translation* parameters $\mathbf{x}_2 \in \mathbb{R}^{3\times 1}$ (column vector) and a matrix valued *rotation* parameters $\mathbf{X}_3 \in O^+(3) := \{\mathbf{X}_3 \in \mathbb{R}^{3\times 3} \mid \mathbf{X}_3^*\mathbf{X}_3 = \mathbf{I}_3, \mid \mathbf{X}_3 \mid= +1\}$, which in total constitute the 7-dimensional *parameter space*. x_1 represents the *dilatation unknown* (scale factor), $\mathbf{x}_2$ the *translation vector unknown* (3 parameters) and $\mathbf{X}_3$ the *unknown orthonormal matrix* (rotation matrix) which is an element of the special orthogonal group in three dimension. In other words, the $O^+(3)$ differentiable manifold can be coordinated by three parameters. In (10.23) on p. 148, relative position vectors are used to form the two matrices $\mathbf{A}$ and $\mathbf{B}$ in the same dimensional space. If the actual coordinates are used instead, the matrix-valued pseudo-observations $\{\mathbf{Y}_1, \mathbf{Y}_2\}$ become

$$\begin{bmatrix} x_1 \; x_2 \; \ldots \; x_n \\ y_1 \; y_2 \; \ldots \; y_n \\ z_1 \; z_2 \; \ldots \; z_n \end{bmatrix}^* =: \mathbf{Y}_1, \quad \mathbf{Y}_2 := \begin{bmatrix} X_1 \; X_2 \; \ldots \; X_n \\ Y_1 \; Y_2 \; \ldots \; Y_n \\ Z_1 \; Z_2 \; \ldots \; Z_n \end{bmatrix}^*, \tag{9.20}$$

with $\{\mathbf{Y}_1$ and $\mathbf{Y}_2\}$ replacing $\{\mathbf{A}$ and $\mathbf{B}\}$. The coordinate matrices of the n points ($n-$dimensional simplex) of a *left three-dimensional Weitzenböck space* as well as a *right three-dimensional Weitzenböck space*, namely $\mathbf{Y}_1 \in \mathbb{R}^{n\times 3}$ and $\mathbf{Y}_2 \in \mathbb{R}^{n\times 3}$ constitute the $6n$ dimensional *observation space*. Left and right matrices $\{\mathbf{Y}_1, \mathbf{Y}_2\}$ are related by means of the *passive 7-parameter conformal group* $\mathbb{C}_7(3)$ in three dimensions (similarity transformation, orthogonal Procrustes transformation) by (cf., 17.1 on p. 304)

$$\mathbf{Y}_1 \doteq F(x_1, \mathbf{x}_2, \mathbf{X}_3 \mid \mathbf{Y}_2) = \mathbf{Y}_2\mathbf{X}_3^* x_1 + \mathbf{1}\mathbf{x}_2^*, \mathbf{1} \in \mathbb{R}^{n\times 1}. \tag{9.21}$$

The nonlinear matrix-valued equation $F(x_1, \mathbf{x}_2, \mathbf{X}_3 \mid \mathbf{Y}_2) \doteq \mathbf{Y}_1$ is inconsistent since the image $\Re(F) \underset{\neq}{\subset} D(\mathbf{Y}_1)$ of F (range space $\Re(F)$) is constrained in the domain $D(\mathbf{Y}_1)$ of $\mathbf{Y}_1 \in \mathbb{R}^{n\times 3}$ (domain space $D(\mathbf{Y}_1)$). First, as a mapping, F is *"not onto, but into"* or *"not surjective"*. Second, by means of the error matrix $\mathbf{E} \in \mathbb{R}^{n\times 3}$ which accounts for errors in the pseudo-observation matrices $\mathbf{Y}_1$ as well as $\mathbf{Y}_2$, respectively, we are able to make the nonlinear matrix-valued equation $F(x_1, \mathbf{x}_2, \mathbf{X}_3 \mid \mathbf{Y}_2) \doteq \mathbf{Y}_1$ as identity. In this case,

$$\mathbf{Y}_1 = F(x_1, \mathbf{x}_2, \mathbf{X}_3 \mid \mathbf{Y}_2) + \mathbf{E} = \mathbf{Y}_2\mathbf{X}_3^* x_1 + \mathbf{1}\mathbf{x}_2^* + \mathbf{E}. \tag{9.22}$$

Furthermore, excluding configuration defect which can be detected a priori we shall assume $\aleph(F) = \{0\}$, the kernel of F (null space $\aleph(F)$) to contain only the zero element (empty null space $\aleph(F)$). A simplex of minimal dimension which allows the computation of the seven parameters of the space $\mathbb{X}$ is constituted by $n = 4$ points, namely a *tetrahedron* which is presented in the next examples.

Example 9.1 (Simplex of minimal dimension, $n = 4$ points, tetrahedron).

$$\mathbf{Y}_1 = \begin{bmatrix} x_1 & x_2 & x_3 & x_4 \\ y_1 & y_2 & y_3 & y_4 \\ z_1 & z_2 & z_3 & z_4 \end{bmatrix}^* \in \mathbb{R}^{n\times 3}, \quad \mathbf{Y}_2 = \begin{bmatrix} X_1 & X_2 & X_3 & X_4 \\ Y_1 & Y_2 & Y_3 & Y_4 \\ Z_1 & Z_2 & Z_3 & Z_4 \end{bmatrix}^* \in \mathbb{R}^{n\times 3},$$

$$\begin{bmatrix} x_1 & y_1 & z_1 \\ x_2 & y_2 & z_2 \\ x_3 & y_3 & z_3 \\ x_4 & y_4 & z_4 \end{bmatrix} = \begin{bmatrix} X_1 & Y_1 & Z_1 \\ X_2 & Y_2 & Z_2 \\ X_3 & Y_3 & Z_3 \\ X_4 & Y_4 & Z_4 \end{bmatrix} \mathbf{X}_3^* x_1 + \mathbf{1x}_2^* + \begin{bmatrix} e_{11} & e_{12} & e_{13} \\ e_{21} & e_{22} & e_{23} \\ e_{31} & e_{32} & e_{33} \\ e_{41} & e_{42} & e_{43} \end{bmatrix}.$$

We will now introduce the weight component and solve it the problem using General Procrustes solution.

*Example 9.2 (**W**eighted **LE**ast **S**quares' **S**olution* **W***-LESS).* We depart from the set up of the pseudo-observation equations given in Example 9.1 (simplex of minimal dimension, $n = 4$ points, tetrahedron). For a diagonal weight $\mathbf{W} = Diag\,(w_1, \ldots, w_4) \in \mathbb{R}^{4\times 4}$ we compute the Frobenius error matrix **W**-semi-norm

$$\| \mathbf{E} \|^2_{\mathbf{W}} := tr(\mathbf{E}^*\mathbf{W}\mathbf{E}) =$$

$$tr\left\{ \begin{bmatrix} e_{11} & e_{21} & e_{31} & e_{41} \\ e_{12} & e_{22} & e_{32} & e_{42} \\ e_{13} & e_{23} & e_{33} & e_{43} \end{bmatrix} \begin{bmatrix} w_1 & 0 & 0 & 0 \\ 0 & w_2 & 0 & 0 \\ 0 & 0 & w_3 & 0 \\ 0 & 0 & 0 & w_4 \end{bmatrix} \begin{bmatrix} e_{11} & e_{12} & e_{13} \\ e_{21} & e_{22} & e_{23} \\ e_{31} & e_{32} & e_{33} \\ e_{41} & e_{42} & e_{43} \end{bmatrix} \right\}$$

$$= tr\left\{ \begin{bmatrix} e_{11}w_1 & e_{21}w_2 & e_{31}w_3 & e_{41}w_4 \\ e_{12}w_1 & e_{22}w_2 & e_{32}w_3 & e_{42}w_4 \\ e_{13}w_1 & e_{23}w_2 & e_{33}w_3 & e_{43}w_4 \end{bmatrix} \begin{bmatrix} e_{11} & e_{12} & e_{13} \\ e_{21} & e_{22} & e_{23} \\ e_{31} & e_{32} & e_{33} \\ e_{41} & e_{42} & e_{43} \end{bmatrix} \right\}$$

$$= \left[\begin{array}{l} w_1 e_{11}^2 + w_2 e_{21}^2 + w_3 e_{31}^2 + w_4 e_{41}^2 \\ + w_1 e_{12}^2 + w_2 e_{22}^2 + w_3 e_{32}^2 + w_4 e_{42}^2 \\ + w_1 e_{13}^2 + w_2 e_{23}^2 + w_3 e_{33}^2 + w_4 e_{43}^2. \end{array} \right.$$

Obviously the coordinate errors (e_{11}, e_{12}, e_{13}) have the same weight w_1, $(e_{21}, e_{22}, e_{23}) \rightarrow w_2$, $(e_{31}, e_{32}, e_{33}) \rightarrow w_3$ and finally $(e_{41}, e_{42}, e_{43}) \rightarrow w_4$. We may also say that the *error weight is pointwise isotropic*, namely weight e_{11} =weight e_{12}=weight e_{13}=weight w_1 etc. But the error weight is *not homogeneous* since w_1 =weight e_{11}≠weight $e_{21} = w_2$. Of course, an ideal *homogeneous and isotropic weight distribution* is guaranteed by the criterion $w_1 = w_2 = w_3 = w_4 = w$.

By means of Solution 9.1 we have summarized the parameter space $(x_1, \mathbf{x}_2, \mathbf{X}_3) \in \mathbb{R} \times \mathbb{R}^3 \times \mathbb{R}^{3\times 3}$. In contrast, Solution 9.2 reviews the pseudo-observation space $(\mathbf{Y}_1, \mathbf{Y}_2) \in \mathbb{R}^{n\times 3} \times \mathbb{R}^{n\times 3}$ equipped with the Frobenius matrix **W**–semi-norm.

Solution 9.1 (The parameter space $\mathbb{X}$).

$x_1 \in \mathbb{R}$	dilatation parameter (scale factor)
$\mathbf{x}_2 \in \mathbb{R}^{3\times 1}$	column vector of translation parameters

$\mathbf{X}_3 \in O^+(3) := \{\mathbf{X}_3 \in \mathbb{R}^{3\times 3} \mid \mathbf{X}_3^*\mathbf{X}_3 = \mathbf{I}_3, \mid \mathbf{X}_3 \mid = +1\}$
orthonormal matrix,
rotation matrix of three
parameters

Solution 9.2 (The observation space $\mathbb{Y}$).

$$\mathbf{Y}_1 = \begin{bmatrix} x_1 \; x_2 \; \ldots \; x_n \\ y_1 \; y_2 \; \ldots \; y_n \\ z_1 \; z_2 \; \ldots \; z_n \end{bmatrix}^* \in \mathbb{R}^{n\times 3}, \quad \mathbf{Y}_2 = \begin{bmatrix} X_1 \; X_2 \; \ldots \; X_n \\ Y_1 \; Y_2 \; \ldots \; Y_n \\ Z_1 \; Z_2 \; \ldots \; Z_n \end{bmatrix}^* \in \mathbb{R}^{n\times 3}$$

left three-dimensional coordinate matrix of an $n-$ dimensional simplex

right three-dimensional coordinate matrix of an $n-$ dimensional simplex

The immediate problem that one is faced with is how to solve the inconsistent matrix-valued nonlinear equation (9.22). Essentially, this is the same problem that we introduced in (17.1) on p. 304. The difference between (17.1) and (9.22) is the incorporation of the error matrix $\mathbf{E}$ in the latter. This takes into consideration the stochasticity of the systems $\mathbf{Y}_1$ and $\mathbf{Y}_2$. In what follows $\mathbf{W}$-LESS (i.e., the **W**eighted **LE**ast **S**quares' **Solution**) is defined and materialized by the *Procrustes algorithm* presented by means of:

- Corollary 9.1 (partial $\mathbf{W}$-LESS for the unknown vector $\mathbf{x}_{2l}$).
- Corollary 9.2 (partial $\mathbf{W}$-LESS for the unknown scalar x_{1l}).
- Corollary 9.3 (partial $\mathbf{W}$-LESS for the unknown matrix $\mathbf{X}_{3l}$).

The partial optimization results are collected in *Theorem* 9.1 ($\mathbf{W}$-LESS of $\mathbf{Y}_1 = \mathbf{Y}_2\mathbf{X}_3^*x_1 + \mathbf{1}\mathbf{x}_2^* + \mathbf{E}$) and Corollary 9.4 ($\mathbf{I}$-LESS of $\mathbf{Y}_1 = \mathbf{Y}_2\mathbf{X}_3^*x_1 + \mathbf{1}\mathbf{x}_2^* + \mathbf{E}$). Solution 9.3 summarizes the general *Procrustes algorithm*.

Definition 9.1 (W-LESS). *The parameter array* $\{x_{1l}, \mathbf{x}_{2l}, \mathbf{X}_{3l}\}$ *is called* **W**-*LESS (least squares solution with respect to* Frobenius matrix $\mathbf{W}$*–semi-norm) of the inconsistent nonlinear matrix-valued system of equations*

$$\mathbf{Y}_2\mathbf{X}_3^*x_1 + \mathbf{1}\mathbf{x}_2^* + \mathbf{E} = \mathbf{Y}_1, \tag{9.23}$$

subject to

$$\mathbf{X}_3^*\mathbf{X}_3 = \mathbf{I}_3, \mid \mathbf{X}_3 \mid = +1, \tag{9.24}$$

if for the parameter array in comparison to all other parameter arrays $\{x_{1l}, \mathbf{x}_{2l}, \mathbf{X}_{3l}\}$, *the inequality*

$$\left[\begin{array}{l} \quad\quad\parallel \mathbf{Y}_1 - \mathbf{Y}_2\mathbf{X}_{3l}^*x_{1l} - \mathbf{1}\mathbf{x}_{2l}^* \parallel_{\mathbf{W}}^2 \\ := tr((\mathbf{Y}_1 - \mathbf{Y}_2\mathbf{X}_{3l}^*x_{1l} - \mathbf{1}\mathbf{x}_{2l}^*)^*\mathbf{W}(\mathbf{Y}_1 - \mathbf{Y}_2\mathbf{X}_{3l}^*x_{1l} - \mathbf{1}\mathbf{x}_{2l}^*) \\ \leq tr((\mathbf{Y}_1 - \mathbf{Y}_2\mathbf{X}_3^*x_1 - \mathbf{1}\mathbf{x}_2^*)^*\mathbf{W}(\mathbf{Y}_1 - \mathbf{Y}_2\mathbf{X}_3^*x_1 - \mathbf{1}\mathbf{x}_2^*) \\ \quad\quad =: \parallel \mathbf{Y}_1 - \mathbf{Y}_2\mathbf{X}_3^*x_1 - \mathbf{1}\mathbf{x}_2^* \parallel_{\mathbf{W}}^2 \end{array}\right. \tag{9.25}$$

holds, in particular if $\mathbf{E}_l := \mathbf{Y}_1 - \mathbf{Y}_2\mathbf{X}_{3l}^* x_{1l} - \mathbf{1}\mathbf{x}_{2l}^*$ *has the minimal Frobenius matrix* $\mathbf{W}$*–semi-norm such that* $\mathbf{W} \in \mathbb{R}^{n \times n}$ *is positive semi-definite.*

Note that $\| \mathbf{E} \|_{\mathbf{W}}^2 := tr(\mathbf{E}^* \mathbf{W} \mathbf{E})$ characterizes the method of least squares tuned to an error matrix $\mathbf{E} \in \mathbb{R}^{n \times 3}$ and a positive semi-definite weight matrix $\mathbf{W}$. Indeed a positive semi-definite weight matrix $\mathbf{W}$ of weights is chosen in-order to have the option to exclude by means of zero weight a particular pseudo-observation, say a particular coordinate row vector $[x_i, y_i, z_i], i \in \mathbb{N}$ arbitrary, but fixed by $w_{ii} = w_i = 0$, which may be an *outlier*. *Example* 9.2 illustrates details of *Definition* 9.1.

In order to construct W-LESS of the inconsistent nonlinear matrix-valued system of equations (9.23) subject to (9.24) we introduce the *Procrustes algorithm.* The *first* algorithmic step is constituted by the *forward* computation of the transformation parameters $\mathbf{x}_{2l}$ from the unconstraint *Lagrangean* $L(x_1, \mathbf{x}_2, \mathbf{X}_3)$ which is twice the value of the Frobenius error matrix $\mathbf{W}$–semi-norm. As soon as the translation parameters $\mathbf{x}_{2l}$ are *backward* substituted we gain a Lagrangean $L(x_1, \mathbf{X}_3)$ which is centralized with respect to the observation matrix $\mathbf{Y}_1 - \mathbf{Y}_2\mathbf{X}_3^* x_1$. In the *second* algorithmic step the scale parameter x_{1l} is *forward* computed from the centralized Lagrangean $L(x_1, \mathbf{X}_3)$. Its *backward* substitution leads to the Lagrangean $L(\mathbf{X}_3)$ which is only dependent on the rotation matrix $\mathbf{X}_3$. Finally the optimization problem $L(\mathbf{X}_3) = min$ subject to $\mathbf{X}_3^*\mathbf{X}_3 = \mathbf{I}_3$, $| \mathbf{X}_3 | = +1$ generates the *third* algorithmic step. This computational step is similar to that of *partial Procrustes algorithm* of Sect. 9-3. By means of *singular value decomposition SVD* the rotation matrix $\mathbf{X}_{3l}$ is *forward* computed and *backward* substituted to gain $(x_1, \mathbf{x}_2, \mathbf{X}_3)$ at the end. The results are collected in Corollaries 9.1 to 9.3.

Corollary 9.1 (Partial W-LESS for the translation vector $\mathbf{x}_{2l}$). *A* 3×1 *vector* $\mathbf{x}_{2l}$ *is* ***partial*** *W-LESS of (9.23) subject to (9.24) if and only if* $\mathbf{x}_{2l}$ *fulfills the system of normal equations*

$$\mathbf{1}^*\mathbf{W}\mathbf{1}\mathbf{x}_{2l} = (\mathbf{Y}_1 - \mathbf{Y}_2\mathbf{X}_3^* x_1)^*\mathbf{W}\mathbf{1}. \qquad (9.26)$$

The translation vector $\mathbf{x}_{2l}$ *always exist and is represented by*

$$\mathbf{x}_{2l} = (\mathbf{1}^*\mathbf{W}\mathbf{1})^{-1}(\mathbf{Y}_1 - \mathbf{Y}_2\mathbf{X}_3^* x_1)^*\mathbf{W}\mathbf{1}. \qquad (9.27)$$

For the special case $\mathbf{W} = \mathbf{I}_n$*, i.e., the weight matrix is unit, the translational parameter vector* $\mathbf{x}_{2l}$ *is given by*

$$\mathbf{x}_{2l} = \frac{1}{n}(\mathbf{Y}_1 - \mathbf{Y}_2\mathbf{X}_3^* x_1)^*\mathbf{1}.$$

proof

W-LESS is constructed by *unconstraint Lagrangean*

$$\left[\begin{array}{l} L(x_1,\mathbf{x}_2,\mathbf{X}_3) := \frac{1}{2} \parallel \mathbf{E} \parallel^2_{\mathbf{W}} = \parallel \mathbf{Y}_1 - \mathbf{Y}_2\mathbf{X}_3^* x_1 - \mathbf{1}\mathbf{x}_2^* \parallel^2_{\mathbf{W}} \\ = \frac{1}{2} tr(\mathbf{Y}_1 - \mathbf{Y}_2\mathbf{X}_3^* x_1 - \mathbf{1}\mathbf{x}_2^*)^* \mathbf{W}(\mathbf{Y}_1 - \mathbf{Y}_2\mathbf{X}_3^* x_1 - \mathbf{1}\mathbf{x}_2^*) = min, \\ subject\, to \{x_1 \geq 0, \mathbf{x}_2 \in R^{3\times 1}, \mathbf{X}_3^*\mathbf{X}_3 = \mathbf{I}_3\} \\ \frac{\partial L}{\partial \mathbf{x}_2^*}(\mathbf{x}_{2l}) = (\mathbf{1}^*\mathbf{W}\mathbf{1})\mathbf{x}_2 - (\mathbf{Y}_1 - \mathbf{Y}_2\mathbf{X}_3^* x_1)^* \mathbf{W}\mathbf{1} = \mathbf{0}, \end{array}\right. \tag{9.28}$$

constitutes the *first necessary condition.* Basics of vector-valued differentials are as given in *Table* 9.1, p. 117. For more details on matrix properties and manipulations, we refer to [180, pp. 439–451]. As soon as we back-substitute the translation parameter $\mathbf{x}_{2l}$, we are led to the *centralized Lagrangean*

$$\left[\begin{array}{c} L(x_1,\mathbf{X}_3) = \\ \frac{1}{2} tr\{\left[\mathbf{Y}_1 - (\mathbf{Y}_2\mathbf{X}_3^* x_1 + (\mathbf{1}^*\mathbf{W}\mathbf{1})^{-1}\mathbf{1}\mathbf{1}^*\mathbf{W}(\mathbf{Y}_1 - \mathbf{Y}_2\mathbf{X}_3^* x_1))\right]^* \mathbf{W} \\ ** \left[\mathbf{Y}_1 - (\mathbf{Y}_2\mathbf{X}_3^* x_1 + (\mathbf{1}^*\mathbf{W}\mathbf{1})^{-1}\mathbf{1}\mathbf{1}^*\mathbf{W}(\mathbf{Y}_1 - \mathbf{Y}_2\mathbf{X}_3^* x_1))\right]\} \end{array}\right. \tag{9.29}$$

$$\left[\begin{array}{c} L(x_1,\mathbf{X}_3) = \\ \frac{1}{2} tr\{\left[(\mathbf{I} - (\mathbf{1}^*\mathbf{W}\mathbf{1})^{-1}\mathbf{1}\mathbf{1}^*)\mathbf{W}(\mathbf{Y}_1 - \mathbf{Y}_2\mathbf{X}_3^* x_1)\right]^* \mathbf{W} \\ ** \left[(\mathbf{I} - (\mathbf{1}^*\mathbf{W}\mathbf{1})^{-1}\mathbf{1}\mathbf{1}^*)\mathbf{W}(\mathbf{Y}_1 - \mathbf{Y}_2\mathbf{X}_3^* x_1)\right]\} \end{array}\right. \tag{9.30}$$

$$\mathbf{C} := \mathbf{I}_n - (\mathbf{1}^*\mathbf{W}\mathbf{1})^{-1}\mathbf{1}\mathbf{1}^*\mathbf{W} \tag{9.31}$$

is a definition of the *centering matrix*, namely for $\mathbf{W} = \mathbf{I}_n$

$$\mathbf{C} := \mathbf{I}_n - \frac{1}{n}\mathbf{1}\mathbf{1},^* \tag{9.32}$$

being symmetric, in general. Substituting the centering matrix into the *reduced Lagrangean* $L(x_1, \mathbf{X}_3)$, we gain the *centralized Lagrangean*

$$\begin{array}{c} L(x_1,\mathbf{X}_3) = \\ \frac{1}{2} tr\{\left[\mathbf{Y}_1 - \mathbf{Y}_2\mathbf{X}_3^* x_1\right]^* \mathbf{C}^*\mathbf{W}\mathbf{C}\left[\mathbf{Y}_1 - \mathbf{Y}_2\mathbf{X}_3^* x_1\right]\} \end{array} \tag{9.33}$$

♣

Corollary 9.2 (Partial W-LESS for the scale factor x_{1l}). *A scalar x_{1l} is **partial** $\mathbf{W}$-LESS of (9.23) subject to (9.24) if and only if*

$$x_{1l} = \frac{tr\mathbf{Y}_1^*\mathbf{C}^*\mathbf{W}\mathbf{C}\mathbf{Y}_2\mathbf{X}_3^*}{tr\mathbf{Y}_2^*\mathbf{C}^*\mathbf{W}\mathbf{C}\mathbf{Y}_2} \tag{9.34}$$

holds. For special case $\mathbf{W} = \mathbf{I}_n$ the scale parameter vector x_{1l} is given by

$$x_{1l} = \frac{tr\mathbf{Y}_1^*\mathbf{C}^*\mathbf{C}\mathbf{Y}_2\mathbf{X}_3^*}{tr\mathbf{Y}_2^*\mathbf{C}^*\mathbf{C}\mathbf{Y}_2} \tag{9.35}$$

proof

Partial **W**-LESS is constructed by the unconstraint centralized Lagrangean

$$\left[\begin{array}{c} L(x_1, \mathbf{X}_3) = \\ = \frac{1}{2} tr\{[(\mathbf{Y}_1 - \mathbf{Y}_2\mathbf{X}_3^* x_1)]^* \mathbf{C}^* \mathbf{W} \mathbf{C} [\mathbf{Y}_1 - \mathbf{Y}_2 \mathbf{X}_3^* x_1]\} = min_{x_1, \mathbf{X}_3}, \\ subject\ to \{x_1 \geq 0, \mathbf{X}_3^* \mathbf{X}_3 = \mathbf{I}_3\}. \end{array}\right. \tag{9.36}$$

$$\frac{\partial L}{\partial x_1}(x_{1l}) = x_{1l} tr \mathbf{X}_3 \mathbf{Y}_2^* \mathbf{C}^* \mathbf{W} \mathbf{C} \mathbf{Y}_2 \mathbf{X}_3^* - tr \mathbf{Y}_1^* \mathbf{C}^* \mathbf{W} \mathbf{C} \mathbf{Y}_2 \mathbf{X}_3^* = \mathbf{0} \tag{9.37}$$

constitutes the *second necessary condition.* Due to (e.g., cyclic property in Table 9.1, p. 117)

$$tr \mathbf{X}_3 \mathbf{Y}_2^* \mathbf{C}^* \mathbf{W} \mathbf{C} \mathbf{Y}_2 \mathbf{X}_3^* = tr \mathbf{Y}_2^* \mathbf{C}^* \mathbf{W} \mathbf{C} \mathbf{Y}_2 \mathbf{X}_3^* \mathbf{X}_3 = \mathbf{Y}_2^* \mathbf{C}^* \mathbf{W} \mathbf{C} \mathbf{Y}_2,$$

(9.37) leads to (9.34). While the forward computation of $\frac{\partial L}{\partial x_1}(x_{1l}) = \mathbf{0}$ enjoyed a representation of the optimal scale parameter x_{1l}, its *backward* substitution into the Lagrangean $L(x_1, \mathbf{X}_3)$ amounts to

$$\left[\begin{array}{c} L(\mathbf{X}_3) = \\ tr\{\left[\mathbf{Y}_1 - \mathbf{Y}_2 \mathbf{X}_3^* \frac{tr \mathbf{Y}_1^* \mathbf{C}^* \mathbf{W} \mathbf{C} \mathbf{Y}_2 \mathbf{X}_3^*}{tr \mathbf{Y}_2^* \mathbf{C}^* \mathbf{W} \mathbf{C} \mathbf{Y}_2}\right] \mathbf{C}^* \mathbf{W} \mathbf{C} \\ ** \left[\mathbf{Y}_1 - \mathbf{Y}_2 \mathbf{X}_3^* \frac{tr \mathbf{Y}_1^* \mathbf{C}^* \mathbf{W} \mathbf{C} \mathbf{Y}_2 \mathbf{X}_3^*}{tr \mathbf{Y}_2^* \mathbf{C}^* \mathbf{W} \mathbf{C} \mathbf{Y}_2}\right]\} \end{array}\right. \tag{9.38}$$

$$\left[\begin{array}{c} L(\mathbf{X}_3) = \\ \frac{1}{2} tr\{(\mathbf{Y}_1^* \mathbf{C}^* \mathbf{W} \mathbf{C} \mathbf{Y}_1) - tr(\mathbf{Y}_1^* \mathbf{C}^* \mathbf{W} \mathbf{C} \mathbf{Y}_2 \mathbf{X}_3^*) \frac{tr \mathbf{Y}_1^* \mathbf{C}^* \mathbf{W} \mathbf{C} \mathbf{Y}_2 \mathbf{X}_3^*}{tr \mathbf{Y}_2^* \mathbf{C}^* \mathbf{W} \mathbf{C} \mathbf{Y}_2} \\ - tr(\mathbf{X}_3 \mathbf{Y}_2^* \mathbf{C}^* \mathbf{W} \mathbf{C} \mathbf{Y}_1) \frac{tr \mathbf{Y}_1^* \mathbf{C}^* \mathbf{W} \mathbf{C} \mathbf{Y}_2 \mathbf{X}_3^*}{tr \mathbf{Y}_2^* \mathbf{C}^* \mathbf{W} \mathbf{C} \mathbf{Y}_2} \\ + tr(\mathbf{X}_3 \mathbf{Y}_2^* \mathbf{C}^* \mathbf{W} \mathbf{C} \mathbf{Y}_2 \mathbf{X}_3^*) \frac{[tr \mathbf{Y}_1^* \mathbf{C}^* \mathbf{W} \mathbf{C} \mathbf{Y}_2 \mathbf{X}_3^*]^2}{[tr \mathbf{Y}_2^* \mathbf{C}^* \mathbf{W} \mathbf{C} \mathbf{Y}_2]^2}\} \end{array}\right. \tag{9.39}$$

$$\left[\begin{array}{c} L(\mathbf{X}_3) = \\ = \frac{1}{2} tr(\mathbf{Y}_1^* \mathbf{C}^* \mathbf{W} \mathbf{C} \mathbf{Y}_1) - \frac{[tr \mathbf{Y}_1^* \mathbf{C}^* \mathbf{W} \mathbf{C} \mathbf{Y}_2 \mathbf{X}_3^*]^2}{[tr \mathbf{Y}_2^* \mathbf{C}^* \mathbf{W} \mathbf{C} \mathbf{Y}_2]} \\ + \frac{1}{2} \frac{[tr \mathbf{Y}_1^* \mathbf{C}^* \mathbf{W} \mathbf{C} \mathbf{Y}_2 \mathbf{X}_3^*]^2}{[tr \mathbf{Y}_2^* \mathbf{C}^* \mathbf{W} \mathbf{C} \mathbf{Y}_2]} \end{array}\right. \tag{9.40}$$

$$\left[\begin{array}{c} L(\mathbf{X}_3) = \\ = \frac{1}{2}tr(\mathbf{Y}_1^*\mathbf{C}^*\mathbf{W}\mathbf{C}\mathbf{Y}_1) - \frac{1}{2}\frac{[tr\mathbf{Y}_1^*\mathbf{C}^*\mathbf{W}\mathbf{C}\mathbf{Y}_2\mathbf{X}_3^*]^2}{[tr\mathbf{Y}_2^*\mathbf{C}^*\mathbf{W}\mathbf{C}\mathbf{Y}_2]} = min, \\ subject\ to\, \{\mathbf{X}_3^*\mathbf{X}_3 = \mathbf{I}_3\} \end{array}\right. \tag{9.41}$$

♣

Corollary 9.3 (Partial W-LESS for the rotation matrix $\mathbf{X}_{3l}$). *A 3×3 orthonormal matrix $\mathbf{X}_3$ is partial* **W***-LESS of (9.41) if and only if*

$$\mathbf{X}_{3l} = \mathbf{U}\mathbf{V}^* \tag{9.42}$$

holds, where $\mathbf{A} := \mathbf{Y}_1^*\mathbf{C}^*\mathbf{W}\mathbf{C}\mathbf{Y}_2 = \mathbf{U}\mathbf{\Sigma}_s\mathbf{V}^*$ is a singular value decomposition with respect to a left orthonormal matrix $\mathbf{U}, \mathbf{U}^*\mathbf{U} = \mathbf{I}_3$, a right orthonormal matrix $\mathbf{V}, \mathbf{V}\mathbf{V}^* = \mathbf{I}_3$, and $\mathbf{\Sigma}_s = Diag(\sigma_1, \sigma_2, \sigma_3)$ a diagonal matrix of singular values $(\sigma_1, \sigma_2, \sigma_3)$. The singular values are the canonical coordinates of the right eigenspace $(\mathbf{A}^*\mathbf{A} - \mathbf{\Sigma}_s^2\mathbf{I})\mathbf{V} = \mathbf{0}$. *The* left eigenspace *is based upon* $\mathbf{U} = \mathbf{A}\mathbf{V}\mathbf{\Sigma}_s^{-1}$.

proof

In (9.41) $L(\mathbf{X}_3)$ subject to $\mathbf{X}_3^*\mathbf{X}_3 = \mathbf{I}_3$ is minimal if

$$tr(\mathbf{Y}_1^*\mathbf{C}^*\mathbf{W}\mathbf{C}\mathbf{Y}_2\mathbf{X}_3^*) = min, \quad subject\ to\, \{x_1 \geq 0, \mathbf{X}_3^*\mathbf{X}_3 = \mathbf{I}_3\}. \tag{9.43}$$

Let $\mathbf{A} := \mathbf{Y}_1^*\mathbf{C}^*\mathbf{W}\mathbf{C}\mathbf{Y}_2 = \mathbf{U}\mathbf{\Sigma}_s\mathbf{V}^*$ *be a singular value decomposition* with respect to a left orthonormal matrix $\mathbf{U}, \mathbf{U}^*\mathbf{U} = \mathbf{I}_3$, a right orthonormal matrix $\mathbf{V}, \mathbf{V}\mathbf{V}^* = \mathbf{I}_3$, and $\mathbf{\Sigma}_s = Diag(\sigma_1, \sigma_2, \sigma_3)$ a diagonal matrix of singular values $(\sigma_1, \sigma_2, \sigma_3)$. Then

$$\left[\begin{array}{c} tr(\mathbf{A}\mathbf{X}_3^*) = tr(\mathbf{U}\mathbf{\Sigma}_s\mathbf{V}^*\mathbf{X}_3^*) \\ = tr(\mathbf{\Sigma}_s\mathbf{V}^*\mathbf{X}_3^*\mathbf{U}) = \sum_{i=1}^3 \sigma_i r_{ii} \leq \sum_{i=1}^3 \sigma_i, \end{array}\right. \tag{9.44}$$

holds since $\mathbf{R} = \mathbf{V}^*\mathbf{X}_3^*\mathbf{U} = [r_{ij}] \in \mathbb{R}^{3\times 3}$ is orthonormal with $| r_{ii} | \leq 1$. The identity $tr(\mathbf{A}\mathbf{X}_3^*) = \sum_{i=1}^{3} \sigma_i$ applies if $\mathbf{V}^*\mathbf{X}_3^*\mathbf{U} = \mathbf{I}_3$, that is $\mathbf{X}_3^* = \mathbf{V}\mathbf{U}^*$, $\mathbf{X}_3 = \mathbf{U}\mathbf{V}^*$, namely if $tr(\mathbf{A}\mathbf{X}_3^*)$ is maximal:

$$\left[\begin{array}{c} tr(\mathbf{A}\mathbf{X}_3^*) = max_{\{\mathbf{X}_3^*\mathbf{X}_3 = \mathbf{I}_3\}} \\ \Leftrightarrow \mathbf{R} = \mathbf{V}^*\mathbf{X}_3^*\mathbf{U} = \mathbf{I}_3. \end{array}\right. \Leftrightarrow tr\mathbf{A}\mathbf{X}_3^* = \sum_{i=1}^{3} \sigma_{\cdot i} \tag{9.45}$$

♣

An alternative proof of *Corollary* 9.3 based on formal differentiation of traces and determinants has been given in Sect. 9-32 of Sect. 9-3. Finally we collect our sequential results in *Theorem* 9.1 identifying the *stationary point* of **W**-LESS of (9.23) specialized for $\mathbf{W} = \mathbf{I}$, i.e., matrix of unit weight in Corollary 9.4. The highlight is the *General Procrustes algorithm* we have developed in Solution 9.3.

Theorem 9.1 (W-LESS of $\mathbf{Y}_1 = \mathbf{Y}_2\mathbf{X}_3^* x_1 + \mathbf{1x}_2^* + \mathbf{E}$).
(i) *The parameter array* $\{x_1, \mathbf{x}_2, \mathbf{X}_3\}$ *is* **W**-*LESS of (9.23) if*

$$x_{1l} = \frac{tr\mathbf{Y}_1^*\mathbf{C}^*\mathbf{WCY}_2\mathbf{X}_{3l}^*}{tr\mathbf{Y}_2^*\mathbf{C}^*\mathbf{WCY}_2} \tag{9.46}$$

$$\mathbf{x}_{2l} = (\mathbf{1}^*\mathbf{W1})^{-1}(\mathbf{Y}_1 - \mathbf{Y}_2\mathbf{X}_{3l}^* x_{1l})^*\mathbf{W1} \tag{9.47}$$

$$\mathbf{X}_{3l} = \mathbf{UV}^*, \tag{9.48}$$

subject to the singular value decomposition of the general 3×3 *matrix*

$$\mathbf{Y}_1^*\mathbf{C}^*\mathbf{WCY}_2 = \mathbf{U}Diag(\sigma_1, \sigma_2, \sigma_3)\mathbf{V}^*, \tag{9.49}$$

namely

$$\left[\begin{array}{c}[(\mathbf{Y}_1^*\mathbf{C}^*\mathbf{WCY}_2)^*(\mathbf{Y}_1^*\mathbf{C}^*\mathbf{WCY}_2) - \sigma_i^2\mathbf{I}]\mathbf{v}_i = \mathbf{0}\ \forall i \in \{1,2,3\} \\ \mathbf{V} = \left[\mathbf{v}_1, \mathbf{v}_2, \mathbf{v}_3\right], \mathbf{VV}^* = \mathbf{I}_3\end{array}\right. \tag{9.50}$$

$$\left[\begin{array}{c}\mathbf{U} = \mathbf{Y}_1^*\mathbf{C}^*\mathbf{WCY}_2\mathbf{V}Diag(\sigma_1^{-1}, \sigma_2^{-1}, \sigma_3^{-1}) \\ \mathbf{U}^*\mathbf{U} = \mathbf{I}_3\end{array}\right. \tag{9.51}$$

and as well as to the centering matrix

$$\mathbf{C} := \mathbf{I}_n - (\mathbf{1}^*\mathbf{W1})^{-1}\mathbf{11}^*\mathbf{W}. \tag{9.52}$$

(ii) *The empirical error matrix of type* **W**-*LESS accounts for*

$$\mathbf{E}_l = [\mathbf{I}_n - \mathbf{11}^*\mathbf{W}(\mathbf{1}^*\mathbf{W1})^{-1}]\left\{\mathbf{Y}_1 - \mathbf{Y}_2\mathbf{VU}^*\frac{tr\mathbf{Y}_1^*\mathbf{C}^*\mathbf{WCY}_2\mathbf{VU}^*}{tr\mathbf{Y}_2^*\mathbf{C}^*\mathbf{WCY}_2}\right\}, \tag{9.53}$$

with the related Frobenius matrix **W**–*semi-norm*

$$\left[\begin{array}{l}\parallel \mathbf{E} \parallel_{\mathbf{W}}^2 := tr(\mathbf{E}_l^*\mathbf{WE}_l) = \\ tr\{(\mathbf{Y}_1 - \mathbf{Y}_2\mathbf{VU}^*\frac{tr\mathbf{Y}_1^*\mathbf{C}^*\mathbf{WCY}_2\mathbf{VU}^*}{tr\mathbf{Y}_2^*\mathbf{C}^*\mathbf{WCY}_2})^*. \\ .[\mathbf{I}_n - \mathbf{11}^*\mathbf{W}(\mathbf{1}^*\mathbf{W1})^{-1}]^*\mathbf{W}[\mathbf{I}_n - \mathbf{11}^*\mathbf{W}(\mathbf{1}^*\mathbf{W1})^{-1}]. \\ .(\mathbf{Y}_1 - \mathbf{Y}_2\mathbf{VU}^*\frac{tr\mathbf{Y}_1^*\mathbf{C}^*\mathbf{WCY}_2\mathbf{VU}^*}{tr\mathbf{Y}_2^*\mathbf{C}^*\mathbf{WCY}_2})\},\end{array}\right. \tag{9.54}$$

and the representative scalar measure of the error of type **W**-*LESS given by*

$$||| E_l |||_{\mathbf{W}} := \sqrt{tr(\mathbf{E}_l^*\mathbf{WE}_l)/3n}. \tag{9.55}$$

Corollary 9.4 (I-LESS of $\mathbf{Y}_1 = \mathbf{Y}_2\mathbf{X}_3^* x_1 + \mathbf{1x}_2^* + \mathbf{E}$).

(i) *The parameter array* $\{x_1, \mathbf{x}_2, \mathbf{X}_3\}$ *is* **I**-*LESS of (9.23) if*

$$x_{1l} = \frac{tr\mathbf{Y}_1^*\mathbf{CY}_2\mathbf{X}_{3l}^*}{tr\mathbf{Y}_2^*\mathbf{CY}_2} \tag{9.56}$$

$$\mathbf{x}_{2l} = \frac{1}{n}(\mathbf{Y}_1 - \mathbf{Y}_2\mathbf{X}_{3l}^* x_{1l})^*\mathbf{1} \tag{9.57}$$

$$\mathbf{X}_{3l} = \mathbf{UV}^*, \tag{9.58}$$

subject to the singular value decomposition of the general 3×3 *matrix*

$$\mathbf{Y}_1^* \mathbf{C} \mathbf{Y}_2 = \mathbf{U} Diag(\sigma_1, \sigma_2, \sigma_3) \mathbf{V}^*, \tag{9.59}$$

namely

$$\left[\begin{array}{c} [(\mathbf{Y}_1^* \mathbf{C} \mathbf{Y}_2)^* (\mathbf{Y}_1^* \mathbf{C} \mathbf{Y}_2) - \sigma_i^2 \mathbf{I}] \mathbf{v}_i = \mathbf{0} \, \forall i \in \{1,2,3\} \\ \mathbf{V} = \left[\mathbf{v}_1, \mathbf{v}_2, \mathbf{v}_3 \right], \mathbf{V}\mathbf{V}^* = \mathbf{I}_3 \end{array} \right. \tag{9.60}$$

$$\begin{array}{c} \mathbf{U} = \mathbf{Y}_1^* \mathbf{C} \mathbf{Y}_2 \mathbf{V} Diag(\sigma_1^{-1}, \sigma_2^{-1}, \sigma_3^{-1}) \\ \mathbf{U}^* \mathbf{U} = \mathbf{I}_3 \end{array} \tag{9.61}$$

and as well as to the centering matrix

$$\mathbf{C} := \mathbf{I}_n - \frac{1}{n} \mathbf{1}\mathbf{1}^*. \tag{9.62}$$

(ii) *The empirical error matrix of type* **I**-*LESS accounts for*

$$\mathbf{E}_l = [\mathbf{I}_n - \frac{1}{n} \mathbf{1}\mathbf{1}^*] \left\{ \mathbf{Y}_1 - \mathbf{Y}_2 \mathbf{V} \mathbf{U}^* \frac{tr \mathbf{Y}_1^* \mathbf{C} \mathbf{Y}_2 \mathbf{V} \mathbf{U}^*}{tr \mathbf{Y}_2^* \mathbf{C} \mathbf{Y}_2} \right\}, \tag{9.63}$$

with the related Frobenius matrix **W**-*semi-norm*

$$\left[\begin{array}{c} \| \mathbf{E} \|_{\mathbf{I}}^2 := tr(\mathbf{E}_l^* \mathbf{E}_l) = \\ tr\{(\mathbf{Y}_1 - \mathbf{Y}_2 \mathbf{V} \mathbf{U}^* \frac{tr \mathbf{Y}_1^* \mathbf{C} \mathbf{Y}_2 \mathbf{V} \mathbf{U}^*}{tr \mathbf{Y}_2^* \mathbf{C} \mathbf{Y}_2})^* \cdot \\ \cdot [\mathbf{I}_n - \frac{1}{n} \mathbf{1}\mathbf{1}^*] . \\ .(\mathbf{Y}_1 - \mathbf{Y}_2 \mathbf{V} \mathbf{U}^* \frac{tr \mathbf{Y}_1^* \mathbf{C} \mathbf{Y}_2 \mathbf{V} \mathbf{U}^*}{tr \mathbf{Y}_2^* \mathbf{C} \mathbf{Y}_2})\} \end{array} \right. \tag{9.64}$$

and the representative scalar measure of the error of type **I**-*LESS*

$$||| \mathbf{E}_l |||_{\mathbf{I}} := \sqrt{tr(\mathbf{E}_l^* \mathbf{E}_l)/3n}.$$

In the proof of Corollary 9.4, we only sketch the results that the matrix $\mathbf{I}_n - \frac{1}{n}\mathbf{1}\mathbf{1}^*$ *is* idempotent*:*

$$\left[\begin{array}{c} (\mathbf{I}_n - \frac{1}{n}\mathbf{1}\mathbf{1}^*)(\mathbf{I}_n - \frac{1}{n}\mathbf{1}\mathbf{1}^*) = \\ = \mathbf{I}_n - \frac{2}{n}\mathbf{1}\mathbf{1}^* + \frac{1}{n^2}(\mathbf{1}\mathbf{1}^*)^2 = \\ = \mathbf{I}_n - \frac{2}{n}\mathbf{1}\mathbf{1}^* + \frac{1}{n^2} n \mathbf{1}\mathbf{1}^* = \mathbf{I}_n - \frac{1}{n}\mathbf{1}\mathbf{1}^*. \end{array} \right.$$

Solution 9.3 (General Procrustes algorithm).

Step 1

<u>*Read*</u> :
$$\mathbf{Y}_1 = \begin{bmatrix} x_1 & y_1 & z_1 \\ . & . & . \\ x_n & y_n & z_n \end{bmatrix} \text{ and } \begin{bmatrix} X_1 & Y_1 & Z_1 \\ . & . & . \\ X_n & Y_n & Z_n \end{bmatrix} = \mathbf{Y}_2$$

Step 2

<u>*Compute*</u> : $\mathbf{Y}_1^*\mathbf{C}\mathbf{Y}_2$ subject to $\mathbf{C} := \mathbf{I}_n - \frac{1}{n}\mathbf{1}\mathbf{1}^*$

Step 3

<u>*Compute*</u> : *SVD* $\mathbf{Y}_1^*\mathbf{C}\mathbf{Y}_2 = \mathbf{U}Diag(\sigma_1, \sigma_2, \sigma_3)\mathbf{V}^*$

<u>3-1</u> $\mid (\mathbf{Y}_1^*\mathbf{C}\mathbf{Y}_2)^*(\mathbf{Y}_1^*\mathbf{C}\mathbf{Y}_2) - \sigma_i^2\mathbf{I} \mid = \mathbf{0} \Rightarrow (\sigma_1, \sigma_2, \sigma_3)$

<u>3-2</u>
$$((\mathbf{Y}_1^*\mathbf{C}\mathbf{Y}_2)^*(\mathbf{Y}_1^*\mathbf{C}\mathbf{Y}_2) - \sigma_i^2\mathbf{I})\mathbf{v}_i = \mathbf{0}\, \forall i \in \{1, 2, 3\}$$
$$\Rightarrow \mathbf{V} = \begin{bmatrix} \mathbf{v}_1, & \mathbf{v}_2, & \mathbf{v}_3 \end{bmatrix} \; right\, eigenvectors$$
$$(right\, eigencoloumns)$$

<u>3-3</u>
$$\mathbf{U} = \mathbf{Y}_1^*\mathbf{C}\mathbf{Y}_2\mathbf{V}Diag(\frac{1}{\sigma_1}, \frac{1}{\sigma_2}, \frac{1}{\sigma_3}) \; left\, eigenvectors$$
$$(left\, eigencoloumns)$$

Step 4

<u>*Compute*</u> : $\mathbf{X}_{3l} = \mathbf{U}\mathbf{V}^*$ (rotation)

Step 5

<u>*Compute*</u> : $x_{1l} = \dfrac{tr\mathbf{Y}_1^*\mathbf{C}\mathbf{Y}_2\mathbf{X}_3^*}{tr\mathbf{Y}_2^*\mathbf{C}\mathbf{Y}_2}$ (dilatation)

Step 6

<u>*Compute*</u> : $\mathbf{x}_{2l} = \frac{1}{n}(\mathbf{Y}_1 - \mathbf{Y}_2\mathbf{X}_3^*x_1)^*\mathbf{1}$ (translation)

Step 7

<u>*Compute*</u> : $\mathbf{E}_l = \mathbf{C}\left\{\mathbf{Y}_1 - \mathbf{Y}_2\mathbf{V}\mathbf{U}^*\dfrac{tr\mathbf{Y}_1^*\mathbf{C}\mathbf{Y}_2\mathbf{V}\mathbf{U}^*}{tr\mathbf{Y}_2^*\mathbf{C}\mathbf{Y}_2}\right\}$ (error matrix)

<u>*"optional control"*</u>

$$\mathbf{E}_l := \mathbf{Y}_1 - (\mathbf{Y}_2\mathbf{X}_{3l}^*x_{1l} + \mathbf{1}\mathbf{x}_{2l}^*)$$

Step 8

<u>*Compute*</u> : $\| \mathbf{E}_l \|_{\mathbf{I}} := \sqrt{tr(\mathbf{E}_l^*\mathbf{E}_l)}$ (error matrix)

Step 9

<u>*Compute*</u> : $||| \mathbf{E}_l |||_{\mathbf{I}} := \sqrt{tr(\mathbf{E}_l^*\mathbf{E}_l)/3n}$ (mean error matrix)

9-5 Extended general Procrustes solution

9-51 Mild anisotropy in scaling

The 3D affine transformation is one possible generalization of the $C_7(3,3)$ *Helmert transformation*, using three different scale (s_1, s_2, s_3) parameters instead of a single one. In this case the scale factors can be modeled by a diagonal matrix ($\mathbf{S}$).

$$\begin{pmatrix} x_i \\ y_i \\ z_i \end{pmatrix} = \mathbf{S} * \mathbf{R} \begin{pmatrix} X_i \\ Y_i \\ Z_i \end{pmatrix} + \begin{pmatrix} X_0 \\ Y_0 \\ Z_0 \end{pmatrix}. \tag{9.65}$$

Recently, an extension of general Procrustes method for solving the 3D affine 9-parameter transformation has been developed by *Awange et al.* [42] and is summarized as follows:

1) Compute the center of gravity of the two systems

$$\mathbf{a}_0 = \frac{1}{N}\mathbf{xyz} \tag{9.66}$$

$$\mathbf{b}_0 = \frac{1}{N}\mathbf{XYZ} \tag{9.67}$$

2) Translate the systems into the center of the coordinate system

$$\mathbf{xyzC} = \mathbf{xyz} - \mathbf{1} * \mathbf{a}_0 \tag{9.68}$$

$$\mathbf{XYZC} = \mathbf{XYZ} - \mathbf{1} * \mathbf{b}_0 \tag{9.69}$$

3) Compute matrix $\mathbf{A}$

$$\mathbf{A} = \mathbf{XYZC}^T * \mathbf{xyzC} \tag{9.70}$$

4) Compute the rotation matrix $\mathbf{R}$ via SVD decomposition of $\mathbf{A}$, getting the matrices $\mathbf{U}$, $\mathbf{V}$ and $diag\,(\lambda_1, \lambda_2, \lambda_3)$, where

$$\mathbf{A} = \mathbf{U}diag(\lambda_1, \lambda_2, \lambda_3)\mathbf{V}$$

and

$$\mathbf{R} = \mathbf{U} * \mathbf{V}^T$$

5) First approximation of the scale matrix considers only the diagonal elements of,

$$\mathbf{S} = \left(\mathbf{R}^T * \mathbf{XYZC}^T * \mathbf{XYZC} * \mathbf{R}\right)^{-1} * \mathbf{R}^T * \mathbf{XYZC}^T * \mathbf{xyzC} \tag{9.71}$$

i.e.,

$$\mathbf{S}_0 = diag(\mathbf{S}) \tag{9.72}$$

5) Iterate to improve the scale matrix

$$\mathbf{w} = \mathbf{xyz} - \mathbf{XYZ} * \mathbf{R} * \mathbf{S}_0 - \mathbf{1} * (\mathbf{a}_0 - \mathbf{b}_0 * \mathbf{R} * \mathbf{S}_0) \tag{9.73}$$

$$\mathbf{dS} = diag\left(\left(\mathbf{R}^T * \mathbf{XYZ}^T * \mathbf{XYZ} * \mathbf{R}\right)^{-1} * \mathbf{R}^T * \mathbf{XYZ}^T * \mathbf{w}\right) \tag{9.74}$$

$$\mathbf{S}_0 = \mathbf{S}_0 + \mathbf{dS} \tag{9.75}$$

6) After the iteration the scale matrix is

$$\mathbf{S} = diag\,(\mathbf{S}_0) \tag{9.76}$$

7) Then the translational vector

$$\mathbf{XYZ}_0 = \mathbf{a}_0 - \mathbf{b}_0 * \mathbf{R} * \mathbf{S}. \tag{9.77}$$

9-52 Strong anisotropy in scalling

The ABC method is very fast, and the computation effort for iteration to improve the **S** scale matrix is negligible. The main problem with ABC method is that after shifting the two systems into the origin of the coordinate system, the rotation matrix is computed via SVD. This means that the physical rotation as well as the deformation (distortion) caused by different scaling in the different directions of the 3 different principal axes will be also involved in the computation of the rotation matrix. This can be approximately allowed only when these scale factors do not differ from each other considerably. In addition, the necessary condition for getting optimal scale matrix is not restricted for diagonal matrix, and the off -diagonal elements are simply deleted in every iteration step, which will not ensure real global minimum for the trace of the error matrix.

The PZ method (see Paláncz et al. [324]) can cure these problems, but there is a price for it! This method start with an initially **guessed** diagonal **S** scale matrix, and first using the inverse of this matrix to eliminate the deformation caused by scaling, and only after the elimination of this "distortion" will be SVD applied to compute the rotation matrix itself. It goes without saying that in this way, the computation of the rotation matrix should be repeated in iterative way in order to decrease the error. That is why that this method is precise and correct for any ratios of scale parameters, but takes longer time to achieve the results, than in case of the ABC method.

The algorithm can be summarized as follows:

1) Guess the elements of the diagonal scale matrix of the transformation **S**

$$\mathbf{S} = \begin{pmatrix} s_1 & 0 & 0 \\ 0 & s_2 & 0 \\ 0 & 0 & s_3 \end{pmatrix} \tag{9.78}$$

The steps for computing the error in case of a given scale matrix **S** are the following.

2)Using equations (9.66) and (9.67) compute the center of gravity of the two system.

3) Translate the system using (9.68) and (9.69)
4) Eliminate rotation (distortion) caused by scaling with different values in different directions in the computation of matrix **A** using

$$\mathbf{A} = \mathbf{XYZC}^T * \mathbf{xyzC} * \mathbf{S}^{-1} \tag{9.79}$$

5) Compute the rotation matrix using SVD decomposition of **A**, getting the matrices **U**, **V** and $diag\,(\lambda_1, \lambda_2, \lambda_3)$, where

$$\mathbf{A} = \mathbf{U}diag(\lambda_1, \lambda_2, \lambda_3)\mathbf{V}$$

and

$$\mathbf{R} = \mathbf{U} * \mathbf{V}^T$$

where $\mathbf{A} = \mathbf{U}diag(\ldots)\mathbf{V}$(see, e.g., 9.59).
6) Compute the translation vector using (9.77).
7) Compute the error matrix

$$\mathbf{E} = \mathbf{xyz} - \mathbf{XYZ} * \mathbf{R}^T * \mathbf{S} - \mathbf{1} * \mathbf{XYZ}_0^T \tag{9.80}$$

8) Compute the square of the norm of the mean error matrix

$$||| \mathbf{E}_l |||_{\mathbf{I}} := \sqrt{tr(\mathbf{E}_l^* \mathbf{W} E_l)/3n} \tag{9.81}$$

If this error is to high, then modify the scale matrix **S**, and repeat the computation. To do that reasonably, this computation is implemented a Newton- algorithm to carry out local minimization of the Frobenius norm of the error matrix. In order to illustrate the proper application area of the two extended Procrustes methods, let us consider two numerical examples. In the case of the 7 parameter similarity transformation (Helmert) the scale value s can be estimated by dividing the sum of length in both systems from the centre of gravity, see Albertz and Kreiling [8].

The center of the gravity of the systems are,

$$\mathbf{a}_0 = \begin{pmatrix} x_s \\ y_s \\ z_s \end{pmatrix} \text{ and } \mathbf{b}_0 = \begin{pmatrix} X_s \\ Y_s \\ Z_s \end{pmatrix} \tag{9.82}$$

Then the estimated scale parameter according to Albertz and Kreimlig [8] in the Helmert similarity transformation is

$$s_{\texttt{priori}} = \frac{\sum_{i=1}^N \sqrt{(x_i - x_s)^2 + (y_i - y_s)^2 + (z_i - z_s)^2}}{\sum_{i=1}^N \sqrt{(X_i - X_s)^2 + (Y_i - Y_s)^2 + (Z_i - Z_s)^2}} \tag{9.83}$$

In case of the 9 parameter affine transformation where 3 different scale values (s_1, s_2, s_3). are applied according to the 3 coordinate axes, a good approach for the scale parameters can be given by modifying the Albertz-Kreiling [8] expression. Instead of the quotient of the two lengths in the centre of gravity system we can use the quotients of the sum of the lengths in the corresponding coordinate axes directions.

The estimated scale parameters according to the modified Albertz-Kreiling [8] expression for the 9 parameter transformation (see also [430]) is

$$s_{O1} = \frac{\sum_{i=1}^{N}\sqrt{(x_i - x_s)^2}}{\sum_{i=1}^{N}\sqrt{(X_i - X_s)^2}},\ s_{O2} = \frac{\sum_{i=1}^{N}\sqrt{(y_i - y_s)^2}}{\sum_{i=1}^{N}\sqrt{(Y_i - Y_s)^2}},\ s_{O3} = \frac{\sum_{i=1}^{N}\sqrt{(z_i - z_s)^2}}{\sum_{i=1}^{N}\sqrt{(Z_i - Z_s)^2}}, \tag{9.84}$$

where (s_{01}, s_{02}, s_{02}) are the initial elements of $\mathbf{S}$ in the iteration.

9-6 Weighted Procrustes transformation

As already stated in Sect. 17-1, other than the problems associated with linearization and iterations, the 7-datum transformation problem (conformal group $\mathbb{C}_7(3)$) is compounded with the problem of incorporating the weights of the systems involved. This section presents Procrustes algorithm II; a reliable means of solving the problem. We have already seen that the problem could be solved using Gauss-Jacobi combinatorial algorithm in Sect. 17-23. Procrustes algorithm II presented in the preceding section offers therefore an alternative that is not computationally intensive as the combinatorial method.

To obtain the weight matrix used in Corollaries 9.1, 9.2 and 9.3 in the *weighted Procrustes problem*, we proceed via the *variance-covariance* matrix of *Theorem* 9.2, whose proof is given in Solution 9.4 where the dispersion matrices of two sets of coordinates in $\mathbb{R}^3$ are presented in (9.86) and (9.88). They are used in (9.90) and (9.91) to obtain the dispersion of the error matrix $\mathbf{E}$ in (9.92). In-order to simplify (9.92), we make use of Corollary 9.5 adopted from [180, Appendix A, p. 419] to express $vec\ x_1\mathbf{X}_3\mathbf{Y}_2^*$ as in (9.93) and substitute it in (9.92) to obtain (9.94). We provide as a summary the following:

Theorem 9.2 (variance-covariance matrix). *Let vec* $\mathbf{E}^*$ *denote the vector valued form of the transposed error matrix* $\mathbf{E} := \mathbf{Y}_1 - \mathbf{Y}_2\mathbf{X}_3^*x_1 - \mathbf{1}\mathbf{x}_2^*$. *Then*

$$\left[\begin{array}{l}\boldsymbol{\Sigma}_{vec\mathbf{E}^*} = \boldsymbol{\Sigma}_{vec\,\mathbf{Y}_1^*} + (\mathbf{I}_n \otimes x_1\mathbf{X}_3)\boldsymbol{\Sigma}_{vec\,\mathbf{Y}_2^*}(\mathbf{I}_n \otimes x_1\mathbf{X}_3)^* \\ \qquad -2\boldsymbol{\Sigma}_{vec\mathbf{Y}_1^*,(\mathbf{I}_n\otimes x_1\mathbf{X}_3)vec\mathbf{Y}_2^*}\end{array}\right. \tag{9.85}$$

is the exact representation of the dispersion matrix (variance-covariance matrix) $\Sigma_{vec\ \mathbf{E}^*}$ *of vec* $\mathbf{E}^*$ *in terms of dispersion matrices (variance-covariance matrices)* $\boldsymbol{\Sigma}_{vec\,\mathbf{Y}_1^*}$ *and* $\boldsymbol{\Sigma}_{vec\,\mathbf{Y}_2^*}$ *of the two coordinates sets vec* $\mathbf{Y}_1^*$ *and vec* $\mathbf{Y}_2^*$ *as well as of their covariance matrix*

$$\boldsymbol{\Sigma}_{vec\mathbf{Y}_1^*,(\mathbf{I}_n\otimes x_1\mathbf{X}_3)vec\mathbf{Y}_2^*}.$$

Proof. By means of Solution 9.4 we define the dispersion matrices, also called variance-covariance matrices, of $vec\,\mathbf{Y}_1^*$ and $vec\,\mathbf{Y}_2^*$ of the two sets of coordinates.

Solution 9.4 (Dispersion matrices of two sets of coordinates in $\mathbb{R}^3$).

$$\boldsymbol{\Sigma}_{vec\,\mathbf{Y}_1^*} = E\{\begin{bmatrix} x_1 - E\{x_1\} \\ x_2 - E\{x_2\} \\ \dots \\ x_n - E\{x_n\} \end{bmatrix} [\,(x_1 - E\{x_1\})^* \dots (x_n - E\{x_n\})^* \,]\} \tag{9.86}$$

$$E\{(vec\mathbf{Y}_1^* - E\{\mathbf{Y}_1^*\})(vec\mathbf{Y}_1^* - E\{\mathbf{Y}_1^*\})^*\} \tag{9.87}$$

$$\boldsymbol{\Sigma}_{vec\,\mathbf{Y}_2^*} = E\{\begin{bmatrix} X_1 - E\{X_1\} \\ X_2 - E\{X_2\} \\ \dots \\ X_n - E\{X_n\} \end{bmatrix} [\,(X_1 - E\{X_1\})^* \dots (X_n - E\{X_n\})^* \,]\} \tag{9.88}$$

Next from the transposed error matrix

$$\mathbf{E}^* := \mathbf{Y}_1^* - (x_1\mathbf{X}_3\mathbf{Y}_2^* + \mathbf{x}_2\mathbf{1}\mathbf{x}_2^*) \tag{9.89}$$

we compute the dispersion matrix (variance-covariance matrix) $\boldsymbol{\Sigma}_{vec\,\mathbf{E}^*}$

$$\left[\begin{array}{l} \boldsymbol{\Sigma}_{vec\,\mathbf{E}^*} := E\{[vec\,\mathbf{E}^* - E\{vec\,\mathbf{E}^*\}][vec\,\mathbf{E}^* - E\{vec\,\mathbf{E}^*\}]^*\} \\ = E\,\{[vec\,\mathbf{Y}_1^* - E\{vec\,\mathbf{Y}_1^*\} - x_1(vec\,\mathbf{X}_3\mathbf{Y}_2^* - E\{vec\,\mathbf{X}_3\mathbf{Y}_2^*\})] \\ \times\, [vec\,\mathbf{Y}_1^* - E\{vec\,\mathbf{Y}_1^*\} - x_1(vec\,\mathbf{X}_3\mathbf{Y}_2^* - E\{vec\,\mathbf{X}_3\mathbf{Y}_2^*\})]^*\}\} \end{array}\right. \tag{9.90}$$

$$\left[\begin{array}{l} \boldsymbol{\Sigma}_{vec\,\mathbf{E}^*} = E\{[vec\,\mathbf{Y}_1^* - E\{vec\,\mathbf{Y}_1^*][vec\,\mathbf{Y}_1^* - E\{vec\,\mathbf{Y}_1^*]^*\} \\ +x_1^2 E\{[vec\,\mathbf{X}_3\mathbf{Y}_2^* - E\{vec\,\mathbf{X}_3\mathbf{Y}_2^*\}][vec\,\mathbf{X}_3\mathbf{Y}_2^* - E\{vec\,\mathbf{X}_3\mathbf{Y}_2^*]^* \\ -2x_1 E\{[vec\,\mathbf{Y}_1^* - E\{vec\,\mathbf{Y}_1^*][vec\,\mathbf{X}_3\mathbf{Y}_2^* - E\{vec\,\mathbf{X}_3\mathbf{Y}_2^*]^*\} \end{array}\right. \tag{9.91}$$

$$\{\, \boldsymbol{\Sigma}_{vec\,\mathbf{E}^*} = \boldsymbol{\Sigma}_{vec\,\mathbf{Y}_1^*} + \boldsymbol{\Sigma}_{vec\,x_1\mathbf{X}_3\mathbf{Y}_2^*} - 2\boldsymbol{\Sigma}_{vec\,\mathbf{Y}_1^*, vec\,x_1\mathbf{X}_3\mathbf{Y}_2^*} \tag{9.92}$$

Corollary 9.5.

$$\left[\begin{array}{l} vec\,\mathbf{AB} = (\mathbf{I}_q \otimes \mathbf{A})vec\,\mathbf{B}\, for\, all\, \mathbf{A} \in \mathbb{R}^{n\times m}, \mathbf{B} \in \mathbb{R}^{m\times q} \\ vec\,\mathbf{X}_3\mathbf{Y}_2^* = (\mathbf{I}_n \otimes x_1\mathbf{X}_3)vec\,\mathbf{Y}_2^* \;\; for\, all\, \mathbf{X}_3 \in \mathbb{R}^{3\times 3},\; \mathbf{Y}_2^* \in \mathbb{R}^{3\times n} \end{array}\right. \tag{9.93}$$

As soon as we implement the *Kronecker-Zehfuß decomposition* of the $vec\,\mathbf{AB}$ we arrive at the general representation of the dispersion matrix $\boldsymbol{\Sigma}_{vec\,\mathbf{E}^*}$, namely

$$\left[\begin{array}{l} \boldsymbol{\Sigma}_{vec\,\mathbf{E}^*} = \boldsymbol{\Sigma}_{vec\,\mathbf{Y}_1^*} + (\mathbf{I}_n \otimes x_1\mathbf{X}_3)\boldsymbol{\Sigma}_{vec\,\mathbf{Y}_2^*}(\mathbf{I}_n \otimes x_1\mathbf{X}_3)^* - \\ \qquad -2\boldsymbol{\Sigma}_{vec\mathbf{Y}_1^*,(\mathbf{I}_n\otimes x_1\mathbf{X}_3)vec\mathbf{Y}_2^*} \end{array}\right. \tag{9.94}$$

♣

The results of *Theorem* 9.2 are interpreted in more detail as follows: The variance-covariance matrix of the vectorized error matrix $\mathbf{E}^*$ depends on;

(i) the variance-covariance matrix $\mathbf{\Sigma}_{vec\,\mathbf{Y}_1^*}$ of the *local* coordinate set $(x_1, y_1, z_1, \dots, x_n, y_n, z_n)$,

(ii) the variance-covariance matrix $\mathbf{\Sigma}_{vec\,\mathbf{Y}_2^*}$ of the *global* coordinate set $(X_1, Y_1, Z_1, \dots, X_n, Y_n, Z_n)$,

(iii) the covariance matrix between $vec\,\mathbf{Y}_1^*$ and $(\mathbf{I}_n \otimes x_1\mathbf{X}_3)vec\mathbf{Y}_2^*$ of the global coordinate set $vec\,\mathbf{Y}_2^*$ as well as

(iv) the nonlinearity of the parameter model on the unknowns x_1, $\mathbf{X}_3$ of type *"scale factor"* and *"rotation matrix"* coupled to $(\mathbf{I}_n \otimes x_1\mathbf{X}_3)$.

So as to take advantage of the equivalence theorem between least squares approximation and best linear uniformly unbiased estimation, e.g., [180, §3, pp. 339–340], which holds for linear Gauss-Markov model, it is tempting to identify the weight matrix $\mathbf{W}$ of $\mathbf{W}$-LESS with $\mathbf{\Sigma}^{-1}_{vec\,\mathbf{E}^*}$ shrunk to a *locally isotropic error situation.* Such a shrinking procedure is outlined in Example 17.3, namely by taking in account *isotropic, but inhomogeneous criterion matrices.*

9-7 Concluding remarks

The *partial* Procrustes algorithm presented in this chapter provides a powerful tool for solving rotation and orientation related problems in general. The approach is straight forward and does not require linearization, which bog down least squares and other techniques commonly used. In Chap. 17, it shall be demonstrated how the general Procrustes approach determines scale and translation parameters of transformation, in addition to the rotation elements.

The 9-parameter Procrustean algorithm considered in this Chapter can thus be used for

(i) quicker and effective generation of 9 transformation parameters given coordinates in two systems as matrix configuration,

(ii) quick checking of the transformation parameters obtained from other methods

(iii) generating three scale parameters which could be useful in correcting distortions following procedures which first determine the rotation and translation parameters independent of scale.

For complete exposition of Procrustes approach, we refer to the works of [79, 86, 87, 93, 110, 111, 116, 120, 121, 130, 151, 153, 157, 191, 197, 198, 293, 294, 333, 355, 356, 378, 397].

Part II

Applications to geodesy and geoinformatics

10 LPS-GNSS orientations and vertical deflections

10-1 Introductory remarks

Since the advent of the Global Navigation Satellite System (GNSS), in particular the Global Positioning System (GPS), many fields within geosciences, such as geodesy, geoinformatics, geophysics, hydrology etc., have undergone tremendous changes. GPS satellites have in fact revolutionized operations in these fields and the entire world in ways that its inventors never imagined. The initial goal of GPS satellites was to provide the capability for the US military to position themselves accurately from space. This way, they would be able to know the positions of their submarines without necessarily relying on fixed ground stations that were liable to enemy attack. Slowly, but surely, the civilian community, led by geodesists, began to devise methods of exploiting the potential of this system. The initial focus of research was on the improvement of positioning accuracies since civilians only have access to the so called *coarse acquisition* or C/A-code of the GPS signal. This code is less precise when compared to the P-code used by the US military and its allies. The other source of error in GPS positioning was the Selective Availability (SA), i.e., intentional degradation of the GPS signal by the US military that would lead to a positioning error of ± 100 m. However, in May 2000, the then president of the United States Bill Clinton, officially discontinued this process.

As research in GPS progressed, so also arose new applications of its use. For example, previous research focussed on modelling or eliminating atmospheric effects such as refraction and multipath on the transmitted signals. In the last decade, however, Melbourne et al. [300] suggested that this negative effect of the atmosphere on GPS signals could be inverted to remote sense the atmosphere for vertical profiles of temperature and pressure. This gave birth to the new field of GPS meteorology, which is currently an active area of research. GPS meteorology has enhanced environmental and atmospheric studies by contributing to weather prediction and forecasting. This new technique is presented in Chap. 15, where the algebraic computations involved are solved.

One would be forgiven to say that the world will soon be unable to operate without GPS satellites. This, however, will not be an understatement either. GPS satellites have influenced our lives such that almost every operation is increasingly becoming GPS dependent! From the use of mobile phones, fertilizer regulation in farming, fish tracking in fisheries, vehicle navigation etc., the word is GPS. These numerous applications of GPS satellites has led the European countries to develop the so-called GALILEO satellites, which are the equivalent of GPS and are currently expected to be operational in 2013. The Russian based Globalnaya Navigationnaya Sputnikovaya Sistema, or GLONASS which were also developed originally for military uses are still operational with modernized versions expected within the next decade.

The direct impact of using these satellites is the requirement that operations be almost entirely *three-dimensional*. The major challenge posed by this requirement

J.L. Awange et al., *Algebraic Geodesy and Geoinformatics*, 2nd ed.,
DOI 10.1007/978-3-642-12124-1_10,

is that of integrating the results from the GNSS satellite system, which operates *globally*, to the traditional techniques that operate *locally*. In geodesy and geoinformatics for example, integrating global and local observations lead to the problem of solving for *3d-orientation* and the *deflection of the vertical*, the subject of this Chapter. These problems have one thing in common: *They require the solution of nonlinear equations that relate the unknowns to the measured values.*

In five sections we introduce positioning systems, global and local, and relate datum problems namely the three-dimensional orientation problem, especially the *procrustes orientation problem* and the vertical deflection. In some detail, we use the example of the test network Stuttgart central.

In short terms, we review the *Global Positioning System* (GPS: Global Problem Solver), the bestseller for *Applied Geodesy*, and the Local Positioning System (LPS: Local Problem Solver) based on "total stations" (theodolites, electronic distance meters EDMs, photogrammetric cameras, *laser scanners*). In some details in Sects. 10-3 and 10-4 we introduce GPS and LPS. A central topic is the relationship between global and local level reference frames based on *vertical deflections* and the classical *orientation unknown*. The classical observation equations for LPS are presented. A modern technique, the *Procrustes solution* of the orientation problem, is illustrated. As a byproduct, we describe the determination of the vertical deflection, the orientation difference between a model orientation and the real orientation (ξ, η). The central part is Sect. 10-6, the example of the test network "Stuttgart Central": GPS coordinates, spherical coordinates of the relative position vector. Unfortunately, the text is very short. An interested reader finds more details in the given references in Sect. 10-7.

10-2 Positioning systems

In daily operations, geodesists and geoinformatists have at their disposal two operating systems namely:

- ***Global Positioning Systems***; in this system, the practitioner operates at a global scale with positions referred to the global reference frame (e.g., World Geodetic System WGS-84). The tools employed comprise mainly satellites with global positioning capabilities. These satellites include; the US based Global Positioning System (GPS), Russian based Globalnaya Navigationnaya Sputnikovaya Sistema (or simply Global Navigation Satellite System) GLONASS and the proposed European Union's proposed Global Navigation Satellite System GALILEO which is expected to be operational in 2013. Unlike GPS satellites which were designed for the US military, GALILEO satellites will be civilian owned. For a brief introduction to GALILEO and other GNSS, we refer to Awange [43].
- ***Local Positioning Systems*** (LPS); which are applicable at national levels. The main positioning tools include; total stations, theodolites, EDMs, photogrammetric cameras, laser scanners etc. Positions in these systems are referred to the local level reference frames. With these systems, for example,

engineers have possibilities of setting horizontal and vertical networks for constructions. Those in geodynamics use them together with GPS for deformation monitoring.

The present chapter discusses these two systems in detail. In particular, for the LPS, the issue of local datum choice is addressed. The test network of "Stuttgart Central" which is applied to test the algorithms of Chaps. 4–7 is also presented.

10-3 Global positioning system (GPS)

Global Positioning System (GPS) are satellites that were primarily designed for use of US military in the early 60's, with a secondary role of civilian navigation. The oscillators aboard the GPS satellites generate a fundamental frequency f_0 of 10.23 MHz. Two carrier signals in the L band denoted L_1 and L_2 are generated by integer multiplication of the fundamental frequency f_0. These carriers are modulated by codes to provide satellite clock readings measured by GPS receivers. Two types of codes; the coarse acquisition C/A and precise acquisition P/A are emitted. C/A code in the L_1 carrier is less precise and is often reserved for civilian use, while the P/A code is reserved for the use of US military and its allies. It is coded on both L_1 and L_2 [227, 43]. The design comprises three segments namely; the space segment, user segment and the control segment. The space segment was designed such that the constellation consisted of 24 satellites (with a spare of four) orbiting at a height of about 20,200 km. The orbits are inclined at an angle of 55° from the equator with an orbiting period of about 12 hours. The user segment consists of a receiver that tracks signals from at least four satellites in-order to position (see e.g., discussion on Chap. 12). The control segment consist 5 ground stations with the master station located at the Air force base in Colorado. The master station measures satellite signals which are incorporated in the orbital models for each satellite. The models compute ephemerids and satellite clock correction parameters which are transmitted to the satellites. The satellites then transmit the orbital data to the receivers.

The results of the *three-dimensional positioning* using GPS satellites are the three-dimensional geodetic coordinates $\{\lambda, \phi, h\}$ of a receiver station. These coordinates comprise the geodetic longitude λ, geodetic latitude ϕ and geodetic height h. When positioning with GPS, the outcome is the geocentric position for an individual receiver or the relative positions between co-observing receivers.

The *global reference frame* $\mathbb{F}^\bullet$ upon which the GPS observations are based is defined by the base vectors $\mathbb{F}_{1\bullet}, \mathbb{F}_{2\bullet}, \mathbb{F}_{3\bullet}$, with the origin being the center of mass. The fundamental vector is defined by the base vector $\mathbb{F}_{3\bullet}$ and coincides with the mean axis of rotation of the Earth and points to the direction of the Conventional International Origin (CIO). $\mathbb{F}_{1\bullet}$ is oriented such that the plane formed by $\mathbb{F}_{1\bullet}$ and $\mathbb{F}_{3\bullet}$ points to the direction of Greenwich in England. $\mathbb{F}_{2\bullet}$ completes the right handed system by being perpendicular to $\mathbb{F}_{1\bullet}$ and $\mathbb{F}_{3\bullet}$. The geocentric Cartesian coordinates of a positional vector $\mathbf{X}$ is given by

$$\mathbf{X} = \mathbb{F}_{1\bullet} X + \mathbb{F}_{2\bullet} Y + \mathbb{F}_{3\bullet} Z, \tag{10.1}$$

where $\{X, Y, Z\}$ are the components of the vector $\mathbf{X}$ in the system $\mathbb{F}_{1\bullet}, \mathbb{F}_{2\bullet}, \mathbb{F}_{3\bullet} \,|o$.

10-4 Local positioning systems (LPS)

Grafarend [164] defines a *local level system* as a three-dimensional reference frame *at the hand* of an experimenter in an engineering network. When one is positioning using a theodolite or a total Station, one first centers the instrument. When the instrument is properly centered and ready for operation, the vertical axis of the instrument at this moment coincides with the direction of the local gravity vector at that particular point, hence the term *direction of local gravity vector*. The vertical axis at the theodolite station however points in the direction opposite to that of the gravity vector (i.e., to the zenith). The instrument can now be used to measure observations of the type horizontal directions T_i, angles, vertical directions B_i or the spatial distances S_i. The triplet $\{S_i, T_i, B_i\}$ are measured in the *local level reference frame* and are used to form the spherical coordinates of a point. These systems as opposed to GPS are only used within the local networks and are referred to as the *Local Positioning Systems* (LPS). When one is operating in these systems, one is faced with two datum choices upon which to operate. The next section elaborates on these datum choices.

10-41 Local datum choice in an LPS 3-D network

When measuring directions in the LPS system, one has two options, namely;

- orienting the theodolite to a station whose azimuth is known or,
- orienting the theodolite to an arbitrary station whose azimuth is unknown.

When the first option is adopted, one operates in the local level reference frame of type $\mathbb{E}^*$ discussed in (a) below. Should the second approach be chosen, then one operates in the local level reference frame of type $\mathbb{F}^*$ discussed in (b).

(a) Local level reference frame of type $\mathbb{E}^*$:
The origin of the $\mathbb{E}^*$ system is a point P whose coordinates

$$\mathbf{X} = \begin{bmatrix} X \\ Y \\ Z \end{bmatrix}_P = \begin{bmatrix} 0 \\ 0 \\ 0 \end{bmatrix} \tag{10.2}$$

are defined by base vectors $\mathbb{E}_{1^*}$, $\mathbb{E}_{2^*}$, $\mathbb{E}_{3^*}$ of type south, east, vertical. $\mathbb{E}_{3^*}$ which points to the direction opposite to that of the local gravity vector $\mathbf{\Gamma}$ at point P. The north direction is defined such that the reference pole agrees with the Geodetic Reference System 2000. $\mathbb{E}_{1^*}$ points south, while $\mathbb{E}_{2^*}$ completes the system by pointing east. The datum spherical coordinates of the direction point P_i in the local level reference frame $\mathbb{E}^*$ are

$$\begin{array}{llll} P & \longrightarrow X^* = Y^* = Z^* = & & 0 \\ PP_i \longrightarrow & \begin{bmatrix} X^* \\ Y^* \\ Z^* \end{bmatrix}_{\mathbb{E}^*} & = & S_i \begin{bmatrix} \cos A_i \cos B_i \\ \sin A_i \cos B_i \\ \sin B_i \end{bmatrix}, \end{array} \tag{10.3}$$

with azimuths A_i, vertical directions B_i, and spatial distances S_i.

(b) Local level reference frame of type $\mathbb{F}^*$:
This system is defined by the base vectors $\mathbb{F}_{1^*}$, $\mathbb{F}_{2^*}$, $\mathbb{F}_{3^*}$, with $\mathbb{F}_{1^*}$ within the local horizontal plane spanned by the base vectors $\mathbb{E}_{1^*}$ and $\mathbb{E}_{2^*}$ directed from P to P_i in vacuo. The angle between the base vectors $\mathbb{E}_{1^*}$ and $\mathbb{F}_{1^*}$ is the *"unknown orientation parameter"* Σ in the horizontal plane. $\mathbb{E}_{1^*}$, $\mathbb{E}_{2^*}$, $\mathbb{E}_{3^*}$ are related to $\mathbb{F}_{1^*}$,$\mathbb{F}_{2^*}$,$\mathbb{F}_{3^*}$ by a *"Karussel-Transformation"* as follows

$$\begin{bmatrix} \mathbb{F}_{1^*} = \mathbb{E}_{1^*} \cos\Sigma + \mathbb{E}_{2^*} \sin\Sigma \\ \mathbb{F}_{2^*} = -\mathbb{E}_{1^*} \sin\Sigma + \mathbb{E}_{2^*} \cos\Sigma \\ \mathbb{F}_{3^*} = \mathbb{E}_{3^*}, \end{bmatrix} \tag{10.4}$$

or

$$[\mathbb{F}_{1^*}, \mathbb{F}_{2^*}, \mathbb{F}_{3^*}] = [\mathbb{E}_{1^*}, \mathbb{E}_{2^*}, \mathbb{E}_{3^*}] \begin{bmatrix} \cos\Sigma & -\sin\Sigma & 0 \\ \sin\Sigma & \cos\Sigma & 0 \\ 0 & 0 & 1 \end{bmatrix}. \tag{10.5}$$

From (10.5), one notes that the local level reference frame of type $\mathbb{F}^*$ is related to the local level reference frame of type $\mathbb{E}^*$ by

$$[\mathbb{E}_{1^*}, \mathbb{E}_{2^*}, \mathbb{E}_{3^*}] = [\mathbb{F}_{1^*}, \mathbb{F}_{2^*}, \mathbb{F}_{3^*}]\ \mathbf{R}_3^{\mathrm{T}}(\Sigma). \tag{10.6}$$

The datum spherical coordinates of point P_i in the local level reference frame $\mathbb{F}^*$ are given as

$$\begin{array}{llll} P \rightarrow X^* = Y^* = Z^* = 0 & & & \\ PP_i \rightarrow & \begin{bmatrix} X^* \\ Y^* \\ Z^* \end{bmatrix}_{\mathbb{F}^*} & = S_i & \begin{bmatrix} \cos T_i \cos B_i \\ \sin T_i \cos B_i \\ \sin B_i \end{bmatrix}, \end{array} \tag{10.7}$$

where T_i and B_i are the horizontal and vertical directions respectively, while S_i are the spatial distances.

The local cartesian coordinates of a point whose positional vector is $\mathbf{x}$ in the $\mathbb{F}^*$ system is given by

$$\mathbf{x} = \mathbb{F}_{1^*} x + \mathbb{F}_{2^*} y + \mathbb{F}_{3^*} z, \tag{10.8}$$

where $\{x, y, z\}$ are the components of the vector $\mathbf{x}$ in the system $\{\mathbb{F}_{1^*}, \mathbb{F}_{2^*}, \mathbb{F}_{3^*} \,|P\}$.

In the chapters ahead, the local level reference frame of type $\mathbb{F}^*$ will be adopted. This system arbitrarily defines the horizontal directions such that the orientation to the system $\mathbb{E}^*$, i.e., Σ, is treated as unknown besides the unknown positions. In case of the three-dimensional orientation problem, it is determined alongside the direction $\{\Lambda_\Gamma, \Phi_\Gamma\}$ of the local gravity vector $\mathbf{\Gamma}$. For position determination using three-dimensional resection method, it is determined alongside unknown coordinates $\{X, Y, Z\}$. This will become clear in Chaps. 13 and 17.

10-42 Relationship between global and local level reference frames

In positioning within the LPS framework, one is interested not only in the geometrical position $\{X, Y, Z\}$, but also in the physical quantities $\{\Lambda_\Gamma, \Phi_\Gamma\}$ which define the direction of the local gravity vector $\mathbf{\Gamma}$ at the instrument station. This direction $\{\Lambda_\Gamma, \Phi_\Gamma\}$ of the local gravity vector $\mathbf{\Gamma}$ together with the unknown orientation Σ relate LPS and GPS systems. They are obtained by solving the *three-dimensional orientation problem.* This is achieved by transforming coordinates from the local level reference frame to the global reference frame (e.g., [10, ITRF2005]). It is conventionally solved by a means of a 3x3 rotation matrix, which is represented by a triplet $\{\Lambda_\Gamma, \Phi_\Gamma, \Sigma_\Gamma\}$ of orientation parameters called the *astronomical longitude* Λ_Γ, *astronomical latitude* Φ_Γ, and the *"orientation unknown"* Σ_Γ in the horizontal plane. With respect to the local gravity vector $\mathbf{\Gamma}$, the triplets $\{\Lambda_\Gamma, \Phi_\Gamma, \Gamma = \|\mathbf{\Gamma}\|\}$ are its spherical coordinates, in particular $\{\Lambda_\Gamma, \Phi_\Gamma\}$ its direction parameters. The *three-dimensional orientation problem* therefore determines;

(i) the 3x3 rotation matrix and,
(ii) the triplet $\{\Lambda_\Gamma, \Phi_\Gamma, \Sigma_\Gamma\}$ of orientation parameters from GPS/LPS measurements.

After the *astronomical longitude* Λ_Γ and *astronomical latitude* Φ_Γ are determined via (i) and (ii) above- no astronomical observations are needed anymore - the vertical deflections with respect to a well-chosen reference frame, e.g., the ellipsoidal normal vector field can be obtained as discussed in Sect. 10-52. When stating that no astronomical observations are needed, we are not advocating that other methods (see Sect. 10-7) should not be used, but rather imply that it is possible to transfer from a GPS reference system in Cartesian coordinates into coordinates Λ_Γ and Φ_Γ of the gravity space.

The three-dimensional orientation problem is formulated by relating the local level reference frame $\mathbb{F}^*$ to the global reference frame $\mathbb{F}^\bullet$ as follows:

$$[\mathbb{F}_{1^*}, \mathbb{F}_{2^*}, \mathbb{F}_{3^*}] = [\mathbb{F}_{1^\bullet}, \mathbb{F}_{2^\bullet}, \mathbb{F}_{3^\bullet}]\, \mathbf{R}_E\left(\Lambda_\Gamma, \Phi_\Gamma, \Sigma_\Gamma\right), \tag{10.9}$$

where the Euler rotation matrix $\mathbf{R}_E$ is parameterized by

$$\boxed{\mathbf{R}_E\left(\Lambda_\Gamma, \Phi_\Gamma, \Sigma_\Gamma\right) := \mathbf{R}_3\left(\Sigma_\Gamma\right) \mathbf{R}_2\left(\tfrac{\pi}{2} - \Phi_\Gamma\right) \mathbf{R}_3(\Lambda_\Gamma),} \tag{10.10}$$

i.e., the three-dimensional orientation parameters; astronomical longitude Λ_Γ, astronomical latitude Φ_Γ, and the orientation unknown Σ in the horizontal plane. In terms of;

(a) Cartesian coordinates $\{x, y, z\}$ of the station point and $\{x_i, y_i, z_i\}$ target points in the local level reference frame $\mathbb{F}$,* and,
(b) Cartesian coordinates $\{X, Y, Z\}$ of the station point and target points $\{X_i, Y_i, Z_i\}$ in the global reference frame $\mathbb{F}^\bullet$,

one writes

$$\begin{bmatrix} x_i - x \\ y_i - y \\ z_i - z \end{bmatrix}_{\mathbb{F}^*} = \mathbf{R}_E(\Lambda_\Gamma, \Phi_\Gamma, \Sigma_\Gamma) \begin{bmatrix} X_i - X \\ Y_i - Y \\ Z_i - Z \end{bmatrix}_{\mathbb{F}^\bullet}, \tag{10.11}$$

with

$$\begin{bmatrix} x_i - x \\ y_i - y \\ z_i - z \end{bmatrix}_{\mathbb{F}^*} = S_i \begin{bmatrix} \cos T_i \cos B_i \\ \sin T_i \cos B_i \\ \sin B_i \end{bmatrix}, \forall_i \in \{1, 2, \ldots, n\}. \tag{10.12}$$

Equation (10.11) contains the orientation parameters $\mathbf{R}_E(\Lambda_\Gamma, \Phi_\Gamma, \Sigma_\Gamma)$ relating the local level reference frame $\mathbb{F}^*$ to the global reference frame $\mathbb{F}^\bullet$. These orientation parameters have been solved by:

1. Determining the direction $(\Lambda_\Gamma, \Phi_\Gamma)$ of the local gravity vector $\mathbf{\Gamma}$ at the origin of the network and the orientation unknown Σ in the horizontal plane from stellar astronomical observations.
2. Solving the three-dimensional resection problem as discussed in Chap. 13. In the approach proposed by [187], directional measurements are performed to the neighbouring 3 points in the global reference frame and used to derive distances by solving the *Grunert's equations*. From these derived distances, a closed form solution of the six unknowns $\{X, Y, Z, \Lambda_\Gamma, \Phi_\Gamma, \Sigma_\Gamma\}$ by means of the *Hamilton-quaternion* procedure is performed.
3. Using the *simple Procrustes algorithm* as discussed in Chap. 9 to determine the three-dimensional orientation parameters $\{\Lambda_\Gamma, \Phi_\Gamma, \Sigma_\Gamma\}$ and the deflection of the vertical for a point whose geometrical positional quantities $\{X, Y, Z\}$ are known.
4. By first determining the geometrical values $\{X, Y, Z\}$ of the unknown station using resection approach as discussed in Chap. 13. Once these geometrical values have been determined, they are substituted back in (10.11) to obtain the Euler rotation matrix $\mathbf{R}_E(\Lambda_\Gamma, \Phi_\Gamma, \Sigma_\Gamma)$. The Euler rotation angles can then be deduced via an inverse map presented in Lemma 17.1 on p. 307 of Chap. 17.

10-43 Observation equations

Let us now have a look at the equations that we often encounter when positioning with a stationary theodolite. Elaborate exposition of three-dimensional observations is given by [159]. Stationed at the point $P_0 \in \mathbb{E}^3$, and with the theodolite properly centered, one sights the target points $P_i \in \mathbb{E}^3$, where i = 1,2,3,.....,n. There exist three types of measurements that will be taken from $P_0 \in \mathbb{E}^3$ to $P_i \in \mathbb{E}^3$ in the LPS system (i.e., local level reference frame $\mathbb{F}^*$). These are:

- Horizontal directions T_i whose equation is given by

$$T_i = \arctan\left(\frac{\Delta y_i}{\Delta x_i}\right)_{\mathbb{F}^*} - \Sigma_\Gamma(P_0), \tag{10.13}$$

  where $\Sigma_\Gamma(P_0)$ is the unknown orientation in the horizontal plane after setting the zero reading of the theodolite in the direction $P \rightarrow P_i$.

- Vertical directions B_i given by

$$B_i = \arctan\left(\frac{\Delta z_i}{\sqrt{\Delta x_i^2 + \Delta y_i^2}}\right)_{\mathbb{F}^*}. \quad (10.14)$$

- Spatial distances S_i, i.e.,

$$S_i = \sqrt{\Delta x_i^2 + \Delta y_i^2 + \Delta z_i^2}\,|_{\mathbb{F}}^*, \quad (10.15)$$

and $\Delta x_i = (x_i - x)$, $\Delta y_i = (y_i - y)$, $\Delta z_i = (z_i - z)$ denote the coordinate difference in the local level reference frame $\mathbb{F}^*$.

The relationship between the local level reference frame $\mathbb{F}^*$ and the global reference frame $\mathbb{F}^\bullet$ is then given by (e.g., 10.11)

$$\begin{bmatrix} \Delta x_i \\ \Delta y_i \\ \Delta z_i \end{bmatrix}_{\mathbb{F}^*} = \mathbf{R}_E(\Lambda_\Gamma, \Phi_\Gamma, 0) \begin{bmatrix} \Delta X_i \\ \Delta Y_i \\ \Delta Z_i \end{bmatrix}_{\mathbb{F}^\bullet}, \quad (10.16)$$

with

$$\mathbf{R}_E(\Lambda_\Gamma, \Phi_\Gamma, 0) = \begin{bmatrix} \sin\Phi_\Gamma \cos\Lambda_\Gamma & \sin\Phi_\Gamma \sin\Lambda_\Gamma & -\cos\Phi_\Gamma \\ -\sin\Lambda_\Gamma & \cos\Lambda_\Gamma & 0 \\ \cos\Phi_\Gamma \cos\Lambda_\Gamma & \cos\Phi_\Gamma \sin\Lambda_\Gamma & \sin\Phi_\Gamma \end{bmatrix}. \quad (10.17)$$

Observations in (10.13), (10.14) and (10.15) are now expressed in the global reference frame as

$$T_i = \arctan\left\{\frac{-\sin\Lambda_\Gamma \Delta X_i + \cos\Lambda_\Gamma \Delta Y_i}{\sin\Phi_\Gamma \cos\Lambda_\Gamma \Delta X_i + \sin\Phi_\Gamma \sin\Lambda_\Gamma \Delta Y_i - \cos\Phi_\Gamma \Delta Z_i}\right\} - \Sigma_\Gamma(P), \quad (10.18)$$

and

$$B_i = \arctan\left\{\frac{\cos\Phi_\Gamma \cos\Lambda_\Gamma \Delta X_i + \cos\Phi_\Gamma \sin\Lambda_\Gamma \Delta Y_i + \sin\Phi_\Gamma \Delta Z_i}{\sqrt{(\sin\Phi_\Gamma \cos\Lambda_\Gamma \Delta X_i + \sin\Phi_\Gamma \sin\Lambda_\Gamma \Delta Y_i - \cos\Phi_\Gamma \Delta Z_i)^2 + D_2}}\right\}, \quad (10.19)$$

where $D_2 = (\cos\Lambda_\Gamma \Delta Y_i - \sin\Lambda_\Gamma \Delta X_i)^2$, $\Delta X_i = (X_i - X)$, $\Delta Y_i = (Y_i - Y)$, $\Delta Z_i = (Z_i - Z)$ in the global reference frame $\mathbb{F}^\bullet$ and $\{\Sigma_\Gamma(P_0), \Lambda_\Gamma(P_0), \Phi_\Gamma(P_0)\}$ are the three unknown orientation parameters at the unknown theodolite station P_0.

10-5 Three-dimensional orientation problem

The transformation of coordinates from the local level reference frame to the global terrestrial reference frame (e.g., ITRF97) is a key, contemporary problem. In carrying out coordinate transformations, some of the sought parameters are those of orientation. Orientations are normally sought for; theodolites, cameras, and CCD sensors, etc. Procedures for solving explicitly the three-dimensional orientation problems in geoinformatics are presented in the works of [360, 382, 383, 434]. In geodesy, attempts to find closed form solution to the orientation problem have been carried out by [16, 168, 187] who proved that the three-dimensional orientation problem could be solved in a closed form through the integration of GPS and LPS systems.

The orientation problem is formulated by expressing (10.11) in Chap. 10 relating the two configurations, i.e., the local level reference frame and the global reference frame, with the left-hand-side in terms of spherical coordinates, as

$$s_i \begin{bmatrix} \cos T_i \cos B_i \\ \sin T_i \cos B_i \\ \sin B_i \end{bmatrix}_{\mathbb{F}^*} = \mathbf{R}(\varLambda_\Gamma, \varPhi_\Gamma, \varSigma_\Gamma^i) \begin{bmatrix} X_i - X \\ Y_i - Y \\ Z_i - Z \end{bmatrix}_{\mathbb{F}^\bullet} \quad (10.20)$$

$$with$$

$$s_i = \sqrt{(X_i - X)^{\bullet 2} + (Y_i - Y)^{\bullet 2} + (Z_i - Z)^{\bullet 2}}.$$

In (10.20), $X, Y, Z, X_i, Y_i, Z_i \ \forall_i \in N$ are GPS coordinates in the global reference frame $\mathbb{F}^\bullet$, while the spherical coordinates $T_i, B_i \ \forall_i \in N$ are used to derive the left-hand-side of (10.20) in the local level reference frame $\mathbb{F}^*$. The orientation problem (10.20) is conventionally solved by means of a 3x3 rotation matrix $\mathbf{R}$, which is represented by the triplet $\{\varLambda_\Gamma, \varPhi_\Gamma, \varSigma_\Gamma\}$ of orientation parameters called the astronomical longitude $\varLambda_\Gamma$, astronomical latitude $\varPhi_\Gamma$, and the "orientation unknown" $\varSigma_\Gamma$ in the horizontal plane. With respect to the local gravity vector $\mathbf{\Gamma}$, the triplet $\{\varLambda_\Gamma, \varPhi_\Gamma, \Gamma = \|\mathbf{\Gamma}\|\}$ are its spherical coordinates, in particular $\{\varLambda_\Gamma, \varPhi_\Gamma\}$ are its direction parameters. Here we solve the problem of determining;

(a) the 3x3 rotation matrix $\mathbf{R}$ and,
(b) the triplet $\{\varLambda_\Gamma, \varPhi_\Gamma, \varSigma_\Gamma\}$ of orientation parameters from GPS/LPS measurements by means of the partial Procrustes algorithm.

10-51 Procrustes solution of the orientation problem

Consider coordinates to be given in two configurations with the same three-dimensional space in the local level reference frame $\mathbb{F}^*$ and global reference frame $\mathbb{F}^\bullet$. For such a three-dimensional space, where $i = 3$ (i.e., 3 target points), the relationship in (10.20) between the two systems is expressed as

$$\begin{bmatrix} x_1 - x & x_2 - x & x_3 - x \\ y_1 - y & y_2 - y & y_3 - y \\ z_1 - z & z_2 - z & z_3 - z \end{bmatrix}_{\mathbb{F}^*} = \mathbf{R} \begin{bmatrix} X_1 - X & X_2 - X & X_3 - X \\ Y_1 - Y & Y_2 - Y & Y_3 - Y \\ Z_1 - Z & Z_2 - Z & Z_3 - Z \end{bmatrix}_{\mathbb{F}^\bullet} . \quad (10.21)$$

For n target points, (10.21) becomes

$$\underbrace{\begin{bmatrix} x_1 - x & x_2 - x & \ldots & x_n - x \\ y_1 - y & y_2 - y & \ldots & y_n - y \\ z_1 - z & z_2 - z & \ldots & z_n - z \end{bmatrix}_{\mathbb{F}^*}}_{3\text{x}n} = \underbrace{\mathbf{R}}_{3\text{x}3} \underbrace{\begin{bmatrix} X_1 - X & X_2 - X & \ldots & X_n - X \\ Y_1 - Y & Y_2 - Y & \ldots & Y_n - Y \\ Z_1 - Z & Z_2 - Z & \ldots & Z_n - Z \end{bmatrix}_{\mathbb{F}^\bullet}}_{3\text{x}n}, \tag{10.22}$$

with their respective dimensions given below them. The transpose of (10.22) is expressed as

$$\underbrace{\begin{bmatrix} x_1 - x & y_1 - y & z_1 - z \\ x_2 - x & y_2 - y & z_2 - z \\ \cdot & \cdot & \cdot \\ \cdot & \cdot & \cdot \\ \cdot & \cdot & \cdot \\ x_n - x & y_n - y & z_n - z \end{bmatrix}_{\mathbb{F}^*}}_{n\text{x}3} = \underbrace{\begin{bmatrix} X_1 - X & Y_1 - Y & Z_1 - Z \\ X_2 - X & Y_2 - Y & Z_2 - Z \\ \cdot & \cdot & \cdot \\ \cdot & \cdot & \cdot \\ \cdot & \cdot & \cdot \\ X_n - X & Y_n - Y & Z_n - Z \end{bmatrix}_{\mathbb{F}^\bullet}}_{n\text{x}3} \underbrace{\mathbf{R}'}_{3\text{x}3}. \tag{10.23}$$

Equation (10.23) contains the relative position vectors of corresponding points in two reference frames. Let us indicate the matrix on the left-hand-side by **A**, the one on the right-hand-side by **B**, and denote the rotation matrix $\mathbf{R}'$ by **T**. The partial Procrustes problem is now concerned with fitting the configuration of **B** into **A** as close as possible. The problem reduces to that of determination of the rotation matrix **T**. The operations involved in the solution of the orientation problem, therefore, are:

- Solution of $\mathbf{T}^* = \mathbf{V}\mathbf{U}'$.
- Obtaining the rotation elements from $\mathbf{R} = (\mathbf{T}^*)'$.

The rotation matrix $\mathbf{T}^*$ is the best possible matrix out of the set of all orthogonal matrices **T** which are obtained by imposing the restriction $\mathbf{T}\mathbf{T}' = \mathbf{T}'\mathbf{T} = \mathbf{I}$. The matrix **T** could otherwise be any matrix, which means, geometrically, that **T** is some linear transformation which in general may not preserve the shape of **B**. A summary of the computational procedure for the three-dimensional orientation parameters based on Example 10.1 is given in Fig. 10.1.

Example 10.1 (Computation of the three-dimensional orientation problem). The partial Procrustes approach discussed in Sect. 9-2 is applied to the Test network of Stuttgart Central presented in Sect. 10-6. 8 GPS stations are used to determine the three-dimensional orientation parameters $\{\Lambda_\Gamma, \Phi_\Gamma, \Sigma_\Gamma\}$. From the observations of Table 10.3 on p. 152, the matrix **A** in (9.1) is computed in terms of the spherical coordinates using (10.20). The Matrix **B** is obtained by subtracting the coordinates of station $K1$ from those of other stations in Table 10.1. The rotation matrix **T** is then computed using partial Procrustes algorithm, i.e., (9.1) to (9.10). For this network, the computed three-dimensional orientation parameters $\{\Lambda_\Gamma, \phi_\Gamma, \Sigma_\Gamma\}$ gave the values $\phi_\Gamma = 48^o46'54''.3$ and $\Lambda_\Gamma = 9^o10'30''.1$, which when compared to $\phi_\Gamma = 48^o46'54''.9$ and $\Lambda_\Gamma = 9^o10'29''.8$ in [257, p. 46] deviates by $\Delta\Lambda_\Gamma = -0''.3$ and $\Delta\Phi_\Gamma = 0''.6$.

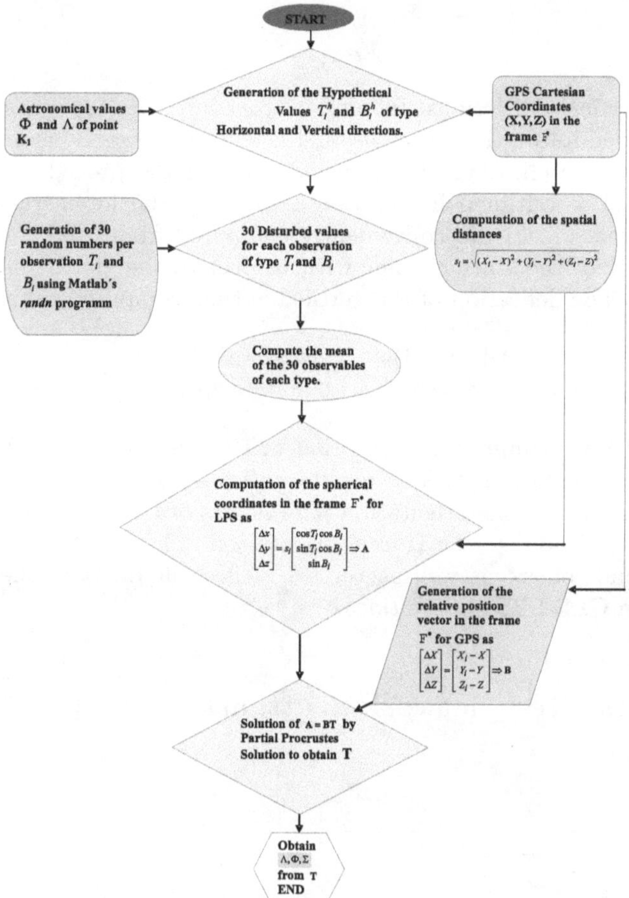

Figure 10.1. Flow chart for computing three-dimensional orientation parameters

10-52 Determination of vertical deflection

As soon as we have determined the astronomical longitude Λ_Γ and astronomical latitude Φ_Γ, the deflection of the vertical can be computed with respect to a well chosen reference frame, e.g., the ellipsoidal normal vector field. Traditionally, orientation parameters $\{\Lambda_\Gamma, \Phi_\Gamma\}$ have been obtained from stellar observations and related to geodetic coordinates $\{\lambda, \phi\}$ to obtain the deflection of the vertical. Through the integration of GPS and LPS systems however, the astronomical observations of type $\{\Lambda_\Gamma, \Phi_\Gamma\}$ are obtained from the three-dimensional orientation solutions as discussed in Sect. 10-5. Such pioneering approach in geodesy can be traced to the works of [16, 52, 168, 187].

To determine the vertical deflection, the reference direction is parameterized in terms of "surface normal"; ellipsoidal longitude λ and ellipsoidal latitude ϕ. These are then subtracted from the local vertical parameterized in terms of astronomical longitude Λ_Γ and astronomical latitude Φ_Γ as

$$\begin{array}{c} \Lambda_\Gamma - \lambda \\ \Phi_\Gamma - \phi, \end{array} \tag{10.24}$$

to access the vertical deflections. In such a procedure, the topographical surface which is embedded into a three-dimensional Euclidean space $\mathbb{R}^3$ is mapped pointwise into a reference ellipsoid of revolution through the procedure discussed in Chap. 11. Indeed as outlined in Solution 11.5 on p. 165 for instance, those direction parameters $\{\Lambda, \Phi\}$ are conveniently computed from GPS Cartesian coordinates $\{\mathrm{X}, \mathrm{Y}, \mathrm{Z}\}$ of the station point with respect to the global reference frame $\{\mathbb{F}_{1\bullet}, \mathbb{F}_{2\bullet}, \mathbb{F}_{3\bullet}\}$. The deflection of the vertical is then computed from (10.24) as

$$\begin{array}{c} \delta\Lambda_\Gamma := \Lambda_\Gamma - \lambda, \delta\Phi_\Gamma := \Phi_\Gamma - \phi \\ \eta := \delta\Lambda_\Gamma \cos\Phi, \xi := \delta\Phi_\Gamma. \end{array} \tag{10.25}$$

Equation (10.25) are simple representation of the east vertical deflection η and the north vertical deflection ξ. The results in Table (3.1) of [168] document the precise determination of the orientation parameters of type astronomic longitude Λ_Γ, astronomic latitude Φ_Γ, horizontal orientation unknown Σ_Γ in the range of fraction of seconds of arc as well as vertical deflection $\{\xi, \eta\}$ in the same range exclusively from GPS-LPS observations.

10-6 Example: Test network Stuttgart Central

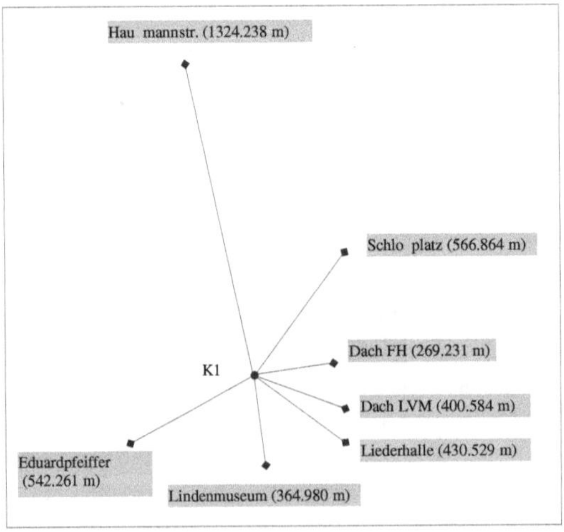

Figure 10.2. Graph of the Test network *"Stuttgart Central"*

Observations and experiment

The following experiment was performed at the center of Stuttgart on one of the pillars of Stuttgart University's building along Kepler Strasse 11 as depicted by Fig. 10.2. The test network *"Stuttgart Central"* consisted of 8 GPS points listed in *Table* 10.1. A theodolite was stationed at pillar $K1$ whose astronomical longitude Λ_Γ as well as astronomic latitude Φ_Γ were known from previous astrogeodetic observations made by the Department of Geodesy and Geoinformatics, Stuttgart University. Since theodolite observations of type horizontal directions T_i as well as vertical directions B_i from the pillar $K1$ to the target points i, $i = 1, 2, \ldots, 6, 7$, were only partially available, the horizontal and vertical directions were simulated from the given values of $\{\Lambda_\Gamma, \Phi_\Gamma\}$ as well as the Cartesian coordinates of the station point $\{X, Y, Z\}$ and target points $\{X_i, Y_i, Z_i\}$ using (10.18) and (10.19). The relationship between the observations of type horizontal directions T_i, vertical directions B_i, values of $\{\Lambda_\Gamma, \Phi_\Gamma\}$ and the Cartesian coordinates of the station point $\{X, Y, Z\}$ and target points $\{X_i, Y_i, Z_i\}$ enabled generation of the observation data sets in *Table* 10.3. Such a procedure had also an advantage in that we had full control of the algorithms that will be tested later in the book. In detail, the directional parameters $\{\Lambda_\Gamma, \Phi_\Gamma\}$ of the local gravity vector were adopted from the astrogeodetic observations $\phi_\Gamma = 48°46^{'}54^{''}.9$ and $\Lambda_\Gamma = 9°10^{'}29^{''}.8$ reported by [257, p. 46] with a root-mean-square error $\sigma_\Lambda = \sigma_\Phi = 10''$. Table 10.1 contains the $\{X, Y, Z\}$ coordinates obtained from a GPS survey of the test network Stuttgart Central, in particular with root-mean-square errors $(\sigma_X, \sigma_Y, \sigma_Z)$ neglecting the covariances $(\sigma_{XY}, \sigma_{YZ}, \sigma_{ZX})$. The spherical coordinates of the relative position vector, namely of the coordinate differences $\{x_i - x, y_i - y, z_i - z\}$, are called horizontal directions T_i, vertical directions B_i and spatial distances S_i and are given in Table 10.2. The standard deviations/root-mean-square errors were fixed to $\sigma_T = 6$", $\sigma_B = 6''$. Such root mean square errors can be obtained on the basis of a proper refraction model. Since the horizontal and vertical directions of Table 10.2 were simulated, with zero noise level, we used a random generator *randn* in Matlab e.g., [207, p. 84, p. 144] to produce additional observational data sets within the framework of the given root-mean-square errors. For each observable of type T_i and B_i, 30 randomly simulated data were obtained and the mean taken. Let us refer to the observational data sets $\{T_i, B_i\}, i = 1, 2, \ldots, 6, 7$, of *Table* 10.3 which were enriched by the root-mean-square errors of the individual randomly generated observations as well as by the differences $\Delta T_i := T_i - T_i(\text{generated})$, $\Delta B_i := B_i - B_i(\text{generated})$. Such differences $(\Delta T_i, \Delta B_i)$ indicate the difference between the ideal values of *Table* 10.2 and those randomly generated.

Observations are thus designed such that by observing the other seven GPS stations, the orientation of the local level reference frame $\mathbb{F}^*$ whose origin is station $K1$, to the global reference frame $\mathbb{F}^\bullet$ is obtained. The direction of Schlossplatz was chosen as the zero direction of the theodolite leading to the determination of the third component Σ_Γ of the three-dimensional orientation parameters. To each of the GPS target points i, the observations of type horizontal directions T_i and the vertical directions B_i are measured. The spatial distances $S_i^2(\mathbf{X}, \mathbf{X}_i) = \|\mathbf{X}_i - \mathbf{X}\|$ are readily obtained from the observation of type horizontal directions T_i and vertical directions B_i. The following symbols have been used: $\sigma_X, \sigma_Y, \sigma_Z$ are the standard errors of the GPS Cartesian coordinates. Covariances $\sigma_{XY}, \sigma_{YZ}, \sigma_{ZX}$ are neglected. σ_T, σ_B are the standard deviation of horizontal and vertical directions respectively after an adjustment, Δ_T, Δ_B are the magnitude of the noise on the horizontal and vertical directions, respectively.

Table 10.1. GPS Coordinates in the global reference frame $\mathbb{F}^{\bullet}(X,Y,Z)$, (X_i,Y_i,Z_i), $i = 1,2,\ldots,7$

Station[1]	$X(m)$	$Y(m)$	$Z(m)$	σ_X mm	σ_X mm	σ_X mm
Dach $K1$	4157066.1116	671429.6655	4774879.3704	1.07	1.06	1.09
1	4157246.5346	671877.0281	4774581.6314	0.76	0.76	0.76
2	4156749.5977	672711.4554	4774981.5459	1.77	1.59	1.61
3	4156748.6829	671171.9385	4775235.5483	1.93	1.84	1.87
4	4157066.8851	671064.9381	4774865.8238	1.38	1.29	1.38
5	4157266.6181	671099.1577	4774689.8536	1.29	1.28	1.34
6	4157307.5147	671171.7006	4774690.5691	0.20	0.10	0.30
7	4157244.9515	671338.5915	4774699.9070	2.80	1.50	3.10

Table 10.2. Ideal spherical coordinates of the relative position vector in the local level reference frame $\mathbb{F}^*$: Spatial distances, horizontal directions, vertical directions

Station Observed from $K1$	Distances (m)	Horizontal directions(gon)	Vertical directions(gon)
Schlossplatz (1)	566.8635	52.320062	-6.705164
Haussmanstr. (2)	1324.2380	107.160333	0.271038
Eduardpfeiffer (3)	542.2609	224.582723	4.036011
Lindenmuseum (4)	364.9797	293.965493	-8.398004
Liederhalle (5)	430.5286	336.851237	-6.941728
Dach LVM (6)	400.5837	347.702846	-1.921509
Dach FH (7)	269.2309	370.832476	-6.686951

Table 10.3. Randomly generated spherical coordinates of the relative position vector: horizontal directions T_i and vertical directions $B_i, i = 1,2,\ldots,6,7$, root-mean-square errors of individual observations, differences $\Delta T_i := T_i - T_i(\text{generated})$, $\Delta B_i := B_i - B_i(\text{generated})$ with respect to (T_i, B_i) ideal data of Table 10.2

St.	H/dir.(gon)	V/dir.(gon)	$\sigma_T(gon)$	$\sigma_B(gon)$	$\Delta_T(gon)$	$\Delta_B(gon)$
1	0.000000	-6.705138	0.0025794	0.0024898	-0.000228	-0.000039
2	54.840342	0.271005	0.0028756	0.0027171	-0.000298	0.000033
3	172.262141	4.035491	0.0023303	0.0022050	0.000293	0.000520
4	241.644854	-8.398175	0.0025255	0.0024874	0.000350	0.000171
5	284.531189	-6.942558	0.0020781	0.0022399	-0.000024	0.000830
6	295.382909	-1.921008	0.0029555	0.0024234	0.000278	-0.000275
7	318.512158	-6.687226	0.0026747	0.0024193	-0.000352	0.000500

10-7 Concluding remarks

What is presented here is just a nutshell of GNSS. For more exposition of its operations and techniques, we refer to related publications, e.g., [124, 192, 227, 228, 264, 296, 361, 372, 419, 43]. For LPS systems, more insight can be found in [170, 175] and [345, p. 28]. In particular, for cases where the theodolite moves from point to point, i.e., moving horizontal triad, [158, 161, 164] presents interesting materials.

We point out that there are other methods capable of providing vertical deflections e.g.,GPS and levelling [398], gravimetric geoid models [132] and modern CCD-Zenith cameras [222, 223, 224, 225, 226]. References to non-astronomical methods capable of determining deflections are presented in Hirt [226]. Further tests are required to assess the real efforts and the accuracy level related to the application of the GPS and LPS method for determination of vertical deflections.

11 Cartesian to ellipsoidal mapping

11-1 Introductory remarks

In establishing a proper reference frame of geodetic point positioning, namely by the Global Positioning System (GPS) - the Global Problem Solver - we are in need to establish a proper model for the *Topography* of the Earth, the Moon, the Sun or planets.By the theory of equilibrium figures, we are informed that an ellipsoid, two-axes or three-axes is an excellent approximation of the *Topography*. For planets similar to the Earth the biaxial ellipsoid, also called "*ellipsoid-of-revolution*" is the best approximation.

It was C.F. Gauss in his work on "*Gauss-Krulger Maps*" or "*Universal Marcator Projection*" relating to the "*ellipsoid-of-revolution*" who designed the elegant method of mapping a reference ellipsoid onto a topographic surface. He used the orthogonal projection of a topographic point onto the reference ellipsoid. *Million times per day in any GPS receiver the Gauss projection* onto the reference ellipsoid of type formula (11.1), (11.2) is used to compact Cartesian coordinates (X, Y, Z) from determined (geodetic longitude, geodetic latitude, surface height) and its inverse.

Here we meet the problem of solving a biquadratic equation in closed form. Various solutions exist which we present in Table 11.1 in short. Of course there are many more and other approaches. The *minimum distance mapping* of a star-shaped surface onto the reference ellipsoid is given by *Solution* 11.3. The constrained minimum distance mapping is analyzed by *Lemma* 11.1, its solution is presented by (i) the *Grafarend-Lohse mapping* and (ii) the *Groebner Basis mapping*, and (iii) the *Extended Newton-Raphson mapping* enriched by numerical examples.

11-2 Mapping topographical points onto reference ellipsoid

The projection of points from the topographical surface to their equivalent on the reference ellipsoid remains one of the fundamental tasks undertaken in geodesy and geoinformatics. This is because the reference ellipsoid of revolution is the mathematical representation of the geoid. Geoid is the surface that approximates mean sea level, and provides vertical datum for heights. It is of prime importance in engineering and geosciences in general. From it, geophysicists can infer on processes taking place below and above the Earth such as earthquakes and rise in sea level. Hydrologists need it to infer on water table, while engineers need it for height determination during roads and structural constructions.

Measurements are normally related to the geoid for computation via its mathematical form, the reference ellipsoid of revolution. There exist two ways of projecting points from a topographical surface onto the reference ellipsoid of revolution. One approach projects a point P onto the geoid p_g and then finally onto the reference ellipsoid of revolution p. This type of projection is called the Pizetti's

J.L. Awange et al., *Algebraic Geodesy and Geoinformatics*, 2nd ed.,
DOI 10.1007/978-3-642-12124-1_11,

projection. The other approach directly projects a topographical point P through the ellipsoidal normal onto a point p on the reference ellipsoid of revolution. The distance between the topographical point P and the ellipsoidal point p gives the geometrical height H above the ellipsoid. The topographical position of point P would therefore be referred by the ellipsoidal height H and the geographical coordinates L, B. In this case, the geographical coordinate L is the longitude and B the latitude. The set of coordinates $\{L, B, H\}$ defining the point P are called geodetic or ellipsoidal coordinates. This second projection is called the Helmert's projection which will be considered in this chapter. The two projections are discussed in detail in [220, pp. 178–184].

The *forward transformation* from ellipsoid to Cartesian coordinates, i.e., $\{L, B, H\} \rightarrow \{X, Y, Z\}$, is demonstrated by Solutions 11.1 and 11.2. The challenge is the inverse transformation which projects topographical points to the ellipsoid. One way of achieving this projection is by first converting topographical Cartesian coordinates into ellipsoidal cartesian coordinates. Once this is done, the ellipsoidal Cartesian coordinates are then converted to their equivalent geodetic coordinates. The problem is formulated as follows: Given topographical coordinates $\{X, Y, Z\}$ of a point P, obtain the geodetic coordinates $\{L, B, H\}$. This problem is a one-to-one mapping of

$$\left[\begin{array}{lcl}\{X, Y, Z\} & \longrightarrow & \{L, B, H\} \\ \textit{Topography} & & \textit{Ellipsoid}\end{array}\right. \tag{11.1}$$

Table 11.1 outlines the existing methods by other authors to convert Cartesian coordinates $\{X, Y, Z\}$ to Gauss ellipsoidal coordinates $\{L, B, H\}$ in (11.1). The target of this chapter is to invert algebraically $\{X, Y, Z\} \rightarrow \{L, B, H\}$ by means of *minimum distance mapping* through the map in (11.2) as

$$\left[\begin{array}{lclcl}\{X, Y, Z\} & \longrightarrow & \{x_1, x_2, x_3\} & \longrightarrow & \{L, B, H\} \\ \textit{Topography} & & \textit{Ellipsoid} & & \textit{Ellipsoid.}\end{array}\right. \tag{11.2}$$

Grafarend [165] already constructed surface normal coordinates with respect to the *international reference ellipsoid*. In this chapter, we will be interested with setting up an algebraic minimum distance mapping to relate a point on the Earth's topographical surface uniquely (one-to-one) to a point on the international reference ellipsoid. The solution to such an optimization problem generates *projective ellipsoidal heights* and the standard transformation of the Gauss ellipsoidal coordinates $\{L, B, H\}$ to geocentric Cartesian coordinates $\{X, Y, Z\}$. The *inverse transformation* of geocentric Cartesian coordinates $\{X, Y, Z\}$ to Gauss ellipsoidal coordinates $\{L, B, H\}$ is here solved algebraically and examples presented.

Table 11.1. Characteristics of inverse transformation Cartesian coordinates to Gauss ellipsoidal coordinates

Author	Publication year	Characteristic
Awange et al. [39]	in press	closed (similar to [14])
Bartelme N, Meissl P [58]	1975	iterative
Benning W [70]	1974	closed
Benning W [71]	1987	iterative first point curve
Borkowski KM [88]	1987	iterative
Borkowski KM [89]	1989	iterative
Bowring BR [90]	1976	approximate "closed"
Bowring BR [91]	1985	approximate
Croceto N [119]	1993	iterative
Fitzgibbon A et al. [136]	1999	iterative
Fotiou A [139]	1998	approximate "closed"
Fröhlich H, Hansen HH [142]	1976	closed
Fukushima T [144]	1999	"fast" iterative
Gander W et al. [145]	1994	iterative
Grafarend EW [166]	2001	closed
Grafarend EW, Lohse P [172]	1991	closed form 4th order equation reduced to 3rd order
Grafarend EW et al. [190]	1995	closed form
Heck B [217]	1987	iterative
Heikkinen M [218]	1982	closed
Heiskannen WA, Moritz H [220]	1976	iterative
Hirvonen R, Moritz H [221]	1963	iterative
Hofman-Wellenhof B et al. [227]	2001	Identical to Bowring [90]
Lapaine M [260]	1990	algebraic equations of higher order
Lin KC, Wang J [274]	1995	iterative
Loskowski P [278]	1991	"simply iterative"
Ozone MI [316]	1985	3rd order equation
Paul MK [327]	1973	iterative
Penev P [328]	1978	Angular variable 3rd order equation
Pick M [330]	1985	approximate "closed"
Sjöberg LE [364]	1999	iterative
Soler T, Hothem LD [365]	1989	iterative "closed" Jacobi ellipsoidal coordinates
Sünkel H [376]	1976	series expansion
Torge W [384]	1991	iterative
Vaniceck P, Krakiwski E [390]	1982	higher order algebraic equation
Vincenty T [395]	1978	Iterative
Vincenty T [396]	1980	approximate "closed"
You RJ [427]	2000	iterative

Solution 11.1 (Forward transformation of Gauss ellipsoidal coordinates).

$$\mathbf{X}(L,B,H) = \mathbf{e}_1\left[\frac{a}{\sqrt{1-e^2\sin^2 B}} + H(L,B)\right]\cos B\cos L +$$
$$\mathbf{e}_2\left[\frac{a}{\sqrt{1-e^2\sin^2 B}} + H(L,B)\right]\cos B\sin L +$$
$$\mathbf{e}_2\left[\frac{a(1-e^2)}{\sqrt{1-e^2\sin^2 B}} + H(L,B)\right]\sin B, \tag{11.3}$$

$$\begin{bmatrix} X \\ Y \\ Z \end{bmatrix} = \begin{bmatrix} \left[\frac{a}{\sqrt{1-e^2\sin^2 B}} + H(L,B)\right]\cos B\cos L \\ \left[\frac{a}{\sqrt{1-e^2\sin^2 B}} + H(L,B)\right]\cos B\sin L \\ \left[\frac{a(1-e^2)}{\sqrt{1-e^2\sin^2 B}} + H(L,B)\right]\sin B \end{bmatrix}, \tag{11.4}$$

with $\{(L(X,Y,Z), B(X,Y,Z), H(X,Y,Z)\}$ as unknowns.

Solution 11.2 (Forward transformation of Gauss complex ellipsoidal coordinates). Consider

$$X + \mathrm{i}Y = \left[\frac{a}{\sqrt{1-e^2\sin^2 B}} + H(L,B)\right]\cos B(\cos L + \mathrm{i}\sin L), \tag{11.5}$$

and

$$Z = \left[\frac{a(1-e^2)}{\sqrt{1-e^2\sin^2 B}} + H(L,B)\right]\sin B, \tag{11.6}$$

then

$$\mathbf{X} = \begin{bmatrix} X + iY & Z \\ -Z & X - \mathrm{i}Y \end{bmatrix} \quad \in \quad \mathbb{C}^{2\times 2} \tag{11.7}$$

11-3 Mapping geometry

In [165], Gauss surface normal coordinates with respect to the international reference ellipsoid $\mathbb{E}^2_{a,a,b}$ are introduced and called $\{l, b\}$. The Gauss surface normal longitude is represented by l (geodetic longitude) and the Gauss surface normal latitude by b (geodetic latitude). Such a coordinate system build up the proper platform for introducing surface normal coordinates $\{L, B, H\}$ for mapping the Earth's topographical surface $\mathbb{T}^2$ with respect to the international reference ellipsoid. In particular, the minimum distance mapping which maps a topographic point $P \in \mathbb{T}^2$ onto a *nearest point* $p \in \mathbb{E}^2_{a,a,b}$ on the international reference ellipsoid is implemented. Such mapping, initiated by *C. F. Gauss*, is *isozenithal* since $\{l = L, b = B\}$: The orthogonal projection of $P \in \mathbb{T}^2$ onto $p \in \mathbb{E}^2_{a,a,b}$ as the nearest point is along the *surface normal* of $\mathbb{E}^2_{a,a,b}$. The minimum distance from point p to P, i.e., $\overline{pP}$ is called accordingly ellipsoidal height H ("geodetic height") complemented by *surface normal longitude* $l = L$ and *surface normal latitude* $b = B$.

In-order to gain a unique solution of *minimum distance mapping*, the assumption that the Earth's topographical surface is *starshaped* has to be made. Figure 11.1 illustrates a topographical surface which is *starshaped*, while Fig. 11.2 illustrates that which is *not*. With respect to these figures, the notion of a *starshaped compact* (closed and bounded) *topographical surface* may be obvious.

Definition 11.1 (Starshaped surface). *A region* $\mathbb{M} \in \mathbb{R}^3$ *is starshaped with respect to a point* $P \in \mathbb{R}^3$*, if the straight line which connects an arbitrary point* $Q \in \mathbb{M}$ *with* P *lies in* $\mathbb{M}$ *. We call a surface starshaped, if it forms the boundary of a starshaped region.*

It is understood that the shape of a star guarantees that the minimum distance mapping of topographical surfaces in $\mathbb{R}^3$ onto the average sphere $\mathbb{S}^2_R$ (*"Bjerhammer sphere"*) is *one-to-one*. If the minimum distance mapping would *not* be one-to-one, it might happen that a point on the *average sphere* $\mathbb{S}^2_R$ *has more than one image* on the topographical surface. Here the condition of *starshaped* has to be relaxed if the topographic surface $\mathbb{T}^2 \subset \mathbb{R}^3$ is mapped onto the international reference ellipsoid $\mathbb{E}^2_{a,a,b}$, an ellipsoid of revolution of semi-major axis a and semi-minor axis b. If any surface normal to the ellipsoid of revolution $\mathbb{E}^2_{a,a,b}$ intersects the topographical surface *only once*, the topographical surface is *ellipsoidal starshaped*. Indeed this condition is not met by any arbitrary topographical surface like the Earth's. Instead we shall assume that we have properly regularized the Earth's topographical surface to meet our requirement. Otherwise the Gauss ellipsoidal coordinates $\{L, B, H\}$ would break down! Figures 11.3 and 11.4 gives a better insight into the notion of *ellipsoidal starshaped* and *anti-ellipsoidal starshaped*.

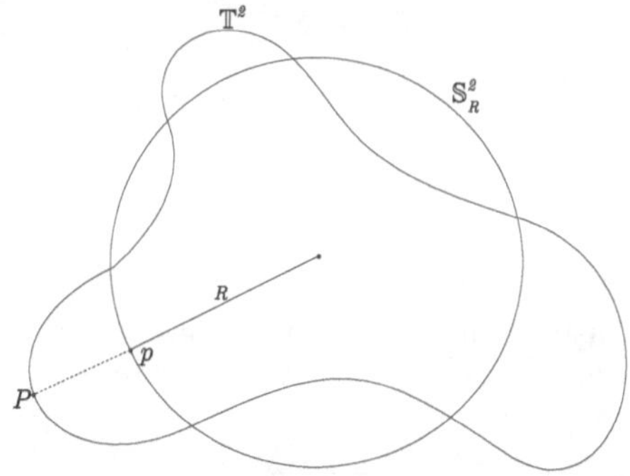

Figure 11.1. Minimum distance mapping, starshaped topographic surface (orthogonal projection of $P \in \mathbb{T}^2$ onto $p \in \mathbb{S}^2_R$)

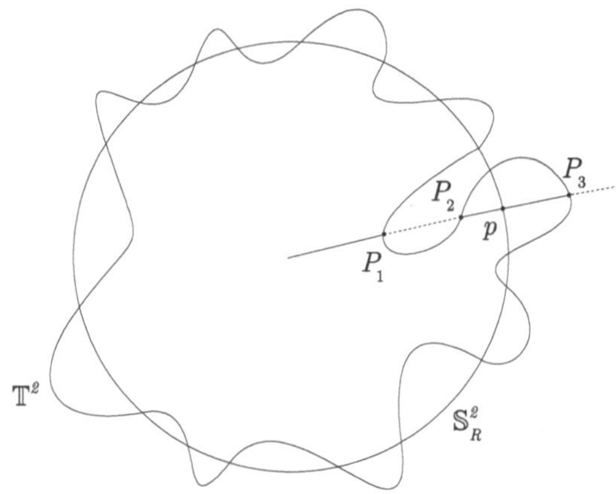

Figure 11.2. Minimum distance mapping, non-starshaped topographic surface (orthogonal projection of $P \in \mathbb{T}^2$ onto $p \in \mathbb{S}^2_R$)

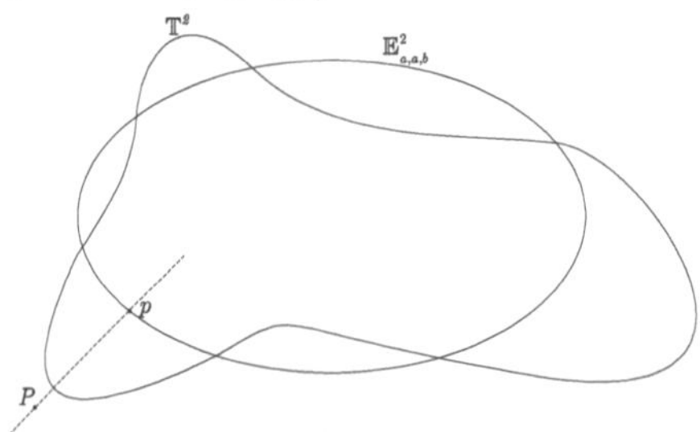

Figure 11.3. Minimum distance mapping, ellipsoidal starshaped topographic surface (orthogonal projection of $P \in \mathbb{T}^2$ onto $p \in \mathbb{E}^2_{a,a,b}$)

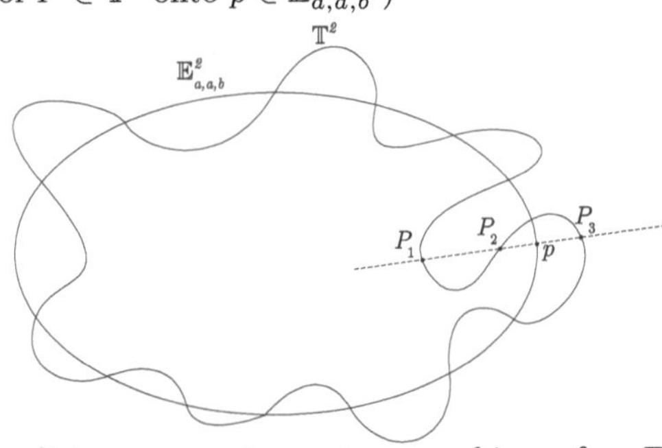

Figure 11.4. Minimum distance mapping, a topographic surface $\mathbb{T}^2$ which is *not* ellipsoidal starshaped (orthogonal projection of $P \in \mathbb{T}^2$ onto $p \in \mathbb{E}^2_{a,a,b}$)

11-4 Minimum distance mapping

In-order to relate a point P on the Earth's topographic surface to a point on the international reference ellipsoid $\mathbb{E}^2_{a,a,b}$, a bundle of half straight lines so called *projection lines* which depart from P and intersect $\mathbb{E}^2_{a,a,b}$ either not at all or in two points are used. There is *one projection line* which is at minimum distance relating P to p. Figure 11.5 is an illustration of such a minimum distance mapping. Let us formulate such an optimization problem by means of the Lagrangean $\pounds(x_1, x_2, x_3, x_4)$ in Solution 11.3.

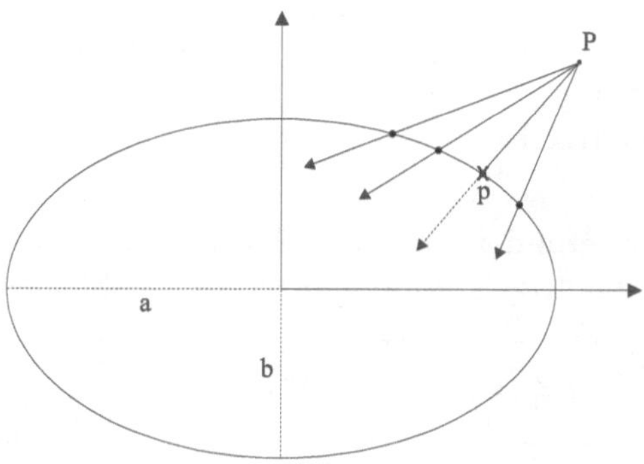

Figure 11.5. Minimum distance mapping of a point P on the Earth's topographic surface to a point p on the *international reference ellipsoid* $\mathbb{E}^2_{a,a,b}$

Solution 11.3 (Constraint minimum distance mapping in terms of Cartesian coordinates).

$$\begin{aligned}\pounds(x_1, x_2, x_3, x_4) &:= \frac{1}{2}\|\mathbf{X} - \mathbf{x}\|^2 + \frac{1}{2}x_4\left[b^2(x_1^2 + x_2^2) + ax_3^2 - a^2b^2\right] \\ &= \frac{1}{2}\{(X - x_1)^2 + (Y - x_2)^2 + (Z - x_3)^2 \\ &\quad + x_4\left[b^2(x_1^2 + x_2^2) + ax_3^2 - a^2b^2\right]\} \\ &= \min_{(x_1, x_2, x_3, x_4)}\end{aligned} \tag{11.8}$$

$$\mathbf{x} \in \mathbb{X} := \{\mathbf{x} \in \mathbb{R}^3 | \frac{x_1^2 + x_2^2}{a^2} + \frac{x_3^2}{b^2} = 1\} =: \mathbb{E}^2_{a,a,b} \tag{11.9}$$

In the first case, the *Euclidean distance* between points P and p in terms of *Cartesian coordinates* of $P(X, Y, Z)$ and of $p(x_1, x_2, x_3)$ is represented. The *Cartesian coordinates* (x_1, x_2, x_3) of the projection point P are unknown. The constraint that the point p is an element of the ellipsoid of revolution

$$\mathbb{E}^2_{a,a,b} := \{\mathbf{x} \in \mathbb{R}^3 | b^2(x_1^2 + x_2^2) + a^2x_2^3 - a^2b^2 = 0, \mathbb{R}^+ \ni a > b \in \mathbb{R}^+\}$$

is substituted into the Lagrangean by means of the Lagrange multiplier x_4, which is unknown too. $\{(x_1^\wedge, x_2^\wedge, x_3^\wedge, x_4^\wedge) = arg\{\pounds(x_1, x_2, x_3, x_4) = min\}$ is the argument of the minimum of the *constrained Lagrangean* $\pounds(x_1, x_2, x_3, x_4)$. The result of the minimization procedure is presented by Lemma 11.1. Equation (11.10) provides the *necessary conditions* to constitute an extremum: The normal equations are of bilinear type. *Products* of the unknowns for instance x_1x_4, x_2x_4, x_3x_4 and *squares* of the unknowns, for instance x_1^2, x_2^2, x_3^2 appear. Finally the matrix of second derivatives $\mathbf{H}_3$ in (11.12) which is *positive definite* constitutes the *sufficient condition* to obtain a minimum. Fortunately the matrix of second derivatives $\mathbf{H}_3$ is *diagonal*. Using (11.11i–11.11iv), together with (11.14) leads to (11.15), which are the eigenvalues of the Hesse matrix $\mathbf{H}_3$. These values are $\Lambda_1 = \Lambda_2 = X\backslash x_1^\wedge, \Lambda_3 = Z\backslash x_3^\wedge$ and must be positive.

Lemma 11.1 (Constrained minimum distance mapping).
The functional $\pounds(x_1, x_2, x_3, x_4)$ is minimal, if the conditions (11.10) and (11.12) hold.

$$\boxed{\frac{\partial \pounds}{\partial x_i}((x_1^\wedge, x_2^\wedge, x_3^\wedge, x_4^\wedge)) = 0 \quad \forall \quad i{=}1,2,3,4.} \tag{11.10}$$

On taking partial derivatives with respect to x_i, we have

$$\left[\begin{array}{ll} (i) & \dfrac{\partial \pounds}{\partial (x_1^\wedge)} = -(X - x_1^\wedge) + b^2 x_1^\wedge x_4^\wedge = 0 \\[2ex] (ii) & \dfrac{\partial \pounds}{\partial (x_2^\wedge)} = -(Y - x_2^\wedge) + b^2 x_2^\wedge x_4^\wedge = 0 \\[2ex] (iii) & \dfrac{\partial \pounds}{\partial (x_3^\wedge)} = -(Z - x_3^\wedge) + a^2 x_3^\wedge x_4^\wedge = 0 \\[2ex] (iv) & \dfrac{\partial \pounds}{\partial (x_4^\wedge)} = \dfrac{1}{2}[b^2(x_1^{\wedge 2} + x_2^{\wedge 2})] + a^2 x_3^{\wedge 2} - a^2 b^2 = 0 \end{array}\right. \tag{11.11}$$

$$\boxed{\frac{\partial^2 \pounds}{\partial x_i \partial x_j}(x_1^\wedge, x_2^\wedge, x_3^\wedge, x_4^\wedge)) > 0 \quad \forall \quad i,j \in \{1,2,3\}.} \tag{11.12}$$

$$\begin{aligned} \mathbf{H}_3 &:= \left[\frac{\partial^2 \pounds}{\partial x_i \partial x_j}(\mathbf{x}^\wedge)\right] \\ &= \begin{bmatrix} 1 + b^2 x_4^\wedge & 0 & \\ 0 & 1 + b^2 x_4^\wedge & 0 \\ 0 & 0 & 1 + a^2 x_4^\wedge \end{bmatrix} \quad \in \quad \mathbb{R}^{3\times 3} \end{aligned} \tag{11.13}$$

"eigenvalues"

$$|\mathbf{H}_3 - \Lambda \mathbf{I}_3| = 0 \quad \Longleftrightarrow \tag{11.14}$$

$$\left[\begin{array}{l} \Lambda_1 = \Lambda_2 := 1 + b^2 x_4^\wedge = \dfrac{X}{x_1^\wedge} = \dfrac{Y}{x_2^\wedge} \\[2ex] \Lambda_3 := 1 + a^2 x_4^\wedge = \dfrac{Z}{x_3^\wedge} \end{array}\right. \tag{11.15}$$

In Sects. 11-41 and 11-42, we present algebraic solutions of the normal equations (11.11).

11-41 Grafarend-Lohse's mapping of $\mathbb{T}^2 \longrightarrow \mathbb{E}^2_{a,a,b}$

Two approaches are proposed by [172] for mapping $\mathbb{T}^2 \longrightarrow \mathbb{E}^2_{a,a,b}$. The *first approach* which is presented in Solution 11.4 is based on substitution technique. The *second approach* is based on degenerate conics and will not be treated in this book. Instead, we refer the reader to [172]. Let us start with the algorithm that solves the normal equations (11.11) in a closed form. Solution 11.4 outlines the *first and second forward steps* of reduction which lead to a univariate polynomial equation (11.20) of *fourth order* (quartic polynomial) in terms of the *Lagrangean multiplier*. First, the solution of the quartic polynomial is implemented. One then continues to determine with the *backward step* the Cartesian coordinates (x_1, x_2, x_3) of the point $p \in \mathbb{E}^2_{a,a,b}$ by means of the minimum distance mapping of the point $P \in \mathbb{T}^2$ to $p \in \mathbb{E}^2_{a,a,b}$.

Solution 11.4 (Grafarend-Lohse MDM solution).

First forward step

Solve $(i), (ii), (iii)$ for x_1, x_2, x_3 respectively.

$$\left[\begin{array}{l} (i) \quad x_1^\wedge(1+b^2x_4^\wedge) = X \Rightarrow x_1^\wedge = \dfrac{X}{1+b^2x_4^\wedge} \\ (ii) \quad x_2^\wedge(1+b^2x_4^\wedge) = Y \Rightarrow x_2^\wedge = \dfrac{Y}{1+b^2x_4^\wedge} \\ (iii) \quad x_3^\wedge(1+a^2x_4^\wedge) = Z \Rightarrow x_3^\wedge = \dfrac{Z}{1+a^2x_4^\wedge}. \end{array}\right. \tag{11.16}$$

Second forward step

Substitute $(x_1^\wedge, x_2^\wedge, x_3^\wedge, x_4^\wedge)$ in (11.11iv)

$$x_1^{\wedge 2} + x_2^{\wedge 2} = \frac{1}{(1+b^2x_4^\wedge)^2}(X^2+Y^2) \tag{11.17}$$

$$x_3^{\wedge 2} = \frac{1}{(1+a^2x_4^\wedge)^2}Z^2 \tag{11.18}$$

$$\left[\begin{array}{l} b^2(x_1^{\wedge 2} + x_2^{\wedge 2}) + a^2x_3^{\wedge 2} - a^2b^2 = 0 \Leftrightarrow \\ \\ \Leftrightarrow b^2\dfrac{X^2+Y^2}{(1+b^2x_4^\wedge)^2} + a^2\dfrac{Z^2}{(1+a^2x_4^\wedge)^2} - a^2b^2 = 0. \end{array}\right. \tag{11.19}$$

Multiply (11.19) by $(1+a^2x_4)^2(1+b^2x_4)^2$ leads to the quartic polynomial (11.20).

$$(11.19i) \left[\begin{array}{l} b^2(1+a^2x_4)^2(X^2+Y^2) + a^2(1+b^2x_4)^2Z^2 \\ -a^2b^2(1+a^2x_4)^2(1+b^2x_4)^2 = 0 \end{array}\right.$$

$$\Longleftrightarrow (1+2a^2x_4+a^4x_4^2)b^2(X^2+Y^2)+(1+2b^2x_4+b^4x_4^2)a^2Z^2$$
$$-a^2b^2(1+2a^2x_4+a^4x_4^2)(1+2b^2x_4+b^4x_4^2)=0$$

$$(11.19ii)\quad \left[\begin{array}{l} -x_4^4a^6b^6-2x_4^3a^4b^4(a^2+b^2) \\ +x_4^2a^2b^2[a^2(X^2+Y^3)+b^2Z^2-4a^2b^2-a^4-b^4]+ \\ 2x_4a^2b^2(X^2+Y^2+Z^2)+b^2(X^2+Y^2)+a^2Z^2-a^2b^2=0 \end{array}\right.$$

$$\left[\begin{array}{l} x_4^4+2x_4^3\dfrac{a^2+b^2}{a^2b^2}+x_4^2\dfrac{4a^2b^2+a^4+b^4-a^2(X^2+Y^2)-b^2Z^2}{a^4b^4} \\ -2x_4\dfrac{X^2+Y^2+Z^2}{a^4b^4}-\dfrac{b^2(X^2+Y^2)+a^2Z^2-a^2b^2}{a^6b^6}=0 \end{array}\right. \tag{11.20}$$

Backward step

Substitute $x_4^\wedge$ into $x_1^\wedge(x_4^\wedge), x_2^\wedge(x_4^\wedge), x_3^\wedge(x_4^\wedge)$

$$x_1^\wedge=(1+b^2x_4^\wedge)^{-1}X,\quad x_2^\wedge=(1+b^2x_4^\wedge)^{-1}Y,\quad x_3^\wedge=(1+a^2x_4^\wedge)^{-1}Z \tag{11.21}$$

Test

$$\Lambda_1=\Lambda_2=1+b^2x_4^\wedge>0,\quad \Lambda_3=1+a^2x_4^\wedge>0 \tag{11.22}$$

if $\Lambda_1=\Lambda_2>0$ and $\Lambda_3>0$ then end.

11-42 Groebner basis' mapping of $\mathbb{T}^2 \longrightarrow \mathbb{E}^2_{a,a,b}$

Without the various forward and backward reduction steps, we could automatically generate an equivalent algorithm for solving the normal equations (11.11i)–(11.11iv) in a closed form by means of Groebner basis approach. Let us write the *Ideal* of the polynomials in *lexicographic order* "$x_1 > x_2 > x_3 > x_4$" (read: x_1 before x_2 before x_3 before x_4) as

$$Ideal\ I:=<\left\{\begin{array}{c} x_1+b^2x_1x_4-X, \\ x_2+b^2x_2x_4-Y, \\ x_3+a^2x_3x_4-Z, \\ b^2x_1^2+b^2x_2^2-a^2x_3^2-a^2b^2 \end{array}\right\}>. \tag{11.23}$$

Expressing the generators of Ideal (11.23) as

$$\left[\begin{array}{l} f_1:=x_1+b^2x_1x_4-X, \\ f_2:=x_2+b^2x_2x_4-Y, \\ f_3:=x_3+a^2x_3x_4-Z, \\ f_4:=b^2x_1^2+b^2x_2^2-a^2x_3^2-a^2b^2, \end{array}\right. \tag{11.24}$$

the Groebner basis of these generators, characteristic for the minimum distance mapping problem, are computed using (4.36) on p. 45 as

$$GroebnerBasis[\{f_1, f_2, f_3, f_4\}, \{x_1, x_2, x_3, x_4\}]. \tag{11.25}$$

Groebner basis computation (11.25) leads to 14 elements presented in Solution 11.5 interpreted as follows: The *first expression* is a univariate polynomial of order four (quartic) in the Lagrange multiplier, i.e.,

$$\left[\begin{array}{l} \boxed{c_4 x_4^4 + c_3 x_4^3 + c_2 x_4^2 + c_1 x_4 + c_o = 0} \\ c_4 = a^6 b^6 \\ c_3 = (2a^6 b^4 + 2a^4 b^6) \\ c_2 = (a^6 b^2 + 4a^4 b^4 + a^2 b^6 - a^4 b^2 X^2 - a^4 b^2 Y^2 - a^2 b^4 Z^2) \\ c_1 = (2a^4 b^2 + 2a^2 b^4 - 2a^2 b^2 X^2 - 2a^2 b^2 Y^2 - 2a^2 b^2 Z^2) \\ c_o = (a^2 b^2 - b^2 X^2 - b^2 Y^2 - a^2 Z^2), \end{array}\right. \tag{11.26}$$

and is identical to (11.19ii). With the admissible values x_4 substituted in linear equations (4),(8),(12) of the computed Groebner basis, i.e.,

$$\left[\begin{array}{l} (1 + a^2 x_4)x_3 - Z \\ (1 + b^2 x_4)x_2 - Y \\ (1 + b^2 x_4)x_1 - X, \end{array}\right. \tag{11.27}$$

the values $(x_1, x_2, x_3) = (x, y, z)$ are finally produced.

Solution 11.5 (Groebner basis MDM solution).

$$(1)\left[\begin{array}{l} a^2b^2x_4^4 + (2a^6b^4 + 2a^4b^6)x_4^3 + (a^6b^2 + 4a^4b^4 + a^2b^6 - a^4b^2X^2 - a^4b^2Y^2 - \\ a^2b^4Z^2)x_4^2 + +(2a^4b^2 + 2a^2b^4 - 2a^2b^2X^2 - 2a^2b^2Y^2 - 2a^2b^2Z^2)x_4 \\ +(a^2b^2 - b^2X^2 - b^2Y^2 - a^2Z^2). \end{array}\right.$$

$$(2)\quad\left[\begin{array}{l} (a^4Z - 2a^2b^2Z + b^4Z)x_3 - a^6b^6x_4^3 - (2a^6b^4 + a^4b^6)x_4^2 \\ -(a^6b^2 + 2a^4b^4 - a^4b^2X^2 - a^4b^2Y^2 - a^2b^4Z^2)x_4 \\ -a^2b^4 + a^2b^2X^2 + a^2b^2Y^2 + 2a^2b^2Z^2 - b^4Z^2. \end{array}\right.$$

$$(3)\quad\left[\begin{array}{l} (2b^2Z + b^4x_4Z - a^2Z)x_3 + a^4b^6x_4^3 + (2a^4b^4 + a^2b^6)x_4^2 \\ +(a^4b^2 + 2a^2b^4 - a^2b^2X^2 - a^2b^2Y^2 - b^4Z^2)x_4 \\ +a^2b^2 - b^2X^2 - b^2Y^2 - 2b^2Z^2. \end{array}\right.$$

$$(4)\quad (1 + a^2x_4)x_3 - Z$$

$$(5)\quad\left[\begin{array}{l} (a^4 - 2a^2b^2 + b^4)x_3^2 + (2a^2b^2Z - 2b^4Z)x_3 \\ -a^4b^6x_4^2 - 2a^4b^4x_4 - a^4b^2 + a^2b^2X^2 + a^2b^2Y^2 + b^4Z^2). \end{array}\right.$$

$$(6)\quad\left[\begin{array}{l} (2b^2 - a^2 + b^4x_4)x_3^2 - a^2Zx_3 + a^4b^6x_4^3 + (2a^4b^4 + 2a^2b^6)x_4^2 \\ + + (a^4b^2 + 4a^2b^4 - a^2b^2X^2 - a^2b^2Y^2 - b^4Z^2)x_4 \\ +2a^2b^2 - 2b^2X - 2bY^2 - 2b^2Z^2. \end{array}\right.$$

$$(7)\quad\left[\begin{array}{l} (X^2 + Y^2)x_2 + a^2b^4Yx_4^2 + Y(a^2b^2 - b^2x_3^2 - b^2Zx_3)x_4 \\ +Yx_3^2 - Y^3 - YZx_3 - YX^2. \end{array}\right.$$

(8) $$(1+b^2x_4)x_2 - Y$$

(9) $$a^2x_3 - b^2x_3 + b^2Z)x_2 - a^2x_3Y$$

(10) $$Yx_1 - Xx_2$$

(11) $$Xx_1 + a^2b^4x_4^2 + (a^2b^2 + b^2x_3^2 - b^2Zx_3)x_4 + x_3^2 - Zx_3 + Yx_2 - X^2 - Y^2.$$

(12) $$(1+b^2x_4)x_1 - X$$

(13) $$(a^2x_3 - b^2x_3 + b^2Z)x_1 - a^2Xx_3$$

(14) $$x_1^2 + a^2b^4x_4^2 + (2a^2b^2 + b^2x_3^2 - b^2Zx_3)x_4 + 2x_3^2 - 2Zx_3 + x_2^2 - X^2 - Y^2.$$

11-43 Extended Newton-Raphson's mapping of $\mathbb{T}^2 \longrightarrow \mathbb{E}^2_{a,a,b}$

Instead of transforming the constrained optimization problem into an unconstrained one as it is usual, we shall solve it as an underdetermined system via *Extended Newton- Raphson method.* Let us introduce new variables,

$$\alpha = X - x, \beta = Y - y, \gamma = Z - z \tag{11.28}$$

Now, our constrain is,

$$-1 + \frac{(X-\alpha)^2 + (Y-\beta)^2}{a^2} + \frac{(Z-\gamma)^2}{b^2} = 0 \tag{11.29}$$

or

$$-a^2b^2 + b^2X^2 + b^2Y^2 + a^2Z^2 - 2b^2X\alpha + b^2\alpha^2 - 2b^2Y\beta + b^2\beta^2 - 2a^2Z\gamma + a^2\gamma^2 = 0 \tag{11.30}$$

Let us consider the input of a topographical point as

$$X = 3770667.9989, Y = 446076.4896, Z = 5107686.2085,$$

$$a = 6378136.602, b = 6356751.860$$

Then our equation in numerical form after normalization becomes,

$$1 - 0.0130735\alpha + 1.7335845788851814 \times 10^{-9}\alpha^2 - 0.00154662\beta +$$

$$+1.7335845788851814 \times 10^{-9}\beta^2 - 0.0178286\gamma + 1.745268098124392 \times 10^{-9}\gamma^2 = 0$$

Now, we have a single equation with 3 variables (α, β, γ). This *underdetermined problem* has infinite solutions. In order to select the proper solution we are looking for the solution with *minimal norm*, but this is what we want, since the distance to be minimized is just $d = \alpha^2 + \beta^2 + \gamma^2$ A good initial guess is $(\alpha, \beta, \gamma) = \{0, 0, 0\}$.

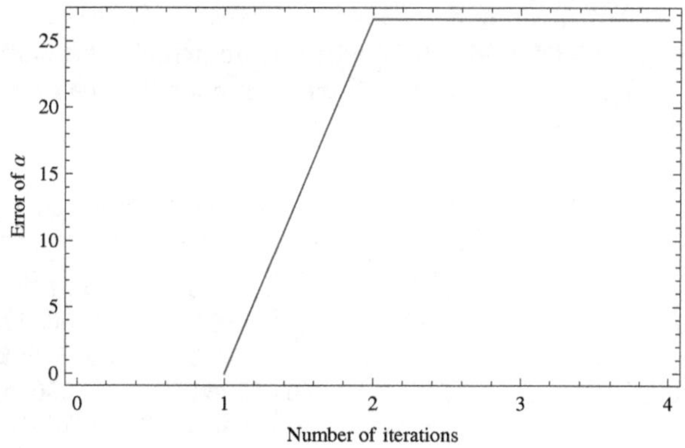

Figure 11.6. Convergence of the method in case of the α = X - x relative coordinates

Let us employ *Extended Newton-Raphson* method. Fig. 11.6 shows the absolute error of α as function of the number of iterations. The convergence is fast. The result are

$$\alpha = 26.6174, \beta = 3.14888, \gamma = 36.2985.$$

Then the absolute coordinates are,

$$x = 3770641.3815, y = 446073.3407, z = 5107649.9100.$$

Once the ellipsoidal Cartesian coordinates $\{x_1, x_2, x_3\}$ have been computed using either Solutions 11.4 or 11.5, they are transformed into their equivalent Gauss ellipsoidal coordinates $\{L, B, H\}$ using (11.31), (11.32) and (11.33) in Solution 11.6.

Solution 11.6 (Coordinates transformation from Cartesian to Gauss ellipsoidal).

$$\{X, Y, Z\} \in \mathbb{T}^2 \quad \{x_1, x_2, x_3\} \in \mathbb{E}^2_{a,a,b}\, to\, \{L, B, H\}$$

"Pythagoras in three dimension"

$$H := \sqrt{(X - x_1)^2 + (Y - x_2)^2 + (Z - x_3)^2} \tag{11.31}$$

"convert $\{x_1, x_2, x_3\}$ and $\{X, Y, Z\}$ to $\{L, B\}$"

$$tanL = \frac{Y - x_2}{X - x_1} = \frac{Y - y}{X - x} \tag{11.32}$$

$$tanB = \frac{Z - x_3}{\sqrt{(X - x_1)^2 + (Y - x_2)^2}} = \frac{Z - x_3}{\sqrt{(X - x)^2 + (Y - y)^2}} \tag{11.33}$$

Example 11.1 (Example from [172]). Given are the geometric parameters of the ellipsoid of revolution; *semi-major axis* $a = 6378137.000$m and *first numerical eccentricity* $e^2 = 0.00669437999013$ from which the *semi-minor axis* b is to be computed. The input data are Cartesian coordinates of 8 points on the surface of the Earth presented in Table 11.2.

Table 11.2. Cartesian coordinates of topographic points

Point	$X(m)$	$Y(m)$	$Z(m)$
1	3980192.960	0	4967325.285
2	0	0	6356852.314
3	0	0	-6357252.314
4	4423689.486	529842.355	4555616.169
5	4157619.145	664852.698	4775310.888
6	-2125699.324	6012793.226	-91773.648
7	5069470.828	3878707.846	-55331.828
8	213750.930	5641092.098	2977743.624

Using these data, the coefficients of the univariate polynomial (11.26) are computed and used in the Matlab's roots command, e.g., (4.42) on p. 46 as $x_4 = roots\left[\, c_4\; c_3\; c_2\; c_1\; c_0 \,\right]$. The obtained roots are then substituted in (11.27) to give the values of $\{x_3, x_2, x_1\}$ of the ellipsoidal Cartesian coordinates. The computed results presented in *Table* 11.3 are identical to those obtained by [172, Table 4, p. 108]. Once the ellipsoidal Cartesian coordinates have been derived, the ellipsoidal coordinates (ellipsoidal longitude L, ellipsoidal latitude B and height H) can be computed using (11.31), (11.32) and (11.33) in Solution 11.6.

Table 11.3. Computed ellipsoidal cartesian coordinates and the Lagrange factor

Point	$x_1(m)$	$x_2(m)$	$x_3(m)$	$x_4(m^{-2})$
1	3980099.549	0.000	4967207.921	5.808116e-019
2	0.000	0.000	6356752.314	3.867016e-019
3	0.000	0.000	-6356752.314	1.933512e-018
4	4420299.446	529436.317	4552101.519	1.897940e-017
5	4157391.441	664816.285	4775047.592	1.355437e-018
6	-2125695.991	6012783.798	-91773.503	3.880221e-020
7	5065341.132	3875548.170	-55286.450	2.017617e-017
8	213453.298	5633237.315	2973569.442	3.450687e-017

Example 11.2 (Case study: Baltic sea level project). Let us adopt the *world geodetic datum 2000* with the semi-major axis a=6378136.602 m and semi-minor axis b=6356751.860 m from [167]. Here we take advantage of given Cartesian coordinates of 21 points of the topographic surface of the Earth presented in Table 11.4. Using these data, the coefficients of (11.26) are computed and used to solve for x_4. With the admissible values of x_4 substituted in (11.27), the values of the ellipsoidal Cartesian coordinates $(x_1, x_2, x_3) = (x, y, z)$ are produced and are as presented in Table 11.5. They are finally converted by means of Solution 11.6 to

(L, B, H) in *Table* 11.6. Figure 11.7 depicts the mapping of topographical points onto the reference ellipsoid.

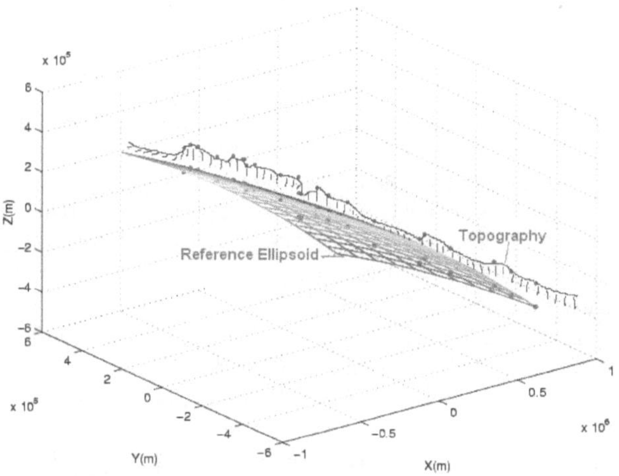

Figure 11.7. Baltic sea level project topographic points mapped on to the *international reference ellipsoid* $\mathbb{E}^2_{a,a,b}$

Table 11.4. Baltic sea level project: Cartesian coordinates of topographic points

Station	$X(m)$	$Y(m)$	$Z(m)$
Borkum (Ger)	3770667.9989	446076.4896	5107686.2085
Degerby (Fin)	2994064.9360	1112559.0570	5502241.3760
Furuoegrund (Swe)	2527022.8721	981957.2890	5753940.9920
Hamina (Fin)	2795471.2067	1435427.7930	5531682.2031
Hanko (Fin)	2959210.9709	1254679.1202	5490594.4410
Helgoland (Ger)	3706044.9443	513713.2151	5148193.4472
Helsinki (Fin)	2885137.3909	1342710.2301	5509039.1190
Kemi (Fin)	2397071.5771	1093330.3129	5789108.4470
Klagshamn (Swe)	3527585.7675	807513.8946	5234549.7020
Klaipeda (Lit)	3353590.2428	1302063.0141	5249159.4123
List/Sylt (Ger)	3625339.9221	537853.8704	5202539.0255
Molas (Lit)	3358793.3811	1294907.4149	5247584.4010
Mäntyluoto (Fin)	2831096.7193	1113102.7637	5587165.0458
Raahe (Fin)	2494035.0244	1131370.9936	5740955.4096
Ratan (Swe)	2620087.6160	1000008.2649	5709322.5771
Spikarna (Swe)	2828573.4638	893623.7288	5627447.0693
Stockholm (Swe)	3101008.8620	1013021.0372	5462373.3830
Ustka (Pol)	3545014.3300	1073939.7720	5174949.9470
Vaasa (Fin)	2691307.2541	1063691.5238	5664806.3799
Visby (Swe)	3249304.4375	1073624.8912	5364363.0732
OElands N. U. (Swe)	3295551.5710	1012564.9063	5348113.6687

Table 11.5. Ellipsoidal Cartesian coordinates (x_1, x_2, x_3) and Lagrange multiplier x_4

Station	$x_1(m)$	$x_2(m)$	$x_3(m)$	$x_4(m^{-2})$
Borkum (Ger)	3770641.3815	446073.3407	5107649.9100	1.746947e-019
Degerby (Fin)	2994054.5862	1112555.2111	5502222.2279	8.554612e-020
Furuoegrund (Swe)	2527009.7166	981952.1770	5753910.8356	1.288336e-019
Hamina (Fin)	2795463.7019	1435423.9394	5531667.2524	6.643801e-020
Hanko (Fin)	2959199.2560	1254674.1532	5490572.5584	9.797001e-020
Helgoland (Ger)	3706019.4100	513709.6757	5148157.7376	1.705084e-019
Helsinki (Fin)	2885126.2764	1342705.0575	5509017.7534	9.533532e-020
Kemi (Fin)	2397061.6153	1093325.7692	5789084.2263	1.028464e-019
Klagshamn (Swe)	3527564.6083	807509.0510	5234518.0924	1.484413e-019
Klaipeda (Lit)	3353562.2593	1302052.1493	5249115.3164	2.065021e-019
List/Sylt (Ger)	3625314.3442	537850.0757	5202502.0726	1.746017e-019
Molas (Lit)	3358777.7367	1294901.3835	5247559.7944	1.152676e-019
Mäntyluoto (Fin)	2831087.1439	1113098.9988	5587146.0214	8.370165e-020
Raahe (Fin)	2494026.5401	1131367.1449	5740935.7483	8.418639e-020
Ratan (Swe)	2620078.1000	1000004.6329	5709301.7015	8.988111e-020
Spikarna (Swe)	2828561.2473	893619.8693	5627422.6007	1.068837e-019
Stockholm (Swe)	3100991.6259	1013015.4066	5462342.8173	1.375524e-019
Ustka (Pol)	3544995.3045	1073934.0083	5174921.9867	1.328158e-019
Vaasa (Fin)	2691299.0138	1063688.2670	5664788.9183	7.577249e-020
Visby (Swe)	3249290.3945	1073620.2512	5364339.7330	1.069551e-019
OElands N. U. (Swe)	3295535.1675	1012559.8663	5348086.8692	1.231803e-019

Table 11.6. Geodetic computed from ellipsoidal Cartesian coordinates in closed form

Station	Longitude L			Latitude B			Ellipsoidal height H
	°	′	″	°	′	″	m
Borkum (Ger)	6	44	48.5914	53	33	27.4808	45.122
Degerby (Fin)	20	23	4.0906	60	1	52.8558	22.103
Furuoegrund (Swe)	21	14	6.9490	64	55	10.2131	33.296
Hamina (Fin)	27	10	47.0690	60	33	52.9819	17.167
Hanko (Fin)	22	58	35.4445	59	49	21.6459	25.313
Helgoland (Ger)	7	53	30.3480	54	10	29.3979	44.042
Helsinki (Fin)	24	57	24.2446	60	9	13.2416	24.633
Kemi (Fin)	24	31	5.6737	65	40	27.7029	26.581
Klagshamn (Swe)	12	53	37.1597	55	31	20.3311	38.345
Klaipeda (Lit)	21	13	9.0156	55	45	16.5952	53.344
List/Sylt (Ger)	8	26	19.7594	55	1	3.0992	45.101
Molas (Lit)	21	4	58.8931	55	43	47.2453	29.776
Mäntyluoto (Fin)	21	27	47.7777	61	35	39.3552	21.628
Raahe (Fin)	24	24	1.8197	64	38	46.8352	21.757
Ratan (Swe)	20	53	25.2392	63	59	29.5936	23.228
Spikarna (Swe)	17	31	57.9060	62	21	48.7645	27.620
Stockholm (Swe)	18	5	27.2528	59	19	20.4054	35.539
Ustka (Pol)	16	51	13.8751	54	35	15.6866	34.307
Vaasa (Fin)	21	33	55.9146	63	5	42.8394	19.581
Visby (Swe)	18	17	3.9292	57	38	21.3487	27.632
OElands N. U. (Swe)	17	4	46.8542	57	22	3.4508	31.823

11-5 Concluding remarks

The chapter has presented a new and direct algebraic approach to the mapping problem that has attracted a lot of research as evidenced in Table 11.1. All that is required is for the user to apply equations (11.26) and (11.27). Probably the best choice is the *Grafarend-Lohse*'s method, but *Linear Homotopy* as a global numerical solver and the Extended Newton-Raphson can also be good candidates. These techniques work well out of the Mathematica, too.

12 Positioning by ranging

12-1 Applications of distances

Throughout history, position determination has been one of the fundamental task undertaken by man on daily basis. Each day, one has to know where one is, and where one is going. To mountaineers, pilots, sailors etc., the knowledge of position is of great importance. The traditional way of locating one's position has been the use of maps or campus to determine directions. In modern times, the entry into the game by Global Navigation Satellite Systems GNSS that comprise the Global Positioning System (GPS), Russian based GLONASS and the proposed European's GALILEO have revolutionized the art of positioning.

In the new field of *GPS meteorology* for example, as well as geodesy, robotics and geoinformatics etc., distances (ranges) play a key role in determining unknown parameters. In the recently developed *Spatial Reference System*[1] designed to check and control the accuracy of three-dimensional coordinate measuring machines and tooling equipments, coordinates of the edges of the instrument are computed from distances of the bars. This signifies that industrial application of distances is fast gaining momentum just as in geosciences. In GPS meteorology that we will discuss in Chap. 15 for example, distances traveled by GPS satellites signals through the atmosphere are measured and related do the would be distances in vacuo (i.e., in the absence of the atmosphere). Since these signals traverse the atmosphere, they enable *accurate global remote sensing* of the atmosphere to retrieve vertical profiles of *temperature*, *pressure* and *water vapour*.

Apart from distances being used to determine the user's position and its application in GPS meteorology, they find use in quick station search in engineering and cadastral surveying operations. Ranging, together with resection and intersection techniques (see e.g., Chaps. 13 and 14) are useful in densifying geodetic networks as illustrated by Fig. 12.1. Densification is vital for extending network control in areas where GPS receivers fail, e.g., in tunnels and forests (see Fig. 12.1). Distances are also used in photogrammetry to determine the perspective center coordinates from measured photo and the ground coordinates. Another area of application is in robotics.

Measured distances (ranges) are normally related to the desired parameters via nonlinear systems of equations that require explicit/exact solutions. Approximate numerical procedures used for solving such nonlinear distance equations are normally iterative in nature, and often require linearization of the nonlinear equations. Where closed form solutions exist, they require differencing and substitution steps which are laborious and time consuming. The desire therefore is to have procedures that can offer direct solutions without linearization, iterations or substitutional steps.

In this chapter, direct procedures for solving nonlinear systems of equations for distances without linearization, iteration, forward and backward substitutions

[1]Metronom US., Inc., Ann Arbor: http://www.metronomus.com

J.L. Awange et al., *Algebraic Geodesy and Geoinformatics*, 2nd ed.,
DOI 10.1007/978-3-642-12124-1_12,

are presented. In particular, the advantages of fast computers with large storage capacities, and computer algebraic software of Mathematica, Maple and Matlab are exploited by the algebraic based approaches. These methods which were presented in Chaps. 4, 5 and 7 directly deliver the position of unknown station from distance measurements. They do so by eliminating variables appearing in the nonlinear systems of equations resulting in univariate polynomials that are solvable using Matlab's *"roots"* command.

The improvements made on measuring instruments has led to Electromagnetic Distance Measuring (EDM) equipments that measure distances to higher accuracies. By measuring distances from an unknown station to two known stations, two nonlinear distance equations, whose geometrical properties have been studied by [178, 179] are formed. They have to be solved for the planar position of the unknown station. If distances are measured from an unknown station to three known stations instead, three nonlinear distance equations have to be solved for the unknown position. In Chaps. 4 and 7, planar distances were encountered in Figs. 4.1 and 7.2 respectively, where they were used to illustrate the concepts that were discussed. The position $\{x_0, y_0\}$ of the unknown station P_0 was related to the measured distances by (4.1) and (4.2) on p. 33.

The term ranging is broadly used in this chapter to incorporate the GPS pseudo-range measurements. For Local Positioning Systems (e.g., using EDMs), distances can be measured directly. For Global Positioning System (GPS) however, distances are not directly measured owing to satellites and receivers' clock uncertainties.

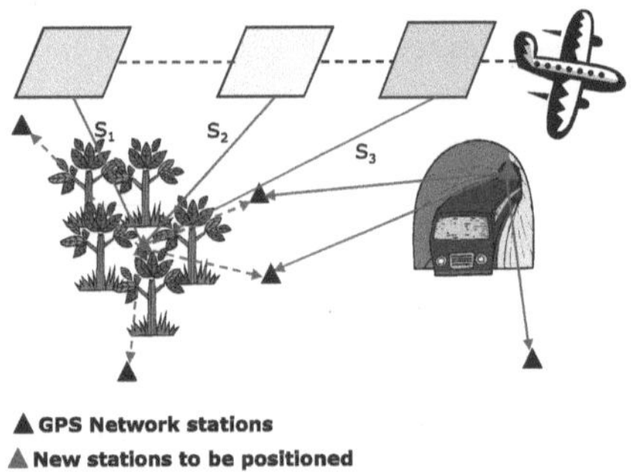

Figure 12.1. Point densification in forest and inside a tunnel

12-2 Ranging by global navigation satellite system (GNSS)

12-21 The pseudo-ranging four-points problem

If one has access to a hand held GPS receiver, a mobile phone or a watch fitted with a GPS receiver, one needs only to press the button to know the position where one is standing. Basically, the operations involve distance measurements to GPS satellites whose properties were discussed in Sect. 10-3. The receiver measures the travel time of the signal transmitted from the satellites. This distance is calculated from the relationship

$$distance = velocity \times time,$$

where velocity is given by the speed of light in vacuum. The distances S_i are then related to the position of the unknown station $\{X_0, Y_0, Z_0\}$ by

$$S_i = \sqrt{(X^i - X_0)^2 + (Y^i - Y_0)^2 + (Z^i - Z_0)^2}, \qquad (12.1)$$

where $\{X^i, Y^i, Z^i\}$ are the position of the satellite i. Geometrically, the three unknowns $\{X_0, Y_0, Z_0\}$ are obtained from the intersection of three spherical cones given by the pseudo-ranging equations. Distance measurements to only one satellite puts the user's position anywhere within the sphere formed by distance S_1 in Fig. 12.2. Measurements to two satellites narrow the position to the intersection of the two spheres S_1 and S_2. 12.2. A third satellite is therefore required to definitely fix the user's position. This is achieved by the intersection of the third sphere S_3 with the other two.

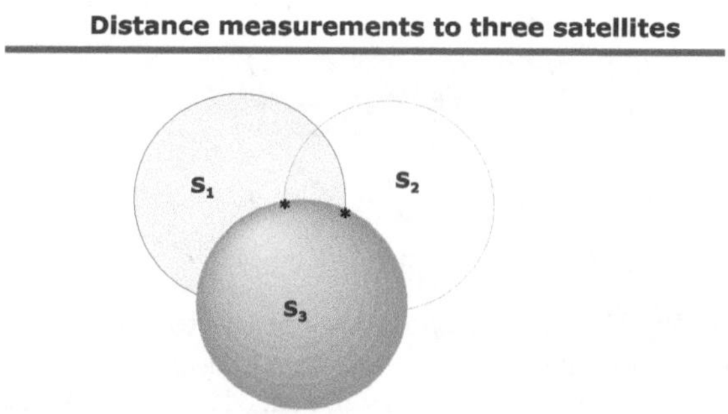

Figure 12.2. Pseudo-ranging geometry

If direct distance measurements to the satellites were possible, (12.1) would have sufficed to provide the user's location. Distance measurements to satellites as already stated are however not direct owing to the satellites and receivers' clock biases. Satellites' clock biases can be modelled while the receivers' clock biases have to be determined as an unknowns. For GPS positioning therefore, in addition to position determination from measured distances, the receiver's clock bias has to be added in the observation equations as unknown. Since distances to the satellites in (12.1) are derived from the transmitted signals that are affected by both satellites and receivers' clock uncertainties, they are normally referred to as pseudo-ranges. What one measures therefore are not the actual distances (ranges) but pseudo-ranges. Pseudo-range measurements lead to GPS pseudo-ranging four-points problem ("pseudo 4P4"), which is the problem of determining the four unknowns. The unknowns comprise the three components of receiver position $\{X_0, Y_0, Z_0\}$ and the stationary receiver *range bias*. Minimum observations required to obtain receiver position and range bias are pseudo-range observations to four satellites as depicted in Fig. 12.3. Besides pseudo-range observations, phase measurements are often used where accurate results are desired.

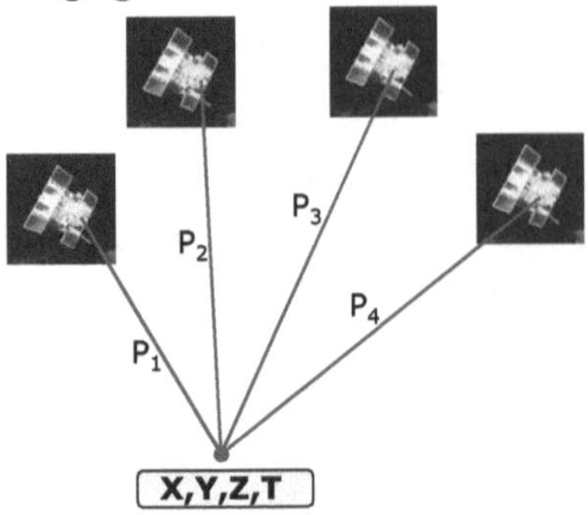

Figure 12.3. Point positioning using GPS satellites

Four pseudo-range equations are formed from (12.1) and expressed algebraically as

$$\left[\begin{array}{c}(x_1-a_0)^2+(x_2-b_0)^2+(x_3-c_0)^2-(x_4-d_0)^2=0\\(x_1-a_1)^2+(x_2-b_1)^2+(x_3-c_1)^2-(x_4-d_1)^2=0\\(x_1-a_2)^2+(x_2-b_2)^2+(x_3-c_2)^2-(x_4-d_2)^2=0\\(x_1-a_3)^2+(x_2-b_3)^2+(x_3-c_3)^2-(x_4-d_3)^2=0\\ where \ x_1,x_2,x_3,x_4\in\\(a_0,b_0,c_0)=(x^0,y^0,z^0)\sim P^0\\(a_1,b_1,c_1)=(x^1,y^1,z^1)\sim P^1\\(a_2,b_2,c_2)=(x^2,y^2,z^2)\sim P^2\\(a_3,b_3,c_3)=(x^3,y^3,z^3)\sim P^3.\end{array}\right. \tag{12.2}$$

In (12.2), $\{P^0, P^1, P^2, P^3\}$ are the positions of the four GPS satellites whose signals are tracked by the receiver at an unknown station P_0. The satellites' positions are given by the coordinates $\{x^i, y^i, z^i | i = 0, 1, 2, 3\}$, where i indicating a particular satellite number. The measured pseudo-ranges to these satellites from a stationary receiver at P_0 are given by $\{d_0, d_1, d_2, d_3\}$. The parameters $\{a_0, b_0, c_0\}$, $\{a_1, b_1, c_1\}$, $\{a_2, b_2, c_2\}$, $\{a_3, b_3, c_3\}$, $\{d_0, d_1, d_2, d_3\}$ are known elements of the spherical cone that intersect at P_0 to give the unknown coordinates $\{x_1, x_2, x_3\}$ of the receiver and the stationary receiver range bias x_4. Several procedures have been put forward to obtain exact solution of (12.2), e.g., [53, 181, 242, 243, 272, 363]. In what follows, we present alternative solutions to (12.2) based on algebraic approaches of Groebner bases and polynomial resultants discussed in Chaps. 4 and 5 respectively. Equation (12.2) is expanded and arranged in the lexicographic order $\{x_1 > x_2 > x_3 > x_4\}$ as

$$\left[\begin{array}{c}x_1^2-2a_0x_1+x_2^2-2b_0x_2+x_3^2-2c_0x_3-x_4^2+2d_0x_4+a_0^2+b_0^2+c_0^2-d_0^2=0\\x_1^2-2a_1x_1+x_2^2-2b_1x_2+x_3^2-2c_1x_3-x_4^2+2d_1x_4+a_1^2+b_1^2+c_1^2-d_1^2=0\\x_1^2-2a_2x_1+x_2^2-2b_2x_2+x_3^2-2c_2x_3-x_4^2+2d_2x_4+a_2^2+b_2^2+c_2^2-d_2^2=0\\x_1^2-2a_3x_1+x_2^2-2b_3x_2+x_3^2-2c_3x_3-x_4^2+2d_3x_4+a_3^2+b_3^2+c_3^2-d_3^2=0,\end{array}\right. \tag{12.3}$$

where the unknown variables to be determined are $\{x_1, x_2, x_3, x_4\}$. The other terms are known constants. Equation (12.3) is written with the linear terms on the right-hand-side and the nonlinear terms on the left-hand-side as

$$\begin{bmatrix} x_1^2+x_2^2+x_3^2-x_4^2 = 2a_0x_1+2b_0x_2+2c_0x_3-2d_0x_4+d_0^2-a_0^2-b_0^2-c_0^2 \\ x_1^2+x_2^2+x_3^2-x_4^2 = 2a_1x_1+2b_1x_2+2c_1x_3-2d_1x_4+d_1^2-a_1^2-b_1^2-c_1^2 \\ x_1^2+x_2^2+x_3^2-x_4^2 = 2a_2x_1+2b_2x_2+2c_2x_3-2d_2x_4+d_2^2-a_2^2-b_2^2-c_2^2 \\ x_1^2+x_2^2+x_3^2-x_4^2 = 2a_3x_1+2b_3x_2+2c_3x_3-2d_3x_4+d_3^2-a_3^2-b_3^2-c_3^2. \end{bmatrix} \quad (12.4)$$

Subtracting the last expression (12.4iv) from the first three expressions (12.4i), (12.4ii), and (12.4iii) leads to

$$\begin{bmatrix} a_{03}x_1+b_{03}x_2+c_{03}x_3+d_{30}x_4+e_{03}=0 \\ a_{13}x_1+b_{13}x_2+c_{13}x_3+d_{31}x_4+e_{13}=0 \\ a_{23}x_1+b_{23}x_2+c_3x_3+d_{32}x_4+e_{23}=0, \end{bmatrix} \quad (12.5)$$

where

$$\begin{bmatrix} a_{03}=2(a_0-a_3),\ b_{03}=2(b_0-b_3),\ c_{03}=2(c_0-c_3),\ d_{30}=2(d_3-d_0), \\ a_{13}=2(a_1-a_3),\ b_{13}=2(b_1-b_3),\ c_{13}=2(c_1-c_3),\ d_{31}=2(d_3-d_1), \\ a_{23}=2(a_2-a_3),\ b_{23}=2(b_2-b_3),\ c_{23}=2(c_2-c_3),\ d_{32}=2(d_3-d_2), \\ e_{03}=(d_0^2-a_0^2-b_0^2-c_0^2)-(d_3^2-a_3^2-b_3^2-c_3^2), \\ e_{13}=(d_1^2-a_1^2-b_1^2-c_1^2)-(d_3^2-a_3^2-b_3^2-c_3^2), \\ e_{23}=(d_2^2-a_2^2-b_2^2-c_2^2)-(d_3^2-a_3^2-b_3^2-c_3^2). \end{bmatrix}$$

We note immediately that (12.5) comprises three equations which are linear with four unknowns leading to an underdetermined system of equations. This is circumvented by treating one variable, say x_4, as a constant thereby leading to a system of three equations in three unknowns. We then apply either Groebner basis or polynomial resultants techniques to solve the linear system of equation for $x_1=g(x_4), x_2=g(x_4), x_3=g(x_4)$, where $g(x_4)$ is a linear function.

12-211 Sturmfels' approach

The Sturmfels' [375] approach discussed in Sect. 5-32 is applied to solve (12.5). Depending on which variable one wants, (12.5) is rewritten such that this particular variable is hidden (i.e., is treated as a constant). If our interest is to solve $x_1 = g(x_4)$ for instance, (12.5) is first homogenized using x_5 (see Definition 5.1 on p. 49) and then written by hiding x_1 as

$$\begin{bmatrix} f_1 := (a_{03}x_1+d_{30}x_4+e_{03})x_5+b_{03}x_2+c_{03}x_3 \\ f_2 := (a_{13}x_1+d_{31}x_4+e_{13})x_5+b_{13}x_2+c_{13}x_3 \\ f_3 := (a_{23}x_1+d_{32}x_4+e_{13})x_5+b_{23}x_2+c_{23}x_3. \end{bmatrix} \quad (12.6)$$

The Jacobian determinant of (12.6) then becomes

$$J_{x_1} = det \begin{bmatrix} \frac{\partial f_1}{\partial x_2} & \frac{\partial f_1}{\partial x_3} & \frac{\partial f_1}{\partial x_5} \\ \frac{\partial f_2}{\partial x_2} & \frac{\partial f_2}{\partial x_3} & \frac{\partial f_2}{\partial x_5} \\ \frac{\partial f_3}{\partial x_2} & \frac{\partial f_3}{\partial x_3} & \frac{\partial f_3}{\partial x_5} \end{bmatrix} = det \begin{bmatrix} b_{03} & c_{03} & (a_{03}x_1+d_{30}x_4+e_{03}) \\ b_{13} & c_{13} & (a_{13}x_1+d_{31}x_4+e_{13}) \\ b_{23} & c_{23} & (a_{23}x_1+d_{32}x_4+e_{23}) \end{bmatrix}. \quad (12.7)$$

The determinant obtained in (12.7) gives the expression for $x_1 = g(x_4)$ as

$x_1 = -(e_{03}b_{13}c_{23}+d_{32}x_4b_{03}c_{13}+d_{30}x_4b_{13}c_{23}-d_{30}x_4c_{13}b_{23}-d_{31}x_4b_{03}c_{23}-e_{03}c_{13}b_{23}-e_{13}b_{03}c_{23}+e_{13}c_{03}b_{23}+e_{23}b_{03}c_{13}+d_{31}x_4c_{03}b_{23}-d_{32}x_4c_{03}b_{13}-e_{23}c_{03}b_{13})/(a_{23}c_{13}b_{03}+a_{13}b_{23}c_{03}-a_{13}c_{23}b_{03}-a_{23}b_{13}c_{03}-a_{03}c_{13}b_{23}+a_{03}c_{23}b_{13})$.

For $x_2 = g(x_4)$, we have

$$\begin{bmatrix} f_4 := (b_{03}x_2 + d_{30}x_4 + e_{03})x_5 + a_{03}x_1 + c_{03}x_3 \\ f_5 := (b_{13}x_2 + d_{31}x_4 + e_{13})x_5 + a_{13}x_1 + c_{13}x_3 \\ f_6 := (b_{23}x_2 + d_{32}x_4 + e_{23})x_5 + a_{23}x_1 + c_{23}x_3, \end{bmatrix} \tag{12.8}$$

whose Jacobian determinant is given by

$$J_{x_2} = det \begin{bmatrix} \frac{\partial f_4}{\partial x_1} & \frac{\partial f_4}{\partial x_3} & \frac{\partial f_4}{\partial x_5} \\ \frac{\partial f_5}{\partial x_1} & \frac{\partial f_5}{\partial x_3} & \frac{\partial f_5}{\partial x_5} \\ \frac{\partial f_6}{\partial x_1} & \frac{\partial f_6}{\partial x_3} & \frac{\partial f_6}{\partial x_5} \end{bmatrix} = det \begin{bmatrix} a_{03} & c_{03} & (b_{03}x_2 + d_{30}x_4 + e_{03}) \\ a_{13} & c_{13} & (b_{13}x_2 + d_{31}x_4 + e_{13}) \\ a_{23} & c_{23} & (b_{23}x_2 + d_{32}x_4 + e_{23}) \end{bmatrix}. \tag{12.9}$$

The determinant obtained in (12.9) gives the expression for $x_2 = g(x_4)$ as

$x_2 = -(a_{23}c_{13}d_{30}x_4 + a_{03}c_{23}d_{31}x_4 + a_{03}c_{23}e_{13} - a_{23}c_{03}d_{31}x_4 - a_{03}c_{13}d_{32}x_4 - a_{03}c_{13}e_{23} + a_{13}c_{03}d_{32}x_4 - a_{13}c_{23}d_{30}x_4 - a_{13}c_{23}e_{03} - a_{23}c_{03}e_{13} + a_{23}c_{13}e_{03} + a_{13}c_{03}e_{23})/(a_{23}c_{13}b_{03} + a_{13}b_{23}c_{03} - a_{13}c_{23}b_{03} - a_{23}b_{13}c_{03} - a_{03}c_{13}b_{23} + a_{03}c_{23}b_{13})$.

Finally $x_3 = g(x_4)$ leads to

$$\begin{bmatrix} f_7 := (c_{03}x_3 + d_{30}x_4 + e_{03})x_5 + a_{03}x_1 + b_{03}x_2 \\ f_8 := (c_{13}x_3 + d_{31}x_4 + e_{13})x_5 + a_{13}x_1 + b_{13}x_2 \\ f_9 := (c_{23}x_3 + d_{32}x_4 + e_{23})x_5 + a_{23}x_1 + b_{23}x_2, \end{bmatrix} \tag{12.10}$$

whose Jacobian determinant is given by

$$J_{x_3} = det \begin{bmatrix} \frac{\partial f_7}{\partial x_1} & \frac{\partial f_7}{\partial x_2} & \frac{\partial f_7}{\partial x_5} \\ \frac{\partial f_8}{\partial x_1} & \frac{\partial f_8}{\partial x_2} & \frac{\partial f_8}{\partial x_5} \\ \frac{\partial f_9}{\partial x_1} & \frac{\partial f_9}{\partial x_2} & \frac{\partial f_9}{\partial x_5} \end{bmatrix} = det \begin{bmatrix} a_{03} & b_{03} & (c_{03}x_3 + d_{30}x_4 + e_{03}) \\ a_{13} & b_{13} & (c_{13}x_3 + d_{31}x_4 + e_{13}) \\ a_{23} & b_{23} & (c_{23}x_3 + d_{32}x_4 + e_{23}) \end{bmatrix}. \tag{12.11}$$

The determinant obtained in (12.7) gives the expression for $x_3 = g(x_4)$ as

$x_3 = -(a_{23}b_{03}d_{31}x_4 + a_{03}b_{13}d_{32}x_4 + a_{03}b_{13}e_{23} - a_{23}b_{13}d_{30}x_4 - a_{03}b_{23}d_{31}x_4 - a_{03}b_{23}e_{13} + a_{13}b_{23}d_{30}x_4 - a_{13}b_{03}d_{32}x_4 - a_{13}b_{03}e_{23} - a_{23}b_{13}e_{03} + a_{23}b_{03}e_{13} + a_{13}b_{23}e_{03})/(a_{23}b_{03}c_{13} + a_{13}b_{23}c_{03} - a_{13}b_{03}c_{23} - a_{23}b_{13}c_{03} - a_{03}b_{23}c_{13} + a_{03}b_{13}c_{23})$.

On substituting the obtained expressions of $x_1 = g(x_4), x_2 = g(x_4)$ and $x_3 = g(x_4)$ in (12.3i), we obtain a quadratic function in x_4. The structure of the quadratic equation is given in [14, Box 3-12, p. 54].

12-212 Groebner basis approach

Using (4.36) on p. 45, the Groebner basis of (12.5) is computed as

$$GroebnerBasis\left[\begin{array}{c}\{a_{03}x_1+b_{03}x_2+c_{03}x_3+d_{30}x_4+e_{03},\\ a_{13}x_1+b_{13}x_2+c_{13}x_3+d_{31}x_4+e_{13},\\ a_{23}x_1+b_{23}x_2+c_3x_3+d_{32}x_4+e_{23}\},\{x_1,x_2,x_3,x_4\}\end{array}\right], \quad (12.12)$$

leading to Solution 12.1.

Solution 12.1. [Computed Groebner basis for GPS pseudo-ranging equations.]
$g_1 := (-a_{23})b_{13}e_{03}+a_{13}b_{23}e_{03}+a_{23}b_{03}e_{13}-a_{03}b_{23}e_{13}-a_{13}b_{03}e_{23}+a_{03}b_{13}e_{23}-a_{23}b_{13}c_{03}x_3+a_{13}b_{23}c_{03}x_3+a_{23}b_{03}c_{13}x_3-a_{03}b_{23}c_{13}x_3-a_{13}b_{03}c_{23}x_3+a_{03}b_{13}c_{23}x_3-a_{23}b_{13}d_{30}x_4+a_{13}b_{23}d_{30}x_4+a_{23}b_{03}d_{31}x_4-a_{03}b_{23}d_{31}x_4-a_{13}b_{03}d_{32}x_4+a_{03}b_{13}d_{32}x_4.$
$g_2 := (-a_{23})e_{13}+a_{13}e_{23}-a_{23}b_{13}x_2+a_{13}b_{23}x_2-a_{23}c_{13}x_3+a_{13}c_{23}x_3-a_{23}d_{31}x_4+a_{13}d_{32}x_4.$
$g_3 := (-a_{23})e_{03}+a_{03}e_{23}-a_{23}b_{03}x_2+a_{03}b_{23}x_2-a_{23}c_{03}x_3+a_{03}c_{23}x_3-a_{23}d_{30}x_4+a_{03}d_{32}x_4.$
$g_4 := (-a_{13})e_{03}+a_{03}e_{13}-a_{13}b_{03}x_2+a_{03}b_{13}x_2-a_{13}c_{03}x_3+a_{03}c_{13}x_3-a_{13}d_{30}x_4+a_{03}d_{31}x_4.$
$g_5 := e_{23}+a_{23}x_1+b_{23}x_2+c_{23}x_3+d_{32}x_4.$
$g_6 := e_{13}+a_{13}x_1+b_{13}x_2+c_{13}x_3+d_{31}x_4.$
$g_7 := e_{03}+a_{03}x_1+b_{03}x_2+c_{03}x_3+d_{30}x_4.$

From Solution 12.1, one notes that g_1 is a polynomial in the variables x_3 and x_4. With g_1 expressed as $x_3 = g(x_4)$, it is substituted in g_2 to obtain $x_2 = g(x_4)$, which together with $x_3 = g(x_4)$ are substituted in g_5 to give $x_1 = g(x_4)$. On substituting the obtained expressions of $x_1 = g(x_4), x_2 = g(x_4)$ and $x_3 = g(x_4)$ in (12.3i), a quadratic equation in x_4 (i.e., $h_2x_4^2 + h_1x_4 + h_o = 0$) is obtained. The coefficients are as given in [14, Box 3-14, p. 55]. The desired variables $x_1 = g(x_4)$, $x_2 = g(x_4)$ and $x_3 = g(x_4)$ could also be obtained directly using the reduced Groebner basis (4.38) on p. 45. If one desired $x_3 = g(x_4)$ for example, (12.12) could be formulated as

$$GroebnerBasis\left[\begin{array}{c}\{a_{03}x_1+b_{03}x_2+c_{03}x_3+d_{30}x_4+e_{03},\\ a_{13}x_1+b_{13}x_2+c_{13}x_3+d_{31}x_4+e_{13},\\ a_{23}x_1+b_{23}x_2+c_3x_3+d_{32}x_4+e_{23}\},\{x_1,x_2,x_3,x_4\},\\ \{x_1,x_2,x_4\}\end{array}\right], \quad (12.13)$$

giving only the value of g_1 in Solution 12.1. This is repeated for $x_1 = g(x_4)$ and $x_2 = g(x_4)$. The algorithms for solving the unknown value x_4 of the receiver range bias from the quadratic equation $\{h_2x_4^2 + h_1x_4 + h_o = 0\}$ and the respective stationary receiver coordinates are;

- Awange-Grafarend Groebner basis algorithm and,
- Awange-Grafarend Multipolynomial resultants algorithm.

They can be accessed in the GPS toolbox[2] and are discussed in detail in [24]. The distinction between the polynomial resultants method and the approach proposed by [181] is that the former does not have to invert the coefficient matrix. It instead uses the *necessary* and *sufficient* conditions requiring the determinant to vanish if the four equations have a nontrivial solution. With the coefficients h_1, h_2 and h_3, the value of x_4 could also be solved from (3.8) or (3.9) on p. 24. Let us consider the example in [181, 242].

Example 12.1 (Ranging to four satellites). From the coordinates of four GPS satellites given in Table 12.1, we apply the Awange-Grafarend algorithms listed above to compute coordinates of a stationary GPS receiver and the receiver range bias term. tiny

Table 12.1. Geocentric coordinates of <u>four</u> GPS satellites and the pseudo-range observations

i	$x^i = a_i$	$y^i = b_i$	$z^i = c_i$	d_i
0	1.483230866e+7	-2.046671589e+7	-7.42863475e+6	2.4310764064e+7
1	-1.579985405e+7	-1.330112917e+7	1.713383824e+7	2.2914600784e+7
2	1.98481891e+6	-1.186767296e+7	2.371692013e+7	2.0628809405e+7
3	-1.248027319e+7	-2.338256053e+7	3.27847268e+6	2.3422377972e+7

The computed coefficients using either of the algorithms are:

$$\begin{bmatrix} h_2 = -9.104704113943708e - 1 \\ h_1 = 5.233385578536521e7 \\ h_0 = -5.233405293375e9. \end{bmatrix}$$

Once these coefficients have been computed, the algorithms proceed to solve the roots x_4 of the quadratic equation $\{h_2 x_4^2 + h_1 x_4 + h_0 = 0\}$ giving the stationary receiver range bias term. The admissible value of the stationary receiver range bias term is then substituted in the expressions $x_1 = g(x_4), x_2 = g(x_4), x_3 = g(x_4)$ in Solution 12.1 to give the values of stationary receiver coordinates $\{x_1 = X, x_2 = Y, x_3 = Z\}$ respectively. With x_4^- =-57479918.164 m or x_4^+ =-100.0006 m, the complete pair of solutions with units in meters are

$$\begin{array}{l} X = -2892123.412, Y = 7568784.349, Z = -7209505.102 \,|\, x_4^- \\ \text{or} \quad X = 1111590.460, Y = -4348258.631, Z = 4527351.820 \,|\, x_4^+ \end{array}$$

The results indicate that the solution space is non unique. In-order to decide on the admissible solution from the pair above, we compute the *norm* (radial distance from the center of the Earth) of the positional vector $\{X, Y, Z\} \,|\, x_4^-$ and $\{X, Y, Z\} \,|\, x_4^+$ using

$$norm = \sqrt{(X^2 + Y^2 + Z^2)}.$$

If the receiver coordinates are in the global reference frame (see Sect. 10-3), the norm of the positional vector of the receiver station will approximate the value

[2]http://www.ngs.noaa.gov/gps-toolbox/awange.htm

of the Earth's radius. The norm of the other solution pair will be in space. The computed norms are

$$\begin{bmatrix} \{X,Y,Z\}\,|x_4^- = 10845636.826\,m \\ \{X,Y,Z\}\,|x_4^+ = 6374943.214\,m, \end{bmatrix}$$

thus clearly giving the second solution $\{X,Y,Z\}\,|x_4^+$ as the admissible solution of the receiver position.

12-22 Ranging to more than four GPS satellites

In Sect. 12-21, we have looked at the case where ranging can be performed to only four satellites (minimum case). In this section, we will extend the concept to the case where more than four GPS satellites are in view as is usually the case in practice. Using Gauss-Jacobi combinatorial, homotopy and ALESS approaches, it is demonstrated how one can obtain the stationary receiver position and range bias without reverting to iterative and linearization procedures such as Newton's or least squares approach.

The common features with the non-algebraic approaches in solving nonlinear problems are that they all have to do with some starting values, linearization of the observation equations and iterations as we have pointed out before. Although the issue of approximate starting values has been addressed in the works of [423, 424], the algebraic approach of Gauss-Jacobi combinatorial enjoys the advantage that all the requirements of non-algebraic approaches listed above are immaterial. The nonlinear problem is solved in an exact form with linearization permitted only during the formation of the variance-covariance matrix to generate the weight matrix of the pseudo-observations (see also [25]). The fact to note is that one has to be able to solve in a closed (exact) form nonlinear systems of equations, a condition already presented in Sect. 12-2.

Let us consider next the example of [372]. The algorithm is used to solve without linearization or iteration the overdetermined pseudo-range problem. The results are then compared to those of linearized least squares solutions.

Example 12.2 (Ranging to more than four satellites). Pseudo-ranges d_i are measured to six satellites whose coordinates $\{x^i, y^i, z^i\}$ are given in *Table* 12.2. From the data in *Table* 12.2 and using (7.28) on p. 93, 15 possible combinations listed in Table 12.3 are obtained. The Position Dilution of Precision (PDOP) are computed as suggested in [227] and presented in *Table* 12.3. From the computed PDOP, it is noticed that the 10th combination had a poor geometry, a fact validated by the plot of the PDOP values versus the combination numbers in Fig. 12.4. Using Gauss-Jacobi combinatorial algorithm, this weaker geometry is accounted for during the adjustment process. Variance-covariance matrix computed through nonlinear error propagation for that respective set is used. Groebner basis or polynomial resultants are used as computing engine (see Fig. 7.5 on p. 98) to compute the minimal combinatorial set as discussed in Sect. 7-331. The computed coefficients are presented in *Table* 12.4.

From the computed coefficients in *Table* 12.4, the 10th combination is once again identified as having significantly different values from the rest. This fact highlights the power of the Gauss-Jacobi combinatorial algorithm in identifying poor geometry. Using the coefficients of *Table* 12.4, the solution of receiver position $\{X, Y, Z\}$ and the range bias $\{cdt\}$ for each minimal combinatorial set is carried out as discussed in Sect. 12-2. The results are presented in *Table* 12.5. The final adjusted position is obtained using linear Gauss-Markov model (7.12) on p. 87. The random pseudo-observation values of *Table* 12.5 are placed in the vector of observation $\mathbf{y}$ and the dispersion matrix $\mathbf{\Sigma}$ obtained by nonlinear error propagation using (7.33) on p. 94. The coefficients of the unknowns $\{X, Y, Z, cdt\}$ form the design matrix $\mathbf{A}$. The dispersion of the estimated parameters are then obtained from (7.13).

Figure 12.9 gives the plot of the scatter of the 15 combinatorial solutions (shown by points) around the adjusted value (indicated by a star). Figure 12.10 is a magnification of Fig. 12.9 for the scatter of 14 solutions (shown by points) that are very close to the adjusted value (indicated by a star). The outlying point in Fig. 12.9 is ignored.

Table 12.2. Geocentric coordinates of *six* GPS satellites and pseudo-range observations

PRN	$x^i = a_i$	$y^i = b_i$	$z^i = c_i$	d_i
23	14177553.47	-18814768.09	12243866.38	21119278.32
9	15097199.81	-4636088.67	21326706.55	22527064.18
5	23460342.33	-9433518.58	8174941.25	23674159.88
1	-8206488.95	-18217989.14	17605231.99	20951647.38
21	1399988.07	-17563734.90	19705591.18	20155401.42
17	6995655.48	-23537808.26	-9927906.48	24222110.91

Table 12.3. Possible combinations and the computed PDOP

Combination Number	Combination	Computed PDOP
1	23-9-5-1	4.8
2	23-9-5-21	8.6
3	23-9-5-17	4.0
4	23-9-1-21	6.5
5	23-9-1-17	3.3
6	23-9-21-17	3.6
7	23-5-1-21	6.6
8	23-5-1-17	6.6
9	23-5-21-17	4.8
10	23-1-21-17	137.8
11	9-5-1-21	5.6
12	9-5-1-17	14.0
13	9-5-21-17	6.6
14	9-1-21-17	5.2
15	5-1-21-17	6.6

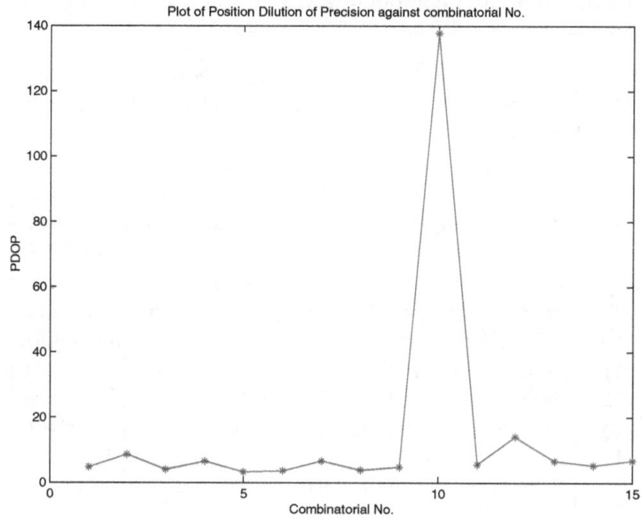

Figure 12.4. A plot of PDOP for respective combinations

Table 12.4. Computed coefficients of the combinations

C/No.	c_2	c_1	c_0
1	-0.914220949236445	52374122.9848733	49022682.3125
2	-0.934176403102736	50396827.4998945	7915541824.84375
3	-0.921130625833683	51741826.0147786	343282824.25
4	-0.865060899130107	54950460.2842167	-10201105114.5
5	-0.922335616484969	51877166.0451888	280298481.625
6	-0.919296962706157	51562232.9601199	1354267366.4375
7	-0.894980063579044	53302005.6927825	-3642644147.5625
8	-0.917233949644576	52194946.1124139	132408747.46875
9	-0.925853049262193	51140847.6331213	3726719112.1875
10	3369.83293928593	-1792713339.80277	6251615074927.06
11	-0.877892756651551	54023883.5656926	-6514735288.13762
12	-0.942581538318523	50793361.5303674	784684294.241371
13	-0.908215141659006	52246642.0794924	-2499054749.05572
14	-0.883364070549387	53566554.3869961	-5481411035.37882
15	-0.866750765656126	54380648.2092251	-7320871488.80859

Table 12.5. Computed combinatorial solution points in a polyhedron

C/No.	$X(m)$	$Y(m)$	$Z(m)$	$cdt(m)$
1	596925.3485	-4847817.3618	4088206.7822	-0.9360
2	596790.3124	-4847765.7637	4088115.7092	-157.0638
3	596920.4198	-4847815.4785	4088203.4581	-6.6345
4	596972.8261	-4847933.4365	4088412.0909	185.6424
5	596924.2118	-4847814.5827	4088201.8667	-5.4031
6	596859.9715	-4847829.7585	4088228.8277	-26.2647
7	596973.5779	-4847762.4719	4088399.8670	68.3398
8	596924.2341	-4847818.6302	4088202.3205	-2.5368
9	596858.7650	-4847764.5341	4088221.8468	-72.8716
10	596951.5275	-4852779.5675	4088758.6420	3510.4002
11	597004.7562	-4847965.2225	4088300.6135	120.5901
12	596915.8657	-4847799.7045	4088195.5770	-15.4486
13	596948.5619	-4847912.9549	4088252.1599	47.8319
14	597013.7194	-4847974.1452	4088269.3206	102.3292
15	597013.1300	-4848019.6766	4088273.9565	134.6230

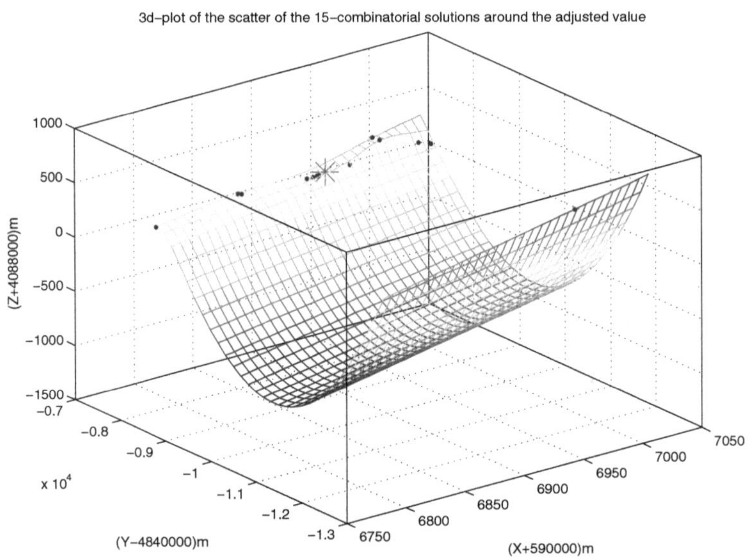

Figure 12.5. Scatter of the *15* combinatorial solutions (•) around the adjusted value (⋆)

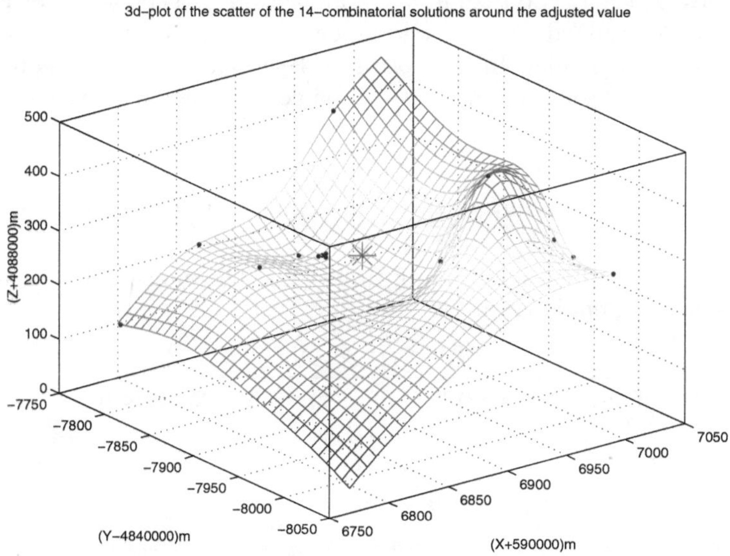

Figure 12.6. Magnification of the scatter of *14* solutions (•) around the adjusted value (⋆) of Fig. 12.9

12-221 Extended Newton-Raphson solution

The distance of the receiver from the i-th satellite, the pseudo- range observation, d_i is related to the unknown position of the receiver, $\{x_1, x_2, x_3\}$,

$$d_i = \sqrt{(x_1 - a_i)^2 + (x_2 - b_i)^2 + (x_3 - c_i)^2} + x_4 \tag{12.14}$$

where $\{a_i, b_i, c_i\}, i = 1, \ldots, m$ are the coordinates of the i-th satellite. The distance is influenced also by the satellite and receiver' clock biases. The satellite clock biases can be modelled while the receiver' clock biases have to be considered as an unknown variable, x_4. This means, we have four unknowns, consequently we need four satellite signals as minimum observation. The general form of the equation for the i-th satellite is

$$f_i = (x_1 - a_i)^2 + (x_2 - b_i)^2 + (x_3 - c_i)^2 - (x_4 - d_i)^2 \tag{12.15}$$

The residual of this type of equation represents the error implicitly. However in geodesy the explicit distance error definition is usual, namely,

$$g_i = d_i - \sqrt{(x_1 - a_i)^2 + (x_2 - b_i)^2 + (x_3 - c_i)^2} - x_4 \tag{12.16}$$

The relation between the two expressions,

$$g_i = d_i - \sqrt{f_i + (x_4 - d_i)^2} - x_4 \tag{12.17}$$

which implies that if $f_i = 0$ then $g_i = 0$ and vice versa. Therefore, in case of *four* observations, determined system, we employ the first expression, which is easy to handle as a polynomial. In case of $m > 4$ satellites, the two representations the pseudo- range observation models will be not equivalent in least square sense, namely

$$\min_{x_1,x_2,x_3,x_4} \sum_{i=1}^{m} f_i \neq \min_{x_1,x_2,x_3,x_4} \sum_{i=1}^{m} g_i^2 \tag{12.18}$$

Let us consider $m = 6$ satellites of Table 12.2. The numerical equations of the overdetermined system is given by

$$g_0 = 2.11193 \times 10^7 - \sqrt{(-1.41776 \times 10^7 + x_1)^2 + (1.88148 \times 10^7 + x_2)^2 + (-1.22439 \times 10^7 + x_3)^2} - x_4 = 0$$

$$g_1 = 2.25271 \times 10^7 - \sqrt{(-1.50972 \times 10^7 + x_1)^2 + (4.63609 \times 10^6 + x_2)^2 + (-2.13267 \times 10^7 + x_3)^2} - x_4 = 0$$

$$g_2 = 2.36742 \times 10^7 - \sqrt{(-2.34603 \times 10^7 + x_1)^2 + (9.43352 \times 10^6 + x_2)^2 + (-8.17494 \times 10^6 + x_3)^2} - x_4 = 0$$

$$g_3 = 2.09516 \times 10^7 - \sqrt{(8.20649 \times 10^6 + x_1)^2 + (1.8218 \times 10^7 + x_2)^2 + (-1.76052 \times 10^7 + x_3)^2} - x_4 = 0$$

$$g_4 = 2.01554 \times 10^7 - \sqrt{(-1.39999 \times 10^6 + x_1)^2 + (1.75637 \times 10^7 + x_2)^2 + (-1.97056 \times 10^7 + x_3)^2} - x_4 = 0$$

$$g_5 = 2.42221 \times 10^7 - \sqrt{(-6.99566 \times 10^6 + x_1)^2 + (2.35378 \times 10^7 + x_2)^2 + (9.92791 \times 10^6 + x_3)^2} - x_4 = 0 \tag{12.19}$$

Now, we solve the original system employing one of the solutions of the Gauss-Jacobi subset solution as initial guess. Let us use the worst one, the 10th solution in Table 12.5, i.e.,

$$\begin{aligned} x_1 &= 596951.52753, \\ x_2 &= -4.8527795710 \times 10^6, \\ x_3 &= 4.08875864269 \times 10^6, \\ x_4 &= 3510.4002370. \end{aligned} \tag{12.20}$$

Employing *Extended Newton-Raphson* method, the convergence is very fast. Fig. 12.7 shows the absolute error of x_1 in meter as function of the number of iterations.

The solution is

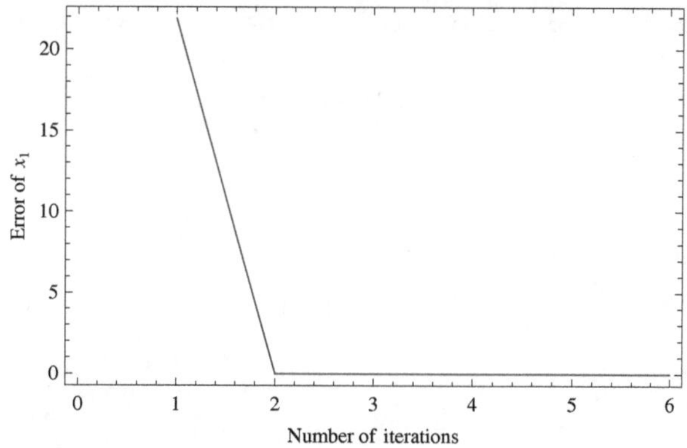

Figure 12.7. Convergence of the method in case of the x_1 coordinate

$$\begin{aligned} x_1 &= 596929.65349, \\ x_2 &= -4.8478515526 \times 10^6, \\ x_3 &= 4.0882267957 \times 10^6, \\ x_4 &= 15.518050625. \end{aligned}$$

12-222 Homotopy solution of GPS N-point problem

In GPS positioning pseudoranges (d) are measured, so we should minimize the error (g) of these measurements (see Eq. (12.16)). Considering the least squares method, we should minimize the following objective function

$$W(x_1, x_2, x_3, x_4) = \sum_{i=1}^{n} g_i^2. \tag{12.21}$$

This minimization problem can be transformed into a square (determined) set of nonlinear equations

$$F_i = \frac{\partial W}{\partial x_i} \tag{12.22}$$

namely

$$F_1 = \sum_{i=1}^{n} \frac{2(x_1 - a_i)\left(x_4 + \sqrt{(x_1 - a_i)^2 + (x_2 - b_i)^2 + (x_3 - c_i)^2} - d_i\right)}{\sqrt{(x_1 - a_i)^2 + (x_2 - b_i)^2 + (x_3 - c_i)^2}} \tag{12.23}$$

$$F_2 = \sum_{i=1}^{n} \frac{2(x_2 - b_i)\left(x_4 + \sqrt{(x_1 - a_i)^2 + (x_2 - b_i)^2 + (x_3 - c_i)^2} - d_i\right)}{\sqrt{(x_1 - a_i)^2 + (x_2 - b_i)^2 + (x_3 - c_i)^2}} \tag{12.24}$$

$$F_3 = \sum_{i=1}^{n} \frac{2(x_3 - c_i)\left(x_4 + \sqrt{(x_1 - a_i)^2 + (x_2 - b_i)^2 + (x_3 - c_i)^2} - d_i\right)}{\sqrt{(x_1 - a_i)^2 + (x_2 - b_i)^2 + (x_3 - c_i)^2}} \tag{12.25}$$

$$F_4 = \sum_{i=1}^{n} 2\left(x_4 + \sqrt{(x_1 - a_i)^2 + (x_2 - b_i)^2 + (x_3 - c_i)^2} - d_i\right) \tag{12.26}$$

For numerical illustration, GPS satellites positions and pseudo-range observations to six satellites in Table 12.2 are used [41]. To solve this nonlinear system with a local method, we need proper initial values. Let us consider the 4 point solution of satellites (23-1-21-17), computed with symbolic-numeric method, as initial values, (i.e., the 10th solution in Table 12.5) in Table 12.6.

Table 12.6. Solution of GPS 4 point problem computed by symbolic-numeric method

$\tilde{x_1} =$	596 951.528
$\tilde{x_2} =$	-4 852 779.568
$\tilde{x_3} =$	4 088 758.642
$\tilde{x_4} =$	3 510.400

These four satellites have a very poor geometry with a Position Dilution of Precision (PDOP) value of 137.8 (a good geometry is indicated by a PDOP value of less than 6). Unfortunately standard Newton-Raphson did not work with these initial values, the line search decreased the specified step size and was unable to find a sufficient decrease in the merit function thereby failing to converge.

In order to ensure convergency for this initial values, we can employ fixed point homotopy, see Section 6-421. Now the start system is

$$\boldsymbol{G} = \begin{pmatrix} x_1 \\ x_2 \\ x_3 \\ x_4 \end{pmatrix} - \begin{pmatrix} \tilde{x_1} \\ \tilde{x_2} \\ \tilde{x_3} \\ \tilde{x_4} \end{pmatrix} \tag{12.27}$$

Then to avoid singularity of the homotopy function, (see [320]), let

$$\gamma = i\{1, 1, 1, 1\}; \tag{12.28}$$

The homotopy function is the linear combination of the target system, Eq.(12.23) - Eq.(12.26) and the start system, Eq.(12.27),

$$H(x_1, x_2, x_3, x_4, \lambda) = \gamma(1 - \lambda)\begin{pmatrix} G_1 \\ G_2 \\ G_3 \\ G_4 \end{pmatrix} + \lambda \begin{pmatrix} F_1 \\ F_2 \\ F_3 \\ F_4 \end{pmatrix} \tag{12.29}$$

and the corresponding system of the differential equations

$$\frac{d}{d\lambda}\begin{pmatrix} x_1(\lambda) \\ x_2(\lambda) \\ x_3(\lambda) \\ x_4(\lambda) \end{pmatrix} = -{H_x}^{-1} H_\lambda \tag{12.30}$$

where the Jacobian

$$(H_x)_{i,j} = \gamma(1-\lambda)\left(\frac{\partial G_i}{\partial x_j}\right)_{i,j} + \lambda\left(\frac{\partial F_i}{\partial x_j}\right)_{i,j} \tag{12.31}$$

and

$$(H_\lambda)_i = F_i\left(x_1, x_2, x_3, x_4\right) - \gamma G_i\left(x_1, x_2, x_3, x_4\right) \tag{12.32}$$

Now employing path tracing by integration with computing inverse, we get 3 solutions, but only one of them is the physically acceptable solution, see [41, 318]. Table 12.7. shows the values of this solution.

Table 12.7. Solution of the GPS N-point problem

$x_1 =$	596 929.653
$x_2 =$	-4 847 851.552
$x_3 =$	4 088 226.795
$x_4 =$	15.518

The trajectories belonging to this solution can be seen in Fig. 12.8.

12-223 Least squares versus Gauss-Jacobi combinatorial

Let us now compare the least squares solution and the Gauss-Jacobi combinatorial approach. Using the combinatorial approach, the stationary receiver position and range bias are computed as discussed in Sect. 12-22. For the least squares approach, the nonlinear observation equations (12.2) are first linearized using Taylor series expansion for the 6 satellites in Table 12.2. This linearization process generates the Jacobi matrix required by the approach. After linearization, the desired values are estimated iteratively using linear models. As approximate starting values for the iterations, let us assign the stationary receiver position and the stationary receiver range bias zero values. Let us also set a convergence limit of 1×10^{-8}, as the difference between values of two successive iterations. With these settings, 6 iterations are required for the threshold condition above to be achieved. In the second case, the values of the combinatorial algorithm are used as approximate starting values for least squares solution. This time round, only two iterations were required to achieve convergence. For users who prefer least squares approach, Gauss-Jacobi combinatorial algorithm can therefore be used to offer quick approximate starting values that lead to faster convergence.

From (12.2) on p. 176, and the results of both procedures, residuals are computed, squared and used to compute the error norm from

$$norm = \sqrt{\left\{\sum_{i=1}^{6}\left(d_i - [\sqrt{(\hat{X}-a_i)^2 + (\hat{Y}-b_i)^2 + (\hat{Z}-c_i)^2} - \hat{x}_4]\right)^2\right\}}. \tag{12.33}$$

In (12.33), $\{\hat{X}, \hat{Y}, \hat{Z}, \hat{x}_4\}$ are the computed values of the stationary receiver position and range bias. The entities $\{a_i, b_i, c_i\} \mid \forall i = \{1, \ldots, 6\}$ are the coordinates of the six satellites in Table 12.2 and $\{d_i\} \mid \forall i = \{1, \ldots, 6\}$ the measured pseudo-ranges.

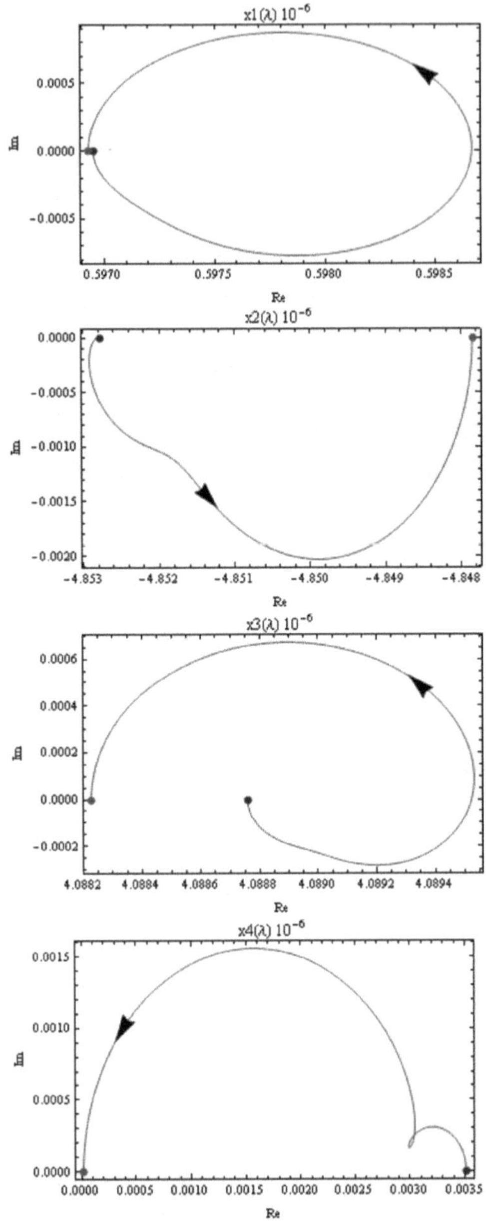

Figure 12.8. Homotopy solution paths of the GPS positioning problem

Table 12.8 compares the results from the Gauss-Jacobi combinatorial algorithm and those obtained from least squares approach. *Table* 12.9 presents the *root-mean-square-errors*. In Table 12.10, we present the computed residuals, their sum of squares and the computed error norm from (12.33). The computed error norm are identical for both procedures. Further comparison of the two procedures will be given in Chap. 17 where they are used to compute the 7-parameter datum transformation problem.

Table 12.8. Computed stationary receiver position and range bias

	$X\,(m)$	$Y\,(m)$	$Z\,(m)$	$cdt(m)$
Combinatorial approach	596929.6542	-4847851.5021	4088226.7858	-15.5098
Least squares	596929.6535	-4847851.5526	4088226.7957	-15.5181
Difference	0.0007	0.0505	-0.0098	0.0083

Table 12.9. Computed root-mean-square errors

	$\sigma_X\,(m)$	$\sigma_Y\,(m)$	$\sigma_Z\,(m)$	$\sigma_{cdt}\,(m)$
Combinatorial Approach	6.4968	11.0141	5.4789	8.8071
Least Squares	34.3769	58.2787	28.9909	46.6018

Table 12.10. Computed residuals, squares of residuals and error norm

PRN	Combinatorial approach (m)	Least squares (m)
23	-16.6260	-16.6545
9	-1.3122	-1.3106
5	2.2215	2.2189
1	-16.4369	-16.4675
21	26.8623	26.8311
17	5.4074	5.3825
Sum of squares	1304.0713	1304.0680
Error norm	36.1119	36.1119

12-3 Ranging by local positioning systems (LPS)

As opposed to GPS ranging where the targets being observed are satellites in space and in motion, Local Positioning Systems' targets are fixed on the surface of the Earth as illustrated in Fig. 12.1 on p. 174. We present both planar and three-dimensional ranging within the LPS system. Planar ranging can be used for quick point search during engineering and cadastral surveying.

12-31 Planar ranging

12-311 Conventional approach

Consider two distances $\{S_1, S_2\}$ measured from an unknown station $P_0 \in \mathbb{E}^2$ to two known stations $P_1 \in \mathbb{E}^2$ and $P_2 \in \mathbb{E}^2$ as shown in Fig. 4.1 in p.34. The two dimensional distance ranging problem involves the determination of the planar coordinates $\{X_0, Y_0\}_{P0}$ of the unknown station $P_0 \in \mathbb{E}^2$ given;

- the observed distances $\{S_1, S_2\}$,
- the planar coordinates $\{X_1, Y_1\}_{P_1}$ of station $P_1 \in \mathbb{E}^2$ and $\{X_2, Y_2\}_{P_2}$ of stations $P_2 \in \mathbb{E}^2$.

The nonlinear distance equations relating the given values above with the coordinates of unknown station are expressed (see e.g., (4.1) and (4.2) on p. 33) as

$$\begin{bmatrix} (X_1 - X_0)^2 + (Y_1 - Y_0)^2 = S_1^2 \\ (X_2 - X_0)^2 + (Y_2 - Y_0)^2 = S_2^2, \end{bmatrix} \tag{12.34}$$

which on expanding leads to

$$\begin{bmatrix} X_1^2 + Y_1^2 - 2X_1X_0 - 2Y_1Y_0 + X_0^2 + Y_0^2 = S_1^2 \\ \\ X_2^2 + Y_2^2 - 2X_2X_0 - 2Y_2Y_0 + X_0^2 + Y_0^2 = S_2^2. \end{bmatrix} \tag{12.35}$$

The conventional analytic approach solves (12.35) by subtracting the first expression, i.e., (12.35i) from the second one, and expressing one unknown in terms of the other. This leads to

$$Y_0 = -\left\{\frac{X_1 - X_2}{Y_1 - Y_2}\right\} X_0 + \frac{S_2^2 - S_1^2 + X_1^2 - X_2^2 + Y_1^2 - Y_2^2}{2(Y_1 - Y_2)}, \tag{12.36}$$

which is substituted for Y_0 in the first expression of (12.35) to give

$$\begin{bmatrix} X_1^2 + Y_1^2 - 2X_1X_0 - 2Y_1\left\{-\left\{\frac{X_1 - X_2}{Y_1 - Y_2}\right\} X_0 + \frac{S_2^2 - S_1^2 + X_1^2 - X_2^2 + Y_1^2 - Y_2^2}{2(Y_1 - Y_2)}\right\} \\ \\ +X_0^2 + \left\{-\left\{\frac{X_1 - X_2}{Y_1 - Y_2}\right\} X_0 + \frac{S_2^2 - S_1^2 + X_1^2 - X_2^2 + Y_1^2 - Y_2^2}{2(Y_1 - Y_2)}\right\}^2 - S_1^2 = 0. \end{bmatrix} \tag{12.37}$$

On expanding and factorizing (12.37) leads to

$$\left[\begin{array}{c} (1+a^2)X_0^2 + (2ab - 2X_1 - 2Y_1 a)X_0 + b^2 - 2Y_1 b + X_1^2 + Y_1^2 - S_1^2 = 0, \\ \text{with} \\ a = -\left\{\dfrac{X_1 - X_2}{Y_1 - Y_2}\right\} \\ \text{and} \\ b = \dfrac{S_2^2 - S_1^2 + X_1^2 - X_2^2 + Y_1^2 - Y_2^2}{2(Y_1 - Y_2)}. \end{array}\right. \tag{12.38}$$

The quadratic equation (12.38) is solved for X_0 using the quadratic formulae (3.8) or (3.9) on p. 24 and substituted back in (12.36) to give the values of Y_0.

12-312 Sylvester resultants approach

Whereas the conventional analytical approach presented above involves differencing, e.g., (12.36), and substitution as in (12.37), the Sylvester resultants technique discussed in Sect. 5-2 solves (12.35) directly. In-order to achieve this, (12.35) is first expressed in algebraic form as

$$\left[\begin{array}{l} g_1 := X_1^2 + Y_1^2 - 2X_1X_0 - 2Y_1Y_0 + X_0^2 + Y_0^2 - S_1^2 = 0 \\ g_2 := X_2^2 + Y_2^2 - 2X_2X_0 - 2Y_2Y_0 + X_0^2 + Y_0^2 - S_2^2 = 0. \end{array}\right. \tag{12.39}$$

Next, the *hide variable* technique is applied. By hiding the variable Y_0 (i.e., considering it as a constant), the coefficient matrix of the variable X_0 is formed as shown in Example 5.2 on p. 50. In (12.39), we note that the polynomials g_1 and g_2 are both of degree 2 and thus both i and j (e.g., (5.1) on p. 49) are equal to 2 resulting into a (4×4) matrix. The coefficient matrix of the variable X_0 formed by hiding the variable Y_0 (i.e., considering the coefficients of X_0 to be polynomials in Y_0) is

$$\mathbf{A}_X = \begin{bmatrix} 1 & -2X_1 & (Y_0^2 - 2Y_1Y_0 + X_1^2 + Y_1^2 - S_1^2) & 0 \\ 0 & 1 & -2X_1 & (Y_0^2 - 2Y_1Y_0 + X_1^2 + Y_1^2 - S_1^2) \\ 1 & -2X_2 & (Y_0^2 - 2Y_2Y_0 + X_2^2 + Y_2^2 - S_2^2) & 0 \\ 0 & 1 & -2X_2 & (Y_0^2 - 2Y_2Y_0 + X_2^2 + Y_2^2 - S_2^2) \end{bmatrix}, \tag{12.40}$$

while that of the variable Y_0 formed by hiding X_0 (i.e., considering the coefficients of Y_0 to be polynomials in X_0) is

$$
\mathbf{A}_Y = \begin{bmatrix} 1 & -2Y_1 & (X_0^2 - 2X_1X_0 + X_1^2 + Y_1^2 - S_1^2) & 0 \\ 0 & 1 & -2Y_1 & (X_0^2 - 2X_1X_0 + X_1^2 + Y_1^2 - S_1^2) \\ 1 & -2Y_2 & (X_0^2 - 2X_2X_0 + X_2^2 + Y_2^2 - S_2^2) & 0 \\ 0 & 1 & -2Y_2 & (X_0^2 - 2X_2X_0 + X_2^2 + Y_2^2 - S_2^2) \end{bmatrix}. \tag{12.41}
$$

Sylvester resultants are now obtained from the determinants of the coefficient matrices (12.40) and (12.41) respectively as

$$
\begin{bmatrix} Res(g_1, g_2, X_0) = det(\mathbf{A}_X) \\ Res(g_1, g_2, Y_0) = det(\mathbf{A}_Y), \end{bmatrix} \tag{12.42}
$$

where $Res(g_1, g_2, X_0)$ and $Res(g_1, g_2, Y_0)$ are the Sylvester resultants of algebraic equations in (12.39), with respect to the variables X_0 and Y_0 as in (12.40) and (12.41) respectively. From (12.42) we obtain two quadratic equations (12.43) for solving the variables X_0 and Y_0 which are the planar coordinates of the unknown station P_0. The coefficients of the quadratic equations are given in Solution 12.2.

Solution 12.2 (Sylvester resultants solution of planar coordinates $\{X_0, Y_0\}$).

$$
\begin{array}{l} a_2Y_0^2 + a_1Y_0 + a_0 = 0 \\ b_2X_0^2 + b_1X_0 + b_0 = 0 \end{array} \tag{12.43}
$$

with the coefficients: $a_2 = (4Y_2^2 + 4X_1^2 - 8Y_1Y_2 + 4X_2^2 + 4Y_1^2 - 8X_2X_1)$.

$a_1 = (-4X_2^2Y_1 - 4S_1^2Y_2 - 4X_1^2Y_2 + 8X_1X_2Y_1 + 4Y_1S_1^2 + 4Y_1Y_2^2 + 8X_2X_1Y_2 - 4Y_2^3 + 4Y_1^2Y_2 - 4Y_2X_2^2 - 4Y_1S_2^2 - 4Y_1^3 - 4Y_1X_1^2 + 4Y_2S_2^2)$.

$a_0 = (X_2^4 + Y_2^4 + S_2^4 - 4X_2X_1Y_2^2 + 4X_2X_1S_2^2 - 4X_1X_2Y_1^2 + 4X_1X_2S_1^2 + 2X_2^2Y_2^2 - 2X_2^2S_2^2 - 2Y_2^2S_2^2 - 4X_1X_2^3 + 6X_2^2X_1^2 + 2X_2^2Y_1^2 - 2X_2^2S_1^2 + 2X_1^2Y_2^2 - 2X_1^2S_2^2 - 2Y_1^2Y_2^2 + 2Y_1^2S_2^2 + 2S_1^2Y_2^2 - 2S_1^2S_2^2 - 4X_2X_1^3 + 2X_1^2Y_1^2 - 2X_1^2S_1^2 - 2Y_1^2S_1^2 + X_1^4 + Y_1^4 + S_1^4)$.

$b_2 = (-8Y_1Y_2 + 4X_1^2 + 4Y_2^2 - 8X_2X_1 + 4X_2^2 + 4Y_1^2)$.

$b_1 = (-4X_1^3 + 4X_2S_2^2 + 8Y_2Y_1X_2 - 4X_2Y_2^2 - 4X_1Y_2^2 - 4X_1S_2^2 + 4X_1X_2^2 - 4X_2S_1^2 + 8Y_1Y_2X_1 - 4X_2Y_1^2 + 4X_2X_1^2 - 4X_1Y_1^2 + 4X_1S_1^2 - 4X_2^3)$.

$b_0 = (4Y_2Y_1S_2^2 - 4Y_2X_2^2Y_1 - 2Y_1^2S_1^2 + X_2^4 + Y_2^4 + S_2^4 + X_1^4 + Y_1^4 + S_1^4 - 4Y_1Y_2^3 + 4Y_1S_1^2Y_2 - 4Y_1X_1^2Y_2 + 2X_2^2Y_2^2 - 2X_2^2S_2^2 - 2Y_2^2S_2^2 - 2X_2^2X_1^2 + 2X_2^2Y_1^2 + 2X_2^2S_1^2 + 2X_1^2Y_2^2 + 2X_1^2S_2^2 + 6Y_1^2Y_2^2 - 2Y_1^2S_2^2 - 2S_1^2Y_2^2 - 2S_1^2S_2^2 + 2X_1^2Y_1^2 - 2X_1^2S_1^2 - 4Y_1^3Y_2)$.

12-313 Reduced Groebner basis approach

Reduced Groebner basis (4.38) on p. 45 solves (12.39) directly through

$$\begin{bmatrix} GroebnerBasis[\{g_1, g_2\}, \{X_0, Y_0\}, \{X_0\}] \\ GroebnerBasis[\{g_1, g_2\}, \{X_0, Y_0\}, \{Y_0\}]. \end{bmatrix} \tag{12.44}$$

The first expression of (12.44) ensures that one gets a quadratic equation only in Y_0 with X_0 eliminated, while the second expression ensures a quadratic equation only in X_0 with Y_0 eliminated. Solution 12.3 presents the results of (12.44).

Solution 12.3 (reduced Groebner basis solution of planar coordinates $\{X_0, Y_0\}$).

$$\begin{array}{|c|} e_2 Y_0^2 + e_1 Y_0 + e_0 = 0 \\ \\ f_2 X_0^2 + f_1 X_0 + f_0 = 0 \end{array} \tag{12.45}$$

with the coefficients: $e_2 = (4X_1^2 - 8X_1X_2 - 8Y_1Y_2 + 4X_2^2 + 4Y_2^2 + 4Y_1^2)$.

$e_1 = (-4X_1^2Y_1 + 4S_1^2Y_1 - 4Y_1^3 - 4X_2^2Y_2 - 4X_1^2Y_2 + 4Y_1^2Y_2 + 4Y_1Y_2^2 - 4Y_2^3 + 4S_2^2Y_2 - 4S_2^2Y_1 - 4S_1^2Y_2 - 4X_2^2Y_1 + 8X_1X_2Y_1 + 8X_1X_2Y_2)$.

$e_0 = (S_2^4 + 2X_1^2Y_2^2 + 4S_1^2X_1X_2 + 4S_2^2X_1X_2 - 2S_2^2X_2^2 - 2Y_1^2Y_2^2 + S_1^4 - 2S_1^2X_1^2 + X_2^4 + 2S_1^2Y_2^2 - 2S_1^2S_2^2 + X_1^4 + Y_1^4 + 2X_2^2Y_1^2 - 4X_1X_2Y_2^2 + 6X_1^2X_2^2 - 2S_2^2X_1^2 - 2S_2^2Y_2^2 + 2X_1^2Y_1^2 - 2S_1^2X_2^2 + 2S_2^2Y_1^2 - 2S_1^2Y_1^2 + 2X_2^2Y_2^2 - 4X_1X_2^3 - 4X_1X_2Y_1^2 + Y_2^4 - 4X_1^3X_2)$.

$f_2 = (4X_1^2 - 8X_1X_2 - 8Y_1Y_2 + 4X_2^2 + 4Y_2^2 + 4Y_1^2)$.

$f_1 = (-4X_2Y_1^2 - 4X_1Y_1^2 + 4X_1^2X_2 - 4S_2^2X_1 - 4X_2Y_2^2 - 4X_1^3 + 8X_1Y_1Y_2 + 4S_1^2X_1 + 8X_2Y_1Y_2 + 4X_1X_2^2 - 4X_2^3 - 4X_1Y_2^2 + 4S_2^2X_2 - 4S_1^2X_2)$.

$f_0 = (S_2^4 + 2X_1^2Y_2^2 - 4X_2^2Y_1Y_2 - 2S_2^2X_2^2 + 6Y_1^2Y_2^2 + 4S_1^2Y_1Y_2 + S_1^4 - 4X_1^2Y_1Y_2 - 2S_1^2X_1^2 + ...X_2^4 - 2S_1^2Y_2^2 - 2S_1^2S_2^2 + 4S_2^2Y_1Y_2 + X_1^4 + Y_1^4 + 2X_2^2Y_1^2 - 2X_1^2X_2^2 + 2S_2^2X_1^2 - 4Y_1^3Y_2 - 2S_2^2Y_2^2 + 2X_1^2Y_1^2 + 2S_1^2X_2^2 - 2S_2^2Y_1^2 - 2S_1^2Y_1^2 + 2X_2^2Y_2^2 - 4Y_1Y_2^3 + Y_2^4)$.

With the given values of known stations and measured distances as listed on p. 192, all that is required of the practitioner, therefore, is to compute the coefficients $\{a_2, a_1, a_0, b_2, b_1, b_0\}$ using the Sylvester resultants Solution 12.2 or $\{e_2, e_1, e_0, f_2, f_1, f_0\}$ using the reduced Groebner basis Solution 12.3. Once the coefficients have been computed, the Matlab, Maple, or Mathematica's roots command can be applied to solve the univariate polynomials (12.43) or (12.45) for the position of the unknown station. The admissible position from the computed pair of solution is chosen with the help of prior information e.g., from existing maps. Although the elimination procedure discussed above work, it is also possible to solve the problem with other *direct elimination* methods, e.g., Groebner basis incorporated into "*NSolve*" or the "*solve*" command in Matlab's symbolic package. Among advantages of using direct elimination in-built functions such as NSolve are; they remove the issue of figuring out which univariate solutions in x correspond to which in y (in higher dimensions, or when degree is large, removing this problem saves considerable work), and they also removes room for mistakes (Lichtblau, Priv. Comm.). We therefore recommend that where possible, users should directly apply these functions depending on their operating platforms (Mathematica, Maple or Matlab).

Example 12.3 (Ranging to two known planar stations). Consider the Example of [237, p. 240] where two distances $\{S_1 = 294.330\,m,\ S_2 = 506.420\,m\}$ have been measured from an unknown station $P_0 \in \mathbb{E}^2$ to two known stations $P_1 \in \mathbb{E}^2$ and $P_2 \in \mathbb{E}^2$ (e.g., Fig. 4.1 in p.34). The Cartesian planar coordinates of the two known stations P_1 and P_2 are given as $\{X_1 = 328.760\,m,\ Y_1 = 1207.850\,m\}_{P_1}$ and $\{X_2 = 925.040\,m,\ Y_2 = 954.330\,m\}_{P_2}$ respectively. The planar ranging problem now involves determining the planar coordinates $\{X_0, Y_0\}_{p_0}$ of the unknown station $P_0 \in \mathbb{E}^2$. Using the given values of known stations and measured distances in either Solution 12.2 or 12.3, the coefficients $\{a_2, a_1, a_0, b_2, b_1, b_0\}$ of the quadratic equation (12.43) or $\{e_2, e_1, e_0, f_2, f_1, f_0\}$ of (12.45) are computed. Using these coefficients and applying Matlab's roots command (see e.g., (4.42) in Example 4.11 on p. 45) leads to

$$X_0 = \{1336.940,\ 927.797\}\,m$$
$$Y_0 = \{593.271,\ 419.316\}\,m$$

In a four step procedure, [237, p. 240] obtained the values $\{X_0(m) = 927.90\}$ and $\{Y_0(m) = 419.42\}$. The algebraic approaches are however direct and fast (i.e., avoids forward and backwards substitutions).

Geometrically, the algebraic curves given by (12.34) would result in a conic intersection of two circles with the centers $\{X_1, Y_1\}$ and $\{X_2, Y_2\}$ and radiuses S_1 and S_2 respectively. The applied polynomial approaches decompose these complicated geometries to those of Figs. 12.9 and 12.10 which represent univariate polynomials and are simpler to solve. Figs. 12.9 and 12.10 indicate the solutions of (12.45) for the Example presented above. The intersection of the quadratic curves with the zero line are the solution points. In Solution 12.4, we present the critical configuration of the planar ranging problem. The computed determinants, (12.51) and (12.52) indicate the critical configuration (where solution ceases to exist) to be cases when points $P_0(X,Y)$, $P_1(X_1,Y_1)$ and $P_2(X_2,Y_2)$ all lie on a straight line with gradient $-\frac{c}{b}$ and intercept $-\frac{a}{b}$.

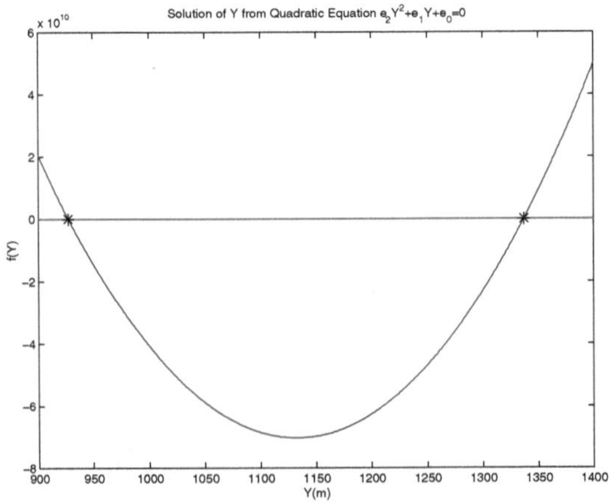

Figure 12.9. Solution of the Y coordinates

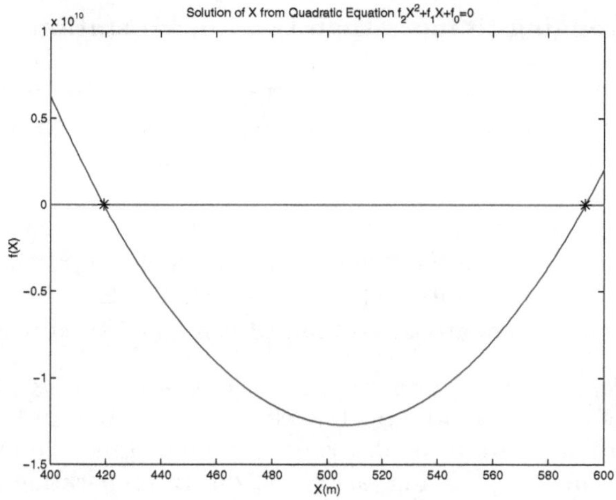

Figure 12.10. Solution of the X coordinates

Solution 12.4 (Critical configuration of the planar ranging problem).

$$\begin{bmatrix} f_1(X,Y;X_1,Y_1,S_1) = (X_1-X)^2+(Y_1-Y)^2-S_1^2 \\ f_2(X,Y;X_2,Y_2,S_2) = (X_2-X)^2+(Y_2-Y)^2-S_2^2 \end{bmatrix} \tag{12.46}$$

$$\begin{bmatrix} \frac{\partial f_1}{\partial X} = -2(X_1-X),\ \frac{\partial f_2}{\partial X} = -2(X_2-X) \\ \\ \frac{\partial f_1}{\partial Y} = -2(Y_1-Y),\ \frac{\partial f_2}{\partial Y} = -2(Y_2-Y), \end{bmatrix} \tag{12.47}$$

$$\begin{bmatrix} D = \left|\frac{\partial f_i}{\partial X_j}\right| = 4\begin{vmatrix} X_1-X & X_2-X \\ Y_1-Y & Y_2-Y \end{vmatrix} \\ D \Leftrightarrow \begin{vmatrix} X_1-X & X_2-X \\ Y_1-Y & Y_2-Y \end{vmatrix} = \begin{vmatrix} X & Y & 1 \\ X_1 & Y_1 & 1 \\ X_2 & Y_2 & 1 \end{vmatrix} = 0, \end{bmatrix} \tag{12.48}$$

$$\begin{bmatrix} \frac{1}{4}D = (X_1-X)(Y_2-Y)-(X_2-X)(Y_1-Y) \\ = X_1Y_2 - X_1Y - XY_2 + XY - X_2Y_1 + X_2Y + XY_1 - XY \\ = X(Y_1-Y_2)+Y(X_2-X_1)+X_1Y_2-X_2Y_1, \end{bmatrix} \tag{12.49}$$

thus

$$\begin{vmatrix} X & Y & 1 \\ X_1 & Y_1 & 1 \\ X_2 & Y_2 & 1 \end{vmatrix} = 2 \times \text{Area of triangle } P(X,Y), P_1(X_1,Y_1), \text{ and } P_2(X_2,Y_2) \tag{12.50}$$

$$D = \begin{vmatrix} X & Y & 1 \\ X_1 & Y_1 & 1 \\ X_2 & Y_2 & 1 \end{vmatrix}\begin{bmatrix} a \\ b \\ c \end{bmatrix} = 0, \tag{12.51}$$

results in a system of homogeneous equations

$$\begin{bmatrix} aX+bY+c=0 \\ aX_1+bY_1+c=0 \\ aX_2+bY_2+c=0. \end{bmatrix} \tag{12.52}$$

12-314 Planar ranging to more than two known stations

In-order to solve the overdetermined two-dimensional ranging problem, the combinatorial algorithm is applied. In the *first step*, combinatorials are formed using (7.28) on p. 93 and solved in a closed form using either (12.43) or (12.45). In the *second step*, the dispersion matrix $\mathbf{\Sigma}$ is obtained from (7.33) on p. 94. Finally the pseudo-observations are adjusted using linear Gauss-Markov model (see e.g., Definition 7.1 on p. 87) in the *third step*, with the unknown parameters estimated via **B**est **L**inear **U**niformly **U**nbiased **E**stimator **BLUUE** (7.12). The dispersion of the estimated parameters are then obtained using (7.13) on p. 87.

Example 12.4 (Planar ranging to more than two known stations). Let us consider the example of [237, pp. 240–241] which is also solved in [32]. In this example, the coordinates of station N are to be determined from distance observations to four stations P_1, P_2, P_3 and P_4 [237, Fig. 6.4.4, p. 229]. In preparation for adjustment, the distances are corrected and reduced geometrically to Gauss-Krueger projection and are as given in *Table* 12.11. Using Gauss-Jacobi combinatorial algorithm, the coordinates of station N are computed and compared to those of least squares in [237, p. 242]. From (7.28), six combinations in the minimal sense are formed and solved for $\{x, y\}_N$ for position of station N using either (12.43) or (12.45). The combinatorial solutions are presented in *Table* 12.12.

Table 12.11. Distance observations to unknown station N

Pt. No.	Easting $x[m]$	Northing $y[m]$	s_i $[m]$
1	48177.62	6531.28	611.023
2	49600.15	7185.19	1529.482
3	49830.93	5670.69	1323.884
4	47863.91	5077.24	1206.524

Table 12.12. Position of station N computed for various combinatorials

Combinatorial No.	combinatorial points	x $[m]$	y $[m]$
1	1-2	48565.2783	6058.9770
2	1-3	48565.2636	6058.9649
3	1-4	48565.2701	6058.9702
4	2-3	48565.2697	6058.9849
5	2-4	48565.3402	6058.9201
6	2-5	48565.2661	6058.9731

The adjusted position of the unknown station N is now obtained either by;

(a) simply taking the arithmetic mean of the combinatorial solutions in columns 3 and 4 of Table 12.12 (an approach which does not take into account full information in terms of the variance-covariance matrix) or,

(b) using *special linear Gauss-Markov model* through the estimation by the **B**est **L**inear **U**niformly **U**nbiased **E**stimator **BLUUE** in (7.12). The dispersion of the estimated parameters are subsequently obtained using (7.13).

The results are presented in *Table* 12.13 and plotted in Fig. 12.11. In *Table* 12.13, we present the coordinates $\{x, y\}$ of station N obtained using the least squares approach in [237], Gauss-Jacobi combinatorial (**BLUUE**) and the Gauss-Jacobi combinatorial (arithmetic mean) in columns 2 and 3, with their respective standard deviations $\{\sigma_x, \sigma_y\}$ in columns 4 and 5. In columns 6 and 7, the deviations $\{\Delta_x, \Delta_y\}$ of the computed coordinates of station N using Gauss-Jacobi combinatorial from the least squares' values of [237] are presented. The deviations of the exact solutions of each combination (columns 3 and 4 of *Table* 12.12) from the adjusted values of **B**est **L**inear **U**niformly **U**nbiased **E**stimator **BLUUE** (i.e., second and third columns of Table 12.13) are plotted in Fig. 12.12.

Table 12.13. Position of station N after adjustments

Approach	$x(m)$	$y(m)$	$\sigma_x(m)$	$\sigma_y(m)$	$\Delta_x(m)$	$\Delta_y(m)$
Least Squares	48565.2700	6058.9750	0.006	0.006	-	-
Gauss-Jacobi (BLUUE)	48565.2709	6058.9750	0.0032	0.0034	-0.0009	0.0000
Gauss-Jacobi (Mean)	48565.2813	6058.9650	-	-	-0.01133	0.0100

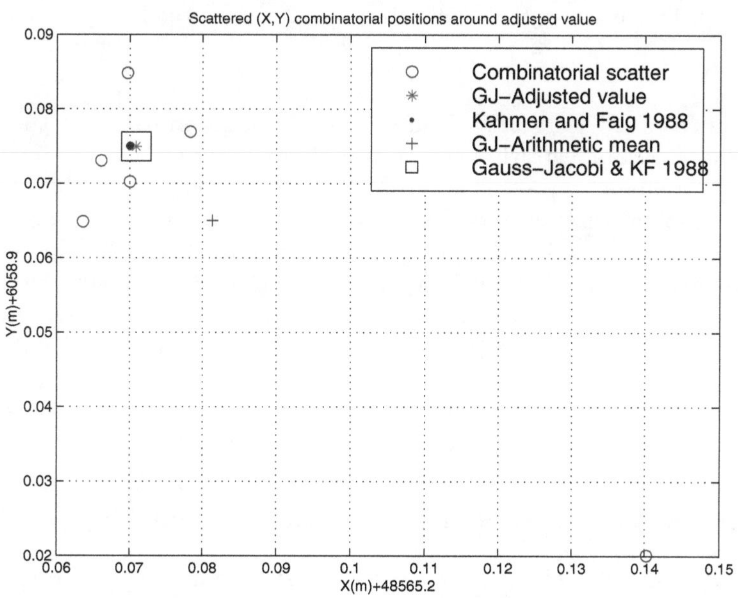

Figure 12.11. Plot of the position of N from various approaches

From the results in *Table* 12.13 and Fig. 12.11, we note that when full information of the observations is taken into account via the *nonlinear error/variance-covariance* propagation, the results of Gauss-Jacobi combinatorial algorithm and

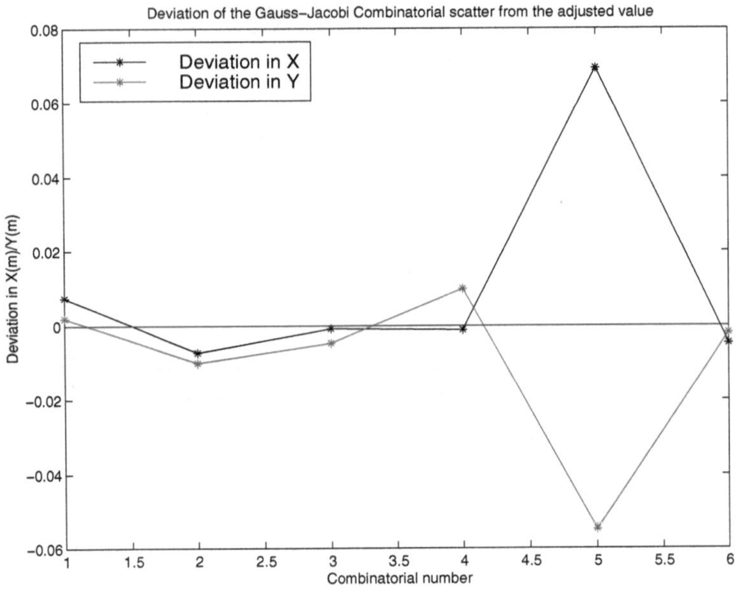

Figure 12.12. Deviations of the combinatorial scatter from the BLUUE adjusted position of N

least squares from [237] are in the same range. In addition to giving the adjusted position, Gauss-Jacobi algorithm can accurately pinpoint a poor combinatorial geometry (e.g., combination 5). This is taken care of through weighting. Fig. 12.11 shows the combinatorial scatter denoted by $\{\circ\}$ and the Gauss-Jacobi combinatorial adjusted value by $\{*\}$. Least squares estimation from [237] is denoted by $\{\bullet\}$ and the arithmetic mean by $\{+\}$. One notes that the estimates from Gauss-Jacobi's BLUUE $\{*\}$ and least squares solution almost coincide. In the Figure, both estimates are enclosed by $\{\square\}$ for clarity purpose. Figure 12.12 indicates the deviations of the combinatorial scatter from the BLUUE adjusted position of N. These results indicate the capability of the Gauss-Jacobi combinatorial algorithm to solve in a closed form the overdetermined planar ranging problems.

12-315 ALESS solution of overdetermined planar ranging problem

Another possibility to solve the overdetermined planar ranging problem is the ALESS method (see Sect. 7-2), defining the objective function as the sum of the square residuals of the equations, and considering the necessary condition for the minimum. The equation for a single measured distance is given by

$$e_i = (X_i - X_0)^2 + (Y_i - Y_0)^2 - S_i^2,$$

where X_0, Y_0 are the coordinates of the unknown station, and X_i, Y_i the coordinates of the known station and S_i the measured distance.

The objective function in case of n measured distance is defined as the sum of the square residuals

$$\Delta = \sum_{i=1}^{n} e_i^2 = \sum_{i=1}^{n} \left[(X_i - X_0)^2 + (Y_i - Y_0)^2 - S_i^2 \right]^2 .$$

The determined square system can easily be created symbolically with Computer Algebra Systems, according to the necessary condition of the minimum,

$$f_1 = \frac{\partial \Delta}{\partial X_0} = J_0 + J_1 X_0 + J_2 X_0^2 + J_3 X_0^3 + J_4 Y_0 + J_5 X_0 Y_0 + J_6 Y_0^2 + J_7 X_0 Y_0^2 = 0$$

$$f_2 = \frac{\partial \Delta}{\partial Y_0} = K_0 + K_1 X_0 + K_2 X_0^2 + K_3 Y_0 + K_4 X_0 Y_0 + K_5 X_0^2 Y_0 + K_6 Y_0^2 + K_7 Y_0^3 = 0$$

where

$$J_0 \to 4\sum_{i=1}^{n} \left(S_i^2 X_i - X_i^3 - X_i Y_i^2 \right), J_1 \to 4\sum_{i=1}^{n} \left(-S_i^2 + 3X_i^2 + Y_i^2 \right),$$

$$J_2 \to -12\sum_{i=1}^{n} X_i, J_3 \to 4n, J_4 \to 8\sum_{i=1}^{n} X_i Y_i, J_5 \to -8\sum_{i=1}^{n} Y_i,$$

$$J_6 \to -4\sum_{i=1}^{n} X_i, J_7 \to 4n$$

$$K_0 \to 4\sum_{i=1}^{n} \left(S_i^2 Y_i - X_i^2 Y_i - Y_i^3 \right), K_1 \to 8\sum_{i=1}^{n} X_i Y_i,$$

$$K_2 \to -4\sum_{i=1}^{n} Y_i, K_3 \to 4\sum_{i=1}^{n} \left(-S_i^2 + X_i^2 + 3Y_i^2 \right), K_4 \to -8\sum_{i=1}^{n} X_i,$$

$$K_5 \to 4n, K_6 \to -12\sum_{i=1}^{n} Y_i, K_7 \to 4n$$

Now we should solve this third order polynomial square system. The complexity of the system is not dependent on the number of redundant data. Only the coefficients change if we have more measured distances. Let use the same data in *Table* 12.11, but in order to increase the precision of the computation the values of the coordinates and the distances are rationalized,

$$X_1 \to \frac{2408881}{50}, Y_1 \to \frac{163282}{25}, S_1 \to \frac{611023}{1000},$$
$$X_2 \to \frac{992003}{20}, Y_2 \to \frac{718519}{100}, S_2 \to \frac{764741}{500},$$
$$X_3 \to \frac{4983093}{100}, Y_3 \to \frac{567069}{100}, S_3 \to \frac{330971}{250},$$
$$X_4 \to \frac{4786391}{100}, Y_4 \to \frac{126931}{25}, S_4 \to \frac{301631}{250}$$

Now our system to solve is the following:

$$f_1 = -\frac{237200}{12500000} + \frac{576246}{50000}X_0 - \frac{586418}{25}X_0^2 + 16X_0^3 + \frac{598320}{625}Y_0 - \frac{978576}{5}X_0Y_0 - \frac{195473}{25}Y_0^2 + 16X_0Y_0^2$$

$$f_2 = -\frac{148689}{6250000} + \frac{598321}{625}X_0 - \frac{489288}{5}X_0^2 + \frac{200121}{50000}Y_0 - \frac{390945}{25}X_0Y_0 + 16X_0^2Y_0 - \frac{1467864}{5}Y_0^2 + 16Y_0^3$$

This system can be solved with Groebner basis using *Mathematica*:

$$GroebnerBasis[\{f_1, f_2\}, \{X_0, Y_0\}];$$

which leads to a univariate and a two-variate polynomials

$$g_1 = L_0 + L_1Y_0 + L_2Y_0^2 + L_3Y_0^3 + L_4Y_0^4 + L_5Y_0^5$$
$$g_2 = M_0 + M_1X_0 + M_2Y_0 + M_3Y_0^2 + M_4Y_0^3 + M_5Y_0^4$$

where

$$L_0 \to -3.93690 \times 10^{55},$$
$$L_1 \to 3.25850 \times 10^{52},$$
$$L_2 \to -1.08236 \times 10^{49},$$
$$L_3 \to 1.80350 \times 10^{45},$$
$$L_4 \to -1.50752 \times 10^{41},$$
$$L_5 \to 5.05731 \times 10^{36}$$

$$M_0 \to 4.22105 \times 10^{85},$$
$$M_1 \to 1.31676 \times 10^{79},$$
$$M_2 \to -2.96087 \times 10^{82},$$
$$M_3 \to 7.67600 \times 10^{78},$$
$$M_4 \to -8.85305 \times 10^{74},$$
$$M_5 \to 3.83418 \times 10^{70}$$

Let us compute the roots of g_1 univariate polynomial (Y_0),

$$Y_0 \to 6058.9782,$$
$$Y_0 \to 5747.6051 - 710.6398\imath,$$
$$Y_0 \to 5747.6051 + 710.6398\imath,$$
$$Y_0 \to 6127.2461 - 873.7323\imath,$$
$$Y_0 \to 6127.2461 + 873.7323\imath$$

There is only one real solution, the first. Knowing Y_0, X_0 can be calculated from the second equation (g_2) which is linear in X_0. Then the solution is,

$$X_0 = 48565.2699 \quad Y_0 = 6058.9782.$$

The result of ALESS is almost the same as Gauss-Jacobi's BLUUE and traditional least squares solution (see *Table* 12.13) (the difference between them is about 3 mm).

12-32 Three-dimensional ranging

12-321 Closed form three-dimensional ranging

Three-dimensional ranging problem differs from the planar ranging in terms of the number of unknowns to be determined. In the planar case, the interest is to obtain from the measured distances the two-dimensional coordinates $\{x_0, y_0\}$ of the unknown station P. For the three-dimensional ranging, the coordinates $\{X, Y, Z\}$ have to be derived from the measured distances. Since three coordinates are involved, distances *must* be measured to at least three known stations for the solution to be determined. If the stations observed are more than three, the case is an overdetermined one. The main task involved is the determination of the unknown position of a station given distance measurements from unknown station $P \in \mathbb{E}^3$, to three known stations $P_i \in \mathbb{E}^3 \mid i = 1, 2, 3$. In general, the three-dimensional closed form ranging problem can be formulated as follows: Given distance measurements from an unknown station $P \in \mathbb{E}^3$ to a minimum of three known stations $P_i \in \mathbb{E}^3 \mid i = 1, 2, 3$, determine the position $\{X, Y, Z\}$ of the unknown station $P \in \mathbb{E}^3$ (see e.g., Fig. 13.4). From the three nonlinear Pythagoras

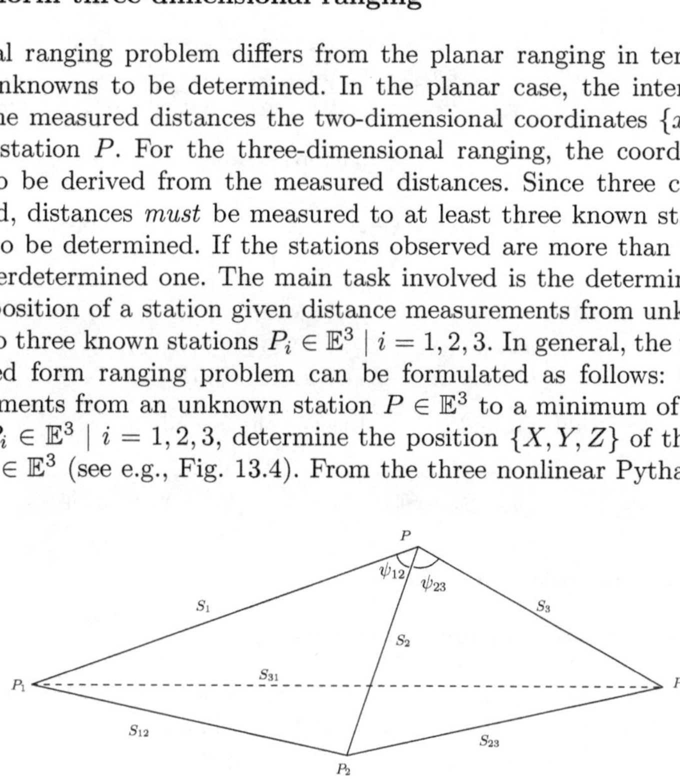

Figure 12.13. Tetrahedron: three-dimensional distance and space angle observations

distance observation equations (12.53) in Solution 12.5, two equations with three unknowns are derived. Equation (12.53) is expanded in the form given by (12.54) and differenced to give (12.55) with the quadratic terms $\{X^2, Y^2, Z^2\}$ eliminated. Collecting all the known terms of (12.55) to the right-hand-side and those relating to the unknowns (i.e., a and b) on the left-hand-side leads to (12.57). The solution of the unknown terms $\{X, Y, Z\}$ now involves solving (12.56), which has two equations with three unknowns. Equation (12.56) is similar to (12.4) on p. 177 which was considered in the case of GPS pseudo-range. Four approaches are considered for solving (12.56), where more unknowns than equations are solved. Similar to the case of GPS pseudo-ranging that we considered, the underdetermined system (12.56) is overcome by determining two of the unknowns in terms of the third unknown (e.g., $X = g(Z), Y = g(Z)$).

Solution 12.5 (Differencing of the nonlinear distance equations).

$$\begin{bmatrix} S_1^2 = (X_1 - X)^2 + (Y_1 - Y)^2 + (Z_1 - Z)^2 \\ S_2^2 = (X_2 - X)^2 + (Y_2 - Y)^2 + (Z_2 - Z)^2 \\ S_3^2 = (X_3 - X)^2 + (Y_3 - Y)^2 + (Z_3 - Z)^2 \end{bmatrix} \tag{12.53}$$

$$\begin{bmatrix} S_1^2 = X_1^2 + Y_1^2 + Z_1^2 + X^2 + Y^2 + Z^2 - 2X_1X - 2Y_1Y - 2Z_1Z \\ S_2^2 = X_2^2 + Y_2^2 + Z_2^2 + X^2 + Y^2 + Z^2 - 2X_2X - 2Y_2Y - 2Z_2Z \\ S_3^2 = X_3^2 + Y_3^2 + Z_3^2 + X^2 + Y^2 + Z^2 - 2X_3X - 2Y_3Y - 2Z_3Z \end{bmatrix} \tag{12.54}$$

differencing above

$$\begin{bmatrix} S_1^2 - S_2^2 = X_1^2 - X_2^2 + Y_1^2 - Y_2^2 + Z_1^2 - Z_2^2 + a \\ S_2^2 - S_3^2 - X_2^2 - X_3^2 + Y_2^2 - Y_3^2 + Z_2^2 - Z_3^2 + b, \end{bmatrix} \tag{12.55}$$

where

$$\begin{bmatrix} a = 2X(X_2 - X_1) + 2Y(Y_2 - Y_1) + 2Z(Z_2 - Z_1) \\ b = 2X(X_3 - X_2) + 2Y(Y_3 - Y_2) + 2Z(Z_3 - Z_2). \end{bmatrix} \tag{12.56}$$

Making a and b the subject of the formula in (12.55) leads to

$$\begin{bmatrix} a = S_1^2 - S_2^2 - X_1^2 + X_2^2 - Y_1^2 + Y_2^2 - Z_1^2 + Z_2^2 \\ b = S_2^2 - S_3^2 - X_2^2 + X_3^2 - Y_2^2 + Y_3^2 - Z_2^2 + Z_3^2. \end{bmatrix} \tag{12.57}$$

12-322 Conventional approaches

Solution by elimination approach-1

In the elimination approach presented in Solution 12.6, (12.56) is expressed in the form (12.58); with two equations and two unknowns $\{X, Y\}$. In this equation, Z is treated as a constant. By first eliminating Y, X is obtained in terms of Z and substituted in either of the two expressions of (12.58) to give the value of Y. The values of $\{X, Y\}$ are depicted in (12.59) with the coefficients $\{c, d, e, f\}$ given by (12.60). The values of $\{X, Y\}$ in (12.59) are substituted in the first expression of (12.53) to give the quadratic equation (12.61) in terms of Z as the unknown. The quadratic formula (3.8) on p. 24 is then applied to obtain the two solutions of Z (see the second expression of (12.61)). The coefficients $\{g, h, i\}$ are given in (12.62). Once we solve (12.61) for Z, we substitute in (12.59) to obtain the corresponding pair of solutions for $\{X, Y\}$.

Solution 12.6 (Solution by elimination).

$$\begin{bmatrix} 2X(X_2 - X_1) + 2Y(Y_2 - Y_1) = a - 2Z(Z_2 - Z_1) \\ 2X(X_3 - X_2) + 2Y(Y_3 - Y_2) = b - 2Z(Z_3 - Z_2) \end{bmatrix} \tag{12.58}$$

$$\begin{bmatrix} X = c - dZ \\ Y = e - fZ \end{bmatrix} \tag{12.59}$$

$$\begin{bmatrix} c = \dfrac{a(Y_3 - Y_2) - b(Y_2 - Y_1)}{2\{(X_2 - X_1)(Y_3 - Y_2) - (X_3 - X_2)(Y_2 - Y_1)\}} \\ \\ d = \dfrac{\{(Z_2 - Z_1)(Y_3 - Y_2) - (Z_3 - Z_2)(Y_2 - Y_1)\}\, Z}{\{(X_2 - X_1)(Y_3 - Y_2) - (X_3 - X_2)(Y_2 - Y_1)\}} \\ \\ e = \dfrac{a(X_3 - X_2) - b(X_2 - X_1)}{2\{(Y_2 - Y_1)(X_3 - X_2) - (Y_3 - Y_2)(X_2 - X_1)\}} \\ \\ f = \dfrac{\{(Z_2 - Z_1)(X_3 - X_2) - (Z_3 - Z_2)(X_2 - X_1)\}\, Z}{\{(Y_2 - Y_1)(X_3 - X_2) - (Y_3 - Y_2)(X_2 - X_1)\}} \end{bmatrix} \tag{12.60}$$

substituting 12.59 in 12.53i

$$\begin{bmatrix} gZ^2 + hZ + i = 0 \\ \\ Z_{1,2} = \dfrac{-h \pm \sqrt{h^2 - 4gi}}{2g}, \end{bmatrix} \tag{12.61}$$

where

$$\begin{bmatrix} g = d^2 + f^2 + 1 \\ h = 2(dX_1 + fY_1 - Z_1 - cd - ef) \\ i = X_1^2 + Y_1^2 + Z_1^2 - 2X_1c - 2Y_1e - S_1^2 + c^2 + e^2. \end{bmatrix} \tag{12.62}$$

12-323 Solution by elimination approach-2

The second approach presented in Solution 12.7 involves first expressing (12.56) in the form (12.63) which can also be expressed in matrix form as in (12.64). We now seek the matrix solution of $\{Y, Z\}$ in terms of the unknown element X as expressed by (12.65), which is written in a simpler form in (12.66). The elements of (12.66) are as given by (12.67). The solution of (12.65) for $\{Y, Z\}$ in terms of X is given by (12.68), (12.69) and (12.70). The coefficients of (12.70) are given by (12.71). Substituting the obtained values of $\{Y, Z\}$ in terms of X in the first expression of (12.53) leads to quadratic equation (12.72) in terms of X as an unknown. Applying the quadratic formula (3.8) on p. 24, two solutions for X are obtained as in the second expression of (12.72). These are then substituted back in (12.70) to obtain the values of $\{Y, Z\}$. The coefficients $\{l, m, n\}$ in (12.72) are given by (12.73).

A pair of solutions $\{X_1, Y_1, Z_1\}$ and $\{X_2, Y_2, Z_2\}$ are obtained. For GPS pseudo-ranging in Sect. 12-2, we saw that the admissible solution could easily be chosen from the pair of solutions. The desired solution was easily chosen as one set of solution was in space while the other set was on the Earth's surface. The solution could therefore be distinguished by computing the radial distances (positional norms). The admissible solution from the pair of the three-dimensional LPS ranging techniques is however difficult to isolate and must be obtained with the help of prior information, e.g., from an existing map.

Solution 12.7 (Solution by matrix approach).

$$\begin{bmatrix} 2Y(Y_2-Y_1)+2Z(Z_2-Z_1)=a-2X(X_2-X_1) \\ 2Y(Y_3-Y_2)+2Z(Z_3-Z_2)=b-2X(X_3-X_2) \end{bmatrix} \quad (12.63)$$

$$\begin{bmatrix} Y_2-Y_1 & Z_2-Z_1 \\ Y_3-Y_2 & Z_3-Z_2 \end{bmatrix}\begin{bmatrix} Y \\ Z \end{bmatrix}=\frac{1}{2}\left\{\begin{bmatrix} a \\ b \end{bmatrix}-2\begin{bmatrix} X_2-X_1 \\ X_3-X_2 \end{bmatrix}X\right\} \quad (12.64)$$

$$\begin{bmatrix} Y \\ Z \end{bmatrix}=\frac{1}{2}d\begin{bmatrix} Z_3-Z_2 & -(Z_2-Z_1) \\ -(Y_3-Y_2) & (Y_2-Y_1) \end{bmatrix}\left\{\begin{bmatrix} a \\ b \end{bmatrix}-2\begin{bmatrix} X_2-X_1 \\ X_3-X_2 \end{bmatrix}X,\right\} \quad (12.65)$$

with

$$d=\{(Y_2-Y_1)(Z_3-Z_2)-(Y_3-Y_2)(Z_2-Z_1)\}^{-1}.$$

$$\begin{bmatrix} Y \\ Z \end{bmatrix}=\{a_{11}a_{22}-a_{12}a_{21}\}^{-1}\begin{bmatrix} a_{22} & -a_{12} \\ -a_{21} & a_{11} \end{bmatrix}\left\{\begin{bmatrix} b_1 \\ b_2 \end{bmatrix}+\begin{bmatrix} c_1 \\ c_2 \end{bmatrix}X,\right\} \quad (12.66)$$

where

$$\begin{bmatrix} a_{11}=Y_2-Y_1,\ a_{12}=Z_2-Z_1,\ a_{21}=Y_3-Y_2,\ a_{22}=Z_3-Z_2 \\ c_1=-(X_2-X_1),\ c_2=-(X_3-X_2),\ b_1=\frac{1}{2}a,\ b_2=\frac{1}{2}b. \end{bmatrix} \quad (12.67)$$

$$\begin{bmatrix} Y=\{a_{11}a_{22}-a_{12}a_{21}\}^{-1}\{a_{22}(b_1+c_1X)-a_{12}(b_2+c_2X)\} \\ Z=\{a_{11}a_{22}-a_{12}a_{21}\}^{-1}\{a_{11}(b_2+c_2X)-a_{21}(b_1+c_1X)\} \end{bmatrix} \quad (12.68)$$

$$\begin{bmatrix} Y=e\left[\{a_{22}b_1-a_{12}b_2\}+\{a_{22}c_1-a_{12}c_2\}X\right] \\ Z=e\left[\{a_{11}b_2-a_{21}b_1\}+\{a_{11}c_2-a_{21}c_1\}X\right] \end{bmatrix} \quad (12.69)$$

$$\begin{bmatrix} Y=e(f+gX) \\ Z=e(h+iX) \end{bmatrix} \quad (12.70)$$

$$\begin{bmatrix} e=(a_{11}a_{22}-a_{12}a_{21})^{-1},\ f=a_{22}b_1-a_{12}b_2,\ g=a_{22}c_1-a_{12}c_2 \\ h=a_{11}b_2-a_{21}b_1,\ i=a_{11}c_2-a_{21}c_1,\ k=X_1^2+Y_1^2+Z_1^2. \end{bmatrix} \quad (12.71)$$

substituting (12.70) in (12.53i)

$$\begin{bmatrix} lX^2+mX+n=0 \\ X_{1,2}=\dfrac{-m\pm\sqrt{m^2-4ln}}{2l}, \end{bmatrix} \quad (12.72)$$

where

$$\begin{bmatrix} l=e^2i^2+e^2g^2+1 \\ m=2(e^2fg+e^2hi-X_1-egY_1-eiZ_1) \\ n=k-S_1^2-2Y_1ef+e^2f^2-2Z_1eh+e^2h^2. \end{bmatrix} \quad (12.73)$$

12-324 Groebner basis approach

Equation (12.56) is expressed in algebraic form (12.74) in Solution 12.8 with the coefficients as in (12.75). Groebner basis of (12.74) is then obtained in (12.76) using (4.36) on p. 45. The obtain Groebner basis solution of the three-dimensional ranging problem is presented in (12.77). The first expression of (12.77) is solved for $Y = g_1(Z)$, and the output presented in (12.78). This value is substituted in the second expression of (12.77) to give $X = g_2(Z)$ in (12.79). The obtained values of Y and X are substituted in the first expression of (12.53) to give a quadratic equation in Z. Once this quadratic equation has been solved for Z using (3.8) on p. 24, the values Y and X are obtained from (12.78) and (12.79) respectively. Instead of solving for $Y = g_1(Z)$ and substituting in the second expression of (12.77) to give $X = g_2(Z)$, direct solution of $X = g(Z)$ in (12.80) could be obtained by computing the reduced Groebner basis (4.38) on p. 45. Similarly we could obtain $Y = g(Z)$ alone by replacing Y with X in the option part of the reduced Groebner basis.

Solution 12.8 (Groebner basis solution).

$$\begin{array}{l} a_{02}X + b_{02}Y + c_{02}Z + f_{02} = 0 \\ a_{12}X + b_{12}Y + c_{12}Z + f_{12} = 0 \end{array} \tag{12.74}$$

$$\begin{bmatrix} a_{02} = 2(X_1 - X_2), b_{02} = 2(Y_1 - Y_2), c_{02} = 2(Z_1 - Z_2) \\ a_{12} = 2(X_2 - X_3), b_{12} = 2(Y_2 - Y_3), c_{12} = 2(Z_2 - Z_3) \\ f_{02} = (S_1^2 - X_1^2 - Y_1^2 - Z_1^2) - (S_2^2 - X_2^2 - Y_2^2 - Z_2^2) \\ f_{12} = (S_2^2 - X_2^2 - Y_2^2 - Z_2^2) - (S_3^2 - X_3^2 - Y_3^2 - Z_3^2). \end{bmatrix} \tag{12.75}$$

$$GroebnerBasis[\{a_{02}X + b_{02}Y + c_{02}Z + f_{02}, a_{12}X + b_{12}Y + c_{12}Z + f_{12}\}, \{X, Y\}] \tag{12.76}$$

$$\begin{bmatrix} g_1 = a_{02}b_{12}Y - a_{12}b_{02}Y - a_{12}c_{02}Z + a_{02}c_{12}Z + a_{02}f_{12} - a_{12}f_{02} \\ g_2 = a_{12}X + b_{12}Y + c_{12}Z + f_{12} \\ g_3 = a_{02}X + b_{02}Y + c_{02}Z + f_{02}. \end{bmatrix} \tag{12.77}$$

$$Y = \frac{\{(a_{12}c_{02} - a_{02}c_{12})Z + a_{12}f_{02} - a_{02}f_{12}\}}{(a_{02}b_{12} - a_{12}b_{02})} \tag{12.78}$$

$$X = \frac{-(b_{12}Y + c_{12}Z + f_{12})}{a_{12}}, \tag{12.79}$$

or

$$X = \frac{\{(b_{02}c_{12} - b_{12}c_{02})Z + b_{02}f_{12} - b_{12}f_{02}\}}{(a_{02}b_{12} - a_{12}b_{02})}. \tag{12.80}$$

12-325 Polynomial resultants approach

The problem is solved in *four steps* as illustrated in Solution 12.9. In the *first step*, we solve for the first variable X in (12.74) by hiding it as a constant and homogenizing the equation using a variable W as in (12.81). In the *second step*, the Sylvester resultants discussed in Sect. 5-2 on p. 49 or the *Jacobian determinant* is obtained as in (12.82). The resulting determinant (12.83) is solved for $X = g(Z)$

and presented in (12.84). The procedure is repeated in *steps three* and *four* from (12.85) to (12.88) to solve for $Y = g(Z)$. The obtained values of $X = g(Z)$ and $Y = g(Z)$ are substituted in the first expression of (12.53) to give a quadratic equation in Z. Once this quadratic has been solved for Z, the values of X and Y are then obtained from (12.84) and (12.88) respectively.

Solution 12.9 (Polynomial resultants solution). Step 1: Solve for X in terms of Z

$$\begin{aligned} f_1 &:= (a_{02}X + c_{02}Z + f_{02})W + b_{02}Y \\ f_2 &:= (a_{12}X + c_{12}Z + f_{12})W + b_{12}Y \end{aligned} \tag{12.81}$$

Step 2: Obtain the Sylvester resultant

$$J_X = det \begin{bmatrix} \frac{\partial f_1}{\partial Y} & \frac{\partial f_1}{\partial W} \\ \frac{\partial f_2}{\partial Y} & \frac{\partial f_2}{\partial W} \end{bmatrix} = det \begin{bmatrix} b_{02} & (a_{02}X + c_{02}Z + f_{02}) \\ b_{12} & (a_{12}X + c_{12}Z + f_{12}) \end{bmatrix} \tag{12.82}$$

$$J_X = b_{02}a_{12}X + b_{02}c_{12}Z + b_{02}f_{12} - b_{12}a_{02}X - b_{12}c_{02}Z - b_{12}f_{02} \tag{12.83}$$

from (12.83)

$$X = \frac{\{(b_{12}c_{02} - b_{02}c_{12})Z + b_{12}f_{02} - b_{02}f_{12}\}}{(b_{02}a_{12} - b_{12}a_{02})} \tag{12.84}$$

Step 3: Solve for Y in terms of Z

$$\begin{aligned} f_3 &:= (b_{02}Y + c_{02}Z + f_{02})W + b_{02}X \\ f_4 &:= (b_{12}Y + c_{12}Z + f_{12})W + a_{12}X \end{aligned} \tag{12.85}$$

Step 4: Obtain the Sylvester resultant

$$J_Y = det \begin{bmatrix} \frac{\partial f_3}{\partial X} & \frac{\partial f_3}{\partial W} \\ \frac{\partial f_4}{\partial X} & \frac{\partial f_4}{\partial W} \end{bmatrix} = det \begin{bmatrix} a_{02} & (b_{02}Y + c_{02}Z + f_{02}) \\ a_{12} & (b_{12}Y + c_{12}Z + f_{12}) \end{bmatrix} \tag{12.86}$$

$$J_Y = a_{02}b_{12}Y + a_{02}c_{12}Z + a_{02}f_{12} - a_{12}b_{02}Y - a_{12}c_{02}Z - a_{12}f_{02} \tag{12.87}$$

from (12.87)

$$Y = \frac{\{(a_{12}c_{02} - a_{02}c_{12})Z + a_{12}f_{02} - a_{02}f_{12}\}}{(a_{02}b_{12} - a_{12}b_{02})} \tag{12.88}$$

Example 12.5 (Three-dimensional ranging to three known stations). Consider distance measurements of Fig. 13.4 as $S_1 = 1324.2380\,m$, $S_2 = 542.2609\,m$ and $S_3 = 430.5286\,m$, the position of P is obtained using either of the procedures above as $X = 4157066.1116\,m$, $Y = 671429.6655\,m$ and $= 4774879.3704\,m$. Figures 12.14, 12.15 and 12.16 indicate the solutions of $\{X, Y, Z\}$ respectively. The stars (intersection of the quadratic curves with the zero line) are the solution points. The critical configuration of the three-dimensional ranging problem is presented in Solution 12.10. Equations (12.94) and (12.95) indicate the critical configuration to be the case where points $P(X, Y, Z)$, $P_1(X_1, Y_1, Z_1)$, $P_2(X_2, Y_2, Z_2)$, and $P_3(X_3, Y_3, Z_3)$ all lie on a plane.

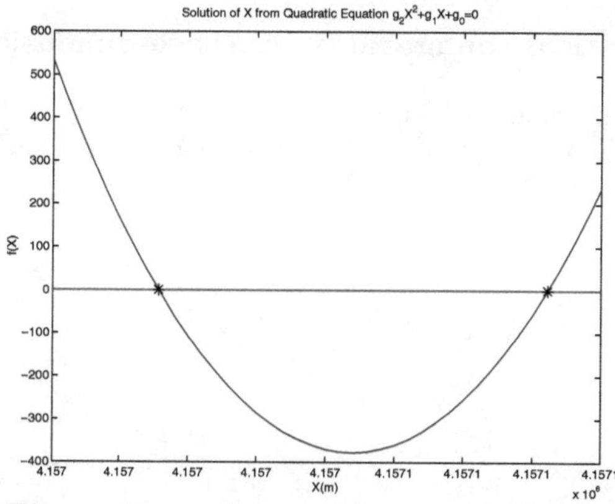

Figure 12.14. Solution of the X coordinates

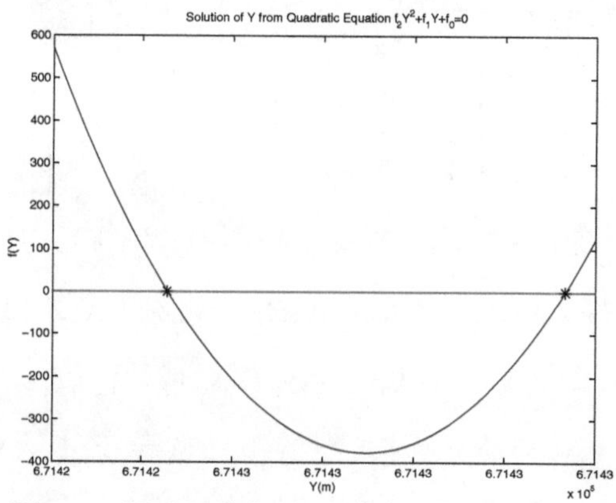

Figure 12.15. Solution of the Y coordinates

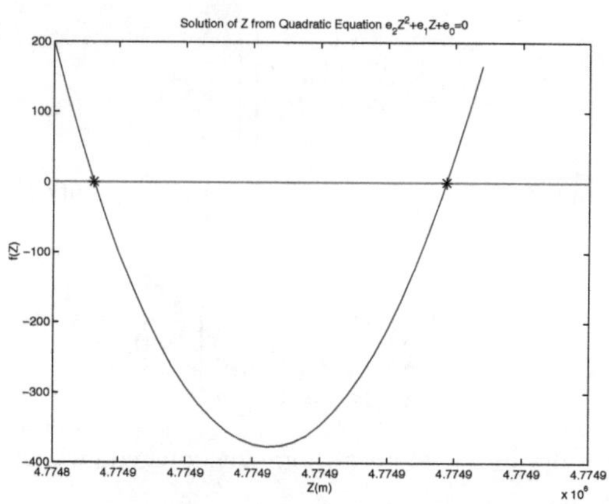

Figure 12.16. Solution of the Z coordinates

Solution 12.10 (Critical configuration of three-dimensional ranging).

$$\begin{bmatrix} f_1(X,Y,Z;X_1,Y_1,Z_1,S_1) = (X_1-X)^2+(Y_1-Y)^2+(Z_1-Z)^2-S_1^2 \\ f_2(X,Y,Z;X_2,Y_2,Z_2,S_2) = (X_2-X)^2+(Y_2-Y)^2+(Z_2-Z)^2-S_2^2 \\ f_3(X,Y,Z;X_3,Y_3,Z_3,S_3) = (X_3-X)^2+(Y_3-Y)^2+(Z_3-Z)^2-S_3^2. \end{bmatrix} \quad (12.89)$$

$$\begin{bmatrix} \frac{\partial f_1}{\partial X} = -2(X_1-X), & \frac{\partial f_2}{\partial X} = -2(X_2-X), & \frac{\partial f_3}{\partial X} = -2(X_3-X) \\ \frac{\partial f_1}{\partial Y} = -2(Y_1-Y), & \frac{\partial f_2}{\partial Y} = -2(Y_2-Y), & \frac{\partial f_3}{\partial Y} = -2(Y_3-Y) \\ \frac{\partial f_1}{\partial Z} = -2(Z_1-Z), & \frac{\partial f_2}{\partial Z} = -2(Z_2-Z), & \frac{\partial f_3}{\partial Zv} = -2(Z_3-Z). \end{bmatrix} \quad (12.90)$$

$$\begin{bmatrix} D = \left|\frac{\partial f_i}{\partial X_j}\right| = -8 \begin{vmatrix} X_1-X & Y_1-Y & Z_1-Z \\ X_2-X & Y_2-Y & Z_1-Z \\ X_3-X & Y_3-Y & Z_1-Z \end{vmatrix} \\ D \Leftrightarrow \begin{vmatrix} X_1-X & Y_1-Y & Z_1-Z \\ X_2-X & Y_2-Y & Z_1-Z \\ X_3-X & Y_3-Y & Z_1-Z \end{vmatrix} = \begin{vmatrix} X & Y & Z & 1 \\ X_1 & Y_1 & Z_1 & 1 \\ X_2 & Y_2 & Z_2 & 1 \\ X_3 & Y_3 & Z_3 & 1 \end{vmatrix} = 0. \end{bmatrix} \quad (12.91)$$

$$\begin{bmatrix} -\tfrac{1}{8}D = \{-Z_1Y_3+Y_1Z_3-Y_2Z_3+Y_3Z_2-Y_1Z_2+Y_2Z_1\}X \\ +\{-Z_1X_2-X_1Z_3+Z_1X_3+X_1Z_2-X_3Z_2+X_2Z_3\}Y \\ +\{Y_1X_2-Y_1X_3+Y_3X_1-X_2Y_3-X_1Y_2+Y_2X_3\}Z \\ +X_1Y_2Z_3-X_1Y_3Z_2-X_3Y_2Z_1+X_2Y_3Z_1-X_2Y_1Z_3+X_3Y_1Z_2, \end{bmatrix} \quad (12.92)$$

thus

$$\begin{vmatrix} X & Y & Z & 1 \\ X_1 & Y_1 & Z_1 & 1 \\ X_2 & Y_2 & Z_2 & 1 \\ X_3 & Y_3 & Z_3 & 1 \end{vmatrix}, \quad (12.93)$$

describes six times volume of the tetrahedron formed by the points $P(X,Y,Z)$, $P_1(X_1,Y_1,Z_1)$, $P_2(X_2,Y_2,Z_2)$, and $P_3(X_3,Y_3,Z_3)$. Therefore

$$D = \begin{vmatrix} X & Y & Z & 1 \\ X_1 & Y_1 & Z_1 & 1 \\ X_2 & Y_2 & Z_2 & 1 \\ X_3 & Y_3 & Z_3 & 1 \end{vmatrix} \begin{bmatrix} a \\ b \\ c \\ d \end{bmatrix} = 0, \quad (12.94)$$

results in a system of homogeneous equations

$$\begin{bmatrix} aX+bY+cZ+d=0 \\ aX_1+bY_1+cY_1+d=0 \\ aX_2+bY_2+cZ_2+d=0 \\ aX_3+bY_3+cZ_3+d=0. \end{bmatrix} \quad (12.95)$$

12-326 N-point three-dimensional ranging

The Gauss-Jacobi combinatorial algorithm is here applied to solve the overdetermined three-dimensional ranging problem. An example based on the test network Stuttgart Central in Fig. 10.2 is considered.

Example 12.6 (Three-dimensional ranging to more than three known stations). From the test network Stuttgart Central in Fig. 10.2 of Sect. 10-6, the three-dimensional coordinates $\{X, Y, Z\}$ of the unknown station $K1$ are desired. One proceeds in three steps as follows:

Step 1 (combinatorial solution):
From Fig. 10.2 on p. 150 and using (7.28) on p. 93, 35 combinatorial subsets are formed whose systems of *nonlinear distance equations* are solved for the position $\{X, Y, Z\}$ of the unknown station $K1$ in closed form. Use is made of either Groebner basis derived equations (12.78) and (12.79) or polynomial resultants derived (12.84) and (12.88). 35 different positions $X, Y, Z|_{K1}$ of the same station $K1$, totalling to 105 (35 × 3) values of X, Y, Z are obtained and treated as pseudo-observations.

Step 2 (determination of the dispersion matrix $\mathbf{\Sigma}$):
The variance-covariance matrix is computed for each of the combinatorial set $j = 1, \ldots, 35$ using error propagation. The closed form observational equations are written algebraically as

$$\begin{bmatrix} f_1 := (X_1 - X)^2 + (Y_1 - Y)^2 + (Z_1 - Z)^2 - S_1^2 \\ f_2 := (X_2 - X)^2 + (Y_2 - Y)^2 + (Z_2 - Z)^2 - S_2^2 \\ f_3 := (X_3 - X)^2 + (Y_3 - Y)^2 + (Z_3 - Z)^2 - S_3^2, \end{bmatrix} \tag{12.96}$$

where $S_i^j \,|i \in \{1, 2, 3\} \mid j = 1$ are the distances between known GPS stations $P_i \in \mathbb{E}^3 \,|i \in \{1, 2, 3\}$ and the unknown station $K1 \in \mathbb{E}^3$ for first combination set $j = 1$. Equation (12.96) is used to obtain the dispersion matrix $\mathbf{\Sigma}$ in (7.33) as discussed in Example 7.4 on p. 95.

Step 3 (rigorous adjustment of the combinatorial solution points in a polyhedron):
For each of the 35 computed coordinates of point $K1$ in step 2, we write the observation equations as

$$\begin{bmatrix} X^j = X + \varepsilon_X^j \,|, j \in \{1, 2, 3, 4, 5, 6, 7, \ldots, 35\} \\ Y^j = Y + \varepsilon_Y^j \,|j \in \{1, 2, 3, 4, 5, 6, 7, \ldots, 35\} \\ Z^j = Z + \varepsilon_Z^j \,|, j \in \{1, 2, 3, 4, 5, 6, 7, \ldots, 35\}. \end{bmatrix} \tag{12.97}$$

The values $\{X^j, Y^j, Z^j\}$ are treated as pseudo-observation and placed in the vector of observation $\mathbf{y}$, while the coefficients of the unknown positions $\{X, Y, Z\}$ are placed in the design matrix $\mathbf{A}$. The vector $\boldsymbol{\xi}$ comprise the unknowns $\{X, Y, Z\}$. The solutions are obtained via (7.12) and the root-mean-square errors of the estimated parameters through (7.13). In the experiment above, the computed position of station $K1$ is given in Table 12.14. The deviations of the combinatorial solutions from the *true (measured) GPS value* are given in Table 12.15. Figure 12.17 indicates the plot of the combinatorial scatter {•} around the adjusted values {∗}.

Table 12.14. Position of station $K1$ computed by *Gauss-Jacobi combinatorial algorithm*

$X(m)$	$Y(m)$	$Z(m)$	σ_X	σ_Y	σ_Z
4157066.1121	671429.6694	4774879.3697	0.00005	0.00001	0.00005

Table 12.15. Deviation of the computed position of $K1$ in Table (12.14) from the real measured GPS values

$\Delta X(m)$	$\Delta Y(m)$	$\Delta Z(m)$
-0.0005	-0.0039	0.0007

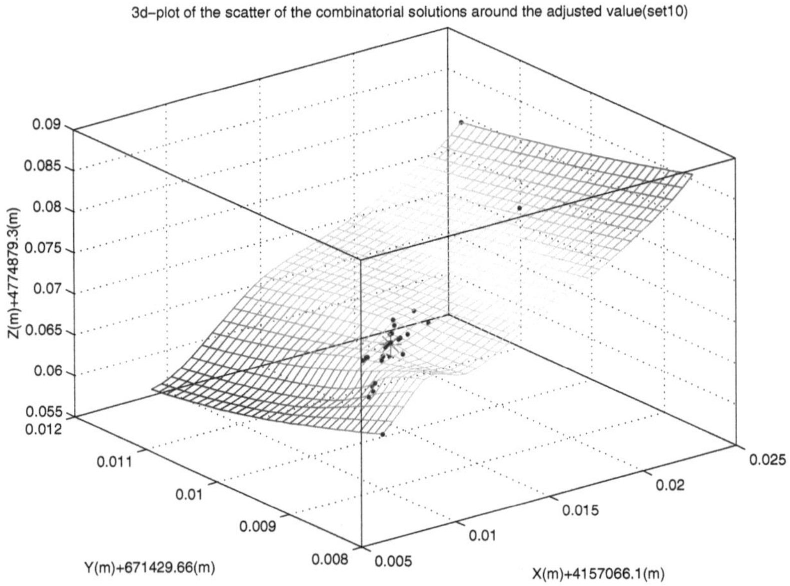

Figure 12.17. Scatter of combinatorial solutions

12-327 ALESS solution

Another possibility to solve the overdetermined three-dimensional ranging problem is the ALESS method (see Sect. 7-2), defining the objective function as the sum of the square residuals of the equations, and considering the necessary condition for the minimum.

A single equation for distance is given by

$$e_i = (X_i - X_0)^2 + (Y_i - Y_0)^2 + (Z_i - Z_0)^2 - S_i^2,$$

where X_0, Y_0, Z_0 are the coordinates of the *unknown* station, X_i, Y_i, Z_i the coordinates of the *known* station, and S_i the measured distances. The objective function in case of n measured distances is defined as the sum of the square residuals

$$\Delta = \sum_{i=1}^{n} e_i^2 = \sum_{i=1}^{n} \left[(X_i - X_0)^2 + (Y_i - Y_0)^2 + (Z_i - Z_0)^2 - S_i^2\right]^2.$$

The determined square system can be created symbolically easily with Computer Algebra Systems, according to the necessary condition of the minimum,

$$\begin{aligned}
f_1 = \frac{\partial \Delta}{\partial X_0} =& J_0 + J_1 X_0 + J_2 X_0^2 + J_3 X_0^3 + J_4 Y_0 + J_5 X_0 Y_0 + J_6 Y_0^2 + J_7 X_0 Y_0^2 \\
&+ J_8 Z_0 + J_9 X_0 Z_0 + J_{10} Z_0^2 + J_{11} X_0 Z_0^2 = 0 \\
f_2 = \frac{\partial \Delta}{\partial Y_0} =& K_0 + K_1 X_0 + K_2 X_0^2 + K_3 Y_0 + K_4 X_0 Y_0 + K_5 X_0^2 Y_0 + K_6 Y_0^2 \\
&+ K_7 Y_0^3 + K_8 Z_0 + K_9 Y_0 Z_0 + K =_{10} Z_0^2 + K_{11} Y_0 Z_0^2 0 \\
f_3 = \frac{\partial \Delta}{\partial Z_0} =& L_0 + L_1 X_0 + L_2 X_0^2 + L_3 Y_0 + L_4 Y_0^2 + L_5 Z_0 + L_6 X_0 Z_0 + L_7 X_0^2 Z_0 \\
&+ L_8 Y_0 Z_0 + L_9 Y_0^2 Z_0 + L_{10} Z_0^2 + L_{11} Z_0^3 = 0
\end{aligned}$$

where

$$\begin{aligned}
&J_0 \to \sum_{i=1}^{n} \left(4S_i^2 X_i - 4X_i^3 - 4X_i Y_i^2 - 4X_i Z_i^2\right), J_1 \to \sum_{i=1}^{n} \left(-4S_i^2 + 12X_i^2 + 4Y_i^2 + 4Z_i^2\right), \\
&J_2 \to \sum_{i=1}^{n} -12X_i, J_3 \to 4n, J_4 \to \sum_{i=1}^{n} 8X_i Y_i, J_5 \to \sum_{i=1}^{n} -8Y_i, J_6 \to \sum_{i=1}^{n} -4X_i, J_7 \to 4n, \\
&J_8 \to \sum_{i=1}^{n} 8X_i Z_i, J_9 \to \sum_{i=1}^{n} -8Z_i, J_{10} \to \sum_{i=1}^{n} -4X_i, J_{11} \to 4n
\end{aligned}$$

$$\begin{aligned}
&K_0 \to \sum_{i=1}^{n} \left(4S_i^2 Y_i - 4X_i^2 Y_i - 4Y_i^3 - 4Y_i Z_i^2\right), K_1 \to \sum_{i=1}^{n} 8X_i Y_i, K_2 \to \sum_{i=1}^{n} -4Y_i, \\
&K_3 \to \sum_{i=1}^{n} \left(-4S_i^2 + 4X_i^2 + 12Y_i^2 + 4Z_i^2\right), K_4 \to \sum_{i=1}^{n} -8X_i, K_5 \to 4n, K_6 \to \sum_{i=1}^{n} -12Y_i, \\
&K_7 \to 4n, K_8 \to \sum_{i=1}^{n} 8Y_i Z_i, K_9 \to \sum_{i=1}^{n} -8Z_i, K_{10} \to \sum_{i=1}^{n} -4Y_i, K_{11} \to 4n
\end{aligned}$$

$$L_0 \to \sum_{i=1}^{n}\left(4S_i^2 Z_i - 4X_i^2 Z_i - 4Y_i^2 Z_i - 4Z_i^3\right), L_1 \to \sum_{i=1}^{n} 8X_i Z_i, L_2 \to \sum_{i=1}^{n} -4Z_i,$$

$$L_3 \to \sum_{i=1}^{n} 8Y_i Z_i, L_4 \to \sum_{i=1}^{n} -4Z_i, L_5 \to \sum_{i=1}^{n}\left(-4S_i^2 + 4X_i^2 + 4Y_i^2 + 12Z_i^2\right),$$

$$L_6 \to \sum_{i=1}^{n} -8X_i, L_7 \to 4n, L_8 \to \sum_{i=1}^{n} -8Y_i, L_9 \to 4n, L_{10} \to \sum_{i=1}^{n} -12Z_i, L_{11} \to 4n$$

Now this third order polynomial square system is to be solved. The complexity of this system is independent of the number of redundant data. Only the coefficients will change if more measured distances are available. To demonstrate the method, let us revisit the test network Stuttgart Central (see data on *Tables* 10.1 and 10.2) with seven distances to seven known stations. Substituting these data for f_1, f_2, f_3 leads to the following system for solving the unknown station K1:

$$\begin{aligned} f_1 = \; & -4.71777 \times 10^{21} + 2.10263 \times 10^{15} X_0 - 3.49196 \times 10^8 X_0^2 + 28X_0^3 \\ & +1.56321 \times 10^{14} Y_0 - 3.76035 \times 10^7 X_0 Y_0 - 1.16399 \times 10^8 Y_0^2 + 28X_0 Y_0^2 \\ & +1.11156 \times 10^{15} Z_0 - 2.6739 \times 10^8 X_0 Z_0 - 1.16399 \times 10^8 Z_0^2 + 28X_0 Z_0^2 \end{aligned}$$

$$\begin{aligned} f_2 = \; & -7.62057 \times 10^{20} + 1.56321 \times 10^{14} X_0 - 1.88017 \times 10^7 X_0^2 \\ & +1.16012 \times 10^{15} Y_0 - 2.32797 \times 10^8 X_0 Y_0 + 28X_0^2 Y_0 - 5.64052 \times 10^7 Y_0^2 \\ & +28Y_0^3 + 1.7955 \times 10^{14} Z_0 - 2.6739 \times 10^8 Y_0 Z_0 - 1.88017 \times 10^7 Z_0^2 + 28Y_0 Z_0^2 \end{aligned}$$

$$\begin{aligned} f_3 = \; & -5.41882 \times 10^{21} + 1.11156 \times 10^{15} X_0 - 1.33695 \times 10^8 X_0^2 + 1.7955 \times 10^{14} Y_0 \\ & -1.33695 \times 10^8 Y_0^2 + 2.41161 \times 10^{15} Z_0 - 2.32797 \times 10^8 X_0 Z_0 + 28X_0^2 Z_0 \\ & -3.76035 \times 10^7 Y_0 Z_0 + 28Y_0^2 Z_0 - 4.01085 \times 10^8 Z_0^2 + 28Z_0^3 \end{aligned}$$

Groebner Basis methods is then applied to solve this system of equation. Let us use high precision data in *Mathematica*:

$$F = SetPrecision[\{f1, f2, f3\}, 300];$$

To solve this system with *reduced Groenber basis* for variable X_0 in *Mathematica*, one writes:

$$gbX_0 = GroebnerBasis[F, \{X_0\}, \{Y_0, Z_0\}];$$

where F is the system of polynomials $\{f_1, f_2, f_3\}$ using high precision data, $\{Y_0, Z_0\}$ are the variables to be eliminated from the system and $\{X_0\}$ the remaining variable. The result is a *univariate polynomial* of order seven for X_0. Its roots are,

$$\{X_0 \to 4.1570478280643144805 \times 10^6 - 17.7824555537030i\},$$
$$\{X_0 \to 4.1570478280643144805 \times 10^6 + 17.7824555537030i\},$$
$$\{X_0 \to 4.1570661115296523158 \times 10^6\},$$
$$\{X_0 \to 4.1571001585939538923 \times 10^6 - 169.0166013474395i\},$$
$$\{X_0 \to 4.1571001585939538923 \times 10^6 + 169.0166013474395i\},$$
$$\{X_0 \to 4.1571279473843755008 \times 10^6 - 270.2601421331170i\},$$
$$\{X_0 \to .1571279473843755008 \times 10^6 + 270.2601421331170i\},$$

where the only real root is $X_0 = 4157066.1115$. Similarly, the variables Y_0 and Z_0 can be calculated by

$$gbY_0 = GroebnerBasis[F, \{Y_0\}, \{X_0, Z_0\}];$$
$$gbZ_0 = GroebnerBasis[F, \{Z_0\}, \{X_0, Y_0\}].$$

For both variables we get a seven order univariate polynomial using reduced Groebner basis, which has only one real root. The computed position of station $K1$ and the deviations of the ALESS solution from the *true (measured) GPS value* are given in Table 12.16. The result of the ALESS method coincides with the measured GPS coordinates.

Table 12.16. Position of station $K1$ computed by *ALESS method*

$X(m)$	$Y(m)$	$Z(m)$	$\Delta X(m)$	$\Delta Y(m)$	$\Delta Z(m)$
4 157 066.1115	671 429.6655	4 774 879.3703	0.0001	0.0000	0.0001

12-328 Extended Newton-Raphson's solution

With more than 3 known stations with the corresponding distances, our system is overdetermined, $m > n$, where $n = 3$ and $m = 7$! The prototype of the equations,

$$(x_i - x_0)^2 + (y_i - y_0)^2 + (z_i - z_0)^2 - s_i^2 = 0, \ i = 1..7, \tag{12.98}$$

where $(x_i \ , \ y_i \ , \ z_i)$ are the known coordinates of the ith station, s_i the known distance of the ith station, and $(x_0 \ , \ y_0 \ , \ z_0)$ the unknown coordinates of the K1 station, see Fig. 10.2 in p.150 for the *Stuttgart Central Test Network.*

Using the GPS coordinates in *Table* 10.1 and the distances indicated in Fig. 10.2, and consider the initial value as the results of the 3-point problems solved via computer algebra (see Awange-Grafarend [41]), one can employ the Extended Newton-Raphson method. Let us consider the solution of the {1 - 2 - 7) combination,

$$x_0 = 4.15707550749 \times 10^6, y_0 = 671431.405189, z_0 = 4.7748874158 \times 10^6 \tag{12.99}$$

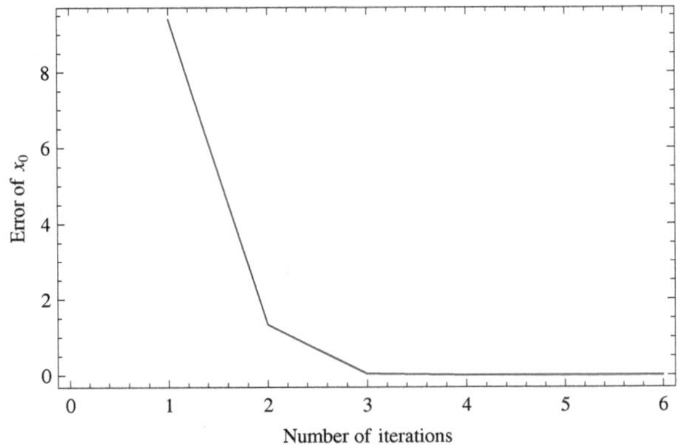

Figure 12.18. Convergence of the method in case of Stuttgart Central Test Network

Employing Extended Newton-Raphson method, the convergence is fast. Fig. 12.18 shows the absolute error of x_0 in meter as function of the number of iterations where $n = 3$ and $m = 7$.
The solution is

$$x_0 = 4.15706611153 \times 10^6, y_0 = 671429.665479, z_0 = 4.77487937031 \times 10^6 \tag{12.100}$$

12-4 Concluding remarks

In cases where positions are required from distance measurements such as point location in engineering and cadastral surveying, the algorithms presented in this chapter are handy. Users need only to insert measured distances and the coordinates of known stations in these algorithms to obtain their positions. In essence, one does not need to re-invent the wheel by going back to the Mathematica software! Additional literature on the topic are [1, 54, 105, 204, 253, 346].

13 Positioning by resection methods

13-1 Resection problem and its importance

In Chap. 12, ranging method for positioning was presented where distances were measured to known targets. In this chapter, an alternative positioning technique which uses direction measurements as opposed to distances is presented. This positioning approach is known as the resection. Unlike in ranging where measured distances are affected by atmospheric refraction, resection methods have the advantage that the measurements are angles or directions which are not affected by refraction.

Resection methods find use in densification of GPS networks. In Fig. 12.1 for example, if the station inside the tunnel or forest is a GPS station, a GPS receiver can not be used due to signal blockage. In such a case, horizontal and vertical directions are measured to three known GPS stations using a theodolite or total station operating in the local positioning systems (LPS). These angular measurements are converted into global reference frame's equivalent using (10.18) and (10.19). The coordinates of the unknown tunnel or forest station is finally computed using resection techniques that we will discuss later in the chapter. A more recent application of resection is demonstrated by [154] who applies it to find the position and orientation of scanner head in object space (Fig. 13.1[1]). The scanner is then used to monitor deformation of a steep hillside in Fig. 13.2[1] which was inaccessible. The only permissible deformation monitoring method was through remote sensing scanning technique.

To understand the resection problem, consider Fig. 13.3. The planar (two-dimensional) resection problem is formulated as follows: Given horizontal direction measurements T_i from unknown station $P_0 \in \mathbb{E}^2$ to three known stations $P_i|i = 1,2,3 \in \mathbb{E}^2$ in Fig. 13.3, determine the position $\{x_0, y_0\}$ and orientation $\{\sigma\}$ of P_0. For the three-dimensional resection, the unknown position $\{X_0, Y_0, Z_0\}$ of point $P_0 \in \mathbb{E}^3$ and the orientation unknown $\mathbf{\Sigma}$ have to be determined. In this case therefore, in addition to horizontal directions T_i, vertical directions B_i have to be measured. In photogrammetry, image coordinates on the photographs are used instead of direction measurements.

Equations relating unknowns and the measurements are nonlinear and the solution has been by linearized numerical iterative techniques. This has been mainly due to the difficulty of solving in closed form the underlying nonlinear systems of equations. Procedures for solving planar nonlinear resection are reported by [85] to exceed 500! Several procedures put forward as early as 1900 concentrated on the solution of the overdetermined version as evidenced in the works of [206, 349, 408, 409, 410]. Most of these works were based on the graphical approaches. Procedures to solve *closed form planar resection* were put forward by [92] and later by [15, 35, 78, 187, 237].

---

[1]Courtesy of Survey Review: Gordon and Lichti (2004)

J.L. Awange et al., *Algebraic Geodesy and Geoinformatics*, 2nd ed.,
DOI 10.1007/978-3-642-12124-1_13,

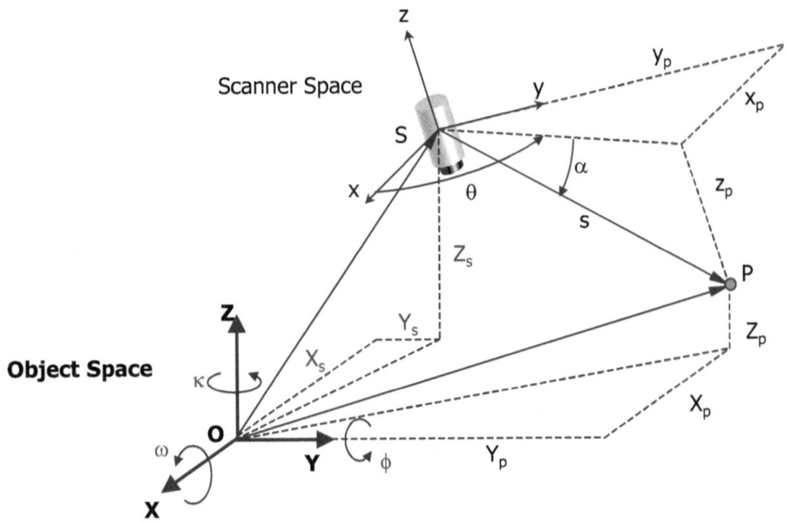

Figure 13.1. Position and orientation of scanner head
©Survey Review: Gordon and Lichti (2004)

Figure 13.2. Slope and Lower Walkway at Kings Park
©Survey Review: Gordon and Lichti (2004)

The search towards the solution of the three-dimensional resection problem traces its origin to the work of a German mathematician J. A. Grunert [194] whose publication appeared in the year 1841. Grunert (1841) solved the three-dimensional

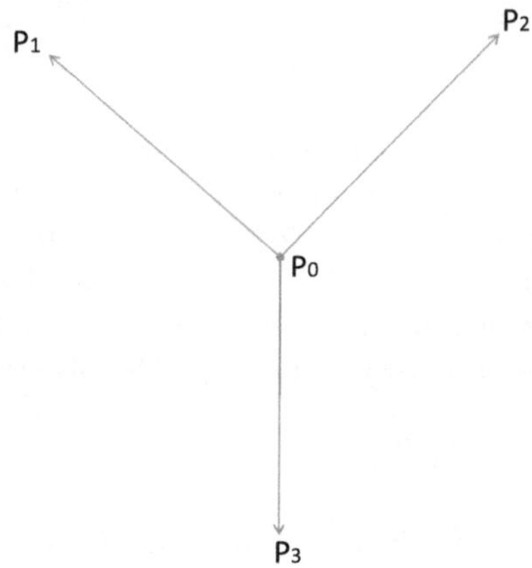

Figure 13.3. Planar distance observations

resection problem – what was then known as the *"Pothenot's"* problem – in a closed form by solving an algebraic equation of degree four. The problem had hitherto been solved by iterative means mainly in photogrammetry and computer vision. Procedures developed later for solving the three-dimensional resection problem revolved around improvements of the approach of [194] with the aim of searching for optimal means of distances determination. Whereas [194] solved the problem by substitution approach in *three steps*, more recent desire has been to solve the distance equations in less steps as exemplified in the works of [133, 135, 187, 275, 276, 301]. In [210, 211, 307], extensive review of these procedures are presented. Other solutions of three-dimensional resection include the works of [14, 27, 28, 154] among others. The closed form solution of overdetermined three-dimensional resection is presented in [29] and elaborate literature on the subject presented in [14].

In this chapter, *Grunert's distance equations* for three-dimensional resection problem are solved using the algebraic techniques of Groebner basis and polynomial resultants. The resulting quartic polynomial is solved for the unknown distances and the admissible solution substituted in any equation of the original system of polynomial equations to determine the remaining two distances. Once distances have been obtained, the position $\{X_0, Y_0, Z_0\}$ are computed using the ranging techniques discussed in Chap. 12. The three-dimensional orientation unknown $\mathbf{\Sigma}$ is thereafter solved using partial Procrustes algorithm of Chap. 9.

13-2 Geodetic resection

13-21 Planar resection

For planar resection, if the horizontal directions are oriented arbitrarily, the unknown orientation in the horizontal plane σ has to be determined in addition to position $\{x, y\}$ of the observing unknown station. The coordinates $X_i, Y_i \mid i \in \{1, 2, 3\}$ of the known target stations $P_i \in \mathbb{E}^2 \mid i \in \{1, 2, 3\}$ are given in a particular reference frame. Horizontal directions $T_i \mid i \in \{1, 2, 3\}$ are observed from an unknown station to the three known target stations. The task at hand as already stated in Sect. 13-1 is to determine the unknowns $\{x, y, \sigma\}$. The observation equation is formulated as

$$tan(T_i + \sigma) = \frac{y_i - y}{x_i - x} \mid \forall_i = 1, 2, 3. \tag{13.1}$$

Next, we present three approaches which can be used to solve (13.1) namely; conventional analytical solution, Groebner basis and Sylvester resultants methods.

13-211 Conventional analytical solution

Using trigonometric additions theorem as suggested by [92], (13.1) is expressed as

$$\frac{tan\, T_i + tan\, \sigma}{1 - tan\, T_i\, tan\, \sigma} = \frac{y_i - y}{x_i - x} \mid \forall_i = 1, 2, 3, \tag{13.2}$$

leading to

$$(tan\, T_i + tan\, \sigma)(x_i - x) = (1 - tan\, T_i\, tan\, \sigma)(y_i - y). \tag{13.3}$$

Expanding (13.3) gives

$$y(tan\, T_i\, tan\, \sigma) - y_i(tan\, T_i\, tan\, \sigma) - y + y_i = x_i tan\, T_i + x_i tan\, \sigma - x tan\, T_i - x tan\, \sigma. \tag{13.4}$$

Equation (13.4) leads to a nonlinear system of equations in the unknowns $\{x, y, \sigma\}$ as

$$\begin{bmatrix} y(tanT_1 tan\, \sigma) - y_1(tanT_1 tan\, \sigma) - y + y_1 = x_1 tanT_1 + x_1 tan\, \sigma - x tanT_1 - x tan\, \sigma \\ y(tanT_2 tan\, \sigma) - y_2(tanT_2 tan\, \sigma) - y + y_2 = x_2 tanT_2 + x_2 tan\, \sigma - x tanT_2 - x tan\, \sigma \\ y(tanT_3 tan\, \sigma) - y_3(tanT_3 tan\, \sigma) - y + y_3 = x_3 tanT_3 + x_3 tan\, \sigma - x tanT_3 - x tan\, \sigma, \end{bmatrix} \tag{13.5}$$

which is solved in three steps for σ and then substituted in the first two equations of (13.5) to obtain the unknowns $\{x, y\}$. The procedure is performed stepwise as follows:

Step 1 (elimination): In this step, the variable y and the term $xtan\,\sigma$ are eliminated from the three equations by subtracting the second and third expressions of (13.5) from the first. This results in

$$\begin{bmatrix} ytan\sigma(tanT_1 - tanT_2) = (tanT_2 - tanT_1)x + (x_1 - x_2 + y_1 tanT_1 \\ \quad -y_2 tanT_2)tan\sigma + x_1 tanT_1 - x_2 tanT_2 - y_1 + y_2 \\ \\ ytan\sigma(tanT_1 - tanT_3) = (tanT_3 - tanT_1)x + (x_1 - x_3 + y_1 tanT_1 \\ \quad -y_3 tanT_3)tan\sigma + x_1 tanT_1 - x_3 tanT_3 - y_1 + y_3. \end{bmatrix} \tag{13.6}$$

Step 2 (division): The first expression of (13.6) is divided by $(tanT_1 - tanT_2)$ and the second expression by $(tanT_1 - tanT_3)$. This is done in-order to make $ytan\,\sigma$ appearing on the left-hand-side of both equations the subject of the formula. The net results are:

$$\begin{bmatrix} ytan\sigma = -x + \dfrac{x_1 - x_2 + y_1 tanT_1 - y_2 tanT_2}{tanT_1 - tanT_2} tan\sigma \\ \quad + \dfrac{x_1 tanT_1 - x_2 tanT_2 - y_1 + y_2}{tanT_1 - tanT_2} \\ \\ ytan\sigma = -x + \dfrac{x_1 - x_3 + y_1 tanT_1 - y_3 tanT_3}{tanT_1 - tanT_3} tan\sigma \\ \quad + \dfrac{x_1 tanT_1 - x_3 tanT_3 - y_1 + y_3}{tanT_1 - tanT_3}. \end{bmatrix} \tag{13.7}$$

Step 3 (elimination): In (13.7), we note that $ytan\,\sigma$ and x appear in both expressions. They are eliminated by subtracting the second expression from the first. On re-arranging the resulting expression leads to $tan\sigma$ on the left-hand-side as

$$\begin{bmatrix} tan\,\sigma = \dfrac{N}{D} \\ \\ N = \dfrac{x_1 tanT_1 - x_2 tanT_2 - y_1 + y_2}{tanT_1 - tanT_2} + \dfrac{x_3 tanT_3 - x_1 tanT_1 - y_3 + y_1}{tanT_1 - tanT_3} \\ \\ D = \dfrac{x_1 - x_3 + y_1 tanT_1 - y_3 tanT_3}{tanT_1 - tanT_3} + \dfrac{x_2 - x_1 + y_2 tanT_2 - y_1 tanT_1}{tanT_1 - tanT_2} \end{bmatrix} \tag{13.8}$$

Step 4 (solution of $\{x, y\}$): Once we have solved for σ in (13.8), the first and the second expressions of (13.5) are re-written in the final step with x, y on the left-hand-side as

$$\begin{bmatrix} a_{11}y + a_{12}x = b_{11} \\ a_{21}y + a_{22}x = b_{22}, \end{bmatrix} \tag{13.9}$$

where;

$$\begin{bmatrix} a_{11} = (tanT_1 tan\sigma - 1), \\ a_{12} = (tanT_1 + tan\sigma), \\ a_{21} = (tanT_2 tan\sigma - 1), \\ a_{22} = (tanT_2 + tan\sigma), \\ b_{11} = y_1 tanT_1 tan\sigma - y_1 + x_1 tanT_1 + x_1 tan\sigma, \\ b_{22} = y_2 tanT_2 tan\sigma - y_2 + x_2 tan\sigma + x_2 tanT_2. \end{bmatrix}$$

In matrix form, (13.9) is expressed as

$$\begin{bmatrix} a_{11} & a_{12} \\ a_{21} & a_{22} \end{bmatrix} \begin{bmatrix} y \\ x \end{bmatrix} = \begin{bmatrix} b_{11} \\ b_{22} \end{bmatrix}, \tag{13.10}$$

giving the solutions as

$$\begin{bmatrix} y \\ x \end{bmatrix} = (a_{11}a_{22} - a_{21}a_{12})^{-1} \begin{bmatrix} a_{22} & -a_{12} \\ -a_{21} & a_{11} \end{bmatrix} \begin{bmatrix} b_{11} \\ b_{22} \end{bmatrix} \tag{13.11}$$

or

$$\boxed{\begin{array}{l} y = (a_{11}a_{22} - a_{21}a_{12})^{-1}(a_{22}b_{11} - b_{22}a_{12}) \\ x = (a_{11}a_{22} - a_{21}a_{12})^{-1}(a_{11}b_{22} - b_{11}a_{21}) \end{array}}, \tag{13.12}$$

which completes the conventional analytic solution.

13-212 Groebner basis approach

Denoting $a = tan\,T_1$, $b = tan\,T_2$, $c = tan\,T_3$, and $d = tan\,\sigma$, (13.5) is simplified in lexicographic order $y > x > d$ as

$$\begin{bmatrix} f_1 := -y + ady + ax + xd - y_1 ad - x_1 d - x_1 a + y_1 = 0 \\ f_2 := -y + bdy + bx + xd - y_2 bd - x_2 d - x_2 b + y_2 = 0 \\ f_3 := -y + cdy + cx + xd - y_3 cd - x_3 d - x_3 c + y_3 = 0. \end{bmatrix} \tag{13.13}$$

The reduced Groebner basis (4.38) on p. 45 is then computed as

$$\begin{bmatrix} GroebnerBasis\,[\{f_1, f_2, f_3\}, \{x, y, d\}, \{x, y\}] \\ GroebnerBasis\,[\{f_1, f_2, f_3\}, \{x, y, d\}, \{y\}] \\ GroebnerBasis\,[\{f_1, f_2, f_3\}, \{x, y, d\}, \{x\}]. \end{bmatrix} \tag{13.14}$$

The first expression of (13.14) gives a linear equation in the variable d allowing the computation of the unknown orientation parameter σ. The second and the third expressions respectively give linear equations in x and y in the variable d. The computed reduced Groebner basis re-arranged with the unknown terms on the left-hand-side are presented in Solution 13.1.

Solution 13.1 (reduced Groebner basis computation of planar resection).

$$\begin{bmatrix} d = \dfrac{N1}{D1} \\ \\ x = -\dfrac{N2}{(-cd^2 + a - c + ad^2)} \\ \\ y = -\dfrac{N3}{(b - c + bd^2 - cd^2)}, \end{bmatrix} \tag{13.15}$$

where

$$\begin{bmatrix} N1 = -(abX_1 - acX_1 + aY_2 - abX_2 + bcX_2 - aY_3 + acX_3 \\ -bcX_3 - bY_1 + cY_1 - cY_2 + bY_3) \\ \\ D1 = (bX_1 - aX_2 + aX_3 - bX_3 + abY_1 - acY_1 - cX_1 - abY_2 \\ +bcY_2 + cX_2 + acY_3 - bcY_3), \end{bmatrix}$$

$$N2 = \begin{bmatrix} (-adY_1 - cdY_1 + acd^2Y_1 - Y_3 - aX_1 - dX_1 + acdX_1 + cd^2X_1 \\ +cX_3 + dX_3 - acdX_3 - ad^2X_3 + Y_1 + adY_3 + cdY_3 - acd^2Y_3) \end{bmatrix}$$

and

$$N3 = \begin{bmatrix} (-bcX_2 - bdX_2 - cdX_2 - d^2X_2 + bcX_3 + bdX_3 + cdX_3 + d^2X_3 \\ +cY_2 + dY_2 - bcdY_2 - bd^2Y_2 - bY_3 - dY_3 + bcdY_3 + cd^2Y_3). \end{bmatrix}$$

Once d has been computed from the first expression of (13.15), it is inserted into the second and third expressions to solve the unknowns $\{x, y\}$ respectively. The unknown orientation in the horizontal plane is then computed via $\sigma = tan^{-1}d$.

13-213 Sturmfels' resultant approach

Let z be a homogenizing variable for (13.13). In-order to solve for the variable d in (13.13), we hide d by making it a polynomial of degree zero (i.e., treating it as a constant) as

$$\begin{bmatrix} g_1 := -y + ady + ax + dx + (-y_1ad - x_1d - x_1a + y_1)z = 0 \\ g_2 := -y + bdy + bx + dx + (-y_2bd - x_2d - x_2b + y_2)z = 0 \\ g_3 := -y + cdy + cx + dx + (-y_3cd - x_3d - x_3c + y_3)z = 0, \end{bmatrix} \tag{13.16}$$

which is expressed in the form (5.12) as

$$J_d = \det \begin{bmatrix} \dfrac{\partial g_1}{\partial x} & \dfrac{\partial g_1}{\partial y} & \dfrac{\partial g_1}{\partial z} \\ \\ \dfrac{\partial g_2}{\partial x} & \dfrac{\partial g_2}{\partial y} & \dfrac{\partial g_2}{\partial z} \\ \\ \dfrac{\partial g_3}{\partial x} & \dfrac{\partial g_3}{\partial y} & \dfrac{\partial g_3}{\partial z} \end{bmatrix} = \det \begin{bmatrix} (a+d) & (ad-1) & (y_1 - y_1ad - x_1d - x_1a) \\ (b+d) & (bd-1) & (y_2 - y_2bd - x_2d - x_2b) \\ (c+d) & (cd-1) & (y_3 - y_3cd - x_3d - x_3c) \end{bmatrix}. \tag{13.17}$$

Equation (13.17) leads to

$$\boxed{e_3d^3 + e_2d^2 + e_1d + e_0 = 0},$$
with

$$e_3 = ax_3 - bx_3 - ay_2b - ax_2 + cx_2 - cx_1 + cy_2b + ay_3c - cy_1a + bx_1 + by_1a - by_3c$$

$$e_2 = cy_1 - ax_2b + ay_2 + ax_3c - by_1 - ay_3 + bx_1a + by_3 + cx_2b - cy_2 - cx_1a - bx_3c$$

$$e_1 = ax_3 - bx_3 - ay_2b - ax_2 + cx_2 - cx_1 + cy_2b + ay_3c - cy_1a + bx_1 + by_1a - by_3c$$

$$e_0 = cy_1 - ax_2b + ay_2 + ax_3c - by_1 - ay_3 + bx_1a + by_3 + cx_2b - cy_2 - cx_1a - bx_3c. \tag{13.18}$$

The value of d is then solved from (13.18) using Matlab's roots command (see 4.42 on p. 46). Comparing the expressions for d in (13.15) and (13.18), we note that the reduced Groebner basis in Solution 13.1 gave a linear function while the Sturmfels' approach results in a cubic polynomial. Both expressions however lead to the same numerical results. The advantage of reduced Groebner basis over the Sturmfels' approach, however, is that the solution is uniquely determined. Sturmfels' approach requires prior information to choose the admissible value of d from the three solutions. Once this value has been selected, the coordinates $\{x, y\}$ are then solved in terms of d as follows:

1. Hiding x and solving in terms of d from (f_1, f_2) of (13.13) gives

$$\begin{bmatrix} h_1 := (ad-1)y + (ax + dx - y_1ad - x_1d - x_1a + y_1)z = 0 \\ h_2 := (bd-1)y + (bx + dx - y_2bd - x_2d - x_2b + y_2)z = 0. \end{bmatrix} \tag{13.19}$$

Applying (5.12) leads to

$$J_x = \det \begin{bmatrix} \frac{\partial h_1}{\partial y} & \frac{\partial h_1}{\partial z} \\ \frac{\partial h_2}{\partial y} & \frac{\partial h_2}{\partial z} \end{bmatrix} = \det \begin{bmatrix} (ad-1) & (ax + dx - y_1ad - x_1d - x_1a + y_1) \\ (bd-1) & (bx + dx - y_2bd - x_2d - x_2b + y_2) \end{bmatrix}. \tag{13.20}$$

The Jacobian determinant of (13.20) is

$$\begin{bmatrix} x = -(y_1 + ady_2 - ad^2y_2b - ad^2x_2 - adx_2b + x_1abd - y_2 + y_2bd + x_2d \\ +x_2b + y_1ad^2b - y_1ad - y_1bd - x_1d - x_1a + x_1d^2b)/(a + ad^2 - b - d^2b). \end{bmatrix} \tag{13.21}$$

2. Hiding y and solving in terms of d from (f_2, f_3) of (13.13) gives

$$\begin{bmatrix} k_1 := (b+d)x + (bdy - y - y_2bd - x_2d - x_2b + y_2)z = 0 \\ k_2 := (c+d)x + (cdy - y - y_3cd - x_3d - x_3c + y_3)z = 0, \end{bmatrix} \tag{13.22}$$

whose Jacobian determinant

$$J_y = \det \begin{bmatrix} \frac{\partial k_1}{\partial x} & \frac{\partial k_1}{\partial z} \\ \frac{\partial k_2}{\partial x} & \frac{\partial k_2}{\partial z} \end{bmatrix} = \det \begin{bmatrix} (b+d) & (bdy - y - y_2bd - x_2d - x_2b + y_2) \\ (c+d) & (cdy - y - y_3cd - x_3d - x_3c + y_3) \end{bmatrix}. \quad (13.23)$$

leads to

$$\begin{bmatrix} y = -(-y_2d + by_3 - by_3cd - bx_3d - bx_3c + x_2bc + dy_3 - y_3cd^2 - x_3d^2 \\ -dx_3c + y_2bdc + y_2bd^2 - y^2c + x_2d^2 + x_2bd + x_2dc)/(-bd^2 - b + cd^2 + c) \end{bmatrix} \quad (13.24)$$

Once d has been computed from ((13.18), it is used in (13.21) and (13.24) to obtain x and y respectively. The unknown orientation in the horizontal plane can now be computed via $\sigma = tan^{-1}d$.

Example 13.1. Let us consider the Example given by [237, p. 234] with our axis defined such that the Easting refer to the $Y-$axis and the Northing refer to the $X-$axis. The input data are given in Table 13.1 for the coordinates of three known stations A, B and M which are denoted by P_1, P_2 and P_3 respectively in Fig. 13.3. Table 13.2 gives directional observations $T_i \mid i \in \{1,2,3\}$ from the observing unknown station $P \in \mathbb{E}^2$ (whose unknown x, y coordinates and orientation parameter σ are sought) to three known stations $P_i \in \mathbb{E}^2 \mid i \in \{1,2,3\}$ whose coordinates $X_i, Y_i \mid i \in \{1,2,3\}$ are given in Table 13.1. The obtained results from either reduced Groebner basis or Sturmfels' resultant algebraic approaches are presented in Table 13.3. They are identical to those of [237, p. 234] once we interchange the axes. If Sturmfels' solution is adopted in (13.18), two complex and one real values of d are obtained. The real value, which is identical to that obtained from reduced Groebner basis solution (13.15) is used to solve for σ from $\sigma = tan^{-1}d$.

Table 13.1. Coordinates of known stations $P_i \in \mathbb{E}^2 \mid i \in \{1,2,3\}$

Station	Easting $Y(m)$	Northing $X(m)$
P_1	46867.94	5537.00
P_2	51293.86	6365.89
P_3	49666.56	4448.58

Table 13.2. Directions measured from unknown station $P \in \mathbb{E}^2$ to known stations $P_i \in \mathbb{E}^2 \mid i \in \{1,2,3\}$

Station	Horizontal directions °	′	″
P_1	60	07	50
P_2	265	18	22
P_3	326	33	59

Table 13.3. Position and orientation of station $P \in \mathbb{E}^2$.

Station	Easting $Y(m)$	Northing $X(m)$	Orientation unknown ° ′ ″
P	48613.3384	6361.1690	
σ			4 35 34.7

13-22 Three-dimensional resection

13-221 Exact solution

Closed form solution of three-dimensional resection problem concerns itself with the determination of position and orientation of a point P connected by angular observations of type horizontal directions T_i and vertical directions B_i to three known stations P_1, P_2, P_3 (see e.g., Fig. 13.4 on p. 227). From these angular measurements, distances are derived by solving equations known as *Grunert's equations*. Once the distances have been established, the unknown position P is determined using ranging techniques that we discussed in Sect. 12-32 of Chap. 12. The closed form solution of the three-dimensional resection problem is completed by solving the unknown orientation parameters that relate the global reference frame $\mathbb{F}^\bullet$ to the local level reference frame of type $\mathbb{F}^*$. As we have already pointed out in Sect. 13-1, several procedures have been suggested for solving Grunert's equations. This section presents three alternative algebraic methods for solving explicitly the three-dimensional resection problem namely; Groebner basis, polynomial resultants and Grafarend-Lohse-Schaffrin methods.

13-222 Solution of Grunert's distance equations

We begin in Solution 13.2 by deriving *Grunert*'s distance equations. These equations relate;

(i) known distances $S_{ij},\ i,j = 1,2,3\,|\,i \neq j$ computed from known stations,
(ii) unknown distances $S_i,\ i = 1,2,3$ between the unknown station $P \in \mathbb{E}^3$, and three known stations $P_i \in \mathbb{E}^3 \mid i \in \{1,2,3\}$ and
(iii) the spatial angles $\psi_{ij},\ i,j = 1,2,3\,|\,i \neq j$ derived from measured horizontal directions T_i and vertical directions B_i in the local level reference frame $\mathbb{F}.^*$

In Solution 13.2, multiplying (10.11) on p. 145 by (13.25) leads to (13.26). After manipulations of (13.27),(13.28) and (13.29), space angles ψ_{ij} can be written in terms of spherical coordinates $\{T_i, B_i\}, \{T_j, B_j\}$ of points P_i and P_j with respect to a theodolite orthogonal Euclidean frame $\mathbb{F}^*$ as in (13.30). The Grunert's equations for the three unknown distances S_1, S_2, S_3 are then written in terms of known distances S_{12}, S_{23}, S_{31} and space angles $\psi_{12}, \psi_{23}, \psi_{31}$ (illustrated in Fig. 13.4) as in (13.32). Solution of (13.32) was first proposed by *J. A. Grunert* [194]. Procedures that were later developed sought to optimize the solution of (13.32) in terms of computational steps. In particular, the interest was to reduce the order

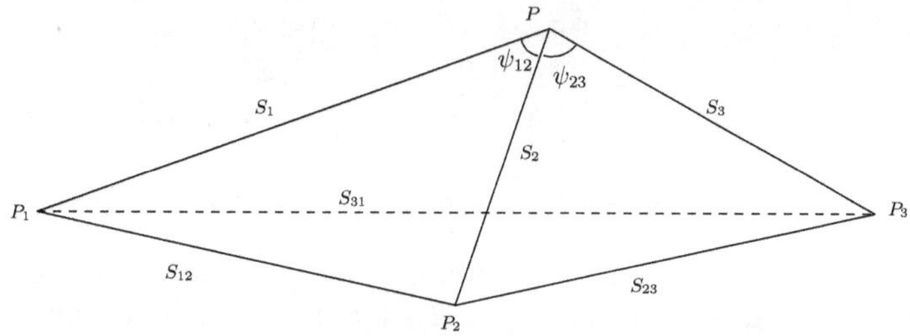

Figure 13.4. Geometrical interpretation of the 3D resection

of the univariate polynomial that resulted following the solution of (13.32). Such procedures were encountered in Sect. 13-1. In what follows, we present algebraic solution of (13.32).

Solution 13.2 (Derivation of Grunert's distance equations).

$$(-2)\,[\cos T_j \cos B_j, \sin T_j \cos B_j, \sin B_j]\, S_j \tag{13.25}$$

$$\left[\begin{array}{c} (-2)[cosT_j cosB_j, sinT_j cosB_j, sinB_j] S_i S_j \begin{bmatrix} cosT_i cosB_i \\ sinT_i cosB_i \\ sinB_i \end{bmatrix} = \\ \\ (-2)[(X_j - X), (Y_j - Y), (Z_j - Z)] \begin{bmatrix} X_i - X \\ Y_i - Y \\ Z_i - Z \end{bmatrix} \end{array}\right. \tag{13.26}$$

$$\left.\begin{array}{c} (X_j - X)(X_i - X) = X_j X_i - X_j X - X_i X + X^2 \\ (X_i - X_j)(X_i - X_j) = X_i^2 - 2X_i X_j + X_j^2 \\ (X_i - X)(X_i - X) = X_i^2 - 2X_i X + X^2 \\ (X_j - X)(X_j - X) = X_j^2 - 2X_j X + X^2 \end{array}\right] \Rightarrow$$
$$\Rightarrow (X_i - X_j)^2 - (X_i - X)^2 - (X_j - X)^2 = -2(X_j - X)(X_i - X) \tag{13.27}$$

$$\left[\begin{array}{c} (-2)[cosT_j cosB_j, sinT_j cosB_j, sinB_j] S_i S_j \begin{bmatrix} cosT_i cosB_i \\ sinT_i cosB_i \\ sinB_i \end{bmatrix} = \\ \\ \left\{\begin{array}{c} (X_i - X_j)^2 + (Y_i - Y_j)^2 + (Z_i - Z_j)^2 - \\ -(X_i - X)^2 - (Y_i - Y)^2 - (Z_i - Z)^2 - \\ -(X_j - X)^2 - (Y_j - Y)^2 - (Z_j - Z)^2 \end{array}\right\} \end{array}\right. \tag{13.28}$$

$$\left[\begin{array}{l} -2\,\{\sin B_j \sin B_i + \cos B_j \cos B_i \cos(T_j - T_i)\}\, S_i S_j = \\ = \left\{\begin{array}{l} (X_i - X_j)^2 + (Y_i - Y_j)^2 + (Z_i - Z_j)^2 - \\ - (X_i - X)^2 - (Y_i - Y)^2 - (Z_i - Z)^2 - \\ - (X_j - X)^2 - (Y_j - Y)^2 - (Z_j - Z)^2 \end{array}\right\} \end{array}\right. \tag{13.29}$$

$$\cos \psi_{ij} = \cos B_i \cos B_j \cos(T_j - T_i) + \sin B_i \sin B_j \tag{13.30}$$

$$\left[\begin{array}{l} -2 \cos \psi_{ij} S_i S_j = S_{ij}^2 - S_i^2 - S_j^2 \\ S_{ij}^2 = S_i^2 + S_j^2 - 2 S_i S_j \cos \psi_{ij} \end{array}\right. \tag{13.31}$$

$$\begin{bmatrix} S_{12}^2 = S_1^2 + S_2^2 - 2S_1S_2\cos\psi_{12} \\ S_{23}^2 = S_2^2 + S_3^2 - 2S_2S_3\cos\psi_{23} \\ S_{31}^2 = S_3^2 + S_1^2 - 2S_3S_1\cos\psi_{31} \end{bmatrix} \tag{13.32}$$

13-223 Groebner basis solution of Grunert's equations

In-order to quicken our understanding of the application of Groebner basis to solve Grunert's distance equations (13.32), let us consider a simple case of a regular tetrahedron. A regular tetrahedron presents a unique case where all the distances and spatial angles of Fig. 13.4 are equal. Instead of computing Groebner basis using 4.36 on p. 45, we will demonstrate by a hand computation how Groebner basis can be computed. Later, we will apply (4.36) to solve the general case of (13.32). We begin by expressing (13.32) in algebraic form

$$\begin{bmatrix} x_1^2 - 2a_{12}x_1x_2 + x_2^2 - a_0 = 0 \\ x_2^2 - 2b_{23}x_2x_3 + x_3^2 - b_0 = 0 \\ x_1^2 - 2c_{31}x_1x_3 + x_3^2 - c_0 = 0, \end{bmatrix} \tag{13.33}$$

where the unknown distances $\{S_1, S_2, S_3\}$ that appear in (13.32) are denoted by $\{x_1, x_2, x_3\}$. The distances between known stations $\{S_{12}, S_{23}, S_{31}\}$ are denoted by $\{a_o, b_o, c_o\}$, while the constants $\{a_{12}, b_{23}, c_{31}\}$ represent $\{cos\,\psi_{12}, cos\,\psi_{23}, cos\,\psi_{31}\}$ respectively. Equation (13.33) therefore has only the distances $\{x_1, x_2, x_3\}$ as unknowns. These are the distances relating the unknown station P_0 to the known stations $P_i|_{\{i=1,2,3\}}$. Grafarend [187] demonstrated that for each of the quadratic equation in (13.33), there exists an elliptical cylinder in the planes $\{x_1, x_2\}$, $\{x_2, x_3\}$ and $\{x_3, x_1\}$ for the first, second and third equations respectively. These cylinders are constrained to their first quadrant since the distances are positive thus $\{x_1 \in \mathbb{R}^+\}$, $\{x_2 \in \mathbb{R}^+\}$ and $\{x_3 \in \mathbb{R}^+\}$. For a regular tetrahedron, the distances $x_1 = x_2 = x_3$ joining the unknown station $P \in \mathbb{E}^3$ to three known stations $P_i \in \mathbb{E}^3|_{\{i=1,2,3\}}$ are all equal to the distances $S_{12} = S_{23} = S_{31}$ between the known stations. Let us give these distances a value $+\sqrt{d}$. The spatial angles are also equal (i.e., $\psi_{12} = \psi_{23} = \psi_{31} = 60°$). In Solution 13.3, a hand computation of Groebner basis of (13.33) is carried out and used to find the Grunert's distances for the regular tetrahedron (i.e., show that the desired solutions for $\{x_1, x_2, x_3\} \in \mathbb{R}^+$ are $x_1 = x_2 = x_3 = +\sqrt{d}$.)

Solution 13.3 (Hand computation of Groebner basis of (13.32) for a regular tetrahedron). For a regular tetrahedron, where $\psi_{ij} = 60°$, and $a_0 = b_0 = c_0 = d$, (13.33) is re-written in lexicographic order $x_1 > x_2 > x_3$ as

$$\begin{bmatrix} x_1^2 - x_1x_2 + x_2^2 - d = 0 \\ x_2^2 - x_2x_3 + x_3^2 - d = 0 \\ x_1^2 - x_1x_3 + x_3^2 - d = 0, \end{bmatrix} \tag{13.34}$$

giving rise to the Ideal I (e.g., 4.14 on p. 37) as

$$\begin{aligned} I = &\langle x_1^2 - x_1x_2 + x_2^2 - d, x_2^2 - x_2x_3 + x_3^2 - d, x_1^2 - x_1x_3 + x_3^2 - d\rangle \\ &\subset \mathbb{R}[x_1, x_2, x_3], \end{aligned} \tag{13.35}$$

whose generators G are

$$\begin{bmatrix} g_1 = x_1^2 - x_1x_2 + x_2^2 - d \\ g_2 = x_1^2 - x_1x_3 + x_3^2 - d \\ g_3 = x_2^2 - x_2x_3 + x_3^2 - d. \end{bmatrix} \tag{13.36}$$

Desired now are the Groebner basis (simplified structure) of the generators (13.36) of the Ideal I in (13.35). Using (4.23) on p. 41, the S pair polynomials $(g_1, g_2), (g_1, g_3), (g_2, g_3)$ are computed from the generators (13.36). From *B. Buchberger's* third criterion explained in Chap. 4, we notice that $\mathrm{LM}(g_2) = x_1^2$ divides the $\mathrm{LCM}(g_1, g_3) = x_1^2x_2^2$. One therefore suppresses (g_1, g_3) and considers only $(g_1, g_2), (g_2, g_3)$ instead. $S(g_1, g_2)$ gives

$$S(g_1, g_2) = -x_1x_2 + x_1x_3 + x_2^2 - x_3^2, \tag{13.37}$$

which is reduced with respect to G by subtracting g_3 to obtain

$$-x_1x_2 + x_1x_3 - 2x_3^2 + x_2x_3 + d. \tag{13.38}$$

Equation (13.38) does not reduce to zero and is added to the original list G of the generating set of the Ideal I as g_4. The S–polynomial pairs to be considered next are $S(g_2, g_3), S(g_2, g_4)$ and $S(g_3, g_4)$ from the new generating set $G = \{g_2, g_3, g_4\}$. Since $LM(g_2)$ and $LM(g_3)$ are relatively prime, $S(g_2, g_3)$ reduces to zero modulo G ($S(g_2, g_3) \rightarrow_G 0$). The S pair polynomials remaining for consideration are (g_2, g_4) and (g_3, g_4). $S(g_2, g_4)$ gives

$$S(g_2, g_4) = x_1^2x_3 + x_1d - 2x_1x_3^2 + x_2x_3^2 - x_2d, \tag{13.39}$$

which is reduced with respect to G by subtracting x_3g_2 to give

$$x_1d - x_1x_3^2 + x_2x_3^2 - x_2d - x_3^3 + x_3d, \tag{13.40}$$

Equation (13.40) does not reduce to zero and is added to the list G of the generating set of the Ideal I as g_5. The S–polynomial pair to be considered next is $S(g_3, g_4)$ from the new generating set $G = \{g_2, g_3, g_4, g_5\}$. $S(g_3, g_4)$ gives

$$S(g_3, g_4) = -x_1x_3^2 + x_1d + 2x_2x_3^2 - x_2^2x_3 - x_2d, \tag{13.41}$$

which is reduced with respect to G by subtracting g_5 and adding x_3g_3 to give

$$2x_3^3 - 2x_3d. \tag{13.42}$$

Equation (13.42) is a univariate polynomial and completes the solution of the set G of Groebner basis summarized as

$$G := \begin{bmatrix} g_2 = x_1^2 - x_1x_3 + x_3^2 - d \\ g_3 = x_2^2 - x_2x_3 + x_3^2 - d \\ g_4 = -x_1x_2 + x_1x_3 - 2x_3^2 + x_2x_3 + d \\ g_5 = x_1d - x_1x_3^2 + x_2x_3^2 - x_2d - x_3^3 + x_3d \\ g_6 = 2x_3^3 - 2x_3d. \end{bmatrix} \tag{13.43}$$

From the computed Groebner basis in (13.43), one notes that the element $g_6 = 2x_3^3 - 2x_3 d$ is a *cubic polynomial* in x_3 and readily gives the values of $x_3 = \left\{0, \pm\sqrt{d}\right\}$. The solutions to the Grunert's distance equations (13.33) for a regular tetrahedron are then deduced as follows: Since $S_3 = x_3 \in \mathbb{R}^+$, the value of $S_3 = +\sqrt{d}$. This is substituted back in $g_3 = x_2^2 - x_2x_3 + x_3^2 - d$ and $g_2 = x_1^2 - x_1x_3 + x_3^2 - d$ to give $x_2 = \left\{0, +\sqrt{d}\right\}$ and $x_1 = \left\{0, +\sqrt{d}\right\}$ respectively. This completes the solution of Grunert's distance equations (13.33) for the unknown distances $x_1 = x_2 = x_3 = +\sqrt{d}$ as we had initially assumed.

Having demonstrated a hand computation of Groebner basis of the Grunert's distance equations (13.32) for a regular tetrahedron, let us consider next the general case. The geometry of the three-dimensional resection problem in practice is hardly a regular tetrahedron. Beginning by expressing (13.32) algebraically as

$$\left[\begin{array}{l} g_1 := x_1^2 + x_2^2 + a_{12}x_1x_2 + a_0 = 0 \\ g_2 := x_2^2 + x_3^2 + b_{23}x_2x_3 + b_0 = 0 \\ g_3 := x_3^2 + x_1^2 + c_{31}x_3x_1 + c_0 = 0, \end{array}\right. \tag{13.44}$$

where

$$\left[\begin{array}{l} S_1 = x_1 \in \mathbb{R}^+, S_2 = x_2 \in \mathbb{R}^+, S_3 = x_3 \in \mathbb{R}^+, \\ -2\cos\psi_{12} = a_{12}, -2\cos\psi_{23} = b_{23}, -2\cos\psi_{31} = c_{31}, \\ -S_{12}^2 = a_0, -S_{23}^2 = b_0, -S_{31}^2 = c_0, \end{array}\right. \tag{13.45}$$

one forms the Ideal

$$I =< x_1^2 + x_2^2 + a_{12}x_1x_2 + a_0, x_2^2 + x_3^2 + b_{23}x_2x_3 + b_0, x_3^2 + x_1^2 + c_{31}x_3x_1 + c_0 > . \tag{13.46}$$

We then seek the Groebner basis of the generators of the Ideal (13.46). Following lexicographic ordering $\{x_1 > x_2 > x_3\}$, (4.36) on p. 45 is applied as

$$GroebnerBasis[\{g_1, g_2, g_3\}, \{x_1, x_2, x_3\}], \tag{13.47}$$

giving the Groebner basis of the Ideal (13.46) expressed in [27, Boxes 3-3a and 3-3b]. Distances can also be derived from (13.44) using reduced Groebner basis. We leave it as an exercise for the reader to try and solve the unknown distances $\{x_1, x_2, x_3\}$ in (13.44) using reduced Groebner basis (4.38) on p. 45.

13-224 Polynomial resultants' solution of Grunert's distance equations

Besides the use of Groebner bases approach demonstrated above, polynomial resultants techniques can also be used to solve Grunert's equations for distances. We illustrate the solution of the problem using F. Macaulay formulation of Sect. 5-31 and B. Sturmfels' formulation presented in Sect. 5-32. We start by expressing (13.44) as

$$\left[\begin{array}{l} R_1 := x_1^2 + x_2^2 + a_{12}x_1x_2 + a_0 = 0 \\ R_2 := x_2^2 + x_3^2 + b_{23}x_2x_3 + b_0 = 0 \\ R_3 := x_1^2 + x_3^2 + c_{31}x_1x_3 + c_0 = 0. \end{array}\right. \tag{13.48}$$

Clearly, (13.48) is not homogeneous (see Definition 5.1 on p. 49). It is therefore homogenized by introducing the fourth variable x_4 and treating the variable which is to be solved, say x_1, as a constant (i.e., hiding it by giving it degree zero). The resulting homogenized polynomial is

$$\begin{bmatrix} R_{11} := x_2^2 + a_{12}x_1x_2x_4 + (a_0 + x_1^2)x_4^2 = 0 \\ R_{21} := x_2^2 + x_3^2 + b_{23}x_2x_3 + b_0x_4^2 \quad\quad = 0 \\ R_{31} := x_3^2 + c_{31}x_1x_3x_4 + (x_1^2 + c_0)x_4^2 = 0, \end{bmatrix} \tag{13.49}$$

which is simplified as

$$\begin{bmatrix} R_{11} := x_2^2 + a_1x_2x_4 + a_2x_4^2 \quad\quad = 0 \\ R_{21} := x_2^2 + x_3^2 + b_1x_2x_3 + b_2x_4^2 = 0 \\ R_{31} := x_3^2 + c_1x_3x_4 + c_2x_4^2 \quad\quad = 0, \end{bmatrix} \tag{13.50}$$

with the coefficients denoted as $a_1 = a_{12}x_1$, $a_2 = (a_0 + x_1^2)$, $b_1 = b_{23}$, $b_2 = b_0$, $c_1 = c_{31}x_1$, $c_2 = (c_0 + x_1^2)$.

Approach 1 (F. Macaulay Formulation):

The *first step* involves the determination of the total degree of (13.50) using (5.7) on p. 52 which gives $d = 4$. In the *second step*, one formulates the general set comprising the monomials of degree 4 in three variables by multiplying the monomials of (13.50) by each other. These monomials form the elements of the set X^d (e.g., 5.8 on p. 52) as

$$X^d = \left\{ \begin{array}{c} x_2^4,\ x_2^3x_4,\ x_2^2x_3^2,\ x_2^3x_3,\ x_2^2x_4^2,\ x_2^2x_3x_4,\ x_2x_3^3 \\ \\ x_2x_4^3,\ x_2x_3^2x_4,\ x_2x_3x_4^2,\ x_3^2x_4^2,\ x_3x_4^3,\ x_4^4,\ x_3^4,\ x_3^3x_4 \end{array} \right\}, \tag{13.51}$$

which is now partitioned in *step 3* according to (5.9) on p. 52 as

$$\begin{bmatrix} X_i^d = \{x^\alpha \mid \alpha_i \geq d_i \, and \, \alpha_j < d_j, \forall_j < i\} \\ \\ X_2^4 = \{x_2^4,\ x_2^3x_4,\ x_2^2x_3^2,\ x_2^3x_3,\ x_2^2x_4^2,\ x_2^2x_3x_4\} \\ \\ X_3^4 = \{x_2x_3^2x_4,\ x_3^2x_4^2,\ x_2x_3^3,\ x_3^4,\ x_3^3x_4\} \\ \\ X_4^4 = \{x_2x_4^3,\ x_2x_3x_4^2,\ x_3x_4^3,\ x_4^4\}. \end{bmatrix} \tag{13.52}$$

In the *fourth step*, the polynomials F_i are formed using the sets in (13.52) according to (5.10) on p. 53 giving rise to

$$\begin{bmatrix} F_1 := \dfrac{X_2^4}{x_2^2} f_1 = \{x_2^2f_1,\ x_2x_4f_1,\ x_3^2f_1,\ x_2x_3f_1,\ x_4^2f_1,\ x_3x_4f_1\} \\ \\ F_2 := \dfrac{X_3^4}{x_3^2} f_2 = \{x_2x_4f_2,\ x_4^2f_2,\ x_2x_3f_2,\ x_3^2f_2,\ x_3x_4f_2\} \\ \\ F_3 := \dfrac{X_4^4}{x_4^2} f_3 = \{x_2x_4f_3,\ x_2x_3f_3,\ x_3x_4f_3,\ x_4^2f_3\}. \end{bmatrix} \tag{13.53}$$

Finally, the matrix **A** of dimension (15 × 15) is formed as discussed on p. 53. Its rows are the coefficients of the f_i in (13.53) and the columns are the monomials $\{c_1 = x_2^4,\ c_2 = x_2^3x_3,\ c_3 = x_2^3x_4,\ c_4 = x_2^2x_3^2,\ c_5 = x_2^2x_3x_4,\ c_6 = x_2^2x_4^2,\ c_7 = x_2x_3^3,\ c_8 = x_2x_3^2x_4,\ c_9 = x_2x_3x_4^2,\ c_{10} = x_2x_4^3,\ c_{11} = x_3^4,\ c_{12} = x_3^3x_4,\ c_{13} = x_3^2x_4^2,\ c_{14} = x_3x_4^3$ and $c_{15} = x_4^4\}$,
elements of the sets formed in (13.52). The matrix **A** is

$$
\mathbf{A} = \begin{bmatrix}
 & c_1 & c_2 & c_3 & c_4 & c_5 & c_6 & c_7 & c_8 & c_9 & c_{10} & c_{11} & c_{12} & c_{13} & c_{14} & c_{15} \\
x_2^2f_1 & 1 & 0 & a_1 & 0 & 0 & a_2 & 0 & 0 & 0 & 0 & 0 & 0 & 0 & 0 & 0 \\
x_3^2f_1 & 0 & 0 & 0 & 1 & 0 & 0 & 0 & a_1 & 0 & 0 & 0 & 0 & a_2 & 0 & 0 \\
x_2x_3f_1 & 0 & 1 & 0 & 0 & a_1 & 0 & 0 & 0 & a_2 & 0 & 0 & 0 & 0 & 0 & 0 \\
x_4^2f_1 & 0 & 0 & 0 & 0 & 0 & 1 & 0 & 0 & 0 & a_1 & 0 & 0 & 0 & 0 & a_2 \\
x_3x_4f_1 & 0 & 0 & 0 & 0 & 1 & 0 & 0 & 0 & a_1 & 0 & 0 & 0 & 0 & a_2 & 0 \\
x_2x_4f_1 & 0 & 0 & 1 & 0 & 0 & a_1 & 0 & 0 & 0 & a_2 & 0 & 0 & 0 & 0 & 0 \\
x_2x_4f_2 & 0 & 0 & 1 & 0 & b_1 & 0 & 0 & 1 & 0 & b_2 & 0 & 0 & 0 & 0 & 0 \\
x_4^2f_2 & 0 & 0 & 0 & 0 & 0 & 1 & 0 & 0 & b_1 & 0 & 0 & 0 & 1 & 0 & b_2 \\
x_3^2f_2 & 0 & 0 & 0 & 1 & 0 & 0 & b_1 & 0 & 0 & 0 & 1 & 0 & b_2 & 0 & 0 \\
x_3x_4f_2 & 0 & 0 & 0 & 0 & 1 & 0 & 0 & b_1 & 0 & 0 & 0 & 1 & 0 & b_2 & 0 \\
x_2x_3f_2 & 0 & 1 & 0 & b_1 & 0 & 0 & 1 & 0 & b_2 & 0 & 0 & 0 & 0 & 0 & 0 \\
x_2x_3f_3 & 0 & 0 & 0 & 0 & 0 & 0 & 1 & c_1 & c_2 & 0 & 0 & 0 & 0 & 0 & 0 \\
x_3x_4f_3 & 0 & 0 & 0 & 0 & 0 & 0 & 0 & 0 & 0 & 0 & 0 & 1 & c_1 & c_2 & 0 \\
x_4^2f_3 & 0 & 0 & 0 & 0 & 0 & 0 & 0 & 0 & 0 & 0 & 0 & 0 & 1 & c_1 & c_2 \\
x_2x_4f_3 & 0 & 0 & 0 & 0 & 0 & 0 & 0 & 1 & c_1 & c_2 & 0 & 0 & 0 & 0 & 0
\end{bmatrix}.
$$

The determinant of this matrix is a univariate polynomial of degree 8 in the variable x_1 given in [28, Box 3-1]. Its roots can be obtained using Matlab's *roots* command. Once these roots have been obtained, the admissible solution is substituted in the third expression of (13.48) on p. 230 to obtain the value of $x_3 \in \mathbb{R}.^+$ The obtained value of $x_3 \in \mathbb{R}^+$ is in turn substituted in the second expression of (13.48) to obtain the last variable $x_2 \in \mathbb{R}.^+$ The admissible values of distances are deduced with the help of prior information.

Approach 2 (B. Sturmfels' Formulation):

From (13.50) on p. 231, the determinant of the Jacobi matrix is computed as

$$
J = det \begin{bmatrix}
\frac{\partial R_{11}}{\partial x_2} & \frac{\partial R_{11}}{\partial x_3} & \frac{\partial R_{11}}{\partial x_4} \\
\frac{\partial R_{21}}{\partial x_2} & \frac{\partial R_{21}}{\partial x_3} & \frac{\partial R_{21}}{\partial x_4} \\
\frac{\partial R_{31}}{\partial x_2} & \frac{\partial R_{31}}{\partial x_3} & \frac{\partial R_{31}}{\partial x_4}
\end{bmatrix}, \tag{13.54}
$$

respectively

$$
J = det \begin{bmatrix}
2x_2 + a_1x_4 & 0 & 2a_2x_4 + a_1x_2 \\
2x_2 + b_1x_3 & 2x_3 + b_1x_2 & 2b_2x_4 \\
0 & 2x_3 + c_1x_4 & 2c_2x_4 + c_1x_3
\end{bmatrix}, \tag{13.55}
$$

which gives a cubic polynomial in x_2, x_3, x_4 as

$$J = 8x_2x_3c_2x_4 + 4x_2c_1x_3^2 + 4b_1x_2^2c_2x_4 + 2b_1x_2^2c_1x_3 - 8x_2b2x_4x_3 - 4x_2b2x_4^2c_1 + 4a_1x_4^2x_3c_2 + 2a_1x_4c_1x_3^2 + 2a_1x_4^2b_1x_2c_2 + 2a_1x_4b_1x_2c_1x_3 - 4a_1x_4^2b2x_3 - 2a_1x_4^3b2c_1 + 8x_2a_2x_4x_3 + 4x_2a_2x_4^2c_1 + 4a_1x_2^2x_3 + 2a_1x_2^2c_1x_4 + 4b_1x_3^2a_2x_4 + 2b_1x_3a_2x_4^2c_1 + 2b_1x_3^2a_1x_2,$$

whose partial derivatives with respect to x_2, x_3, x_4 can be written in the form (5.14) on p. 54. The coefficients b_{ij} and a_{ij} are given as in [28]. The computation of the resultant of the matrix using (5.15) on p. 54 leads to a univariate polynomial in x_1 of degree eight, e.g., [28, Box 3-2].

Fischler and Bolles [135, pp. 386-387, Fig. 5] have demonstrated that because every term in (13.32) is either a constant or of degree 2, for every real positive solution, there exist a geometrically isomorphic negative solution. Thus there are at most four positive solutions to (13.32). This is because (13.32) has eight solutions according to [112, p. 415] who states that for n independent polynomial equations in n unknowns, there can be no more solution than the product of their respective degrees. Since each equation of (13.32) is of degree 2 there can only be up to eight solutions.

Finally, in comparing the polynomial resultants approach to Groebner basis method, the latter in most cases is slow and there is always a risk of the computer breaking down during computations. Besides, the Groebner basis approach computes unwanted intermediary elements which occupy more space and thus lead to storage problems. The overall speed of computation is said to be proportional to twice exponential the number of variables [286, 287, 288, 290]. This has led to various studies advocating for the use of the alternate method; the resultant and specifically multipolynomial resultant approach. Groebner bases can be made faster by computing the reduced Groebner bases as explained in Chap. 4. For the special cases used throughout this book, however, this bound is not so tight since the size of the solution sets are finite. In such cases, there exist single exponential bounds.

Polynomial resultants on the other hand involve computing with larger matrices which may require a lot of work. For linear systems and ternary quadrics, Sturmfels' approach offers a remedy through the application of the Jacobi determinants. Once the distances have been computed, they are subjected to the ranging techniques (Chap. 12) to compute positions. Finally, the three-dimensional orientation parameters are computed from (9.10) on p. 118.

13-225 Linear homotopy solution

Example 13.2 (3D resection problem). Let us express the Grunert's equations (13.32) in the following form,

$$x_1^2 - 2f_{12}x_1x_2 + x_2^2 - d_{12} = 0 \tag{13.56}$$

$$x_2^2 - 2f_{23}x_2x_3 + x_3^2 - d_{23} = 0 \tag{13.57}$$

$$x_3^2 - 2f_{31}x_1x_3 + x_1^2 - d_{31} = 0. \tag{13.58}$$

Numerical data from *Awange-Grafarend* [41] are used in the following, see Table 13.4.

Table 13.4. Data for 3D resection problem

$\varphi_{1,2}$	$\varphi_{2,3}$	$\varphi_{3,1}$	$S_{1,2}$	$S_{2,3}$	$S_{3,1}$
1.843620	1.768989	2.664537	1560.3302	755.8681	1718.1090

Employing the homotopy solution we need to search 8 paths, because the Bezout bound in Eq (13.56)-Eq (13.58) is $d = 2^3 = 8$ ($d_1 = 2, \quad d_2 = 2$ and $d_3 = 2$). Now we have three start equations in our start system generated randomly, using Eq (6.29)-Eq (6.30),

$$g_1 = (0.988262 + 0.152767\boldsymbol{i})\left(-0.613157 + 0.789961\boldsymbol{i} + x_1^2\right) \tag{13.59}$$

$$g_2 = (0.367759 - 0.929921\boldsymbol{i})\left(0.390737 - 0.920503\boldsymbol{i} + x_2^2\right) \tag{13.60}$$

$$g_3 = (-0.530598 + 0.847624\boldsymbol{i})\left(0.62502 + 0.780609\boldsymbol{i} + x_3^2\right). \tag{13.61}$$

Table 13.5. shows the solutions of this start system. These values will be the initial values of the 8 paths of the homotopy function in Eq (13.62).

Table 13.5. Solutions of the start system

i	x_{1i}	x_{2i}	x_{3i}
1	$-0.898097 + 0.439797i$	$0.551935 + 0.833887i$	$0.433001 - 0.901393i$
2	$-0.898097 + 0.439797i$	$0.551935 + 0.833887i$	$-0.433001 + 0.901393i$
3	$-0.898097 + 0.439797i$	$-0.551935 - 0.833887i$	$0.433001 - 0.901393i$
4	$-0.898097 + 0.439797i$	$-0.551935 - 0.833887i$	$-0.433001 + 0.901393i$
5	$0.898097 - 0.439797i$	$0.551935 + 0.833887i$	$0.433001 - 0.901393i$
6	$0.898097 - 0.439797i$	$0.551935 + 0.833887i$	$-0.433001 + 0.901393i$
7	$0.898097 - 0.439797i$	$-0.551935 - 0.833887i$	$0.433001 - 0.901393i$
8	$0.898097 - 0.439797i$	$-0.551935 - 0.833887i$	$-0.433001 + 0.901393i$

The homotopy function is the linear combination of the target system, Eq (13.56)-Eq (13.58) and the start system, Eq (13.59)-Eq (13.61),

$$H(x_1, x_2, x_3, \lambda) = (1-\lambda)\begin{pmatrix} g_1(x_1, x_2, x_3) \\ g_2(x_1, x_2, x_3) \\ g_3(x_1, x_2, x_3) \end{pmatrix} + \lambda \begin{pmatrix} f_1(x_1, x_2, x_3) \\ f_2(x_1, x_2, x_3) \\ f_3(x_1, x_2, x_3) \end{pmatrix} \tag{13.62}$$

The homotopy paths will be the solution of the following differential equation system (see Eq (6.15)),

$$\frac{d}{d\lambda}\begin{pmatrix} x_1(\lambda) \\ x_2(\lambda) \\ x_3(\lambda) \end{pmatrix} = -H_x^{-1}H_\lambda \tag{13.63}$$

where the Jacobian

$$(H_x)_{i,j} = (1-\lambda)\left(\frac{\partial g_i}{\partial x_j}\right)_{i,j} + \lambda\left(\frac{\partial f_i}{\partial x_j}\right)_{i,j} \tag{13.64}$$

and

$$(H_\lambda)_i = f_i(x_1, x_2, x_3) - g_i(x_1, x_2, x_3)\,. \tag{13.65}$$

The end points of these 8 homotopy paths belonging to the eight initial values in Table 13.5 will give us the eight solutions of the target system Eq (13.56)-Eq (13.58), provided in Table 13.6.

Table 13.6. End points of the 8 homotopy paths

i	x_{1i}	x_{2i}	x_{3i}
1	$-22456.5 - 1735.3i$	$4375.48 + 22037.5i$	$20757.3 - 8626.43i$
2	-1580.11	770.958	-153.711
3	$-22456.5 + 1735.3i$	$4375.48 - 22037.5i$	$20757.3 + 8626.43i$
4	-1324.24	-542.261	-430.529
5	1324.24	542.261	430.529
6	$22456.5 - 1735.3i$	$-4375.48 + 22037.5i$	$-20757.3 - 8626.43i$
7	1580.11	-770.958	153.711
8	$22456.5 + 1735.3i$	$-4375.48 - 22037.5i$	$-20757.3 + 8626.43i$

The corresponding solutions for distances in this example should be all positive real numbers, consequently only the fifth solution is acceptable, see Table 13.7.

Table 13.7. Solution of the 3D resection problem

$x_1 =$	1324.240
$x_2 =$	542.261
$x_3 =$	430.529

The paths corresponding to this single solution are illustrated in Fig. 13.5.

The computation of the 8 paths takes 0.172 sec with *Mathematica*. The same solution was achieved using the reduced Groebner basis, eliminating x_2 and x_3, resulted in a univariate polynomial of order 8 for x_1 taking 6.297 sec also in *Mathematica*. However, if NSolve employing numerical Groebner basis is used, the computation time is 0.078 sec. Dixon resultant method improved by [239] and implemented into *Mathematica* by Nakos and Williams [308] cannot solve the problem in practicable time (less than 500 sec). However, Dixon method implemented in the computer algebra system Fermat by [267, 269] can solve the problem in about 0.02 seconds!

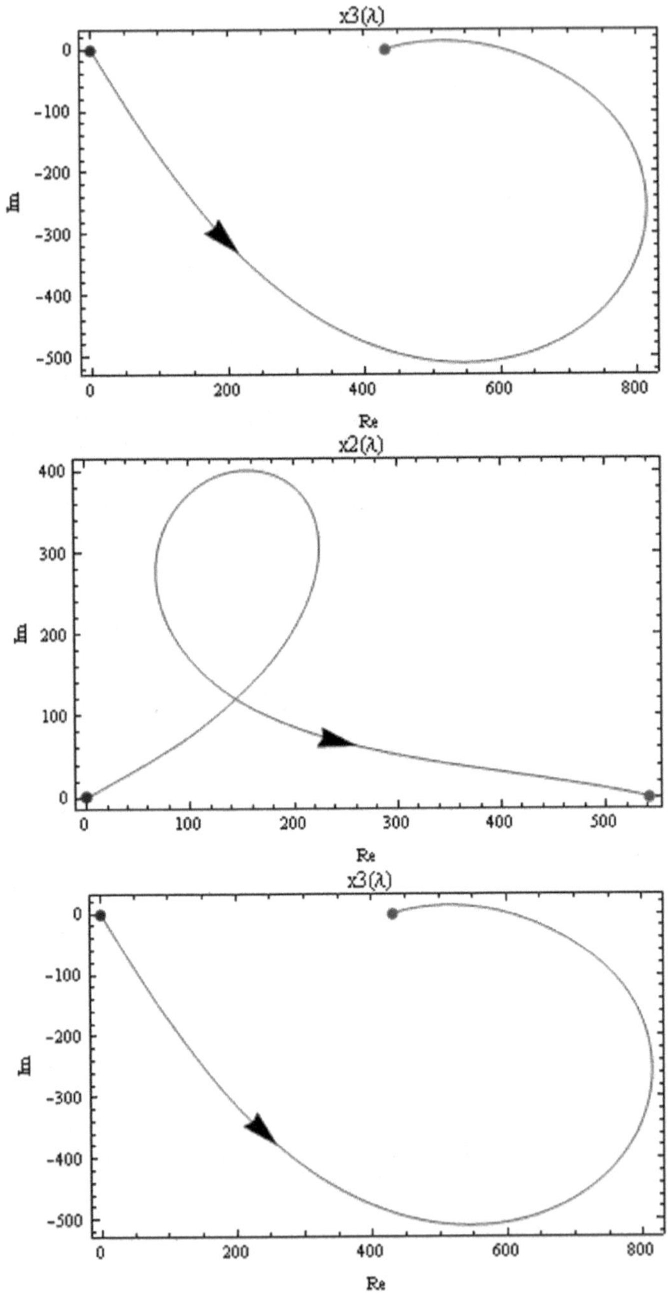

Figure 13.5. Trajectories of the homotopy path providing positive, real solutions for the 3D resection problem

13-226 Grafarend-Lohse-Schaffrin approach

In this approach, [187] begin by first setting up rigorous projective equations of resection problem in three-dimensional Euclidean space. They classify the equations as six dimensional algebraic system of nonlinear equations of cubic type. In the second part, a three step procedure is adopted for solving the Grunert's distance equations. The nonlinear system of distance equations are projected into linear equations by means of the technique of degenerate quadrics called the stencil method. The stencil method gives the solution of Grunert's equation (13.44) as

$$\begin{bmatrix} x_1^2 = S_1^2 = -\dfrac{a_{00}}{1+2a_{12}p+p^2} \\ \\ x_2^2 = S_2^2 = -\dfrac{b_{00}p^2}{p^2+2b_{23}pq+q^2} \\ \\ x_3^2 = S_3^2 = -\dfrac{c_{00}}{1+2c_{31}q+q^2}. \end{bmatrix} \quad (13.66)$$

The solution for p and q are as discussed in [187]. Once the distances have been solved from (13.66), the three orientation parameters and the cartesian coordinates of the unknown stations are solved from a 6×6 system of linear equations. The linear system of equations are formed using the normalized Hamilton-quaternion (see e.g., p. 11). For a complete discussion on the approach and a numerical example, we refer to papers by [187]. Lohse [276] extends the approach by proposing an alternative solution of the Grunert's distance equations.

Example 13.3 (Three-dimensional resection given three known stations). In-order to position station $K1$ (see Fig. 10.2 on p. 150) by resection method, horizontal directions T_i and vertical directions B_i are measured to three known stations Haussmanstr., Eduardpfeiffer, and Liederhalle. The computation is performed in three steps as follows:

- In the *first* step, the spatial distances are computed. This involves solving the Grunert's distance equations.
- The *second step* is the computation of the GPS Cartesian coordinates $\{X, Y, Z\}$ of the unknown station $K1 \in \mathbb{E}^3$ in the global reference frame using the algebraic ranging techniques of Chap. 12.
- The three-dimensional orientation parameters are computed in the *final step* using the partial Procrustes approach (see Chap. 9).

Using the computed univariate polynomials in [27, Boxes 3-3a and 3-3b] or [28, Box 3-2], and the observations in Tables 10.1 and 10.3 on p. 152, the distances $S_i = x_i \in \mathbb{R},^+ \; i = \{1,2,3\} \in \mathbb{Z}_+^3$ between the unknown station $K1 \in \mathbb{E}^3$ and the known stations $P_i \in \mathbb{E}^3$ are determined. For control purposes, these distances are as expressed in Fig. 10.2. The unknown station $K1$ is located on top of one of the University's building at Kepler Strasse 11. Points $\{P_1, P_2, P_3\}$ of the tetrahedron

$\{PP_1P_2P_3\}$ in Fig. 13.4 correspond to the chosen known GPS stations Haussmannstr., Eduardpfeiffer, and Liederhalle. The distance from $K1$ *to Haussmannstr.* is designated $S_1 = x_1 \in \mathbb{R},^+$ $K1$ *to Eduardpfeiffer* $S_2 = x_2 \in \mathbb{R},^+$ while that of $K1$ *to Liederhalle* is designated $S_3 = x_3 \in \mathbb{R}.^+$ The distances between the known stations $\{S_{12}, S_{23}, S_{31}\} \in \mathbb{R}^+$ are computed from their respective GPS coordinates as indicated in Solution 13.4. Their corresponding space angles $\psi_{12}, \psi_{23}, \psi_{31}$ are computed from (13.30) on p. 227.

In-order to control the computations, the Cartesian GPS coordinates of station $K1$ are also known. Solution 13.4 gives the unknowns distances $\{x_1, x_2, x_3\} \in \mathbb{R}^+$ computed using Groebner basis. The univariate polynomial in x_3 has eight roots, four of which are complex and four real. Of the four real roots two are positive and two are negative. The desired distance $x_3 \in \mathbb{R}^+$ is thus chosen amongst the two positive roots with the help of prior information and substituted in [27, g_{11} in Box 3-3b] to give two solutions of x_1, one of which is positive. Finally the obtained values of $\{x_1, x_3\} \in \mathbb{R}^+$ are substituted in [27, g_5 in Box 3-3b] to obtain the remaining indeterminate x_2. Using this procedure, we have in Solution 13.4 that $S_3 = \{430.5286, 153.7112\}$. Since $S_3 = x_3 \in \mathbb{R},^+$ from prior information (e.g., Fig. 10.2), we choose $S_3 = 430.5286$, leading to $S_1 = 1324.2381$, and $S_2 = 542.2608$. These values compare well with their real values depicted in Fig. 10.2 on p. 150.

Solution 13.4 (Computation of distances for test network Stuttgart Central).
Using the entries of Table 10.1 on p. 152, inter-station distances are computed by Pythagoras $S_{ij} = \sqrt{(X_j - X_i)^2 + (Y_j - Y_i)^2 + (Z_j - Z_i)^2}$, and spatial angles obtained from (13.30). The values are

$$\begin{bmatrix} S_{12} = 1560.3302\,m \\ S_{23} = 755.8681\,m \\ S_{31} = 1718.1090\,m \end{bmatrix} \; and \; \begin{bmatrix} \psi_{12} = 1.843620 \\ \psi_{23} = 1.768989 \\ \psi_{31} = 2.664537 \end{bmatrix}$$

and are substituted in (13.45) on p. 230 to compute the terms $a_{12}, b_{23}, c_{31}, a_0, b_0, c_0$ which are needed to determine the coefficients of the Groebner basis element g_1 in [27, Box 3-3a]. Expressing the univariate polynomial g_1 as $A_8x_3^8 + A_6x_3^6 + A_4x_3^4 + A_2x_3^2 + A_0 = 0$, the computed coefficients are:

$$\begin{bmatrix} \mathrm{A}_0 = 4.833922266706213\mathrm{e}+023 \\ \mathrm{A2} = -2.306847176510587\mathrm{e}+019 \\ \mathrm{A4} = 1.104429253262719\mathrm{e}+014 \\ \mathrm{A6} = -3.083017244255380\mathrm{e}+005 \\ \mathrm{A8} = 4.323368172460818\mathrm{e}-004. \end{bmatrix}$$

The solutions to the univariate polynomial equation are then obtained using Matlab's roots command (e.g., 4.42 on p. 46) as

$$\begin{bmatrix} c = [A_8\, A_7\, A_6\, A_5\, A_4\, A_3\, A_2\, A_1\, A_0] \\ x_3 = roots(c), \end{bmatrix}$$

where A_7, A_5, A_3, A_1 are all zero. The obtained values of x_3 are:

$$x_3 = \begin{bmatrix} -20757.2530734872 + 8626.43262759353i \\ -20757.2530734872 - 8626.43262759353i \\ 20757.2530734872 + 8626.4326275935i \\ 20757.2530734872 - 8626.4326275935i \\ 430.528578109464 \\ -430.528578109464 \\ 153.711222705295 \\ -153.711222705295. \end{bmatrix}$$

Alternatively, the polynomial resultants techniques can be used to solve the Grunert's distance equations. They proceed as follows:

(a) The *F. Macaulay* formulation discussed in Sect. 5-31 solves for the determinant of the matrix **A** leading to a univariate polynomial in x_1. The solution of the obtained univariate polynomial equation expressed in [28, Box 3-1] leads to similar results as those of Groebner basis, i.e.,

$$\begin{bmatrix} det(A) = A_8x_1^8 + A_6x_1^6 + A_4x_1^4 + A_2x_1^2 + A_0 \\ A_0 = -4.87154987980622^{26},\ A_2 = 4.74815547158708^{20} \\ A_4 = -113109755605017 \\ A_8 = -0.000432336817247789,\ A_6 = 435283.472057364 \\ \\ x_1 = -22456.4891074245 + 1735.29702574406i \\ -22456.4891074245 - 1735.29702574406i \\ 22456.4891074245 + 1735.29702574406i \\ 22456.4891074245 - 1735.29702574406i \\ 1580.10924379877 \\ -1580.10924379877 \\ 1324.23808451944 \\ -1324.23808451944 \\ x_3 = 430.528578109536, -2783.30427366986 \\ x_2 = 542.260767703823,\ -711.800947103387. \end{bmatrix}$$

(b) The *B. Sturmfels* formulation discussed in Sect. 5-32 solves the determinant of a 6×6 matrix leading to a univariate polynomial in x_1. The solution of the obtained univariate polynomial equation expressed in [28, Box 3-2] gives identical results as those of Groebner basis, i.e.,

$$\begin{bmatrix} det(A) = A_8 x_1^8 + A_6 x_1^6 + A_4 x_1^4 + A_2 x_1^2 + A_0 \\ A_0 = -1.94861995192249^{27},\ A_2 = 1.89926218863483^{21} \\ A_4 = -452439022420067 \\ A_8 = -0.00172934726897456,\ A_6 = 1741133.88822977 \\ \\ x_1 = -22456.4891075064 + 1735.29702538544i \\ -22456.4891075064 - 1735.29702538544i \\ 22456.4891075064 + 1735.29702538544i \\ 22456.4891075064 - 1735.29702538544i \\ 1580.10924379877 \\ -1580.10924379877 \\ 1324.23808451944 \\ -1324.23808451944 \\ x_3 = 430.528578109535, -2783.30427366986 \\ x_2 = 542.260767703824, -711.800947103388. \end{bmatrix}$$

The computed distances from *F. Macaulay* and *B. Sturmfels'* approaches above tally. The required solutions $\{x_1, x_2, x_3\}$ obtained from Groebner basis computation and those of multipolynomial resultants are the same {i.e, 1324.2381 m, 542.2608 m, 430.5286 m} respectively. The computed distances are then used to determine the position of $K1$ using ranging techniques discussed in Sect. 12-32. The unknown orientation elements are computed from (9.10) on p. 118.

13-227 3d-resection to more than three known stations

In the preceding section, only three known stations were required to solve in a closed form the three-dimension resection problem for the position and orientation of the unknown station $K1$. If superfluous observations are available, due to the availability of several known stations, as in the case of the test network Stuttgart Central, closed form three-dimensional resection procedures give way to Gauss-Jacobi combinatorial approach. We illustrate this by means of Example 13.4.

Example 13.4 (Three-dimensional resection given more than three known stations). From the test network Stuttgart Central in Fig. 10.2 of Sect. 10-6, the three-dimensional coordinates $\{X, Y, Z\}$ of the unknown station $K1$ are sought. Using observations in Tables 10.2 and 10.3 on p. 152, the algorithm is applied in four steps as follows:

Step 1 (combinatorial solution):
From Fig. 10.2, 35 minimal combinatorials are formed using (7.28) on p. 93. The systems of nonlinear Grunert's distance equations (13.32) for each of the 35 combinatorials is solved in a closed form to give the distances linking the unknown station $K1$ to the 7 known stations. Use is made of either Groebner basis or polynomial resultants approaches as already discussed in Sect. 13-22. Each combinatorial minimal subset results in 3 distances, thus giving rise to a total of 105 (3 × 35) which are used in the next steps as pseudo-observations.

Step 2 (error propagation to determine the dispersion matrix $\boldsymbol{\Sigma}$):
The variance-covariance matrix is computed for each of the combinatorial set $j = 1, \ldots, 35$ using error propagation. Equation (13.32) on p. 228 is applied to obtain the dispersion matrix $\boldsymbol{\Sigma}$ using (7.33) as discussed in Example 7.4 on p. 95.

Step 3 (rigorous adjustment of the combinatorial solution points):
The 105 combinatorial distances from step 1 are finally adjusted using the linear Gauss-Markov model (7.9) on p. 87. Each of the 105 pseudo-observations is expressed as

$$S_i^j = S_i + \varepsilon_i^j \,|i \in \{1,2,3,4,5,6,7\}, j \in \{1,2,3,4,5,6,7,\ldots,35\},$$

and placed in the vector of observation $\mathbf{y}$. The coefficients of the unknown distances S_i are placed in the design matrix $\mathbf{A}$. The vector $\boldsymbol{\xi}$ comprises the unknowns S_i. The solutions are obtained via (7.12) and the root-mean-square errors of the estimated parameters through (7.13) on p. 87. The results of the adjusted distances, root-mean-square-errors and the deviations in distances are presented in Table 13.8. These deviations are obtained by subtracting the combinatorial derived distance S_i from its ideal value S in Table 10.2 on p. 152. The adjusted distances in Table 13.8 were: $K1$-Haussmanstr.(S_1) , $K1$-Eduardpfeiffer (S_2), $K1$-Lindenmuseum (S_3) , $K1$-Liederhalle (S_4), $K1$-Dach LVM (S_5) , $K1$-Dach FH (S_6) and $K1$-Haussmanstr.(S_7).

Table 13.8. Gauss-Jacobi combinatorial derived distances

Distance	Value (m)	Root mean square (m)	Deviation $\Delta(m)$
S_1	1324.2337	0.0006	0.0042
S_2	542.2598	0.0006	0.0011
S_3	364.9832	0.0006	-0.0035
S_4	430.5350	0.0008	-0.0063
S_5	400.5904	0.0007	-0.0067
S_6	269.2346	0.0010	-0.0037
S_7	566.8608	0.0005	0.0027

Step 4 (determination of position by ranging method):
The derived distances in Table 13.8 are then used as in Example 12.6 on p. 211 to determine the position of $K1$.

Example 13.5 (Comparison between exact and overdetermined 3d-resection solutions).
In this example, we are interested in comparing the solutions of the position of station $K1$ obtained from;

- closed form procedures of either Groebner basis or polynomial resultants,
- closed form solution of Gauss-Jacobi combinatorial for the overdetermined 3d-resection to the 7-stations.

To achieve this, 11 sets of experiments were carried out. For each experiment, the position of $K1$ was determined using Groebner basis and the obtained values subtracted from the known position in Table 10.1 on p. 152. The experiments were then repeated for the Gauss-Jacobi combinatorial approach. In Table 13.9, the deviation of the positions computed using Gauss-Jacobi combinatorial approach from the real values, for the 11 sets of experiments are presented. In Figs. 13.6, 13.7 and 13.8, the plot of the deviations of the X, Y, Z coordinates respectively are presented.

Table 13.9. Deviation of position of station $K1$ from the real value in Table 10.1

Set No.	$\Delta X(m)$	$\Delta Y(m)$	$\Delta Z(m)$
1	-0.0026	0.0013	-0.0001
2	-0.0034	-0.0001	0.0009
3	0.0016	0.0005	0.0028
4	0.0076	0.0007	0.0016
5	0.0027	0.0020	0.0005
6	-0.0011	0.0004	0.0020
7	0.0027	-0.0000	0.0005
8	0.0014	0.0012	-0.0016
9	0.0010	0.0006	0.0005
10	-0.0005	-0.0039	0.0007
11	0.0016	0.0001	-0.0001

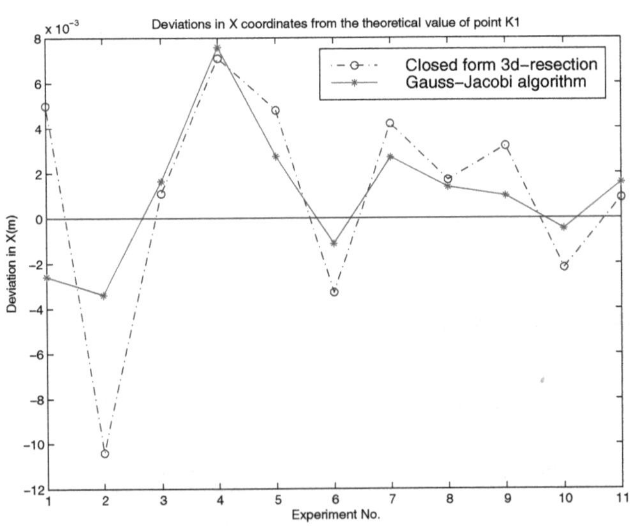

Figure 13.6. Deviations of $X - Coordinates$ of station $K1$

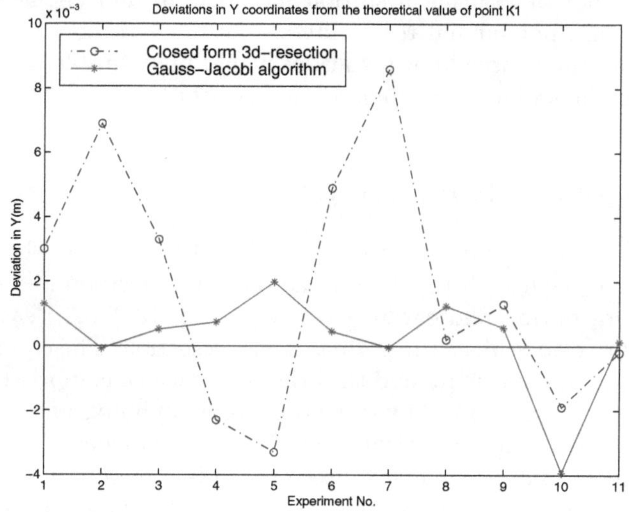

Figure 13.7. Deviations of $Y - Coordinates$

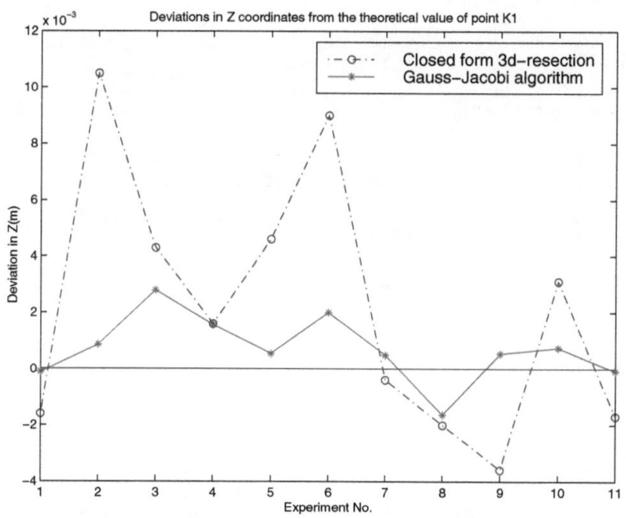

Figure 13.8. Deviations of $Z - Coordinates$

From the plots of Figs. 13.6, 13.7 and 13.6, it is clearly seen that closed form solutions with more than three known stations yield better results. For less accurate results such as that of locating a station in cadastral and engineering surveys, Groebner basis and polynomial resultants are useful. For more accurate results, resecting to more than three known stations would be desirable. In this case, one could apply the Gauss-Jacobi combinatorial algorithm.

13-3 Photogrammetric resection

Similar to the case of the scanner resection in Fig. 13.1 on p. 218, photogrammetric resection concerns itself with the determination of the position and orientation of the aerial camera during photography (see e.g., Fig. 13.9). At least three scene objects are required to achieve three-dimensional resection. The coordinates of the perspective center of the camera and the orientation which comprises the elements of exterior orientation are solved by three-dimensional photogrammetric resection. Once the coordinates and the orientation of the camera have been established, they are used in the intersection step (see Sect. 14-3) to compute coordinates of the pass points. Besides, they also find use in transformation procedures where coordinates in the photo plane are transformed to ground system and vice versa. The three-dimensional photogrammetric resection is formulated as follows: Given image coordinates of at least three points $\{p_i, p_j\}$, respectively, $i \neq j$ (e.g., in Fig. 13.9), determine the position $\{X_0, Y_0, Z_0\}$ and orientation $\{\omega, \phi, \kappa\}$ of the perspective center p.

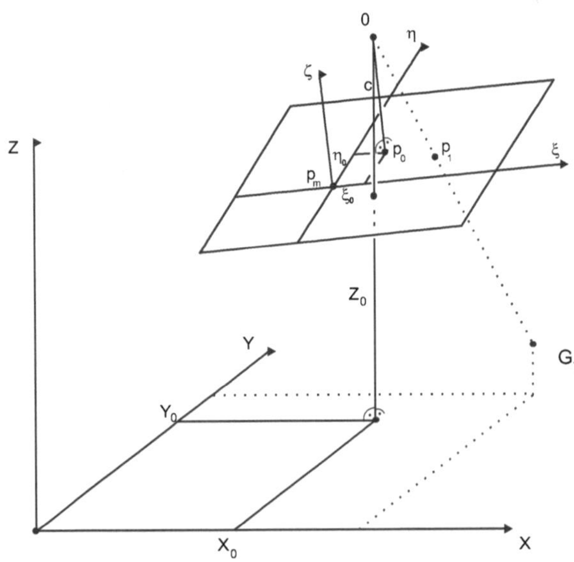

Figure 13.9. Photogrammetric three-dimensional resection

In practice, using stereoplotters etc., the bundle of rays projected from the perspective center to the ground are normally translated and rotated until there exist a match between points on the photograph and their corresponding points on the ground. The mathematical relationship between the points $\{\xi_i, \eta_i, f\}$ on the photographs and their corresponding ground points $\{X_i, Y_i, Z_i\}$ are related by

$$\begin{bmatrix} \xi_i = \xi_0 - f\dfrac{r_{11}(X_i - X_0) + r_{21}(Y_i - Y_0) + r_{31}(Z_i - Z_0)}{r_{13}(X_i - X_0) + r_{23}(Y_i - Y_0) + r_{33}(Z_i - Z_0)} \\ \\ \eta_i = \eta_0 - f\dfrac{r_{12}(X_i - X_0) + r_{22}(Y_i - Y_0) + r_{32}(Z_i - Z_0)}{r_{13}(X_i - X_0) + r_{23}(Y_i - Y_0) + r_{33}(Z_i - Z_0)}, \end{bmatrix} \tag{13.67}$$

where f is the focal length, $\{\xi_0, \eta_0\}$ the perspective center coordinates on the photo plane and r_{ij} the elements of the rotation matrix (see Sect. 17-21 p. 305 for details)

$$R = \begin{bmatrix} r_{11} & r_{12} & r_{13} \\ r_{21} & r_{22} & r_{23} \\ r_{31} & r_{32} & r_{33} \end{bmatrix} |_{\omega,\phi,\kappa}. \tag{13.68}$$

The solution of (13.67) for the unknown position $\{X_0, Y_0, Z_0\}$ and orientation $\{\omega, \phi, \kappa\}$ is often achieved by;

- first linearizing about approximate values of the unknowns,
- application of least squares approach to the linearized observations,
- iterating to convergence.

In what follows, we present algebraic solutions of (13.67) based on the Grafarend-Shan Möbius and Groebner basis/polynomial resultant approaches.

13-31 Grafarend-Shan Möbius photogrammetric resection

In this approach of [183], the measured image coordinates $\{x_i, y_i, f\}$ of point p_i and $\{x_j, y_j, f\}$ of point p_j are converted into space angles by

$$\cos\psi_{ij} = \frac{x_i x_j + y_i y_j + f^2}{\sqrt{x_i^2 + y_i^2 + f^2}\sqrt{x_j^2 + y_j^2 + f^2}}. \tag{13.69}$$

Equation (13.69) is the photogrammetric equivalent of (13.30) on p. 227 for geodetic resection. The Grafarend-Shan algorithm operates in five steps as follows:

Step 1: The space angles ψ_{ij} relating angles to image coordinates of at least four known stations are computed from (13.69).

Step 2: The distances $\mathbf{x}_i - \mathbf{x}_j$ from the given cartesian coordinates of points p_i and p_j are computed using

$$s_{ij} = \sqrt{((x_i - x_j)^2 + (y_i - y_j)^2 + (z_i - z_j)^2)|_{\{i \neq j\}}}. \tag{13.70}$$

Step 3: Using the distances from (13.70) and the space angles from (13.69) in step 1, Grunert's distance equations (13.32) are solved using the Grafarend-Lohse-Schaffrin procedure discussed in Sect. 13-226.

Step 4: Once the distances have been obtained in step 3, they are used to compute the perspective center coordinates using the Ansermet's algorithm [11].

Step 5: The orientation is computed by solving (17.3) on p. 305.

13-32 Algebraic photogrammetric resection

The algebraic algorithms of Groebner basis or polynomial resultants operate in five steps as follows:

Step 1: The space angles ψ_{ij} relating the angles to the image coordinates of at least four known stations are computed from (13.69).
Step 2: The distances $\mathbf{x}_i - \mathbf{x}_j$ from the given cartesian coordinates of points p_i and p_j are computed from (13.70).
Step 3: Using the distances from step 2 and the space angles from step 1, Grunert's distance equations in (13.32) are solved using procedures of Sect. 13-224.
Step 4: Once distances have been solved in step 3, they are used to compute the perspective center coordinates using ranging techniques of Chap. 12.
Step 5: The three-dimensional orientation parameters are computed from (9.10) on p. 118.

Example 13.6 (Three-dimensional photogrammetric resection). In this example, we will use data of two photographs adopted from [414]. From these data, we are interested in computing algebraically the perspective center coordinates of the two photographs. The image coordinates of photographs 1010 and 1020 are given in *Tables* 13.10 and 13.11 respectively. The corresponding ground coordinates are as given in *Table* 13.12. Table 13.14 gives for control purposes the known coordinates of the projection center adopted from [414]. We use four image coordinates (No. 1,2,3,5) to compute algebraically the perspective center coordinates. From these image coordinates, combinatorials are formed using (7.28) on p. 93 and are as given in Table 13.13. For each combination, distances are solved using reduced Groebner basis (4.38) on p. 45. Once the distances have been computed for each combination, the perspective center coordinates are computed using the ranging techniques of Chap. 12. The mean values are then obtained. The results are summarized in *Table* 13.13. A comparison between the Groebner basis derived results and those of *Table* 13.14 is presented in Table 13.15. Instead of the mean value which does not take weights into consideration, the Gauss-Jacobi combinatorial techniques that we have studied can be used to obtain refined solutions. For photo 1020, the first combination 1-2-3 gave complex values for the distances S_2 and S_3. For computation of the perspective center in this *Example*, the real part was adopted.

Table 13.10. Image coordinates in Photo 1010: $f = 153000.000[\mu m]$

Point	No.	$x(\mu m)$	$y(\mu m)$
100201	1	18996.171	-64147.679
100301	2	113471.749	-73694.266
200201	3	16504.609	16331.646
200301	4	128830.826	21085.172
300201	5	13716.588	106386.802
300301	6	120577.473	128214.823

Table 13.11. Image coordinates in Photo 1020: $f = 153000.000[\mu m]$

Point	No.	$x(\mu m)$	$y(\mu m)$
100201	1	-74705.936	-71895.580
100301	2	5436.953	-78524.687
200201	3	-87764.035	7895.436
200301	4	3212.790	10311.144
300201	5	-84849.923	94110.338
300301	6	802.388	106585.613

Table 13.12. Ground coordinates

Point	X(m)	Y(m)	Z(m)
100201	-460.000	-920.000	-153.000
100301	460.000	-920.000	0.000
200201	-460.000	0.000	0.000
200301	460.000	0.000	153.000
300201	-460.000	920.000	-153.000
300301	460.000	920.000	0.000

Table 13.13. Algebraic computed perspective center coordinates

Photo 1010						
Combination	$S_1(m)$	$S_2(m)$	$S_3(m)$	$X(m)$	$Y(m)$	$Z(m)$
1-2-3	1918.043	2008.407	1530.000	-459.999	0.000	1530.000
1-2-5	1918.043	2008.407	1918.043	-459.999	0.000	1530.000
1-3-5	1918.043	1530.000	1918.043	-459.292	0.000	1530.000
2-3-5	2008.407	1530.000	1918.043	-459.999	0.000	1530.000
mean				-459.822	0.000	1530.000
Photo 1020						
Combination	$S_1(m)$	$S_2(m)$	$S_3(m)$	$X(m)$	$Y(m)$	$Z(m)$
1-2-3	2127.273	1785.301	1785.301i	460.001	0.001	1529.999
1-2-5	2127.273	1785.301	2127.273	460.001	0.000	1530.000
1-3-5	2127.273	1785.301	2127.272	460.036	0.002	1529.978
2-3-5	1785.301	1785.301	2127.273	459.999	0.000	1530.000
mean				460.009	0.000	1529.994

Table 13.14. Ground coordinates of the projection centers

Photo	X(m)	Y(m)	Z(m)
1010	-460.000	0.000	1530.000
1020	460.000	0.000	1530.000

Table 13.15. Deviation of the computed mean from the real value

Photo	ΔX(m)	ΔY(m)	ΔZ(m)
1010	-0.178	0.000	0.000
1020	-0.009	0.000	0.006

13-4 Concluding remarks

As evident in the chapter, the problem of resection is still vital in geodesy and geoinformatics. The algebraic algorithms of Groebner basis, polynomial resultants and the Gauss-Jacobi combinatorial are useful where exact solutions are required. They may be used to control the numeric methods used in practice. Instead of the forward and backward steps, straight forward algebraic approaches presented save on computational time. Further references are [11, 62, 155, 156, 162, 171, 229, 240, 389].

14 Positioning by intersection methods

14-1 Intersection problem and its importance

The similarity between resection methods presented in the previous chapter and intersection methods discussed herein is their application of angular observations. The distinction between the two however, is that for resection, the unknown station is *occupied* while for intersection, the unknown station is *observed*. Resection uses measuring devices (e.g., theodolite, total station, camera etc.) which occupy the unknown station. Angular (direction) observations are then measured to three or more known stations as we saw in the preceding chapter. Intersection approach on the contrary measures angular (direction) observations to the unknown station; with the measuring device occupying *each* of the three or more known stations. It has the advantage of being able to position an unknown station which can not be physically occupied. Such cases are encountered for instance during engineering constructions or cadastral surveying. During civil engineering construction for example, it may occur that a station can not be occupied because of swampiness or risk of sinking ground. In such a case, intersection approach can be used. The method is also widely applicable in photogrammetry. In aero-triangulation process, simultaneous resection and intersection are carried out where common rays from two or more overlapping photographs intersect at a common ground point (see e.g., Fig. 12.1).

The applicability of the method has further been enhanced by the Global Positioning System (GPS), which the authors also refer to as GPS: Global Problem Solver. With the entry of GPS system, classical geodetic and photogrammetric positioning techniques have reached a new horizon. Geodetic and photogrammetric directional observations (machine vision, total stations) have to be analyzed in a three-dimensional Euclidean space. The challenge has forced positioning techniques such as resection and intersection to operate three-dimensionally. As already pointed out in Chap. 13, closed form solutions of the three-dimensional resection problem exist in a great number. On the contrary, closed form solutions of three-dimensional intersection problem are very rare. For instance [182, 183] solved the two *P4P* or the combined three-dimensional resection-intersection problem in terms of *Möbius barycentric coordinates* in a closed form. One reason for the rare existence of the closed form solutions of the three-dimensional intersection problem is the nonlinearity of directional observation equations, partially caused by the *external orientation parameters*. One target of this chapter, therefore, is to address the problem of orientation parameters.

The key to overcome the problem of nonlinearity caused by orientation parameters is taken from the *Baarda Doctrine*. Baarda [44, 48] proposed to use *dimensionless quantities* in geodetic and photogrammetric networks: Angles in a three-dimensional *Weitzenböck space*, shortly called *space angles* as well as *distance ratios* are the *dimensionless* structure elements which are *equivalent* under the action of the seven parameter *conformal group*, also called similarity transformation.

J.L. Awange et al., *Algebraic Geodesy and Geoinformatics*, 2nd ed.,
DOI 10.1007/978-3-642-12124-1_14, © Springer-Verlag Berlin Heidelberg 2010

14-2 Geodetic intersection

14-21 Planar intersection

The planar intersection problem is formulated as follows: Given directions or angular measurements from two known stations P_1 and P_2 to an unknown station P_0, determine the position $\{X_0, Y_0\}$. The solution to the problem depends on whether angles or directions are used as discussed in the next section.

14-211 Conventional solution

Closed form solution of planar intersection in terms of angles has a long tradition. Let us consult Fig. 14.1 on p. 255 where we introduce the angles ψ_{12} and ψ_{21} in the planar triangle $\Delta : P_0P_1P_2$, with P_0, P_1, P_2 being the nodes. The Cartesian coordinates $\{X_1, Y_1\}$ and $\{X_2, Y_2\}$ of the points P_1 and P_2 are given while $\{X_0, Y_0\}$ of the point P_0 are unknown. The angles $\psi_{12} = \alpha$ and $\psi_{21} = \beta$ are derived from direction observations by differencing horizontal directions. $\psi_{12} - T_{10} - T_{12}$ or $\psi_{21} = T_{20} - T_{21}$ are examples of observed horizontal directions T_{10} and T_{12} from P_1 to P_0 and P_1 to P_2 or T_{21} and T_{20} from P_2 to P_1 and P_2 to P_0 respectively. By means of taking differences we map direction observations to angles and eliminate orientation unknowns. The solution of the two-dimensional intersection problem in terms of angles, a classical procedure in analytical surveying, is given by (14.1) and (14.2) as

$$X_0 = s_{12}\frac{\cos\alpha \sin\beta}{\sin(\alpha+\beta)} \tag{14.1}$$

$$Y_0 = s_{12}\frac{\sin\alpha \sin\beta}{\sin(\alpha+\beta)}. \tag{14.2}$$

Note: The Euclidean distance between the nodal points is given by

$$s_{12} = \sqrt{(X_2 - X_1)^2 + (Y_2 - Y_1)^2}.$$

In deriving (14.1) and (14.2), use was made of angular observations. In case directions are adopted, measured values from known stations P_1 and P_2 to unknown station P_0 are designated T_{10} and T_{20} respectively. If the theodolite horizontal circle reading from point P_1 to P_0 is set to zero, then the measured angle α is equal to the directional measurement T_{12} from point P_1 to P_2. Likewise if the direction from P_2 to P_1 is set to zero, the measured angle β is equal to the directional measurement T_{20} from point P_2 to P_0. In this way, we make use of both the angles and directions thus introducing two more unknowns, i.e., the unknown orientation σ_1 and σ_2 in addition to the unknown coordinates $\{X_0, Y_0\}$ of point P_0. This leads to four observation equations in four unknowns, written as:

$$\begin{bmatrix} tan(T_{12}+\sigma_1) = \left\{\dfrac{Y_2-Y_1}{X_2-X_1}\right\} \\ \\ tan(T_{10}+\sigma_1) = \left\{\dfrac{Y_0-Y_1}{X_0-X_1}\right\} \\ \\ tan(T_{21}+\sigma_2) = \left\{\dfrac{Y_1-Y_2}{X_1-X_2}\right\} \\ \\ tan(T_{20}+\sigma_2) = \left\{\dfrac{Y_0-Y_2}{X_0-X_2}\right\}, \end{bmatrix} \tag{14.3}$$

where $\{X_1, Y_1, X_2, Y_2\}$ are coordinates of the two known stations $\{P_1, P_2\}$, while $\{T_{12}, T_{10}, T_{21}, T_{20}\}$ are the measured horizontal directions and $\{X_0, Y_0, \sigma_1, \sigma_2\}$ are the desired position and orientation of the unknown station P_0. In (14.3), the first and the third expressions contain the orientation elements σ_1 and σ_2 as the only unknowns. They are solved by obtaining the inverse of the tangents as

$$\begin{bmatrix} \sigma_1 = tan^{-1}\left\{\dfrac{Y_2-Y_1}{X_2-X_1}\right\} - T_{12} \\ \sigma_2 = tan^{-1}\left\{\dfrac{Y_1-Y_2}{X_1-X_2}\right\} - T_{21}. \end{bmatrix} \tag{14.4}$$

Once the unknown orientation elements have been solved in (14.4), they are substituted in the second and fourth expressions of (14.3) to form simultaneous equation whose solution give the values $\{X_0, Y_0\}$. Next, let us see how (14.3) can be solved using reduced Groebner basis (4.38) on p. 45.

14-212 reduced Groebner basis solution

The left-hand-sides of (14.3) are expanded using additions theorem

$$tan(\alpha+\beta) = \frac{tan\,\alpha + tan\,\beta}{1 - tan\,\alpha\, tan\,\beta}, \tag{14.5}$$

to give:

$$\begin{bmatrix} \dfrac{tan\,T_{12} + tan\,\sigma_1}{1 - tan\,T_{12} tan\,\sigma_1} = \left\{\dfrac{Y_2-Y_1}{X_2-X_1}\right\} \\ \\ \dfrac{tan\,T_{10} + tan\,\sigma_1}{1 - tan\,T_{10}\, tan\,\sigma_1} = \left\{\dfrac{Y_0-Y_1}{X_0-X_1}\right\} \\ \\ \dfrac{tan\,T_{21} + tan\,\sigma_2}{1 - tan\,T_{21}\, tan\,\sigma_2} = \left\{\dfrac{Y_1-Y_2}{X_1-X_2}\right\} \\ \\ \dfrac{\tan T_{20} + tan\,\sigma_2}{1 - tan\,T_{20}\, tan\,\sigma_2} = \left\{\dfrac{Y_0-Y_2}{X_0-X_2}\right\}. \end{bmatrix} \tag{14.6}$$

Expanding (14.6) and re-arranging gives trigonometric algebraic expressions

$$\begin{bmatrix} (X_2 - X_1 + Y_2 tanT_{12} - Y_1 tanT_{12}) tan\,\sigma_1 + X_2 tanT_{12} - X_1 tanT_{12} + Y_1 - Y_2 = 0 \\ \\ X_0 tanT_{10} + X_0 tan\,\sigma_1 + Y_0 tanT_{10} tan\,\sigma_1 - Y_0 - X_1 tan\,\sigma_1 - Y_1 tanT_{10} tan\sigma_1 \\ -X_1 tanT_{10} + Y_1 = 0 \\ \\ (X_1 - X_2 + Y_1 tanT_{21} - Y_2 tanT_{21}) tan\,\sigma_2 + X_1 tanT_{21} - X_2 tanT_{21} + Y_2 - Y_1 = 0 \\ \\ X_0 tanT_{20} + X_0 tan\,\sigma_2 + Y_0 tanT_{20} tan\,\sigma_2 - Y_0 - X_2 tan\,\sigma_2 - Y_2 tanT_{20} tan\,\sigma_2 \\ -X_2 tanT_{20} + Y_2 = 0. \end{bmatrix} \tag{14.7}$$

Denoting

$$\begin{bmatrix} a_1 = tanT_{12},\ a_2 = tanT_{21} \\ b = tanT_{10} \\ c = tanT_{20} \\ d_1 = tan\sigma_1,\ d_2 = tan\sigma_2, \end{bmatrix} \tag{14.8}$$

and substituting in (14.7) leads to four algebraic equations which are arranged in the lexicographic order $\{X_0 > Y_0 > d_2 > d_1\}$ as

$$\begin{bmatrix} f_1 = d_1X_2 - d_1X_1 + d_1Y_2a_1 - d_1Y_1a_1 + X_2a_1 - X_1a_1 + Y_1 - Y_2 = 0 \\ f_2 = X_0b + X_0d_1 - Y_0 + Y_0bd_1 - X_1d_1 - Y_1bd_1 - X_1b + Y_1 = 0 \\ f_3 = d_2X_1 - d_2X_2 + d_2Y_1a_2 - d_2Y_2a_2 + X_1a_2 - X_2a_2 + Y_2 - Y_1 = 0 \\ f_4 = X_0c + X_0d_2 - Y_0 + Y_0cd_2 - X_2d_2 - Y_2cd_2 - X_2c + Y_2 = 0. \end{bmatrix} \tag{14.9}$$

Using reduced Groebner basis (4.38) on p. 45, (14.9) is solved as

$$\begin{bmatrix} GroebnerBasis\,[\{f_1, f_2, f_3, f_4\}, \{X_0, Y_0, d_2, d_1\}, \{X_0, Y_0, d_2\}] \\ GroebnerBasis\,[\{f_1, f_2, f_3, f_4\}, \{X_0, Y_0, d_1, d_2\}, \{X_0, Y_0, d_1\}] \\ GroebnerBasis\,[\{f_1, f_2, f_3, f_4\}, \{d_2, d_1, Y_0, X_0\}, \{Y_0, d_2, d_1\}] \\ GroebnerBasis\,[\{f_1, f_2, f_3, f_4\}, \{d_2, d_1, X_0, Y_0\}, \{X_0, d_2, d_1\}]. \end{bmatrix} \tag{14.10}$$

The first and second expressions of (14.10) give linear equations relating the tangents d_1 and d_2 of the unknown orientations σ_1 and σ_2 and the coordinates $\{X_1, Y_1, X_2, Y_2\}$ of the known stations P_1 andP_2. The third and fourth expressions give linear equations relating the coordinates X_0 and Y_0 of unknown station P_0, coordinates $\{X_1, Y_1, X_2, Y_2\}$ of known stations P_1 and P_2, and the orientation terms d_1 and d_2. The computed reduced Groebner basis (linear functions) are

$$
\left[
\begin{array}{l}
d_1 = \dfrac{(-a_1X_1 + a_1X_2 + Y_1 - Y_2)}{(X_1 - X_2 + a_1Y_1 - a_1Y_2)} \\[2ex]
d_2 = \dfrac{(-a_2X_1 + a_2X_2 + Y_1 - Y_2)}{(X_1 - X_2 + a_2Y_1 - a_2Y_2)} \\[2ex]
X_0 = \dfrac{\left\{\begin{array}{l} -(Y_1 - Y_2 - d_1X_1 + d_2X_2 - bX_1 + cX_2 - bd_1Y_1 + bcY_1 - cd_2Y_1 \\ +bd_1Y_2 - bcY_2 + cd_2Y_2 - cd_1Y_2 + cd_1Y_1 + a_2cd_1d_2Y_2 + a_2cd_1X_2 - \\ a_2cd_1X_1 + cd_1d_2X_2 - bd_1d_2X_2 + bcd_2X_2 + a_2bcX_2 - a_2cd_1d_2Y_1 - \\ bcd_1X_2 - a_2bcX_1 - bcd_1d_2Y_2 + bcd_1d_2Y_1 + a_2bcd_2Y_2 - a_2bcd_2Y_1) \end{array}\right\}}{d_1 + bcd_1 - d_2 - bcd_2 + bd_1d_2 + b - c - cd_1d_2} \\[2ex]
Y_0 = \dfrac{\left\{\begin{array}{l} -(a_2bX_1 - a_2bd_2Y_2 + cd_1X_2 - a_2d_1X_2 + bcX_2 - a_2bX_2 - cd_1X_1 + \\ a_2d_1X_1 - bcd_1Y_1 - bcX_1 - a_2d_1d_2Y_2 - bY_1 - bd_1d_2Y_1 + bcd_2Y_2 + \\ cd_1d_2Y_2 + a_2d_1d_2Y_1 + d_2Y_1 + cY_1 + a_2bd_2Y_1 - d_1Y_1) \end{array}\right\}}{d_1 + bcd_1 - d_2 - bcd_2 + bd_1d_2 + b - c - cd_1d_2}
\end{array}
\right.
\tag{14.11}
$$

Example 14.1 (Planar intersection problem). Consider the example of [237, p. 292]. In this example, planar Cartesian coordinates of two known stations $F := P_1$ and $E := P_2$ are given as

$$
\begin{array}{l}
\{X_1 = 2490.50\,m,\ Y_1 = 2480.79\,m\}_{P_1} \\
\{X_2 = 780.67\,m,\ Y_2 = 7394.05\,m\}_{P_2}\,.
\end{array}
$$

The adjusted angles from points $F := P_1$ and $E := P_2$ to the unknown station $G := P_0 \in \mathbb{E}^2$ are $117°\ 11'\ 20.7''$ and $27°\ 35'\ 47.9''$ respectively. Using these angles and Fig. 14.1 on p. 255 one writes the directions as: $T_{10} = 0°\ 00'\ 00.0''$, $T_{12} = 117°\ 11'\ 20.7''$, $T_{21} = 0°\ 00'\ 00.0''$ and $T_{20} = 27°\ 35'\ 47.9''$. These directions are used in (14.8) to compute the constants $\{a_1, a_2, b, c\}$ which are then inserted in the first two expressions of (14.11) to give the values of d_1 and d_2, which are used in the fourth expression of (14.8). This leads to the two unknown orientation parameters σ_1 and σ_2 as; $351°\ 59'\ 56.3''$ and $289°\ 11'\ 17.0''$ respectively. The planar coordinates $\{X_0, Y_0\}_{P_0}$ of the unknown station $G := P_0 \in \mathbb{E}^2$ are then computed from the third and fourth expressions of (14.11) as; $\{X_0 = 6629.0952\,m,\ Y_0 = 1899.0728\,m\}_{P_1}$, which compare well with those of [237, p. 292].

Example 14.2 (Planar intersection problem). Let us consider another example of [237, p. 292] where the planar Cartesian coordinates of two known stations $E := P_1$ and $D := P_2$ are given as

$$
\begin{array}{l}
\{X_1 = 780.67\,m,\ Y_1 = 7394.05\,m\}_{P_1} \\
\{X_2 = 5044.25\,m,\ Y_2 = 7752.70\,m\}_{P_2}\,.
\end{array}
$$

The adjusted angles from points $E := P_1$ and $D := P_2$ to the unknown station $G := P_0 \in \mathbb{E}^2$ are $48^0\ 01'\ 25.3''$ and $100^0\ 20'\ 27.8''$ respectively. Using these angles and Fig. 14.1 as in the previous example, one writes the directions as: $T_{10} =$

$0^\circ\ 00'\ 00.0''$, $T_{12} = 48^\circ\ 01'\ 25.3''$, $T_{21} = 0^\circ\ 00'\ 00.0''$ and $T_{20} = 100^\circ\ 20'\ 27.8''$. These directions are used in (14.8) to compute $\{a_1, a_2, b, c\}$, which are inserted in the first two expressions of (14.11) to give the values of d_1 and d_2. These values of d_1 and d_2 are inserted in the fourth expression of (14.8) to give the two unknown orientation parameters σ_1 and σ_2 as $316^\circ\ 47'\ 04.8''$ and $04^\circ\ 48'\ 30.1''$ respectively. The planar coordinates $\{X_0, Y_0\}P_0$ of the unknown station $G := P_0 \in \mathbb{E}^2$ are then computed from the third and fourth expressions of (14.11) as; $\{X_0 = 6629.1007\,m,\ Y_0 = 1899.0635\,m\}_{P_1}$, which compare well with those of [237, p. 292].

14-22 Three-dimensional intersection

14-221 Closed form solution

In the case of three-dimensional intersection problem, the triple of three points $\{P_1, P_2, P_3\}$ in Fig. 14.1 are given by their three-dimensional Cartesian coordinates $\{X_1, Y_1, Z_1\}$, $\{X_2, Y_2, Z_2\}$, $\{X_3, Y_3, Z_3\}$, but the coordinates $\{X_0, Y_0, Z_0\}$ of point P_0 are unknown. The dimensionless quantities $\{\psi_{12}, \psi_{23}, \psi_{31}\}$ are space angles; $\psi_{12} = \angle P_0P_1P_2$, $\psi_{23} = \angle P_0P_2P_3$, $\psi_{31} = \angle P_1P_3P_0$. This problem is formulated as follows; Given horizontal directions T_i and vertical directions B_i measured from three known stations to an unknown station, determine the position of the unknown station P_0. These directional measurements are transformed into space angles $\{\psi_{12}, \psi_{23}, \psi_{31}\}$ using (13.30) on p. 227 (see e.g., Fig. 14.1). Equation (13.30) is the analytic version of a map of directions to space coordinates. Indeed, the map eliminates the external orientation parameters. The space angles are then used to obtain the unknown distances $\{x_1 = S_i, x_2 = S_2, x_3 = S_3\}$. These distances relate the unknown station $P_0 \in \mathbb{E}^3$ to three known stations $P_i \in \mathbb{E}^3 \mid i = \{1, 2, 3\}$ in the first step. The nonlinear system of equations relating the unknown distances $\{x_1 = S_i, x_2 = S_2, x_3 = S_3\}$ to the space angles $\{\psi_{12}, \psi_{23}, \psi_{31}\}$ are given as

$$\begin{bmatrix} x_2^2 = x_1^2 + S_{12}^2 - 2S_{12}\cos(\psi_{12})x_1 \\ x_3^2 = x_2^2 + S_{23}^2 - 2S_{23}\cos(\psi_{23})x_2 \\ x_1^2 = x_3^2 + S_{31}^2 - 2S_{31}\cos(\psi_{31})x_3. \end{bmatrix} \tag{14.12}$$

In the *second step*, the computed distances from *step* 1 are used in the three-dimensional ranging techniques of Chap. 12 to solve for the unknown position $P_0 \in \mathbb{E}^3$.

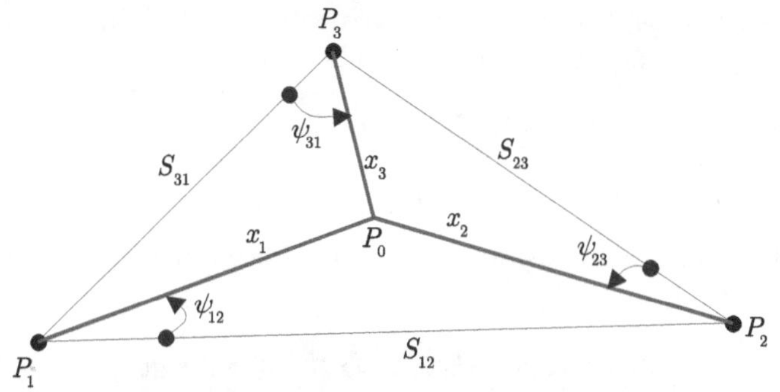

Figure 14.1. 3d-intersection

14-222 Conventional solution

Equation (14.12) is solved by first adding (14.12)i, (14.12)ii and (14.12)iii to eliminate the squared terms. The resulting expression

$$S_{12}^2+S_{23}^2+S_{31}^2-2x_1S_{12}\cos(\psi_{12})-2x_2S_{23}\cos(\psi_{23})-2x_3S_{31}\cos(\psi_{31})=0 \quad (14.13)$$

is linear in x_1, x_2 and x_3. The variable x_1 in (14.13) is then expressed in terms of x_2 and x_3 as

$$x_1=\frac{S_{12}^2+S_{23}^2+S_{31}^2-2x_2S_{23}\cos(\psi_{23})-2x_3S_{31}\cos(\psi_{31})}{2S_{12}\cos(\psi_{12})}, \quad (14.14)$$

and substituted in (14.12)i to give an expression in x_2 and x_3 only. The resulting expression in x_2 and x_3 is solved simultaneously with (14.12)ii to give values of x_2 and x_3. On the other hand, if (14.13) is now written such that x_3 is expressed in terms of x_2 and x_1 and substituted in (14.12)iii, an expression in x_2 and x_1 will be given which together with (14.12)i can be solved for the values of x_2 and x_1.

The setback with this approach is that one variable, in this case x_2, is determined twice with different values being given; which clearly is undesirable. A direct solution to the problem based on algebraic approaches of either Groebner basis or polynomial resultants alleviates the problem.

14-223 Reduced Groebner basis solution

reduced Groebner basis (4.38) on p. 45 is performed in two steps as follows:

- Step 1 (derivation of distances):
  Equation (14.12) is re-written algebraically as

$$\begin{bmatrix} f_1 := x_1^2+b_1x_1-x_2^2+a_0=0 \\ f_2 := x_2^2+b_2x_2-x_3^2+b_0=0 \\ f_3 := x_3^2+b_3x_3-x_1^2+c_0=0, \end{bmatrix} \quad (14.15)$$

with $b_1 = -2S_{12}\cos(\psi_{12}), b_2 = -2S_{23}\cos(\psi_{23}), b_3 = -2S_{31}\cos(\psi_{31})$ and $a_0 = S_{12}^2, b_0 = S_{23}^2, c_0 = S_{31}^2$. The reduced Groebner basis of (14.15) is then computed as

$$\begin{bmatrix} GroebnerBasis[\{f_1, f_2, f_3\}, \{x_1, x_2, x_3\}, \{x_2, x_3\}] \\ GroebnerBasis[\{f_1, f_2, f_3\}, \{x_1, x_2, x_3\}, \{x_1, x_3\}] \\ GroebnerBasis[\{f_1, f_2, f_3\}, \{x_1, x_2, x_3\}, \{x_2, x_3\}], \end{bmatrix} \tag{14.16}$$

which leads to three quartic polynomials for determining the unknown distances $\{x_1 = S_i, x_2 = S_2, x_3 = S_3\}$;

$$\begin{bmatrix} x_1 := d_4x_1^4 + d_3x_1^3 + d_2x_1^2 + d_1x_1 + d_0 = 0 \\ x_2 := e_4x_2^4 + e_3x_2^3 + e_2x_2^2 + e_1x_2 + e_0 = 0 \\ x_3 := f_4x_3^4 + f_3x_3^3 + f_2x_3^2 + f_1x_3 + f_0 = 0. \end{bmatrix} \tag{14.17}$$

The coefficients of (14.17) are as given in [33, Appendix].

- Step 2 (position determination):
In this step, the computed distances from (14.17) are used to determine the unknown position $P_0 \in \mathbb{E}^3$ as discussed in Sect. 12-32.

14-224 Sturmfels' resultants solution

Algorithm presented in Sect. 5-32 proceeds in two steps as follows:

- Step 1 (derivation of distances):
Following (5.12) on p. 54, (14.15) is homogenized using the variable x_4 and re-written for the solutions of x_1, x_2 and x_3 in (14.18), (14.19) and (14.20) respectively as
  - Solving for x_1 by treating it as a constant (polynomial of degree zero)

$$\begin{bmatrix} g_1 := (x_1^2 + b_1x_1 + a_0)x_4^2 - x_2^2 = 0 \\ g_2 := x_2^2 + b_2x_2x_4 - x_3^2 + b_0x_4^2 = 0 \\ g_3 := x_3^2 + b_3x_3x_4 + (c_0 - x_1^2)x_4^2 = 0. \end{bmatrix} \tag{14.18}$$

  - Solving for x_2 by treating it as a constant (polynomial of degree zero)

$$\begin{bmatrix} h_1 := x_1^2 + b_1x_1x_4 + (a_0 - x_2^2)x_4^2 = 0 \\ h_2 := (x_2^2 + b_2x_2 + b_0)x_4^2 - x_3^2 = 0 \\ h_3 := x_3^2 + b_3x_3x_4 - x_1^2 + c_0x_4^2 = 0. \end{bmatrix} \tag{14.19}$$

  - Solving for x_3 by treating it as a constant (polynomial of degree zero)

$$\begin{bmatrix} k_1 := x_1^2 + b_1x_1x_4 - x_2^2 + a_0x_4^2 = 0 \\ k_2 := x_2^2 + b_2x_2x_4 + (b_0 - x_3^2)x_4^2 = 0 \\ k_3 := (x_3^2 + b_3x_3 + c_0)x_4^2 - x_1^2 = 0. \end{bmatrix} \tag{14.20}$$

From (14.18), (14.19) and (14.20), expressing $a_1 = (x_1^2 + b_1x_1 + a_0)$ and $a_2 = (c_0 - x_1^2)$ in (14.18), $a_3 = (x_2^2 + b_2x_2 + b_0)$ and $c_2 = (a_0 - x_2^2)$ in (14.19), and finally $c_3 = (b_0 - x_3^2)$ and $c_1 = (x_3^2 + b_3x_3 + c_0)$ in (14.20), one forms the Jacobian determinant matrices with (5.13) on p. 54 respectively as

$$J_{x1} = det\begin{bmatrix} \frac{\partial g_1}{\partial x_2} & \frac{\partial g_1}{\partial x_3} & \frac{\partial g_1}{\partial x_4} \\ \frac{\partial g_2}{\partial x_2} & \frac{\partial g_2}{\partial x_3} & \frac{\partial g_2}{\partial x_4} \\ \frac{\partial g_3}{\partial x_2} & \frac{\partial g_3}{\partial x_3} & \frac{\partial g_3}{\partial x_4} \end{bmatrix} = det\begin{bmatrix} -2x_2 & 0 & 2a_1x_4 \\ 2x_2 + b_2x_4 & -2x_3 & b_2x_2 + 2b_0x_4 \\ 0 & 2x_3 + b_3x_4 & b_3x_3 + 2a_2x_4 \end{bmatrix}, \quad (14.21)$$

$$J_{x2} = det\begin{bmatrix} \frac{\partial h_1}{\partial x_1} & \frac{\partial h_1}{\partial x_3} & \frac{\partial h_1}{\partial x_4} \\ \frac{\partial h_2}{\partial x_1} & \frac{\partial h_2}{\partial x_3} & \frac{\partial h_2}{\partial x_4} \\ \frac{\partial h_3}{\partial x_1} & \frac{\partial h_3}{\partial x_3} & \frac{\partial h_3}{\partial x_4} \end{bmatrix} = det\begin{bmatrix} 2x_1 + b_1x_4 & 0 & b_1x_1 + 2c_2x_4 \\ 0 & -2x_3 & 2a_3x_4 \\ -2x_1 & 2x_3 + b_3x_4 & b_3x_3 + 2c_0x_4 \end{bmatrix}, \quad (14.22)$$

and

$$J_{x3} = det\begin{bmatrix} \frac{\partial k_1}{\partial x_1} & \frac{\partial k_1}{\partial x_2} & \frac{\partial k_1}{\partial x_4} \\ \frac{\partial k_2}{\partial x_1} & \frac{\partial k_2}{\partial x_2} & \frac{\partial k_2}{\partial x_4} \\ \frac{\partial k_3}{\partial x_1} & \frac{\partial k_3}{\partial x_2} & \frac{\partial k_3}{\partial x_4} \end{bmatrix} = det\begin{bmatrix} 2x_1 + b_1x_4 & -2x_2 & b_1x_1 + 2a_0x_4 \\ 0 & 2x_2 + b_2x_4 & b_2x_2 + 2c_3x_4 \\ -2x_1 & 0 & 2c_1x_4 \end{bmatrix}. \quad (14.23)$$

The resulting determinants are cubic polynomials:

$J_{x1} = 4x_2b_3x_3^2 + 8x_2x_3a_2x_4 + 4b_2x_2^2x_3 + 2b_2x_2^2b_3x_4 + 8x_2b_0x_4x_3 + 4x_2b_0x_4^2b_3 + 8a_1x_4x_2x_3 + 4a_1x_4^2x_2b_3 + 4a_1x_4^2b_2x_3 + 2a_1x_4^3b_2b_3$.

$J_{x2} = -4x_1b_3x_3^2 - 8x_1x_3c_0x_4 - 8x_1a_3x_4x_3 - 4x_1a_3x_4^2b_3 - 2b_1x_4b_3x_3^2 - 4b_1x_4^2x_3c_0 - 4b_1x_4^2a_3x_3 - 2b_1x_4^3a_3b_3 - 4x_1^2x_3b_1 - 8x_1x_3c_2x_4$.

$J_{x3} = 8c_1x_4x_1x_2 + 4c_1x_4^2x_1b_2 + 4c_1x_4^2b_1x_2 + 2c_1x_4^3b_1b_2 + 4x_1b_2x_2^2 + 8x_1x_2c_3x_4 + 4b_1x_1^2x_2 + 2b_1x_1^2b_2x_4 + 8x_1a_0x_4x_2 + 4x_1a_0x_4^2b_2$.

Making use of (5.14) and (5.15) on p. 54 lead to

$$\begin{bmatrix} x_1 := d_4x_1^4 + d_3x_1^3 + d_2x_1^2 + d_1x_1 + d_0 = 0 \\ x_2 := e_4x_2^4 + e_3x_2^3 + e_2x_2^2 + e_1x_2 + e_0 = 0 \\ x_3 := f_4x_3^4 + f_3x_3^3 + f_2x_3^2 + f_1x_3 + f_0 = 0. \end{bmatrix} \quad (14.24)$$

The coefficients of (14.24) are given in [34, Appendix].

- Step 2 (position determination):
  In this step, the computed distances from (14.24) are used to determine the unknown position $P_0 \in \mathbb{E}^3$ as discussed in Sect. 12-32.

Example 14.3 (3d-intersection from three known stations). Using the computed quartic polynomials (14.17) or (14.24), the distances $S_i = x_i \in \mathbb{R},^+ \; i = \{1,2,3\} \in \mathbb{Z}^3_+$ between an unknown station $K1 \in \mathbb{E}^3$ and known stations $P_i \in \mathbb{E}^3$ for the test network Stuttgart Central in *Fig.* 10.2 on p. 150 are determined. Points P_1, P_2, P_3 of the tetrahedron $\{P_0 P_1 P_2 P_3\}$ in Fig. 14.1 correspond to the chosen known GPS stations Schlossplatz, Liederhalle, and Eduardpfeiffer (see *Fig.* 10.2). The distance from $K1$ to Schlossplatz. is designated $S_1 = x_1 \in \mathbb{R},^+$ $K1$ *to Liederhalle* $S_2 = x_2 \in \mathbb{R},^+$ while that of $K1$ *to Eduardpfeiffer* is designated $S_3 = x_3 \in \mathbb{R}.^+$ The distances between the known stations $\{S_{12}, S_{23}, S_{31}\} \in \mathbb{R}^+$ are computed from their respective GPS coordinates in *Table* 10.1 on p. 152. Using the horizontal directions T_i and vertical directions B_i from *Table* 10.3 on p. 152, space angles $\{\psi_{12}, \psi_{23}, \psi_{31}\}$ are computed using (13.30) on p. 227 and presented in *Table* 14.1. From (14.17), we see that $S_1 = x_1$, $S_2 = x_2$ and $S_3 = x_3$ each has four roots. The solutions are real as depicted in *Figs.* 14.2, 14.3 and 17.5. The desired distances are selected with the help of prior information (e.g., from Fig. 10.2) as $S_1 = 566.8635$, $S_2 = 430.5286$, and $S_3 = 542.2609$. These values compare well with their real values in *Fig.* 10.2. Once the distances have been established, they are used to determine the coordinates of the unknown station $K1$ in step 2 via ranging techniques. In this example, the computed Cartesian coordinates of $K1$ are $X = 4157066.1116\,m$, $Y = 671429.6655\,m$ and $Z = 4774879.3704\,m$; which tallies with the GPS coordinates in *Table* 10.1.

Table 14.1. Space angles

Observation from	Space angle (gon)
$K1$-Schlossplatz-Liederhalle ψ_{12}	35.84592
$K1$-Liederhalle-Eduardpfeiffer ψ_{23}	49.66335
$K1$-Eduardpfeiffer-Schlossplatz ψ_{31}	14.19472

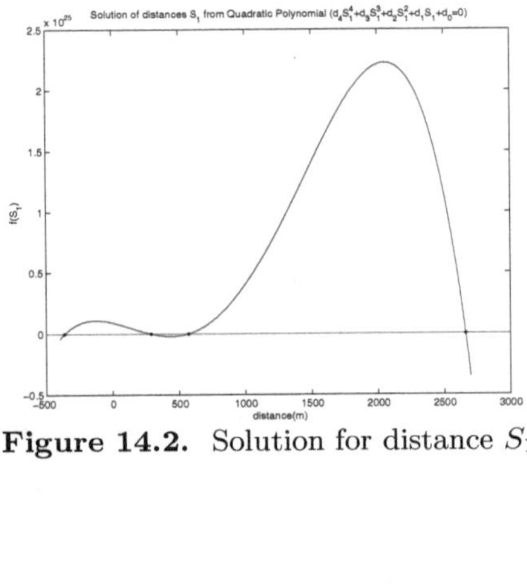

Figure 14.2. Solution for distance S_1

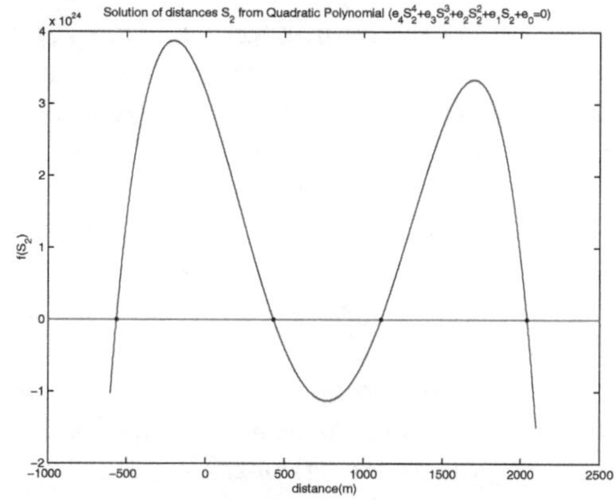

Figure 14.3. Solution for distance S_2

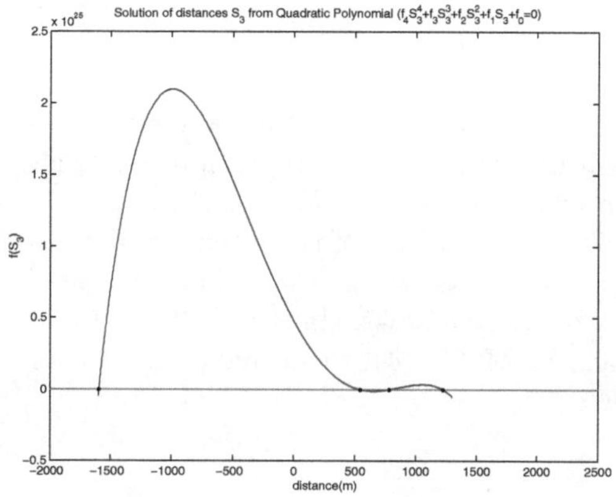

Figure 14.4. Solution for distance S_3

14-225 Intersection to more than three known stations

The formulation of the *overdetermined* three-dimension intersection problem is as follows; given space angles from *more than three known stations*, i.e., $P_1, P_2, P_3, ..., P_n$, determine the unknown position $P_0 \in \mathbb{E}^3$. In this case, the observations will comprise horizontal directions T_i and vertical directions B_i from P_1 to P_0, P_2 to P_0, P_3 to P_0,...,P_n to P_0, with the unknowns being $\{X, Y, Z\}$.

Example 14.4 (Three-dimensional intersection from more than three known stations). For the test network Stuttgart Central in Fig. 10.2, the three-dimensional coordinates $\{X, Y, Z\}$ of the unknown station $K1$ are desired. Using all the observation data of Table 10.2 on p. 152, one proceeds to compute the position of $K1$ in four steps as follows:

Step 1 (combinatorial solution):
From Fig. (10.2) on p. 150, and using (7.28) on p. 93, 35 minimal combinatorials are formed whose nonlinear systems of equations (14.12) are solved for

the distances $\{S_i | i = 1, 2, 3\}$ to the unknown station $K1$ in closed form using either (14.17) or (14.24). Each combinatorial minimal subset results in 3 distances thus giving rise to a total of (3×35) 105 distances which we consider in the subsequent steps as pseudo-observations.

Step 2 (error propagation to determine the dispersion matrix $\boldsymbol{\Sigma}$):
The variance-covariance matrix is computed for each of the combinatorial set $j = 1, \ldots, 35$ using error propagation. Equation (14.12) is used to obtain the dispersion matrix $\boldsymbol{\Sigma}$ in (7.33) as discussed in Example 7.4 on p. 95.

Step 3 (rigorous adjustment of the combinatorial solution points in a polyhedron):
Once the 105 combinatorial solution points in a polyhedron have been obtained in step 1, they are finally adjusted using the linear Gauss-Markov model (7.9) on p. 87 with the dispersion matrix $\boldsymbol{\Sigma}$ obtained via the nonlinear error propagation law in step 2.

Step 4 (position determination by ranging):
The position is then determined from values of steps 1 to 3 as in Example 12.6 on p. 211.

Using the data of *Table* 10.2, space angles for the network are computed and used to determine the position of the unknown station $K1$. Figure 14.5 presents the deviation of the computed scatter of the distance Haussmanstr.-$K1$ around its adjusted value. The plot of deviations of the adjusted distances from those derived from GPS coordinates are presented in Fig. 14.6. The numbers in the X-axis of Fig. 14.6 represent distances as follows; Haussmanstr.-$K1$ (1), Schlossplatz-$K1$ (2), Dach FH-$K1$ (3), Dach LVM-$K1$ (4), Liederhalle-$K1$ (5), Lindenmuseum-$K1$ (6) and Eduardpfeiffer-$K1$ (7).

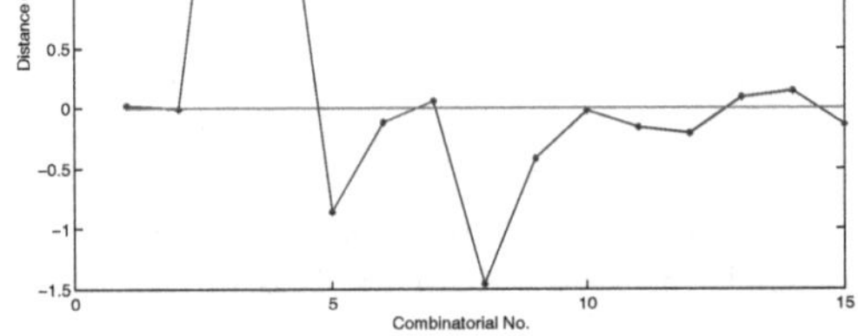

Figure 14.5. Deviation of the scatter solutions of the distance Haussmanstr.-$K1$ from the adjusted value

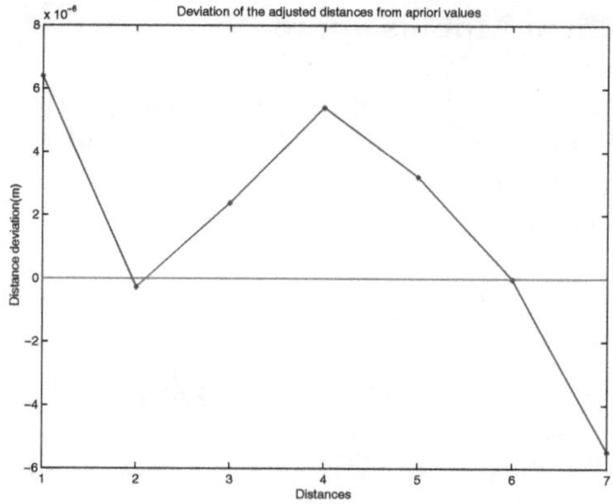

Figure 14.6. Deviation of the 7-adjusted distances from real measured values

14-3 Photogrammetric intersection

In Chap. 13, the exterior elements of orientation were determined as discussed in Sect. 13-3. Using these exterior elements, we demonstrate in this section how they are applied to determine algebraically the ground coordinates of unknown station. The problem of photogrammetric intersection is formulated as follows: Given the position and orientation of two or more photographs, let the conjugate image rays from the photographs intersect at a common ground point (e.g., Fig. 14.7). Determine the ground coordinates of the unknown station P. Let us examine two possible ways of solving this problem algebraically.

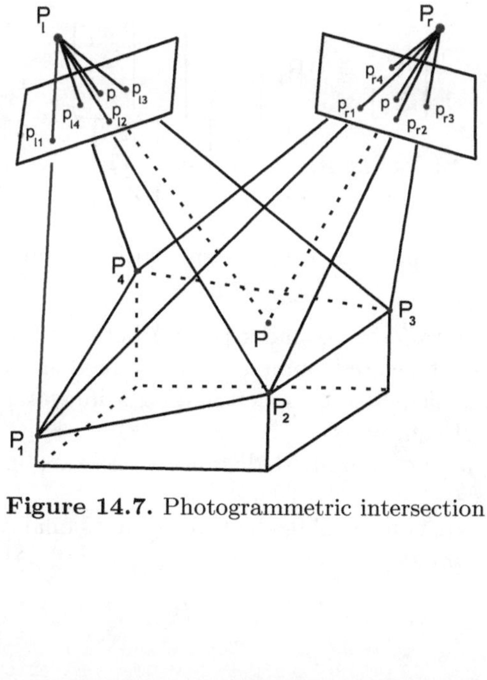

Figure 14.7. Photogrammetric intersection

14-306 Grafarend-Shan Möbius approach

Let us assume that the Cartesian coordinates $\{X_l, Y_l, Z_l\}$ and $\{X_r, Y_r, Z_r\}$, respectively, for the left perspective center P_l and the right perspective center P_r in Fig. 14.7 have been obtained from the photogrammetric resection approach in Sect. 13-3. The perspective center equations are

$$\begin{bmatrix} X - X_l \\ Y - Y_l \\ Z - Z_l \end{bmatrix} = s_l \mathbf{R}_l \begin{bmatrix} x_l \\ y_l \\ -f_l \end{bmatrix}, \quad (14.25)$$

and

$$\begin{bmatrix} X - X_r \\ Y - Y_r \\ Z - Z_r \end{bmatrix} = s_r \mathbf{R}_r \begin{bmatrix} x_r \\ y_r \\ -f_r \end{bmatrix}. \quad (14.26)$$

In (14.25) and (14.26), f_l and f_r are the left and right focal lengths and $\mathbf{R}$ the rotation matrix. The distance ratios s_l and s_r are given respectively by

$$s_l := \frac{\|\overrightarrow{PP_l}\|}{\|\overrightarrow{pp_l}\|} = \frac{\sqrt{(X - X_l)^2 + (Y - Y_l)^2 + (Z - Z_l)^2}}{\sqrt{x_l^2 + y_l^2 + z_l^2}}, \quad (14.27)$$

and

$$s_r := \frac{\|\overrightarrow{PP_r}\|}{\|\overrightarrow{pp_r}\|} = \frac{\sqrt{(X - X_r)^2 + (Y - Y_r)^2 + (Z - Z_r)^2}}{\sqrt{x_r^2 + y_r^2 + z_r^2}}. \quad (14.28)$$

Equations (14.25) and (14.26) could be expanded into the 7-parameter similarity transformation equation

$$\begin{bmatrix} X \\ Y \\ Z \end{bmatrix} = s_l \mathbf{R}_l \begin{bmatrix} x_l \\ y_l \\ -f_l \end{bmatrix} + \begin{bmatrix} X_l \\ Y_l \\ Z_l \end{bmatrix} = s_r \mathbf{R}_r \begin{bmatrix} x_r \\ y_r \\ -f_r \end{bmatrix} + \begin{bmatrix} X_r \\ Y_r \\ Z_r \end{bmatrix}. \quad (14.29)$$

Grafarend and Shan [183] propose a *three-step* solution approach based on Möbius coordinates as follows:

- In the *first step*, the ratio of distances $\{s_l, s_r\}$ between the perspective centers and the unknown intersected point are determined from a linear system of equations. The area elements of the left and right images are employed to form the linear system of equations.
- In the *second step*, the computed distance ratios are used to compute the Möbius coordinates.
- These coordinates are converted to the three-dimensional cartesian coordinates $\{X, Y, Z\}$ in the *third step*.

14-307 Commutative algebraic approaches

Whereas the Grafarend-Shan approach discussed in Sect. 14-307 solves the intersection of rays from two photographs, the algebraic approaches solves the intersection of rays from three photographs. Consider the case in Fig. 12.1, the unknown station is intersected from three photographs. The distances $\{S_1, S_2, S_3\}$ to the unknown stations are determined using either using (14.17) or (14.24). The coordinates of the unknown stations are determined from procedures of Sect. 12-326.

14-4 Concluding remarks

The techniques presented in this chapter could provide direct approaches for obtaining positions from direction (angular) measurements to stations that can not be physically occupied. Intersection techniques discussed could be useful in structural deformation monitoring in industries. For surfaces or structures that are harmful when physical contact is made, intersection techniques come in handy. the methods can also be used for quick station search during cadastral and engineering surveying operations. These methods can be augmented with resection and ranging techniques to offer a wide range of possibilities. Further reference are [17, 30, 40, 163, 173, 183].

15 GNSS environmental monitoring

15-1 Satellite environmental monitoring

In 1997, the Kyoto protocol to the United Nation's framework convention on climate change spelt out measures that were to be taken to reduce the greenhouse gas emission that has contributed to *global warming*. Global warming is just but one of the many challenges facing our environment today. The rapid increase in *desertification* on one hand and *flooding* on the other hand are environmental issues that are increasingly becoming of concern. For instance, the torrential rains that caused havoc and destroyed properties in USA in 1993 is estimated to have totalled to $15 billion, 50 people died and thousands of people were evacuated, some for months [261]. Today, the threat from torrential rains and flooding still remains real as was seen in 1997 El'nino rains that swept roads and bridges in Kenya, the 2000 Mozambique flood disaster, 2002 Germany flood disaster or the Hurricane Isabel in the US coast[1]. The melting of polar ice thus raising the sea level is creating fear of submersion of beaches and cities surrounded by the oceans and those already below sea level. In-order to be able to predict and model these occurrences so as to minimize damages such as those indicated by [261], atmospheric studies have to be undertaken with the aim of improving on mechanism for providing *reliable*, *accurate* and *timely* data. These data are useful in Numerical Weather Prediction (NWP) models for weather forecasting and climatic models for monitoring climatic changes. Besides, *accurate* and *reliable information* on weather is essential for other applications such as agriculture, flight navigation, etc.

Data for NWP and climatic models are normally collected using balloon filled radiosondes, satellites (polar and geostationary) and other sources e.g., flight data from aeroplanes. Whereas [283, p. 94] points out that about 9500 surface based stations and 7000 merchant ships exist that send up weather balloons, [413] noted that most of these data cover the northern hemisphere, with the southern hemisphere (mainly Africa and South America) lacking adequate data due to financial constraints. Lack of radiosonde data is also noted in the oceanic areas hence leading to shortage of adequate data for NWP and climatic models. These models require precise and accurate data for estimating initial starting values in-order to give accurate and reliable weather forecast, and to be of use for climate monitoring. The shortage of radiosonde data is complemented with the polar and geostationary satellite data. Polar satellites include for instance the US owned National Ocean and Atmospheric Administration NOAA-14 and NOAA-15, while the geostationary satellites include US based Geostationary Operational Environmental Satellite (GEOS) and Europe owned METEOrological SATellite (METEOSAT).

Polar and geostationary satellites (e.g., NOAA, GOES and METEOSAT) used for temperature and water vapour profile measurements have their own limitations however. In high altitude winter conditions for instance, use of passive Infra Red (IR) is difficult due to *very cold temperatures*, *common near surface thermal*

[1]BBC 19th Sept. 2003 online report: http://news.bbc.co.uk/

J.L. Awange et al., *Algebraic Geodesy and Geoinformatics*, 2nd ed.,
DOI 10.1007/978-3-642-12124-1_15, © Springer-Verlag Berlin Heidelberg 2010

inversion, and *high percentage of ice cloud* that play a role in limiting the IR sounding [300]. In volcanic areas, low flying remote sensing satellites are also affected by the presence of dust and aerosol. Large-scale volcanic eruption normally injects large amount of aerosols into the lower stratosphere and thus limiting the IR observation of the stratosphere and lower regions. In-order therefore to enhance global weather and climatic prediction, current systems have to be complemented by a system that will provide global coverage and whose signals will be able to penetrate clouds and dust to remote sense the atmosphere. Such system, already proposed as early as 1965 by Fischbach [134], and which is currently an active area of research, is the new field of ***GPS-Meteorology***. It involves the use of GPS satellites to obtain atmospheric profiles of *temperature*, *pressure* and *water vapour/humidity*.

As we saw in Chap. 10, Global Positioning System (GPS) satellites were primarily designed to be used by the US military. their main task was to obtain the position of any point on Earth from space. The signals emitted by GPS satellites traverse the ionosphere and neutral atmosphere to be received by ground based GPS receivers. One of the major obstacles to positioning or navigating with GPS is the signal delay caused by atmospheric refraction. Over the years, research efforts have been dedicated to modelling atmospheric refraction in-order to improve on positioning accuracy. In the last decade however, [300] suggested that this negative effect of the atmosphere on GPS signals could be inverted to remote sense the atmosphere using space borne techniques. Melbourne [300], proposed that Low Earth Orbiting Satellites LEO be fitted with GPS receivers and be used to track the signals of rising or setting GPS satellites (occulting satellites). The signal delay could then be measured and used to infer on the atmospheric profiles of temperature, pressure, water vapour and geopotential heights.

This new technology of GPS atmospheric remote sensing has the advantages of;

(a) being *global*,
(b) *stable* owing to the stable GPS oscillators and
(c) having radio frequencies that can *penetrate* clouds and dusts.

The new technology therefore plays a major role in complementing the existing techniques, e.g., radiosondes. Atmospheric profiles from GPS remote sensing have been tested in NWP models and preliminary results so far are promising [216]. Indeed, [250] have already demonstrated using the data of the pilot project GPS/MET that the accuracy of global and regional analysis of weather prediction can significantly be improved. Also motivating are the results of [370] who showed that high accuracy of measurements and vertical resolution around the tropopause would be relevant to monitor climatic changes in the next decades. Several atmospheric sounding missions have been launched, e.g., the CHAllenging Minisatellite Payload mission (CHAMP), Gravity Recovery And Climate Experiment (GRACE) and SAC-C. Constellation Observing System for Meteorology, Ionosphere and Climate (COSMIC) mission that was launched on 15th of April 2006 by University Corporation of Atmospheric Research UCAR is already recording more than 2500 daily measurements, see e.g., [412]. Its optimum operation is expected to provide up to 3000 occultation data daily [12]. EQUatorial Atmosphere Research Satellite Currently, studies are being undertaken at Jet Propulsion Laboratory (JPL) on possibilities of having future atmospheric sounding missions that will have satellites of the sizes of a laptop with GPS receivers of the sizes of a credit card [429]. The European owned EUropean organization for the exploitation of METeorological SATellites (EUMETSAT) recently installed a GPS occultation receiver GRAS (GNSS Receiver for Atmospheric Sounding). The planned satellite missions, together with the proposed GALILEO satellites anticipated around 2013, the Russian GLONASS and the Chinese Compass [228] promises a brighter future for environmental monitoring. Indeed, that these atmospheric sounding missions promise to provide daily global coverage of thousands of remotely sensed data which will be vital for weather, climatic and atmospheric sciences studies will be a revolution in the history of environmental studies.

Space borne GPS meteorology which we discuss in detail in Sect. 15-21 is just but one part of this new technique. The other component is the ground based GPS meteorology which will be discussed in detail in Sect. 15-22. Collection of articles on this new technique has been presented for instance in [13]. In ground based GPS meteorology, a dense GPS network is used to measure precisely GPS path delays caused by the ionosphere and the neutral troposphere traversed by the GPS signals. These path delays are then converted into Total Electronic Contents (TEC) and Integrated Precipitate Water Vapour IPWV. Conversion to IPWV requires prior information of surface pressure or estimates along the GPS ray path. These create a continuous, accurate, all weather, real time lower and upper atmospheric data with a variety of opportunities for atmospheric research [407].

Clearly, GPS meteorology promises to be a real boost to atmospheric studies with expected improvements on weather forecasting and climatic change monitoring, which directly impact on our day to day lives. In [21], the possible use of IPWV for flood prediction is proposed, while [50] have outlined the potential of water vapour for meteorological forecasting. For environmental monitoring, GPS meteorology will further play the following roles:

1. Precisely derive vertical temperature and pressure profiles: These will be useful in the following ways [300]:
   (a) By combining them with other observations of ozone densities and dynamic models, our understanding of conditions which lead to the formation of polar stratosphere clouds will be improved. We will also be able to understand how particles in which heterogeneous chemical reactions lead to ozone loss are believed to occur.
   (b) The precise measured temperature will enable the monitoring of global warming and the effect of greenhouse gases. This is made possible as the change in surface temperatures caused by an increase in the greenhouse gas densities is generally predicted to be the largest and therefore most apparent at high latitudes. Precise temperature can be used to map the structure of the stratosphere, particularly in the polar region where tem-

perature is believed to be an important factor in the minimum levels of ozone observed in spring.

(c) Accurate high vertical resolution temperature reconstruction in the upper troposphere will increase our understanding on the conditions which cirrus clouds form. The cirrus clouds will generate for instance a positive feed back effect if global warming displaces a given cloud layer to a higher and colder region. The colder cloud will then emit less radiation forcing the troposphere to warm in-order to compensate for the decrease.

(d) Accurate temperature retrievals from GPS meteorological measurements combined with high horizontal resolution temperatures derived from the nadir-viewing microwave radiometers will provide a powerful data set for climate studies of the Earth's lower atmosphere. This can be achieved by using the derived profiles to monitor trends in the upper troposphere and lower stratosphere where the GPS meteorological techniques yield its most accurate results.

(e) The measured pressure is expected to contribute to the monitoring of global warming. This is because pressure versus geometrical height is potentially an interesting diagnostic of troposphere's climatic change since the height of any pressure surface is a function of the integrated temperature below.

(f) The temperature in the upper troposphere/tropopause influences the amount of energy radiated to space. Accurate measurements of temperature in this region over a long period of time will provide data for global warming and climatologic studies.

2. Derive water vapour: Precise analysis of the water vapour will contribute to the data required by hydrologists to enhance the prediction of local torrential rain that normally cause damage and havoc (see e.g., [21]). Besides, the knowledge of water vapour density in the lower troposphere will be useful in;
   - providing data that will be directly assimilated into meteorological models to enhance predictability and forecasting of weather,
   - applicable for creation of distribution of water vapour via tomographic techniques (e.g., [138]),
   - applied to correct the wet delay component in both Synthetic Aperture Radar (SAR) and GPS positioning thus benefiting applications requiring precise positioning such as deformation monitoring,
   - beneficial to low altitude aircraft navigation, since limitation in the mitigation of tropospheric delay is a major source of positioning error,
   - global warming monitoring by determining the latent heat suspended in the atmosphere where water vapour comprise one of the greenhouse gases,
   - the radiative forcing due to vapour and cloud inferred from humidity,
   - improved inputs for weather forecasting, climate and hydrology. Water vapour will be essential for short term (0-24hrs) forecasting of precipitation. Currently, lack of atmospheric water vapour is the major source of error in short term weather forecasting [209].
3. Contribute towards climatic studies: By comparing the observed temperatures against the predicted model values, a method for detecting and characterizing

stratospheric climatic variations as well as a means for evaluating the performance of model behaviour at stratospheric altitudes will be developed and the existing ones tested.

4. Enhance geodynamic studies: The study of the gravitation effects of the atmospheric pressure, water vapour and other phenomenons will contribute towards the determination of high-resolution local geoid, which is vital for monitoring crustal deformation. The transient drift that occurs per week in estimate of crustal deformation from GPS measurement will be corrected.
5. Enhance disaster mitigation measures: Its information will contribute to the much-needed information required to improve forecasting of catastrophic weather around the world.
6. With abundance of GPS remote sensing data, accuracy better than 1 -2K in temperature given by GPS meteorological missions (e.g., CHAMP, GRACE etc.) will be realized.

In-order to fully realize the potential of the GPS atmospheric remote sensing listed above, estimated profiles have to be of high quality. Already, comparative results with the existing models such as European Centre for Medium Weather Forecast (ECMWF) and National Centre for Environmental Prediction (NCEP) are promising as seen from the works of [348, 413] with respect to GPS/MET and CHAMP missions, respectively. Detailed exposition of the application of GNSS remote sensing to environment is presented in Awange [43].

15-2 GNSS remote sensing

15-21 Space borne GNSS meteorology

Radio occultation with GPS takes place when a transmitting GPS satellite, setting or rising behind the Earth's limb, is viewed by a LEO satellite as illustrated in Fig. 15.1[2]. GPS satellites send navigation signals, which passes through successively deeper layer of the Earth's atmosphere and are received by LEO satellites. These signals are bent and retarded causing a delay in the arrival at the Leo (see Fig.15.1[2]). Figure 15.3 indicates the occultation geometry where the signal is sent from GPS to the LEO satellite passing through dispersive layers of the ionosphere and atmosphere remote sensing them. As the signal is bent, the total bending angle α, an impact parameter a and a tangent radius r_t define the ray passing through the atmosphere. *Refraction angle* is accurately measured and related to atmospheric parameters of temperature, pressure and water vapour via the refractive index. Use is made of radio waves where the LEO receiver measures, at the required sampling rate, the dual band carrier phase, the C/A and P-code group delay and the signal strength made by the flight receiver [300]. The data is then processed to remove errors arising from short time oscillator and instabilities in; satellites and receivers. This is achieved by using at least one ground station and

[2]source: http://geodaf.mt.asi.it/html/GPSAtmo/space.html

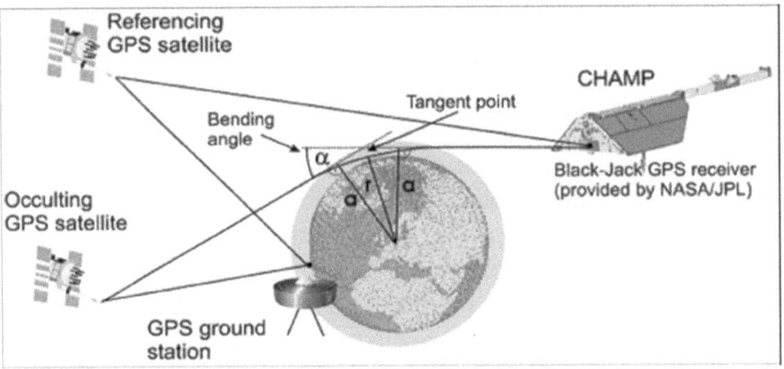

Figure 15.1. GPS Radio occultation (source: GfZ [413])

one satellite that is not being occulted. Once the observations have been corrected for possible sources of errors, the resulting *Doppler shift* is used to determine the refraction angle α (see Fig. 15.3).

The variation of α with a during an occultation depends primarily on the vertical profile of atmospheric refractive index, which is determined globally by *Fermat's principle* of least time and locally by *Snell's law*

$$n sin\phi = constant, \tag{15.1}$$

where ϕ denotes the angle between the gradient of refraction and the ray path. Doppler shift is determined by projecting spacecraft velocities onto the ray paths at the transmitter and receiver, so that atmospheric bending contributes to its measured value. Data from several GPS transmitters and post-processing ground stations are used to establish the precise positions and velocities of the GPS transmitters and LEO satellites. These derived positions and velocities are used to calculate the Doppler shift expected in the absence of atmospheric bending (i.e., were the signal to travel in vacuo). By subtracting the *expected* shift from the measured shift, one obtains the excess Doppler shift. Assuming local symmetry and with Snell's law, the excess Doppler shift together with satellites' *positions* and *velocities* are used to compute the values of the bending angles α with respect to the impact parameters a. In Sect. 15-3, we will present an algebraic approach for computing bending angles and impact parameters. Once computed, these bending (refraction) angles are related to the refractive index by

$$\alpha(a) = 2a \int_{r=r_0}^{r=\infty} \frac{1}{\sqrt{n^2 r^2 - a^2}} \frac{\mathrm{d}In(n)}{\mathrm{d}r} \mathrm{d}r, \tag{15.2}$$

which is inverted using Abel's transformation to give the desired refractive index

$$n(r_0) = \exp\left[\frac{1}{\pi} \int_{a=a_0}^{a=\infty} \frac{\alpha(a)}{\sqrt{a^2 - a_0^2}} da\right]. \tag{15.3}$$

Rather than the refractive index in (15.3), refractivity is used as

$$N = (n-1)10^6 = 77.6\frac{P}{T} + 3.73 \times 10^5 \frac{P_w}{T^2} - 40.3 \times 10^6 \frac{n_e}{f^2} + 1.4w. \quad (15.4)$$

In (15.4), P denotes the atmospheric pressure in {mbar}, T the atmospheric temperature in K, P_w the water vapour in {mbar}, n_e the electron number density per cubic meter {number of electron/m^3}, f the transmitter frequency in Hz and w the liquid water content in g/m^3. Three main contributors to refractivity are:

- The *dry neutral atmosphere* (called the dry component, i.e., the first component on the right-hand-side of (15.4)).
- Water vapour (also called the wet or moist components, i.e., the second component on the right-hand-side of (15.4))
- The free electrons in the ionosphere (i.e., the third component on the right-hand-side of (15.4)).

If the atmospheric temperature T and pressure P are provided from external source, e.g., from models and synoptic meteorological data over tropical oceanic regions, then the vertical water vapour density may be recovered from satellite remote sensing data [300]. The refraction effects on the signals in the ionosphere must be corrected using signals at two frequencies at which these effects are substantially different.

15-22 Ground based GNSS meteorology

Whereas GPS receivers are onboard low flying (LEO) satellites (e.g., CHAMP, GRACE etc.) in space borne GPS remote sensing, they are fixed on ground stations in the case of ground-based GPS meteorology. These receivers track the transmitted signals which have traversed the atmosphere as indicated in Fig.15.2[3]. As the signals travel through the atmosphere from the satellites to the ground based receivers, they are delayed by the troposphere. The tropospheric delay comprise the hydrostatic and the wet parts as seen in (15.4). The contribution of hydrostatic part which can be modeled and eliminated very accurately using surface pressure data or three-dimensional numerical models is about 90% [107, 126]. The wet delay however is highly variable with little correlation to surface meteorological measurements. Assuming that the wet delay can be accurately derived from GPS data, and that reliable surface temperature data are available, the wet delay can be converted into the estimation of the total atmospheric water vapour P_w present along the GPS ray path as already suggested by [67]. This atmospheric water vapour P_w is termed precipitable water in GPS meteorology.

The precipitable water as opposed to the vertical profile is estimated with a correction made for the fact that the radio beams normally are slanted from the zenith. The phase delay along the zenith direction is called the "zenith delay" and is related to the atmospheric refractivity by:

---

[3]Source: http://apollo.lsc.vsc.edu/classes/remote/lecture_notes/gps/theory/ theory-html.htm

$$zenith\,Delay = 10^6 \int_{antenna}^{\infty} N(z)\mathrm{d}z, \tag{15.5}$$

where the integral is in the zenith direction. Substituting (15.4) in (15.5) leads to the calculation of the zenith wet delay which is related to the total amount of water vapour along the zenith direction. The zenith delay is considered to be constant over a certain time interval. It is the average of the individual slant ray path delays that are projected to the zenith using the mapping functions (e.g., [311]) which are dependent on the receiver to satellite elevation angle, altitude and time of the year.

The significant application of GPS satellites in ground based GPS meteorology is the determination of the slant water. If one could condense all the water vapour along the ray path of a GPS signal (i.e., from the GPS satellite to the ground receiver), the column of the liquid water after condensation is termed slant water. By converting the GPS derived tropospheric delay during data processing, slant water is obtained. By using several receivers to track several satellites (e.g., Fig.15.2[3]), a three-dimensional distribution of water vapour and its time variation can be quantified. In Japan, there exist (by 2004) more than 1200 GPS receivers within the framework of GPS Earth Observing NETwork (GEONET) with a spatial resolution of 12-25km dedicated to GPS meteorology (see e.g., [13, 387]). These dense network of GPS receivers are capable of delivering information on water vapour that are useful as already stated in Sect. 15-1.

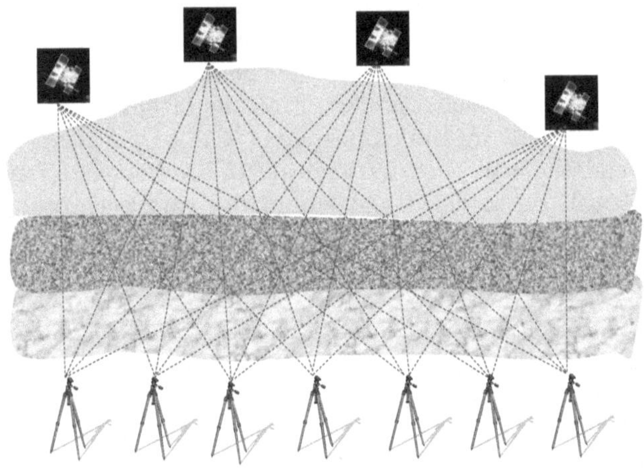

Figure 15.2. Ground based GPS meteorology

15-3 Refraction (bending) angles

In space borne GPS meteorology, the measured quantities are normally the excess path delay of the signal. It is obtained by measuring the excess phase of the signal owing to atmospheric refraction during the traveling period. The determination of the refraction angle α from the measured excess phase therefore marks the beginning of the computational process to retrieve the atmospheric profiles of temperature, pressure, water vapour and geopotential heights. The unknown refraction angle α is related to the measured excess phase by a system of two nonlinear trigonometric equations;

1. an equation relating the doppler shift at the Low Earth Orbiting (LEO) satellite (e.g., CHAMP, GRACE etc.) expressed as the difference in the projected velocities of the two moving satellites on the ray path tangent on one hand, and the doppler shift expressed as the sum of the atmosphere free propagation term and a term due to atmosphere on the other hand,
2. an equation that makes use of Snell's law in a spherically layered medium [369, p. 59].

Equations formed from (1) and (2) are nonlinear e.g., (15.6) and have been solved using iterative numerical methods such as Newton's (see e.g., [203, 256, 369, 413]. In-order to solve the trigonometric nonlinear system of equations (15.6), Newton's approach assumes the refractive angles to be small enough such that the relationship between the doppler shift and the bending angles formed from (1) and (2) are linear. The linearity assumption of the relationship between the doppler shift and refraction angles introduces some small *nonlinearity errors*. Vorob'ev and Krasil'nikova [400] have pointed out that neglecting the nonlinearity in (15.6) causes an error of 2% when the beam perigee is close to the Earth's ground and decrease with the altitude of the perigee. The extent of these errors in the dry part of the atmosphere, i.e., the upper troposphere and lower stratosphere, particularly the height 5-30 km, whose bending angle data are directly used to compute the atmospheric profiles or directly assimilated in Numerical Weather Prediction Models (NWPM) (e.g., [216]) is however not precisely stated. The effects of nonlinearity error on the impact parameters to which the refraction bending angles are related is also not known.

In an attempt to circumvent the nonlinearity of (15.6), [400] expand it into series of V/c, where V is the velocity of the artificial satellite and c the velocity of light in vacuum. This corrects for relativistic effects and introduce the concept of perturbation. The angle between the relative position vectors of the two satellites and the tangent velocity vector at GPS is expressed in quadratic terms of the corresponding angle at LEO (also expanded to the second order). The refraction angle is then obtained by making use of its infinitesimal values that are less than 10^{-2}. Though the approach attempts to provide an analytic (direct) solution to nonlinear system of equations for bending angles, it is still nevertheless "quasi-nonlinear" and as such does not offer a complete, exact solution to the problem. The fact that there existed no direct (exact) solution to the nonlinear system of bending angle's equations of space borne GPS meteorology had already been pointed out by [413].

Motivated by Wickert's [413] observation, we will demonstrate in the next sections how the algebraic techniques of Sylvester resultant and reduced Groebner basis offer direct solution to bending angles' nonlinear system of equations (15.6).

15-31 Transformation of trigonometric equations to algebraic

The system of nonlinear trigonometric equations for determining the refraction angles comprises of two equations given as

$$\left[\begin{array}{l} v_L cos(\beta_L - \phi_L) - v_G cos(\phi_G + \beta_G) = \dfrac{dL_i}{dt} + v_L cos(\beta_L - \psi_L) - v_G cos(\psi_G + \beta_G) \\ r_G sin\phi_G = r_L sin\phi_L, \end{array}\right. \tag{15.6}$$

where v_L, v_G are the projected LEO and GPS satellite velocities in the occultation plane, r_L, r_G the radius of tangent points at LEO and GPS respectively, and $\frac{dL_i}{dt}$, the doppler shift. The angles in (15.6) are as shown in Fig. 15.3.

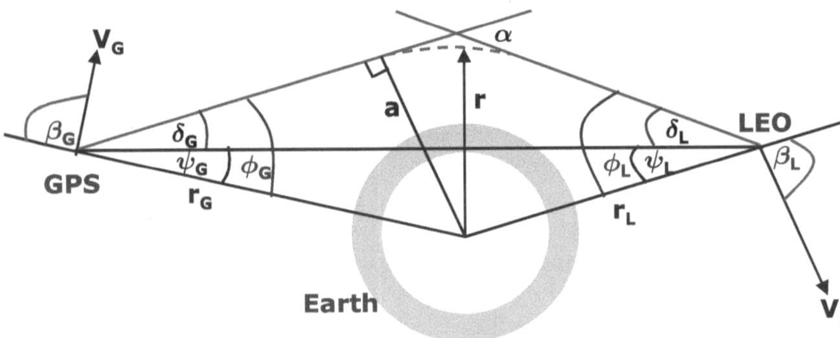

Figure 15.3. Geometry of space borne GPS meteorology

Let us denote

$$\left[\begin{array}{l} x = sin\phi_G, y = sin\phi_L, a_1 = v_L cos\beta_L, a_2 = v_L sin\beta_L \\ a_3 = -v_G cos\beta_G, a_4 = v_G sin\beta_G, a_5 = r_G, a_6 = -r_L, \end{array}\right. \tag{15.7}$$

where the signs of the velocities change depending on the directions of the satellites. Using;

- Theorem 3.1 on p. 19,
- the *trigonometric addition formulae*,
- and (15.7),

(15.6) simplifies to

$$\begin{bmatrix} a_1 cos\phi_L + a_2 y + a_3 cos\phi_G + a_4 x = a \\ a_5 x + a_6 y = 0. \end{bmatrix} \tag{15.8}$$

In (15.8), the right-hand-side of the first expression of (15.6) has been substituted with a. In-order to eliminate the trigonometric terms $cos\phi_L$ and $cos\phi_G$ appearing in (15.8), they are taken to the right-hand-side and the resulting expression squared as

$$(a_2 y + a_4 x - a)^2 = (-a_1 cos\phi_L - a_3 cos\phi_G)^2. \tag{15.9}$$

The squared trigonometric values $cos^2\phi_G$ and $cos^2\phi_L$ from (15.9) are then replaced by variables $\{x, y\}$ from (15.7). This is done following the application of trigonometric Pythagorean theorem of a unit circle $\{cos^2\phi_G + sin^2\phi_G = 1\}$ and $\{cos^2\phi_L + sin^2\phi_L = 1\}$ which convert the cosine terms into sines. The resulting expression has only trigonometric product $\{2a_1 a_3 cos\phi_L cos\phi_G\}$ on the right-hand-side. Squaring both sides of the resulting expression and replacing the squared trigonometric values $cos^2\phi_G$ and $cos^2\phi_L$, with $\{x, y\}$ from (15.7) completes the conversion of (15.6) into algebraic

$$\begin{bmatrix} d_1 x^4 + d_2 x^3 + d_3 x^3 y + d_4 x^2 + d_5 x^2 y^2 + d_6 x^2 y + d_7 x + d_8 x y^3 + d_9 x y^2 + d_{00} = 0 \\ a_5 x + a_6 y = 0, \end{bmatrix} \tag{15.10}$$

where $d_{00} = d_{10} xy + d_{11} y^4 + d_{12} y^3 + d_{13} y^2 + d_{14} y + d_{15}$. The coefficients $d_1, ..., d_{15}$ are:

$$\begin{array}{ll} d_1 = b_4^2 & d_9 = 2b_1 b_5 + 2b_2 b_3 \\ d_2 = 2b_4 b_5 & d_{10} = 2b_3 b_6 + 2b_5 b_2 \\ d_3 = 2b_4 b_3 & d_{11} = b_1^2 \\ d_4 = 2b_6 b_4 + b_5^2 + b_7^2 & d_{12} = 2b_1 b_2 \\ d_5 = 2b_1 b_4 + b_3^2 - b_7^2 & d_{13} = 2b_1 b_6 + b_2^2 + b_7^2 \\ d_6 = 2b_3 b_5 + 2b_2 b_4 & d_{14} = 2b_2 b_6 \\ d_7 = 2b_6 b_5 & d_{15} = b_6^2 - b_7^2, \\ d_8 = 2b_1 b_3 & \end{array}$$

with

$$\begin{array}{l} b_1 = a_1^2 + a_2^2 \\ b_2 = -2aa_2 \\ b_3 = 2a_2 a_4 \\ b_4 = (a_3^2 + a_4^2) \\ b_5 = -2aa_4 \\ b_6 = a^2 - a_1^2 - a_3^2 \\ b_7 = 2a_1 a_3. \end{array}$$

The algebraic equation (15.10) indicates that the solution of the nonlinear bending angle equation (15.6) is given by the intersection of a *quartic polynomial* (e.g., p. 26) and a *straight line* (see e.g., Fig. 15.4).

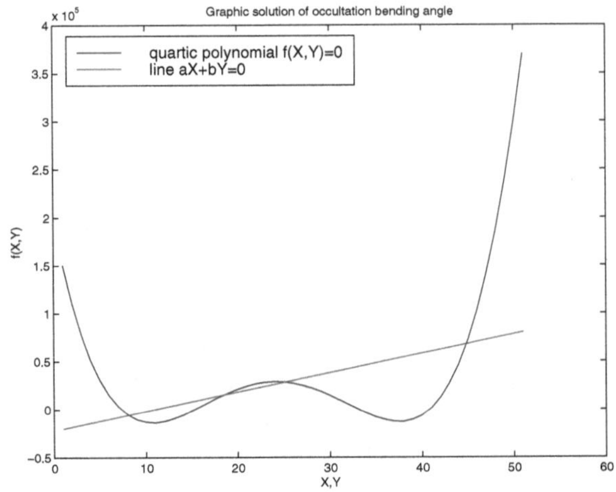

Figure 15.4. Algebraic curves for the solution of nonlinear system of bending angle equations

15-32 Algebraic determination of bending angles

15-321 Application of Groebner basis

Denoting the nonlinear system of algebraic (polynomial) equations (15.10) by

$$\begin{bmatrix} f_1 := d_1x^4 + d_2x^3 + d_3x^3y + d_4x^2 + d_5x^2y^2 + d_6x^2y + d_7x + d_8xy^3 + d_9xy^2 + d_{00} \\ f_2 := a_5x + a_6y, \end{bmatrix} \tag{15.11}$$

reduced Groebner basis (4.38) on p. 45 is computed for x and y as

$$\begin{bmatrix} GroebnerBasis[\{f_1, f_2\}, \{x, y\}, \{y\}] \\ GroebnerBasis[\{f_1, f_2\}, \{x, y\}, \{x\}]. \end{bmatrix} \tag{15.12}$$

The terms $\{f_1, f_2\}$ in (15.12) indicate the polynomials in (15.11), $\{x, y\}$ the variables with lexicographic ordering x comes before y, and $\{y\}$, $\{x\}$ the variables to be eliminated. The first expression of (15.12), i.e., $GroebnerBasis[\{f_1, f_2\}, \{x, y\}, \{y\}]$ gives a quartic polynomial in x (the first expression of 15.13), while the second expression gives a quartic polynomial in y (the second expression of 15.13). The results of (15.12) are:

$$\boxed{\begin{bmatrix} h_4x^4 + h_3x^3 + h_2x^2 + h_1x + h_0 = 0 \\ g_4y^4 + g_3y^3 + g_2y^2 + g_1y + g_0 = 0, \end{bmatrix}} \tag{15.13}$$

with the coefficients as

$$
\begin{aligned}
h_4 &= (a_6^4 d_1 + a_5^4 d_{11} - a_5 a_6^3 d_3 + a_5^2 a_6^2 d_5 - a_5^3 a_6 d_8) \\
h_3 &= (-a_5^3 a_6 d_{12} + a_6^4 d_2 - a_5 a_6^3 d_6 + a_5^2 a_6^2 d_9) \\
h_2 &= (-a_5 a_6^3 d_{10} + a_5^2 a_6^2 d_{13} + a_6^4 d_4) \\
h_1 &= (-a_5 a_6^3 d_{14} + a_6^4 d_7) \\
h_0 &= a_6^4 d_1 5,
\end{aligned}
$$

and

$$
\begin{aligned}
g_4 &= (a_6^4 d_1 + a_5^4 d_{11} - a_5 a_6^3 d_3 + a_5^2 a_6^2 d_5 - a_5^3 a_6 d_8) \\
g_3 &= (a_5^4 d_{12} - a_5 a_6^3 d_2 + a_5^2 a_6^2 d_6 - a_5^3 a_6 d_9) \\
g_2 &= (-a_5^3 a_6 d_{10} + a_5^4 d_{13} + a_5^2 a_6^2 d_4) \\
g_1 &= (a_5^4 d_{14} - a_5^3 a_6 d_7) \\
g_0 &= a_5^4 d_{15}.
\end{aligned}
$$

Four solutions are obtained from (15.13) for both x and y using Matlab's roots command (e.g., 4.42 on p. 46) as $x = roots([\,h_4\ h_3\ h_2\ h_1\ h_0\,])$ and $y = roots([\,g_4\ g_3\ g_2\ g_1\ g_0\,])$. From (15.7) and the roots of (15.13), the required solutions can now be obtained from

$$
\begin{bmatrix} \phi_G = sin^{-1}x, \\ \phi_L = sin^{-1}y. \end{bmatrix} \tag{15.14}
$$

The desired bending angle α in Fig. 15.3 is then obtained by first computing δ_G and δ_L as

$$
\begin{bmatrix} \delta_G = \phi_G - \psi_G \\ \delta_L = \phi_L - \psi_L, \end{bmatrix} \tag{15.15}
$$

leading to

$$
\begin{bmatrix} \alpha = \delta_G + \delta_L \\ p = \frac{1}{2}(r_L sin\phi_L + r_G sin\phi_G), \end{bmatrix} \tag{15.16}
$$

where $\alpha(p)$ is the bending angle and p the impact parameter.

15-322 Sylvester resultants solution

The quartic polynomials (15.13) can also be obtained using Sylvester resultants technique as follows:

- Step 1: From the nonlinear system of equations (15.10), hide y by treating it as a constant (i.e., polynomial of degree zero). Using (5.1) and (5.2) on p. 49, one computes the resultant of a 5×5 matrix

$$
Res\,(f_1,\, f_2, y) = det \begin{bmatrix} a_5 & a_6 y & 0 & 0 & 0 \\ 0 & a_5 & a_6 y & 0 & 0 \\ 0 & 0 & a_5 & a_6 y & 0 \\ 0 & 0 & 0 & a_5 & a_6 y \\ d_1 & d_2 + d_3 y & b_{53} & b_{54} & b_{55} \end{bmatrix}, \tag{15.17}
$$

  with $b_{53} = d_4 + d_5 y^2 + d_6 y$, $b_{54} = d_7 + d_8 y^3 + d_9 y^2 + d_{10} y$ and $b_{55} = d_{00}$. The solution of (15.17) leads to the first expression of (15.13).

- Step 2: Now hide x by treating it as a constant (i.e., polynomial of degree zero) and compute the resultant of a 5×5 matrix as

$$Res\,(f_1, f_2, x) = det \begin{bmatrix} a_6 & a_5x & 0 & 0 & 0 \\ 0 & a_6 & a_5x & 0 & 0 \\ 0 & 0 & a_6 & a_5x & 0 \\ 0 & 0 & 0 & a_6 & a_5x \\ d_{11} & d_{12}+d_8x & c_{53} & c_{54} & c_{55} \end{bmatrix}, \tag{15.18}$$

  with $c_{53} = d_{13} + d_5x^2 + d_9x$, $c_{54} = d_{14} + d_3x^3 + d_6x^2 + d_{10}x$ and $c_{55} = d_{15} + d_1x^4 + d_2x^3 + d_4x^2 + d_7x$. The solution of (15.18) leads to the second expression of (15.13) from which the bending angles and the impact parameters can be solved as already discussed.

In summary, the algebraic solution of refraction angles in space borne GPS meteorology proceeds in five steps as follows:

Step 1 (coefficients computation):
Using (15.7), compute the coefficients $\{h_4\ h_3\ h_2\ h_1\ h_0\}$ and $\{g_4\ g_3\ g_2\ g_1\ g_0\}$ of the quartic polynomials in (15.13).

Step 2 (solution of variables $\{x, y\}$):
Using the coefficients $\{h_i, g_i\}|i = 1, 2, 3, 4$ computed from step 1, obtain the roots of the univariate polynomials in (15.13) for $\{x, y\}$.

Step 3 (determine the angles $\{\phi_G, \phi_L\}$):
With the admissible values of $\{x, y\}$ from step 2, compute the angles $\{\phi_G, \phi_L\}$ using (15.14).

Step 4 (obtain the angles $\{\delta_G, \delta_L\}$):
Using the values of $\{\phi_G, \phi_L\}$ from step 3, compute the angles $\{\delta_G, \delta_L\}$ using (15.15).

Step 5 (determine the angle α and the impact parameter p):
Finally, the bending angle α and the impact parameter p are computed using the values of $\{\phi_G, \phi_L\}$ from step 4 in (15.16).

15-4 Algebraic analysis of some CHAMP data

Let us now apply the algebraic algorithm outlined in steps 1 to 5 to assess the effect of neglecting nonlinearity (i.e., nonlinearity error) in using Newton's iterative approach, which assumes (15.6) to be linear. In-order carry out the analysis, bending angles from CHAMP satellite level 2 data for two satellite occultations were computed and compared to those obtained from iterative approach in [369]. The occultations were chosen at different times of the day and years. Occultation number 133 of 3rd May 2002 occurred past mid-day at 13:48:36. For this period of the day, the solar radiation is maximum and so is the ionospheric noise. In contrast, occultation number 3 of 14th May 2001 occurred past mid-night at 00:39:58.00. For this period, the solar radiation is minimum and the effect of ionospheric noise is also minimum.

The excess phase length data are smoothed using polyfit function of Matlab software and the resulting doppler shift values for L1 and L2 used together with (15.7), (15.13) and (15.14) to obtain the angles ϕ_G and ϕ_L. These angles were then used in (15.16) to compute the refraction angle α and the impact parameter p (also denoted in this analysis as a). Let us examine in detail the computation of occultation number 133 of 3rd May 2002 which occurred during the maximum solar radiation period. The results of occultation number 3 of 14th May 2001 will thereafter be briefly presented. For occultation number 133 of 3rd May 2002, which occurred from the time 13:48:36 to 13:49:51.98, the bending angles were computed using both algebraic and the classical Newton's (e.g., [369]) algorithms. Since the algebraic procedure leads to four solutions of (15.13), a criteria for choosing the admissible solution had to be developed. This was done by using the bending angles from the classical Newton's approach as prior information. Time $t = 24.66\, sec$ was randomly chosen and its solutions from both algebraic and classical Newton's methods for the L1 signal compared. Figures (15.5) and (15.7) show the plot of the four solutions for x and y computed from (15.13) respectively. These solutions are converted into angular values $\{\delta_G, \delta_L\}$ using (15.14) and (15.15) respectively and plotted in Figs. 15.6 and 15.8. From the values of Figs. 15.6 and 15.8, the smallest values (encircled) were found to be close to those of the classical Newton's solution. The algebraic algorithm was then set to select the smallest value amongst the four solutions. Though Newton's approach converged after three iterations, a fixed value of 20 was set for this analysis. The threshold was set such that the difference between two consecutive solutions were smaller than 1×10^{-6}.

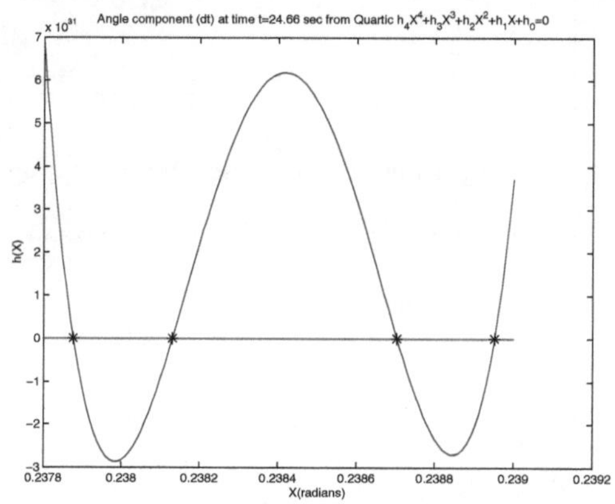

Figure 15.5. x values for computing the bending angle component δ_G for L1 at $t = 24.66 sec$

For the entire occultation, the bending angles $\{\alpha = \delta_G + \delta_L\}$ for both L1 and L2 signals were computed using algebraic algorithm and are plotted in Fig. 15.9. A magnifications of Fig. 15.9 above the height 30 km is plotted in Figs. 15.10 to show the effect of the residual ionospheric errors on bending angles.

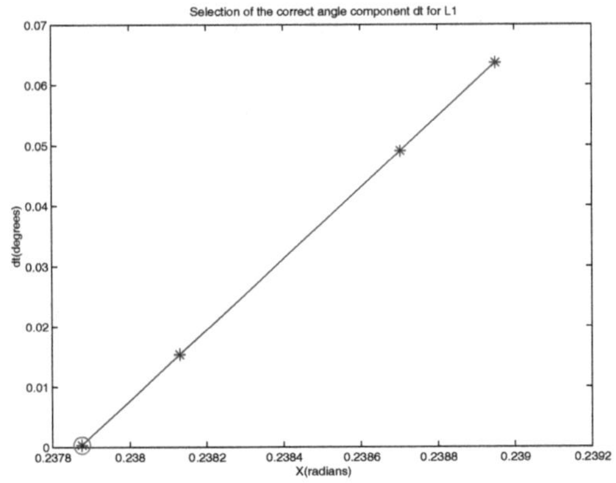

Figure 15.6. Selection of the admissible x value for computing the component δ_G for L1 at $t = 24.66sec$

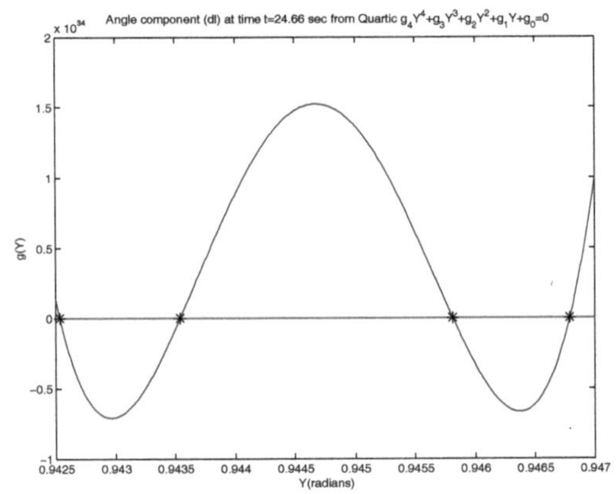

Figure 15.7. y values for computing the bending angle component δ_L for L1 at $t = 24.66sec$

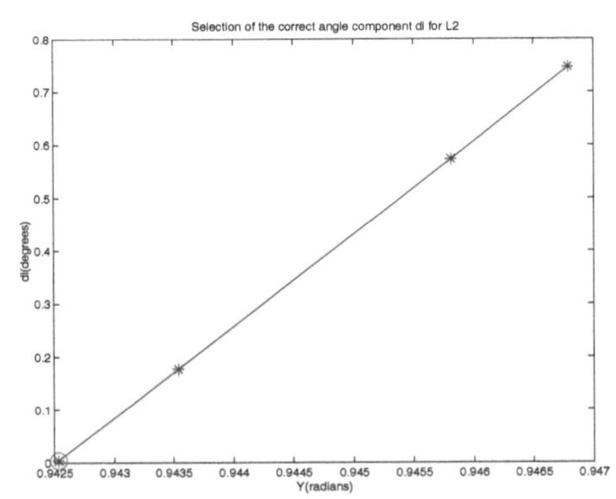

Figure 15.8. Selection of the admissible y value for computing the component δ_L for L1 at $t = 24.66sec$

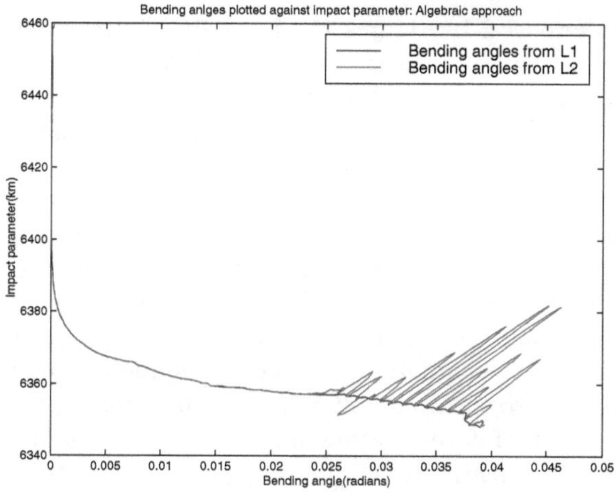

Figure 15.9. Bending angles for L1 and L2 from algebraic algorithm

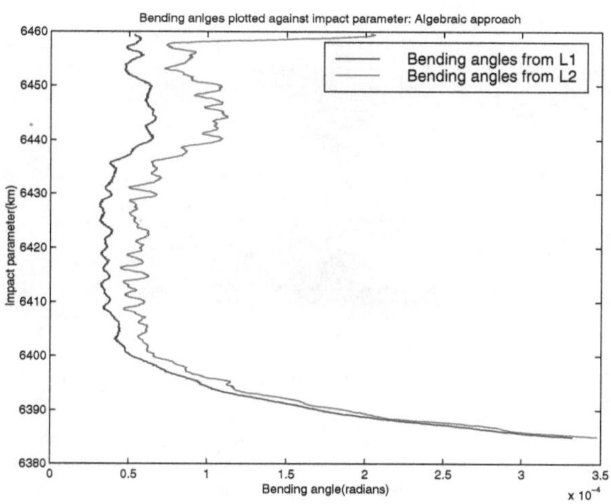

Figure 15.10. Magnification of the bending angles in Fig. 15.9 above 6380 km

Since bending angle's data above 40 km are augmented with model values and those below 5 km are highly influenced by the presence of water vapour (see e.g., Figs. 15.9 and 15.10), we will restricted our analysis to the region between 5-40 km. Data in this region are normally used directly to derive the atmospheric profiles required for Numerical Weather Prediction (NWP) models . In-order to assess the effect of nonlinearity assumptions, we subtract the results of the classical Newton's approach from those of algebraic approach. This is performed for both the bending angles α and the impact parameter p. The computations were carried out separately for both L1 and L2 signals. In-order to compare the results, the computed differences are plotted in Figs. 15.11, 15.12, 15.13 and 15.14. In these Figures, the vertical axes are fixed while the horizontal axes indicate the range of the computed differences. In Figures 15.11 and 15.12, the computed differences in bending angles due to nonlinearity assumption for L1 are in the range $\pm 6 \times 10^{-5}(degrees)$ with the maximum absolute value of $5.14 \times 10^{-5}(degree)$. For L2, they are in the range $\pm 5 \times 10^{-5}(degrees)$, with the maximum absolute value of $4.85 \times 10^{-5}(degree)$. The effects of nonlinearity error on the impact parameters for L1 are in the range $\pm 1.5\,m$ with the maximum absolute value of $1.444\,m$, while those of L2 are in the range $\pm 2\,m$ with the maximum absolute value of $1.534\,m$. The large differences in the impact parameters are due to the large distances of the GPS satellites ($r_G > 20,000\,km$). They are used in the second equation of (15.16) to compute the impact parameters to which the bending angles are related. Any small difference in the computed bending angles due to nonlinearity therefore contributes significantly to the large differences in the impact parameters. For this particular occultation therefore, the bending angles of L1 and L2 signals could probably be related to impact parameters that are off by up to $\pm 2\,m$.

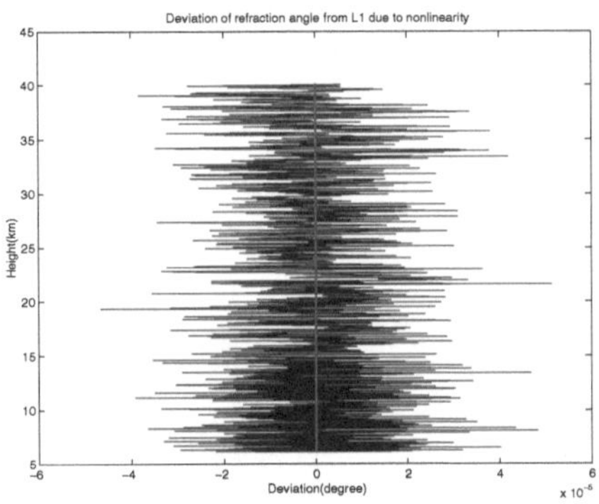

Figure 15.11. Differences in bending angles from L1 due to nonlinearity for occultation 133 of 3rd May 2002.

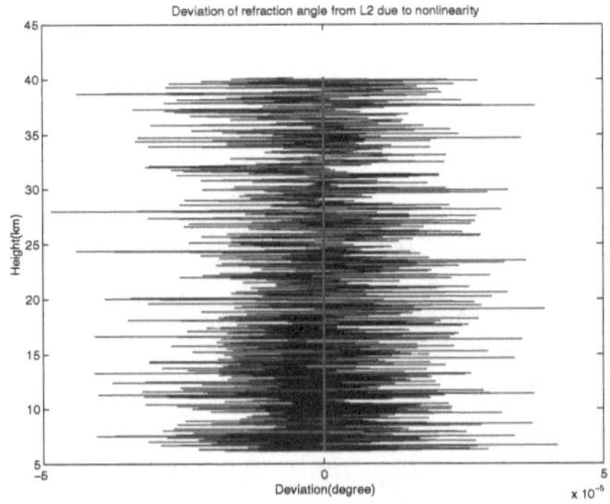

Figure 15.12. Differences in bending angles from L2 due to nonlinearity for occultation 133 of 3rd May 2002.

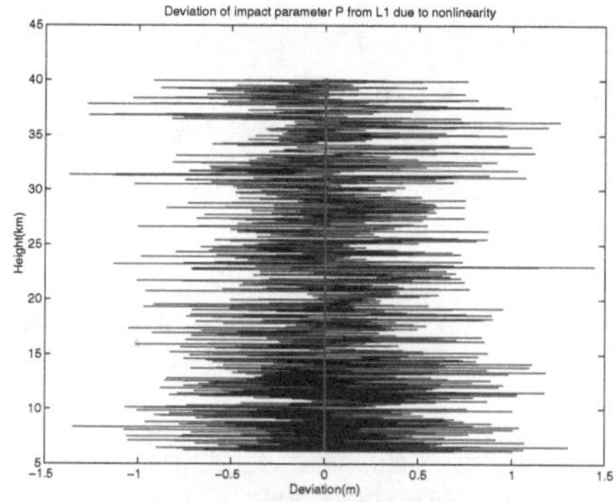

Figure 15.13. Differences in impact parameters from L1 due to nonlinearity for occultation 133 of 3rd May 2002.

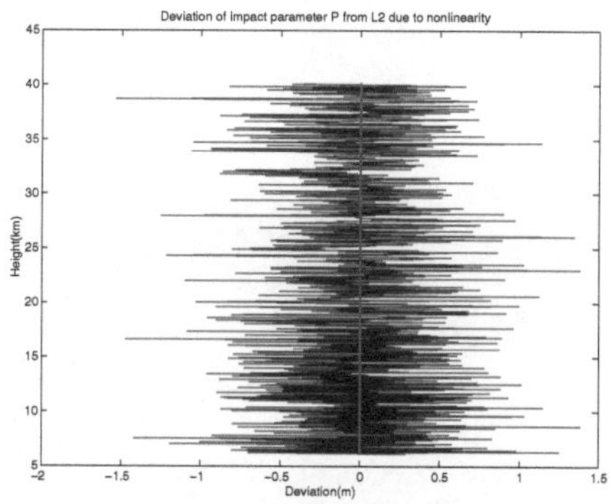

Figure 15.14. Differences in impact parameters from L2 due to nonlinearity for occultation 133 of 3rd May 2002.

In-order to assess the overall effect of nonlinearity on the bending angles, both bending angles from algebraic and Newton's procedures have to be related to the same impact parameters. In this analysis, the bending angles of L2 from algebraic approach and those of L1 and L2 from Newton's approach are all matched through interpolation to the impact parameters P1 of L1 from algebraic approach. The resulting total bending angles from both algebraic and iterative procedures are then obtained by the linear correction method of [400] as

$$\alpha(a) = \frac{f_1^2\alpha_1(a) - f_2^2\alpha_2(a)}{f_1^2 - f_2^2}, \tag{15.19}$$

where f_1, f_2 are the frequencies of L1 and L2 signals respectively and, $\alpha_1(a)$ and $\alpha_2(a)$ the bending angles from L1 and L2 signals respectively. The resulting bending angles $\alpha(a)_i$ from the Newton's approach and $\alpha(a)_a$ from algebraic approach using (15.19) are plotted in Fig. 15.15. The deviation $\nabla\alpha = \alpha(a)_a - \alpha(a)_i$ obtained are plotted in Fig. 15.16 which indicates the nonlinearity error to increase with decreasing atmospheric height. From 40km to 15km, the deviation is within

$\pm 2\times10^{-4}(degrees)$ but increases to $\pm 7\times10^{-4}(degrees)$ for the region below 15km with the maximum absolute deviation of 0.00069° for this particular occultation. This maximum absolute error is less than 1%. Vorob'ev and Krasil'nikova [400] pointed out that the error due to nonlinearity increases downwards to a maximum of about 2% when the beam perigee is close to the Earth's ground. The large difference in computed bending angles with decrease in height is expected as the region below 5km is affected by the presence of water vapour, and as seen from Fig. 15.9, the bending angles due to L2 are highly nonlinear.

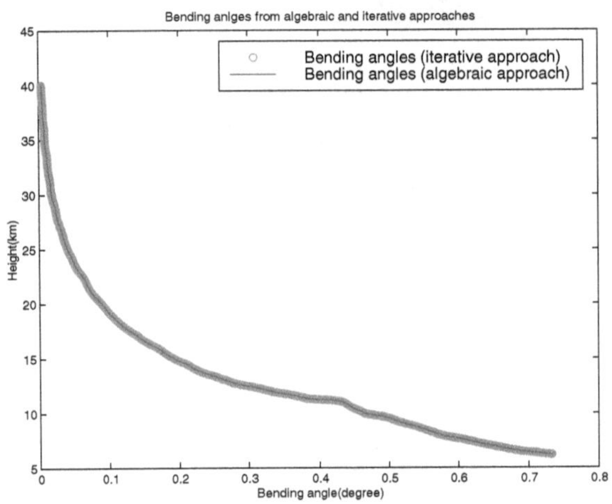

Figure 15.15. Bending angles from iterative and algebraic algorithms matched to the same impact parameters for occultation 133 of 3rd May 2002.

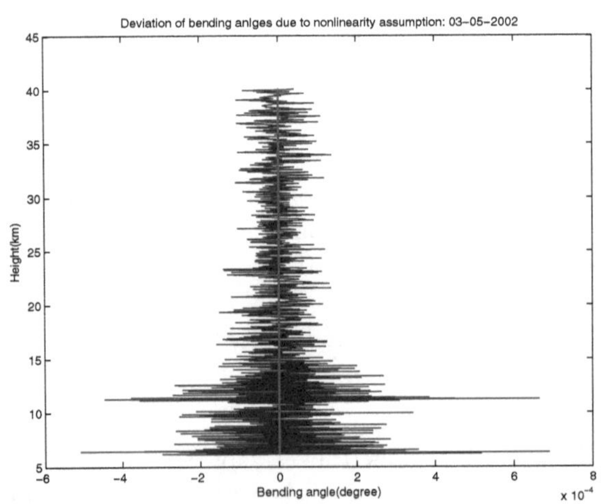

Figure 15.16. Differences of computed bending angles due to nonlinearity for occultation 133 of 3rd May 2002.

The algebraic approach was next used to compute the bending angles of occultation number 3 of 14th May 2001 which occurred past mid-night at 00:39:58.00. For this period, as stated earlier, the solar radiation is minimum and the effect of ionospheric noise is also minimum. The results from this occultation show the differences in bending angles from the algebraic and Newton's methods to be smaller (see Fig. 15.17) compared to those of solar maximum period. The maximum absolute difference value for bending angles was 0.00001°. For the computed impact parameters, the differences were in the range $\pm 5\,cm$ for L1 signal (Fig. 15.18) and $\pm 6\,cm$ for L2 (Fig. 15.19). The maximum absolute values were 4 cm and 5 cm respectively. In comparison to the results of occultation 133 of 3rd May 2002, the results of occultation 3 of 14th May 2001 indicate the effect of ionospheric noise during low solar radiation period to be less. The ionospheric noise could therefore increase the errors due to nonlinearity. In [36], further analysis of nonlinear bending angles have shown that there could exist other factors that influence the nonlinearity error other than the ionospheric noise.

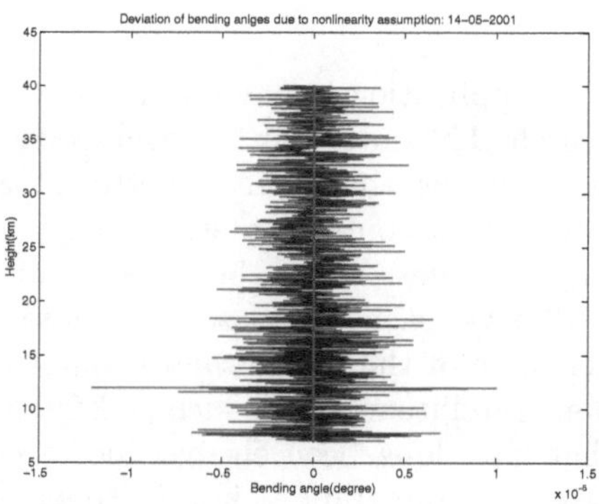

Figure 15.17. Differences in bending angles due to nonlinearity for occultation number 3 of 14th May 2001.

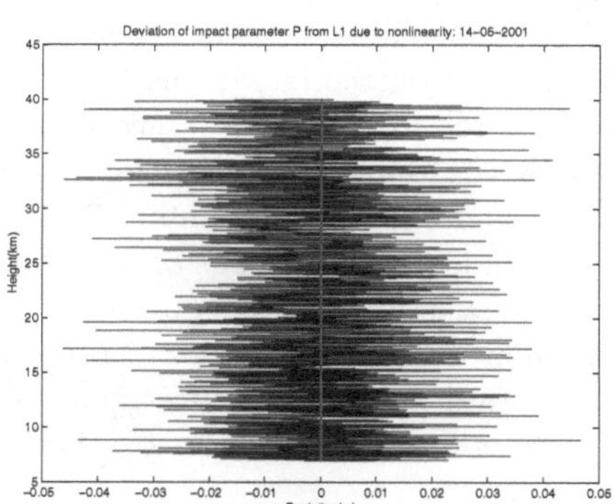

Figure 15.18. Differences in impact parameters from L1 due to nonlinearity for occultation number 3 of 14th May 2001

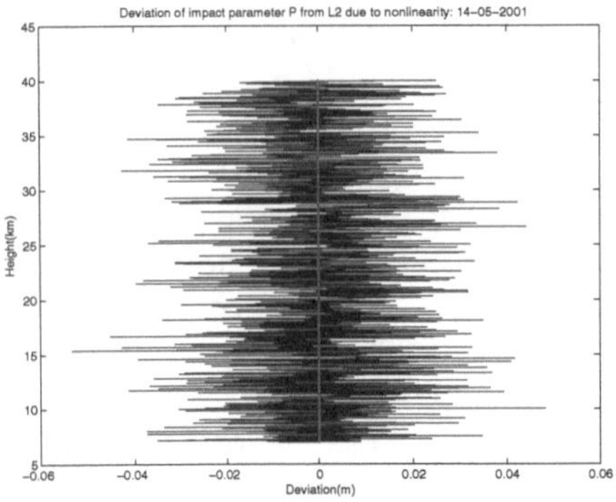

Figure 15.19. Differences in impact parameters from L2 due to nonlinearity for occultation number 3 of 14th May 2001

15-5 Concluding remarks

The new concept of GPS meteorology and its application to environmental monitoring is still new and an active area of research. The data that has been collected so far have unearthed several atmospheric properties that were hitherto difficult to fathom. The new technique clearly promises to contribute significantly and enormously to environmental and atmospheric studies. When the life span of the various missions (e.g., CHAMP, GRACE) will have reached, thousands of data will have been collected which will help unravel some of the hidden and complicated atmospheric and environmental phenomenon. Satellite missions such as EQUARS will contribute valuable equatorial data that have long been elusive due to poor radiosonde coverage. From the analysis of water vapour trapped in the atmosphere and the tropopause temperature, global warming studies will be enhanced a great deal.

We have also successfully presented an independent algebraic algorithm for solving the system of nonlinear bending angles for space borne GPS meteorology and shown that nonlinearity correction should be taken into account if the accuracy of the desired profiles are to be achieved to 1%. In particular, it has been highlighted how the nonlinearity errors in bending angles contribute to errors in the impact parameters to which the bending angles are related. Occultation number 133 of 3rd May 2002 which occurred past noon and occultation number 3 of 14th May 2001 which occurred past mid-night indicated the significance of ionospheric noise on nonlinearity error. When ionospheric noise is minimum, e.g., during midnight, the computed differences in bending angles between the two procedures are almost negligible. During maximum solar radiation in the afternoons with increased ionospheric noise, the computed differences in bending angles between algebraic and classical Newton's methods increases.

The proposed algebraic method could therefore be used to control the results of the classical Newton's method especially when the ionospheric noise is suspected to be great, e.g., for occultations that occur during maximum solar radiation periods. The hurdle that must be overcome however is to concretely identify the criteria for selecting the admissible solution amongst the four algebraic solutions. In this analysis, the smallest values amongst the four algebraic solutions turned out to be the admissible in comparison with values of the classical Newton's approach. Whether this applies in general is still subject to investigation. In terms of computing time, the algebraic approach would probably have an advantage over the classical Newton's iterative procedure in cases where thousands of occultations are to be processed. For single occultations however, the classical Newton' s approach generally converges after few iterations and as such, the advantage of the algebraic approach in light of modern computers may not be so significant. For further literature on GPS meteorology, we refer to [13, 343].

16 Algebraic diagnosis of outliers

16-1 Outliers in observation samples

In Chap. 7, we introduced parameter estimation from observational data sample and defined the models applicable to linear and nonlinear cases. In-order for the estimates to be meaningful however;

(a) proper field observations must be carried out to minimize chances of gross errors,
(b) the observed data sample must be processed to minimize or eliminate the effects of systematic errors,
(c) appropriate estimation procedures that account for random errors have to be applied.

Despite the care taken during observation period and the improved models used to correct systematic errors, observations may still be contaminated with outliers or gross errors. Outliers are those observations that are inconsistent with the rest of the observation sample. They often degrade the quality of the estimated parameters and render them unreliable for any meaningful inferences (deductions). Outliers find their way into observational data sample through:

- Miscopying during data entry, i.e., a wrong value can be entered during data input into the computer or other processing machines.
- Misreading during observation period, e.g., number 6 can erroneously be read as 9.
- Instrumental errors (e.g., problems with centering, vertical and horizontal circles, unstable tripod stands etc.)
- Rounding and truncation errors (e.g., during data processing)
- Poor models applied to correct systematic errors and estimate parameters (e.g., a linear model may be assumed where a nonlinear model could be suitable). This error is also common during data smoothing where a linear fit is used where actually a cubic fit could have been most suitable etc.
- Key punch errors during data input etc.

A special problem faced by users while dealing with outliers is the basis on which to discard observations from a set of data on the grounds that they are contaminated with outliers.

The least squares method used to estimate parameters assume the observational errors to be independent and normally distributed. In the presence of gross errors in the observational data sample, these assumptions are violated and hence render the estimators, such as least squares, ineffective. Earlier attempts to circumvent the problem of outlier involved procedures that would first detect and isolate the outliers before adjusting the remaining data sample. Such procedures were both statistical as seen in the works of [45, 46, 47, 304, 420, 421, 422], and non statistical e.g., [219]. Other outlier detection procedures have been presented by [4, 7].

J.L. Awange et al., *Algebraic Geodesy and Geoinformatics*, 2nd ed.,
DOI 10.1007/978-3-642-12124-1_16,

The detection and isolation approach to the outlier problem comes with its own shortcoming. On one hand, there exists the danger of *false deletion* and *false retention* of the assumed outliers. On the other hand, there exists the problem that the detection techniques are based on the residuals computed initially using the least squares method which has the tendency of masking the outliers by pulling their residuals closer to the regression fit. This makes the detection of outliers difficult. These setbacks had been recognized by the father of robust statistics P. J. Huber [232], [233, p. 5] and also [205, pp. 30–31] who suggested that the best option to deal with the outlier problem was to use *robust estimation procedures*. Such procedures would proceed safely despite the presence of outliers, isolate them and give admissible estimates that could have been achieved in the absence of outliers (i.e., if underlying distribution was normal). Following the fundamental paper by P. J. Huber in 1964 [231] and [233], several robust estimation procedures have been put forward that revolve around the robust M-estimators, L-estimators and R-estimators. In geodesy and geoinformatics, use of robust estimation techniques to estimate parameters has been presented e.g., in [5, 19, 20, 108, 196, 244, 246, 247, 306, 351, 422, 425, 426, 428] among others.

In this chapter, we present a non-statistical algebraic approach to outlier diagnosis that uses the Gauss-Jacobi combinatorial algorithm presented in Chap. 7. The combinatorial solutions are analyzed and those containing falsified observations identified. In-order to test the capability of the algorithm to diagnose outliers, we inject outliers of different magnitudes and signs on planar ranging and GPS pseudo-ranging problems. The algebraic approach is then employed to diagnose the outlying observations.

For GPS pseudo-range observations, the case of multipath effect is considered. Multipath is the error that occurs when the GPS signal is reflected (mostly by reflecting surfaces in built up areas) towards GPS receivers, rather than travelling directly to the receiver. This error still remains a menace which hinders full exploitation of the GPS system. Whereas other GPS observational errors such as ionospheric and atmospheric refractions can be modelled, the error due to multipath still poses some difficulties in being contained thus necessitating a search for procedures that can deal with it. In proposing procedures that can deal with the error due to multipath, [416] have suggested the use of robust estimation approach that is based on iteratively weighted least squares (e.g., a generalization of the Danish method to heterogeneous and correlated observations). Awange [18] proposed the use of algebraic deterministic approach to diagnose outliers of type multipath.

16-2 Algebraic diagnosis of outliers

Let us illustrate by means of a simple linear example how the algebraic algorithm diagnoses outliers.

Example 16.1 (Outlier diagnosis using Gauss-Jacobi combinatorial algorithm). Consider a case where three linear equations have been given for the purpose of solving the two unknowns(x, y) in Fig. 16.1. Three possible combinations, each containing two equations necessary for solving the two unknowns, can be formed as shown in the box labelled "combination". Each of these systems of two linear equations is either solved by substitution, graphically or matrix inversion to give three pairs of solutions $\{x_{1,2}, y_{1,2}\}, \{x_{2,3}, y_{2,3}\}$ and $\{x_{1,3}, y_{1,3}\}$. The final step involves adjusting these pseudo-observations $\{x_{1,2}, y_{1,2}, x_{2,3}, y_{2,3}, x_{1,3}, y_{1,3}\}$ as indicated in the box labelled "adjustment of the combinatorial subsets solutions". The weight matrix $\mathbf{\Sigma}$ or the weight elements $\{\pi_{1,2}, \pi_{2,3}, \pi_{1,3}\}$ are obtained via nonlinear error/variance-covariance propagation. Assuming now that observation y_1 is contaminated by gross error ∂y, then the first two combinatorial sets of the system of equations containing observation y_1 will have their results $\{x_{1,2}, y_{1,2}\}, \{x_{1,3}, y_{1,3}\}$ changed to $\{x^*_{1,2}, y^*_{1,2}\}, \{x^*_{1,3}, y^*_{1,3}\}$ respectively because of the change of observation y_1 to $(y_1 + \partial y)$. The third combination set $\{x_{2,3}, y_{2,3}\}$ without observation $(y_1 + \partial y)$ remains unchanged. If one computes the *combinatorial positional norms*

$$\left[\begin{array}{l} p_1 = \sqrt{(x^{*2}_{1,2} + y^{*2}_{1,2})} \\ p_2 = \sqrt{(x^{*2}_{1,3} + y^{*2}_{1,3})} \\ p_3 = \sqrt{(x^2_{2,3} + y^2_{2,3})}, \end{array}\right.$$

and subtract them from the norms of the adjusted positional values, median or a priori values (say from maps), one can analyze the deviations to obtain the falsified observation y_1 which is common in the first two sets. It will be noticed that the deviation of the first two sets containing the contaminated value is larger than the uncontaminated set. The *median* is here used as opposed to the *mean* as it is less sensitive to extreme scores.

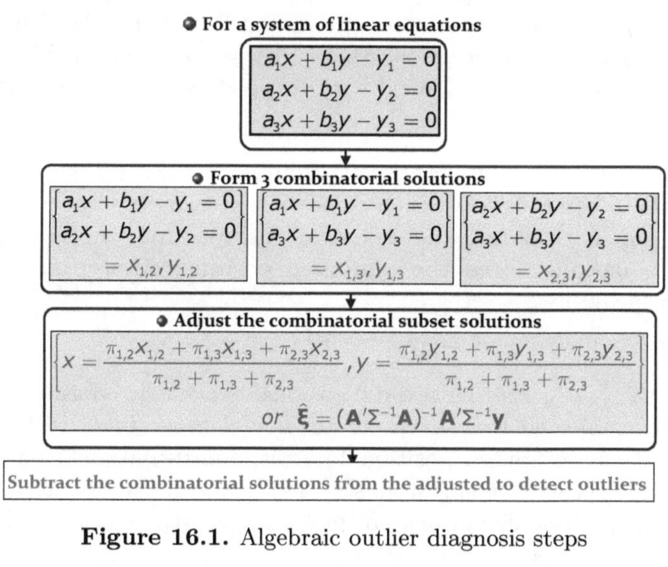

Figure 16.1. Algebraic outlier diagnosis steps

The program operates in the following steps:

Step 1: Given an overdetermined system with n observations in m unknowns, k combinations are formed using (7.28) on p. 93.
Step 2: Each of the minimal combination is solved in closed form using either Groebner basis or polynomial resultant algebraic technique of Chaps. 4 or 5. From the combinatorial solutions, compute the positional norm.
Step 3: Perform the nonlinear error/variance-covariance propagation to obtain the weight matrix of the pseudo-observations resulting from step 2.
Step 4: Using these pseudo-observations and the weight matrix from step 3, perform an adjustment using linear Gauss-Markov model (7.12) on p. 87.
Step 5: Compute the adjusted barycentric coordinate values together with its positional norm and the median positional norm from step 2. Subtract these positional norms from those of the combinatorial solutions to diagnose outliers from the deviations.

16-21 Outlier diagnosis in planar ranging

In Sect. 12-3 of Chap. 12, we discussed the planar ranging problem and presented the solution to the overdetermined case. We demonstrated by means of Example 12.4 on p. 198 how the position of unknown station could be obtained from distance measurements to more than two stations. In this section, we use the same example to demonstrate how the algebraic combinatorial algorithm can be used to diagnose outliers. From observational data of the overdetermined planar ranging problem of [237] given in *Table* 12.11 on p. 198, the position of the unknown station is determined. The algorithm is then applied to diagnose outlying observations. Let us consider three cases as follows; first, the algorithm is subjected to outlier free observations and used to compute the positional norms. Next, an outlier of 0.95 m is injected to the distance observation to station 2 and the algorithm applied to diagnose that particular observation. Finally, the distance observed to station 4 is considered to have been miss-booked with 6 typed as 9, thus leading to an error of 3 m.

Example 16.2 (Outlier free observations). From the values of *Table* 12.11 and using (7.28) on p. 93, 6 combinations, each consisting of two observation equations are formed. The aim is to obtain the unknown position from the nonlinear ranging observations equations. From the computed positions in step 2, the positional norms are given by

$$P_i = \sqrt{X_i^2 + Y_i^2}\,|_{i=1,\dots,6}, \qquad (16.1)$$

where $(X_i, Y_i)\,|_{i=1,\dots,6}$ are the two-dimensional geocentric coordinates of the unknown station computed from each combinatorial pair. *Table* 16.1 indicates the combinations, their computed positional norms, and deviations from the norm of the adjusted value (48941.769 m) from step 4. The results are for the case of outlier free observations. These deviations are plotted against combinatorial numbers in Fig. 16.2.

Table 16.1. Combinatorial positional norms and their deviations from that of the adjusted value (outlier free observations)

Combination	Positional norm (m)	Deviation from adjusted value (m)
1-2	48941.776	0.007
1-3	48941.760	-0.009
1-4	48941.767	-0.002
2-3	48941.769	0.000
2-4	48941.831	0.062
3-4	48941.764	-0.005

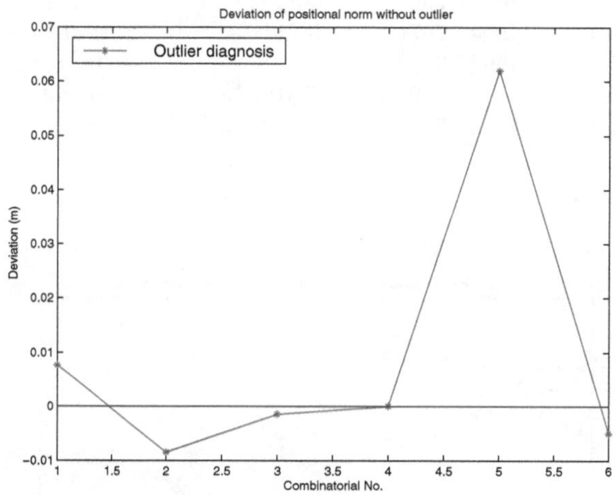

Figure 16.2. Deviations of the combinatorial positional norms from that of the adjusted value (outlier free observation)

Example 16.3 (Outlier of 0.950 m in observation to station 2). Let us now consider a case where the distance observation to station 2 in *Table* 12.11 has an outlier of 0.950 m. For this case, the distance is falsified such that the observed value is recorded as 1530.432 m instead of the correct value appearing in *Table* 12.11. The computed deviations of the combinatorial positional norms from the adjusted value (48941.456 m) and the median value of (48941.549 m) are presented in Table 16.2 and plotted in *Fig.* 16.3.

Given that there exists an outlier in the distance observation to station 2, which appears in combinations (1-2), (2-3) and (2-5), i.e., the *first*, *fourth*, and *fifth* combinations respectively, one expects the deviations in positional norms in these combinations to be larger than those combinations without station 2. Values of *Table* 16.2 columns *three* and *four* indicate that whereas combinations (1-3), (1-4) and (3-4), i.e., the *second*, *third*, and *sixth* combinations respectively have deviations in the same range (c.a. 0.3 m and 0.2 m in columns three and four respectively), the other combinations with outliers are clearly seen to have varying deviations. *Figure* 16.3 clearly indicates the *first*, *fourth*, and *fifth* combinations respectively with outlying observation to have larger deviations as compared to the rest. This is attributed to observation to station 2 containing gross error.

Table 16.2. Combinatorial positional norms and their deviations from that of the adjusted value and median (error of 0.950 m in observation to station 2)

Combination	Positional norm-(m)	Deviation from adjusted value (m)	Deviation from median value (m)
1-2	48940.964	-0.492	-0.585
1-3	48941.760	0.304	0.211
1-4	48941.767	0.311	0.218
2-3	48941.338	-0.118	-0.211
2-4	48936.350	-5.106	-5.199
3-4	48941.764	0.308	0.215

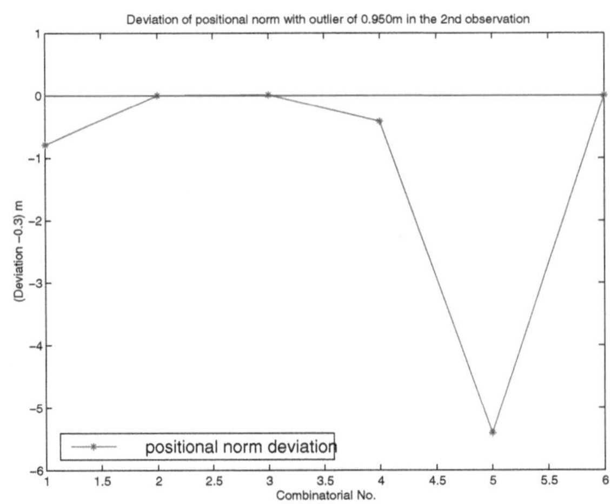

Figure 16.3. Deviations of the combinatorial positional norms from that of the adjusted value (error of 0.950 m in observation to station 2)

Example 16.4 (Outlier of 3 m in observation to station 4). Next, we consider a case where observation to station 4 in *Table* 12.11 has an outlier of 3 m, which erroneously resulted from miss-booking of the number 6 as 9. This falsified the distance such that the recorded value was 1209.524 m. The computed deviations of the combinatorial positional norms from the norm of the adjusted value (48942.620 m) and the median value (48941.772 m) are given as in *Table* 16.3 and plotted in Fig. 16.4. Given that there exists an outlier in the distance observation to station 4, which appears in combinations (1-4), (2-4) and (3-4) i.e., the *third*, *fifth*, and *sixth* combinations, Table 16.3 columns three and four together with Fig. 16.4 clearly indicates the deviations from these combinations to be larger than those of the combinations without observation 4, thus attributing it to observation to station 4 containing gross error.

Table 16.3. Combinatorial positional norms and deviations from the norms of adjusted value and median (error of 3m in observation to station 4)

Combination	Positional norm-(m)	Deviation from adjusted value (m)	Deviation from median value (m)
1-2	48941.776	-0.844	0.004
1-3	48941.760	-0.860	-0.012
1-4	48944.388	1.768	2.616
2-3	48941.769	-0.851	-0.003
2-4	48927.912	-14.708	-13.860
3-4	48943.061	0.441	1.289

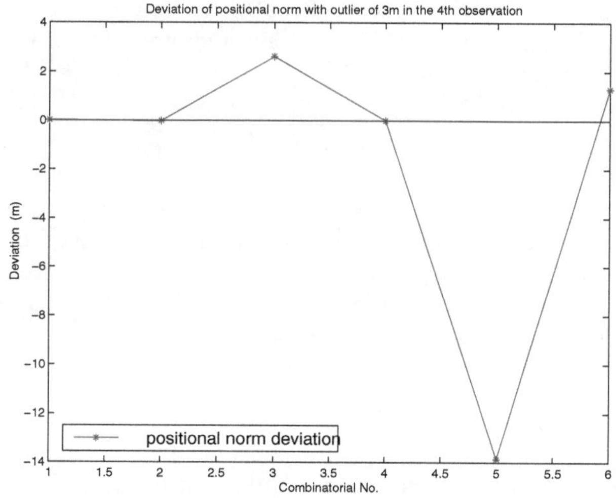

Figure 16.4. Deviations of the combinatorial positional norms from that of the median value (error of 3m in observation to station 4)

16-22 Diagnosis of multipath error in GNSS positioning

For GPS pseudo-ranging, consider that a satellite signal meant to travel straight to the receiver was reflected by a surface as shown in *Fig.* 16.5. The measured pseudo-range reaching the receiver ends up being longer than the actual pseudo-range, had the signal travelled directly. In-order to demonstrate how the algorithm

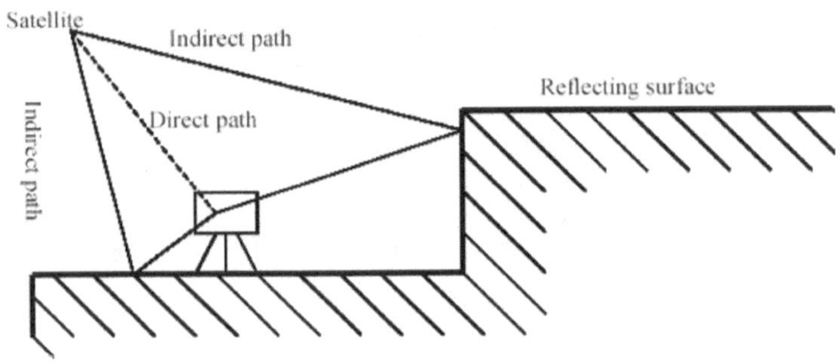

Figure 16.5. Multipath effect

can be used to detect outlier of type multipath, let us make us of Example 12.2 on p. 181. Using the six satellites, 15 combinations are formed whose positional norms are computed using

$$P_i = \sqrt{X_i^2 + Y_i^2 + Z_i^2}\,|_{i=1,\ldots,15}, \tag{16.2}$$

where $(X_i, Y_i, Z_i)\,|_{i=1,\ldots,15}$ are the three-dimensional geocentric coordinates of the unknown station computed from each combinatorial set. The computed positional norm are then used to diagnose outliers. Three cases are presented as follows: In case A, outlier free observations are considered while for cases B and C, outliers of 500 m and 200 m are injected in pseudo-range measurements from satellites 23 and 9 respectively.

Example 16.5 (Case A: Multipath free pseudo-ranges). From the values of *Table* 12.2 and using (7.28) on p. 93, 15 combinations, each consisting of four satellites, are formed with the aim of solving for the unknown position. For each combination, the position of the receiver is computed as discussed in Example 12.2 on p. 181. *Table* 16.4 indicates the combinations, the computed combinatorial positional norms from (16.2) and the deviations from the norms of the adjusted value of 6369. 582 m and the median from step 4, for outlier free case. The combinatorial algorithm diagnoses the poor geometry of the 10th combination. *Figure* 16.6 indicates the plotted deviations versus combinatorials.

Table 16.4. Positional norms and deviations from the norms of adjusted value and median (multipath free)

Combination Number	Combination	Positional norm (km)	Deviation from the norm of the adjusted value (m)	the norm of Deviation from the median (m)
1	23-9-5-1	6369.544	-39.227	-17.458
2	23-9-5-21	6369.433	-149.605	-127.837
3	23-9-5-17	6369.540	-43.255	-21.487
4	23-9-1-21	6369.768	185.342	207.110
5	23-9-1-17	6369.538	-44.603	-22.835
6	23-9-21-17	6369.561	-21.768	0.000
7	23-5-1-21	6369.630	47.449	69.217
8	23-5-1-17	6369.542	-41.229	-19.461
9	23-5-21-17	6369.507	-76.004	-54.235
10	23-1-21-17	6373.678	4094.748	4116.516
11	9-5-1-21	6369.724	140.976	162.744
12	9-5-1-17	6369.522	-60.746	-38.978
13	9-5-21-17	6369.648	64.830	86.598
14	9-1-21-17	6369.712	128.522	150.2908
15	5-1-21-17	6369.749	166.096	187.865

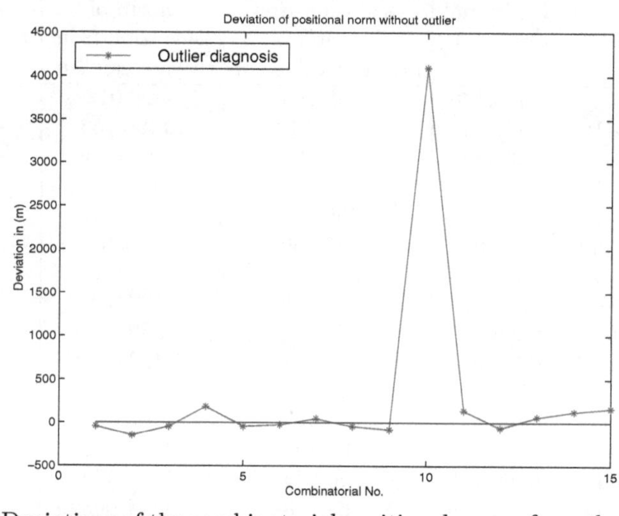

Figure 16.6. Deviations of the combinatorial positional norms from that of the adjusted value (outlier free observation).

Example 16.6 (Case B: Multipath error of 500 m in pseudo-range measurements from satellite 23). Let us assume that satellite number 23 had its pseudo-range longer by 500 m owing to multipath effect. Once the positions have been computed for the various combinations in *Table* 12.2, the positional norms for the 15 combinatorials are then computed via (16.2). The computed deviations of the positional norms from the norm of the adjusted value 6368.785 m, norm of the median value of 6368.638 m and a priori norm from case A are presented in *Table* 16.5. The deviations from a priori norm in case A are plotted in Fig. 16.7.

Given that there exists an outlier in the pseudo-range measurements from satellite 23, which appears in combinations 1 to 10, one expects the deviation in positional norms in these combinations to contain higher fluctuations than the combinations without satellite 23. Values of *Table* 12.2 columns four, five and six indicate that whereas combinations 11, 12, 13, 14 and 15 without satellite number 23 have values with less fluctuation of positional norms, the variation of the first 10 combinations containing satellite 23 were having larger fluctuations. The case is better illustrated by Fig. 16.7, where prior information is available on the desired position (e.g., from the norm of outlier free observations in Example 16.5). In such case, it becomes clearer which combinations are contaminated. From the figure, the first 10 combinations have larger deviations as opposed to the last 5, thus diagnosing satellite 23 as the outlying satellite. In practice, such prior information can be obtained from existing maps.

Table 16.5. Positional norms and deviations from the norms of adjusted value, median norm and the norm of case A (Multipath error of 500 m in satellite 23)

Comb. No.	Combination	Positional norm (km)	Deviation from norm of adjusted value (m)	norm of Deviation from the median (m)	Deviation from a priori norm (m)
1	23-9-5-1	6368.126	-658.763	-512.013	-1456.925
2	23-9-5-21	6367.147	-1637.513	-1490.763	-2435.675
3	23-9-5-17	6368.387	-397.366	-250.616	-1195.528
4	23-9-1-21	6370.117	1332.597	1479.347	534.435
5	23-9-1-17	6368.475	-309.906	-163.155	-1108.068
6	23-9-21-17	6368.638	-146.750	0.000	-944.912
7	23-5-1-21	6368.895	110.069	256.820	-688.093
8	23-5-1-17	6368.256	-528.806	-382.055	-1326.967
9	23-5-21-17	6368.006	-779.197	-632.447	-1577.359
10	23-1-21-17	6368.068	-716.569	-569.818	-1514.730
11	9-5-1-21	6369.724	939.136	1085.888	140.976
12	9-5-1-17	6369.522	737.416	884.166	-60.746
13	9-5-21-17	6369.648	862.991	1009.742	64.829
14	9-1-21-17	6369.712	926.684	1073.434	128.522
15	5-1-21-17	6369.749	964.258	1111.008	166.096

Example 16.7 (Case C: Multipath error of 200 m in satellite 9). Let us now suppose that satellite number 9 appearing in the last 5 combinations in case B (i.e., combinations 11, 12, 13, 14 and 15) has outlier of 200 m. The positional norms for the 15 combinatorials are then computed from (16.2). The computed deviations of the positional norms from the norm of the adjusted value 6369.781 m, norm of the median value 6369.804 m and a priori norm from case A are presented in Table 16.6

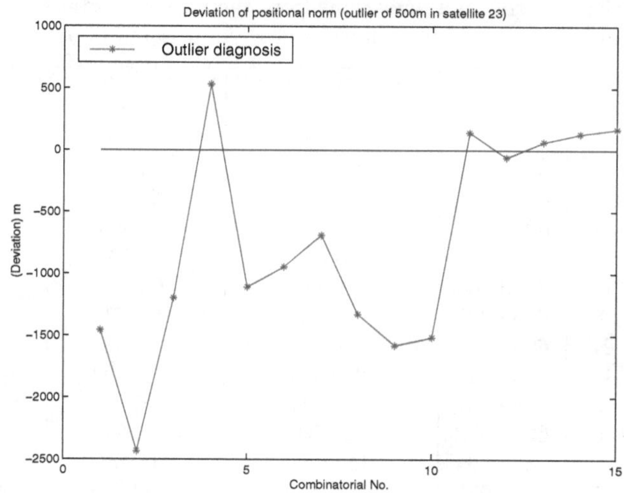

Figure 16.7. Deviations of combinatorial positional norms from the norm of the a priori value in case A (error of 500m in pseudo-range of satellite 23)

and plotted in Fig. 16.8. Satellite 9 appears in combinations 1, 2, 3, 4, 5, 6, 11, 12, 13, 14 and 15, with larger deviations in positional norms as depicted in Table 16.6 columns four, five and six and plotted in Fig. 16.8. Whereas combinations 7, 8, 9 and 10 without satellite number 9 have values with less deviations of positional norms, the deviations of the first 6 combinations and those of combinations 11-15 containing satellite 9 were larger. With prior information as shown in Fig. 16.8, satellite 9 can then be isolated to be the satellite with outlier. The value of combinatorial 10 is due to poor geometry as opposed to outlier since this particular combination does not contain satellite 9. This can be confirmed by inspecting the coefficients of the quadratic equations used to solve the unknown pseudo-range equations as already discussed in Example 12.2 on p. 181.

The diagnosed outliers in planar ranging observations as well as the pseudo-ranges of satellites 23 and 9 could therefore either;

- be eliminated from the various combinations and the remaining observations used to estimate the desired parameters or,
- the effect of the outlier could be managed using robust estimation techniques such as those discussed in [19, 20, 416], but with the knowledge of the contaminated observations.

Table 16.6. Positional norms and deviations from the norms of adjusted value, median norm and the norm of case A (Multipath error of 200m in satellite 9)

Comb. No.	Combination	Positional norm (km)	Deviation from norm of adjusted value (m)	norm of Deviation from the median (m)	Deviation from a priori norm (m)
1	23-9-5-1	6369.386	-394.656	-417.522	-196.748
2	23-9-5-21	6369.075	-705.644	-728.511	-507.737
3	23-9-5-17	6369.699	-81.796	-104.6639	116.111
4	23-9-1-21	6370.019	238.062	215.195	435.969
5	23-9-1-17	6369.804	22.866	0.000	220.774
6	23-9-21-17	6369.825	44.237	21.371	242.145
7	23-5-1-21	6369.630	-150.459	-173.325	47.449
8	23-5-1-17	6369.542	-239.137	-262.003	-41.229
9	23-5-21-17	6369.507	-273.911	-296.778	-76.004
10	23-1-21-17	6373.678	3896.840	3873.974	4094.748
11	9-5-1-21	6369.894	113.076	90.209	310.983
12	9-5-1-17	6371.057	1276.444	1253.578	1474.352
13	9-5-21-17	6370.333	552.230	529.364	750.138
14	9-1-21-17	6369.966	184.890	162.024	382.798
15	5-1-21-17	6369.749	-31.811	-54.678	166.096

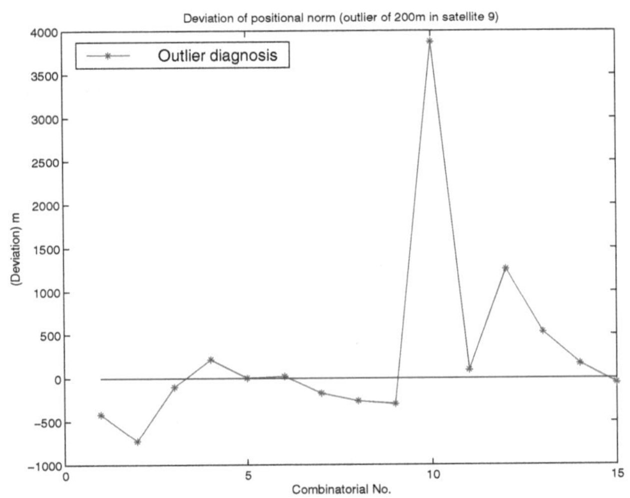

Figure 16.8. Deviations of combinatorial positional norms from the norm of the a priori value (error of 200m in pseudo-range of satellite 9)

16-3 Concluding remarks

The success of the algebraic Gauss-Jacobi combinatorial algorithm to diagnose outliers in the cases considered is attributed to its computing engine. The capability of the powerful algebraic tools of Groebner basis and polynomial resultants to solve in a close form the nonlinear systems of equations is the key to the success of the algorithm. With prior information from e.g., existing maps, the method can further be enhanced. For the 7-parameter datum transformation problem discussed in the next chapter, Procrustes algorithm II could be used as the computing engine instead of Groebner basis or polynomial resultants. The algebraic approach presented could be developed to further enhance the statistical approaches for detecting outliers.

17 Datum transformation problems

17-1 The 7-parameter datum transformation and its importance

The 7-parameter datum transformation $\mathbb{C}_7(3)$ problem involves the determination of seven parameters required to transform coordinates from one system to another. The transformation of coordinates is a computational procedure that maps one set of coordinates in a given system onto another. This is achieved by translating the given system so as to cater for its origin with respect to the final system, and rotating the system about its own axes so as to orient it to the final system. In addition to the translation and rotation, scaling is performed in order to match the corresponding baseline lengths in the two systems. The *three translation parameters*, *three rotation parameters* and the *scale element* comprise the *7 parameters of the datum transformation* $\mathbb{C}_7(3)$ *problem*, where one understands $\mathbb{C}_7(3)$ to be the notion of the *seven parameter conformal group in* $\mathbb{R}^3$, leaving "*space angles*" and "*distance ratios*" *equivariant* (invariant). A mathematical introduction to conformal field theory is given by [141, 357], while a systematic approach of geodetic datum transformation, including geometrical and physical terms, is presented by [188]. For a given network, it suffices to compute the transformation parameters using three or more coordinates in both systems. These parameters are then later used for subsequent conversions.

In geodesy and geoinformatics, the 7-parameter datum transformation problem has gained significance following the advent of Global Navigation Satellites Systems (GNSS), and particularly GPS. Since satellite positioning operates on a global reference frame (see e.g., Chap. 10), there often exists the need to transform coordinates from local systems onto GPS's World Geodetic System 84 (WGS-84). Specifically, coordinates can be transformed;

- from map systems to digitizing tables (e.g., in Geographical Information System GIS),
- from photo systems (e.g., photo coordinates) to ground systems (e.g., WGS-84),
- from local (national) systems to global reference systems (e.g., WGS-84) as in (14.29) on p. 262,
- from regional (e.g., European Reference Frame EUREF system) to global reference systems (e.g., WGS-84),
- from local (national) systems to regional reference systems, and
- from one local system onto another local system. In some countries, there exist different systems depending on political boundaries.

This problem, also known as 7-parameter similarity transformation, has its 7 unknown transformation parameters related to the known coordinates in the two systems by nonlinear equations. These equations are often solved using numerical methods which, as already pointed out in the preceding chapters, rely on

J.L. Awange et al., *Algebraic Geodesy and Geoinformatics*, 2nd ed.,
DOI 10.1007/978-3-642-12124-1_17, © Springer-Verlag Berlin Heidelberg 2010

linearization, approximate starting values and iterations. In this chapter, we solve the problem algebraically using;

(a) Procrustes,
(b) Groebner basis,
(c) Gauss-Jacobi combinatorial algorithms, and
(d) Dixon's resultant.

Before we present the usage of these algebraic algorithms, let us see how the 7-parameter datum transformation problem is formulated.

Formulation of the problem

Consider a case where coordinates have been given in two systems, A and B. For clarity purposes, let us assume the two coordinate systems to be e.g., photo image coordinates in system A and ground coordinates in system B (see e.g., Fig. 14.7 on p. 261). The ground coordinates $\{X_i, Y_i, Z_i | i, \ldots, n\}$ of the objects are obtained from, say, GPS measurements. Given the photo coordinates $\{a_i = x_i, b_i = y_i, c_i = -f | i, \ldots, n\}$ and their equivalent ground coordinates $\{X_i, Y_i, Z_i | i, \ldots, n\}$, the 7-parameter datum transformation problem concerns itself with determining;

(1) the scale parameter $x_1 \in \mathbb{R}$,
(2) three translation parameters $\mathbf{x}_2 \in \mathbb{R}^3$, and
(3) the rotation matrix $\mathbf{X}_3 \in \mathbb{R}^{3\times3}$ comprising three rotation elements.

Once the listed unknowns have been determined, coordinates can subsequently be transformed from one system onto another. The nonlinear equations relating these unknowns and coordinates from both systems are given by (cf., equation 14.29 on p. 262)

$$\begin{bmatrix} a_i \\ b_i \\ c_i \end{bmatrix} = x_1 \mathbf{X}_3 \begin{bmatrix} X_i \\ Y_i \\ Z_i \end{bmatrix} + \mathbf{x}_2 \mid i = 1, 2, 3, \ldots, n, \tag{17.1}$$

subject to

$$\boxed{\mathbf{X}_3' \mathbf{X}_3 = \mathbf{I}_3}. \tag{17.2}$$

In (17.1), $\{a_i, b_i, c_i\}$ and $\{X_i, Y_i, Z_i\}$ are the coordinates of the same points, e.g., in both photo and ground coordinate systems respectively. The determination of the unknowns $x_1 \in \mathbb{R}$, $\mathbf{x}_2 \in \mathbb{R}^3$, $\mathbf{X}_3 \in \mathbb{R}^{3\times3}$ require a minimum of *three points* in both systems whose coordinates are known. Owing to the nonlinearity of (17.1), the solutions have always been obtained using a least squares approach iteratively. With this approach, (17.1) is first linearized and some initial approximate starting values of the unknown parameters used. The procedure then iterates, each time improving on the solutions of the preceding iteration step. This is done until a convergence criteria is achieved.

Where the rotation angles are small e.g., in photogrammetry, the starting values of zeros are normally used. In other fields such as geodesy, the rotation angles are unfortunately not small enough to be initialized by zeros, thereby making the solution process somewhat difficult and cumbersome. Bad choices of initial starting values often lead to many iterations being required before the convergence criteria is achieved. In some cases, where the initial starting values are far from those of the unknown parameters, iteration processes may fail to converge. With these uncertainties in the initial starting values, the cumbersomeness of the linearization and iterations, procedures that would offer an exact solution to the 7-parameter datum transformation problem would be desirable. To answer this challenge, we propose algebraic approaches whose advantages over the approximate numerical methods have already been mentioned.

Apart from the computational difficulties associated with numerical procedures, the 7-parameter datum transformation problem poses another challenge to existing algorithms. This is, the incorporation of the variance-covariance (weight) matrices of the two systems involved. Communications between [245, 266, 344] on the subject, following the work of [265], provides an insight to this problem. In practice, users have been forced to rely on iterative procedures and linearized least squares solutions which are incapable of incorporating the variance-covariance matrices of both systems in play. We will attempt to address this challenge algebraically in this chapter.

17-2 Algebraic solution of the 7-parameter transformation problem

17-21 Groebner basis transformation

By making use of the *skew-symmetric* matrix $\mathbf{S}$, the rotation matrix $\mathbf{X}_3 \in \mathbb{R}^{3\times 3}$ in (17.1) is expressed as

$$\mathbf{X}_3 = (\mathbf{I}_3 - \mathbf{S})^{-1}(\mathbf{I}_3 + \mathbf{S}), \tag{17.3}$$

where $\mathbf{I}_3$ is the identity matrix and the *skew-symmetric* matrix $\mathbf{S}$ given by

$$\mathbf{S} = \begin{bmatrix} 0 & -c & b \\ c & 0 & -a \\ -b & a & 0 \end{bmatrix}. \tag{17.4}$$

The rotation matrix $\mathbf{X}_3 \in \mathbb{R}^{3\times 3}$ is parameterized using *Euler* or *Cardan* angles. With Cardan angles, we have:

Solution 17.1 (Parametrization of the rotation matrix by Cardan angles).

$$\mathbf{X}_3 = \mathbf{R}_1(\alpha)\mathbf{R}_2(\beta)\mathbf{R}_3(\gamma) \tag{17.5}$$

with

$$\mathbf{R}_1 = \begin{bmatrix} 1 & 0 & 0 \\ 0 & cos\alpha & sin\alpha \\ 0 & -sin\alpha & cos\alpha \end{bmatrix}, \mathbf{R}_2 = \begin{bmatrix} cos\beta & 0 & -sin\beta \\ 0 & 1 & 0 \\ sin\beta & 0 & cos\beta \end{bmatrix}, \mathbf{R}_3 = \begin{bmatrix} cos\gamma & sin\gamma & 0 \\ -sin\gamma & cos\gamma & 0 \\ 0 & 0 & 1 \end{bmatrix},$$

leading to

$$\mathbf{R}_1(\alpha)\mathbf{R}_2(\beta)\mathbf{R}_3(\gamma) = \begin{bmatrix} cos\beta cos\gamma & cos\beta sin\gamma & -sin\beta \\ sin\alpha sin\beta cos\gamma - cos\alpha sin\gamma & sin\alpha sin\beta sin\gamma + cos\alpha cos\gamma & sin\alpha cos\beta \\ cos\alpha sin\beta cos\gamma + sin\alpha sin\gamma & cos\alpha sin\beta sin\gamma - sin\alpha cos\gamma & cos\alpha cos\beta \end{bmatrix}. \tag{17.6}$$

The Cardan angles are then obtained from the rotation matrix $\mathbf{X}_3 \in \mathbb{R}^{3\times3}$ through:

$$\begin{bmatrix} \alpha = tan\left\{\dfrac{r_{23}}{r_{33}}\right\} \\ \\ \gamma = tan\left\{\dfrac{r_{12}}{r_{11}}\right\} \\ \\ \beta = tan\left\{\dfrac{-r_{31}}{\sqrt{r_{11}^2 + r_{12}^2}}\right\} = tan\left\{\dfrac{-r_{31}}{\sqrt{r_{23}^2 + r_{33}^2}}\right\}. \end{bmatrix} \tag{17.7}$$

For parametrization using Euler angles, we have:

Solution 17.2 (Parametrization of the rotation matrix by Euler angles $\Lambda_\Gamma, \Phi_\Gamma, \Sigma_\Gamma$).

$$\mathbf{R}_{\mathrm{E}}(\Lambda_\Gamma, \Phi_\Gamma, \Sigma_\Gamma) := \mathbf{R}_3(\Sigma_\Gamma)\mathbf{R}_2(\frac{\pi}{2} - \Phi_\Gamma)\mathbf{R}_3(\Lambda_\Gamma) \tag{17.8}$$

$$\mathbf{R}_1 := \begin{bmatrix} 1 & 0 & 0 \\ 0 & \cos 1 & \sin 1 \\ 0 & -\sin 1 & \cos 1 \end{bmatrix}, \quad \mathbf{R}_2 := \begin{bmatrix} \cos 2 & 0 & -\sin 2 \\ 0 & 1 & 0 \\ \sin 2 & 0 & \cos 2 \end{bmatrix}, \quad \mathbf{R}_3 := \begin{bmatrix} \cos 3 & \sin 3 & 0 \\ -\sin 3 & \cos 3 & 0 \\ 0 & 0 & 1 \end{bmatrix} \tag{17.9}$$

$$\mathbf{R}_3(\Lambda_\Gamma) = \begin{bmatrix} \cos\Lambda_\Gamma & \sin\Lambda_\Gamma & 0 \\ -\sin\Lambda_\Gamma & \cos\Lambda_\Gamma & 0 \\ 0 & 0 & 1 \end{bmatrix}, \quad \mathbf{R}_2(\frac{\pi}{2} - \Phi_\Gamma) = \begin{bmatrix} \sin\Phi_\Gamma & 0 & -\cos\Phi_\Gamma \\ 0 & 1 & 0 \\ \cos\Phi_\Gamma & 0 & \sin\Phi_\Gamma \end{bmatrix} \tag{17.10}$$

$$\mathbf{R}_2(\frac{\pi}{2} - \Phi_\Gamma)\mathbf{R}_3(\Lambda_\Gamma) = \begin{bmatrix} \sin\Phi_\Gamma\cos\Lambda_\Gamma & \sin\Phi_\Gamma\sin\Lambda_\Gamma & -\cos\Phi_\Gamma \\ -\sin\Lambda_\Gamma & \cos\Lambda_\Gamma & 0 \\ \cos\Phi_\Gamma\cos\Lambda_\Gamma & \cos\Phi_\Gamma\sin\Lambda_\Gamma & \sin\Phi_\Gamma \end{bmatrix} \tag{17.11}$$

$$\mathbf{R} := \mathbf{R}_3(\Sigma_\Gamma)\mathbf{R}_2(\tfrac{\pi}{2} - \Phi_\Gamma)\mathbf{R}_3(\Lambda_\Gamma) =$$

$$\begin{bmatrix} \cos\Sigma_\Gamma \sin\Phi_\Gamma \cos\Lambda_\Gamma - \sin\Sigma_\Gamma \sin\Lambda_\Gamma & \cos\Sigma_\Gamma \sin\Phi_\Gamma \sin\Lambda_\Gamma + \sin\Sigma_\Gamma \cos\Lambda_\Gamma & -\cos\Sigma_\Gamma \cos\Phi_\Gamma \\ -\sin\Sigma_\Gamma \sin\Phi_\Gamma \cos\Lambda_\Gamma - \cos\Sigma_\Gamma \sin\Lambda_\Gamma & -\sin\Sigma_\Gamma \sin\Phi_\Gamma \sin\Lambda_\Gamma + \cos\Sigma_\Gamma \cos\Lambda_\Gamma & \sin\Sigma_\Gamma \cos\Phi_\Gamma \\ \cos\Phi_\Gamma \cos\Lambda_\Gamma & \cos\Phi_\Gamma \sin\Lambda_\Gamma & \sin\Phi_\Gamma \end{bmatrix} \tag{17.12}$$

$$0 \le \Lambda_\Gamma < 2\pi, -\tfrac{\pi}{2} < \Phi_\Gamma < +\tfrac{\pi}{2}, 0 \le \Sigma_\Gamma < 2\pi$$

The inverse map of

$$\mathbf{R} = [\mathrm{r}_{kl}]\,, k, l \in \{1,2,3\},$$

to

$$(\Lambda_\Gamma, \Phi_\Gamma, \Sigma_\Gamma)$$

is given by Lemma 17.1.

Lemma 17.1 (Inverse map $\mathbf{R} \mapsto \Lambda_\Gamma, \Phi_\Gamma, \Sigma_\Gamma$). *Let the direct Euler map of the rotation matrix be given by (17.12), namely*

$$\mathbf{R} := \mathbf{R}_3(\Sigma_\Gamma)\mathbf{R}_2(\frac{\pi}{2} - \Phi_\Gamma)\mathbf{R}_3(\Lambda_\Gamma),$$

$$(\Lambda_\Gamma, \Phi_\Gamma, \Sigma_\Gamma) \in \left\{\mathbf{R^3} \,\middle|\, \mathbf{0 \le \Lambda_\Gamma < 2\pi, -\frac{\pi}{2} < \Phi_\Gamma < +\frac{\pi}{2}, 0 \le \Sigma_\Gamma < 2\pi}\right\}.$$

The inverse Euler map is parameterized by

$$\left[\begin{array}{l} \tan\Lambda_\Gamma = \dfrac{r_{32}}{r_{31}} \Rightarrow \Lambda_\Gamma = \arctan\dfrac{r_{32}}{r_{31}} \\[2ex] \tan\Phi_\Gamma = \dfrac{r_{33}}{\sqrt{r_{31}^2 + r_{32}^2}} \Rightarrow \Phi_\Gamma = \arctan\dfrac{r_{33}}{\sqrt{r_{31}^2 + r_{32}^2}} \\[2ex] \tan\Sigma_\Gamma = -\dfrac{r_{23}}{r_{13}} \Rightarrow \Sigma_\Gamma = \arctan -\dfrac{r_{23}}{r_{13}}. \end{array}\right. \tag{17.13}$$

The properties of the rotation matrix $\mathbf{X}_3 \in \mathbb{R}^{3\times 3}$ expressed in (17.3) have been examined by [434] and shown to fulfill (17.2). Only a minimum of three corresponding points in both systems are required for the transformation parameters to be obtained. For these points, (17.1) is now written for $i = 1, 2, 3$ using (17.3) as

$$\begin{bmatrix} 1 & c & -b \\ -c & 1 & a \\ b & -a & 1 \end{bmatrix} \begin{bmatrix} a_i \\ b_i \\ c_i \end{bmatrix} = x_1 \begin{bmatrix} 1 & -c & b \\ c & 1 & -a \\ -b & a & 1 \end{bmatrix} \begin{bmatrix} X_i \\ Y_i \\ Z_i \end{bmatrix} + \begin{bmatrix} 1 & c & -b \\ -c & 1 & a \\ b & -a & 1 \end{bmatrix} \begin{bmatrix} X_0 \\ Y_0 \\ Z_0 \end{bmatrix}, \tag{17.14}$$

with $\{X_0, Y_0, Z_0\} \in \mathbf{x}_2$ being the translation parameters. For these three corresponding points in both systems, the observation equations for solving the 7 transformation parameters are expressed from (17.14) as:

$$\begin{bmatrix} f_1 := x_1X_1 - x_1cY_1 + x_1bZ_1 + X_0 + cY_0 - bZ_0 - a_1 - cb_1 + bc_1 = 0 \\ f_2 := x_1cX_1 + x_1Y_1 - x_1aZ_1 - cX_0 + Y_0 + aZ_0 + ca_1 - b_1 - ac_1 = 0 \\ f_3 := -x_1bX_1 + x_1aY_1 + x_1Z_1 + bX_0 - aY_0 + Z_0 - ba_1 + ab_1 - c_1 = 0 \\ f_4 := x_1X_2 - x_1cY_2 + x_1bZ_2 + X_0 + cY_0 - bZ_0 - a_2 - cb_2 + bc_2 = 0 \\ f_5 := x_1cX_2 + x_1Y_2 - x_1aZ_2 - cX_0 + Y_0 + aZ_0 + ca_2 - b_2 - ac_2 = 0 \\ f_6 := -x_1bX_2 + x_1aY_2 + x_1Z_2 + bX_0 - aY_0 + Z_0 - ba_2 + ab_2 - c_2 = 0 \\ f_7 := x_1X_3 - x_1cY_3 + x_1bZ_3 + X_0 + cY_0 - bZ_0 - a_3 - cb_3 + bc_3 = 0 \\ f_8 := x_1cX_3 + x_1Y_3 - x_1aZ_3 - cX_0 + Y_0 + aZ_0 + ca_3 - b_3 - ac_3 = 0 \\ f_9 := -x_1bX_3 + x_1aY_3 + x_1Z_3 + bX_0 - aY_0 + Z_0 - ba_3 + ab_3 - c_3 = 0, \end{bmatrix} \tag{17.15}$$

where $\{x_i, y_i, z_i\} := \{a_i, b_i, c_i\} \mid i \in \{1, 2, 3\}$ are coordinates of the three points in one of the systems (e.g., local system) and $\{X_i, Y_i, Z_i\} \mid i \in \{1, 2, 3\}$ are the corresponding coordinates in the other system (e.g., global system). In (17.15), $\{f_1, f_2, f_3\}$ are algebraic expressions formed from the *first point* with coordinates in both systems, $\{f_4, f_5, f_6\}$ from the *second point* and $\{f_7, f_8, f_9\}$ from the *third point*. From (17.15), considering the unknowns (see Eqn. 17.14) as $x_1, a, b, c, X_0, Y_0, Z_0$, one requires only seven equations for a closed form solution of the 7-parameter transformation problem.

Let us consider the system of nonlinear equations extracted from (17.15) to be formed by the 7 polynomials $\{f_1, f_2, f_3, f_4, f_5, f_6, f_9\}$. Our target now is to solve algebraically this nonlinear system of equations using Groebner basis approach to provide symbolic solutions. We proceed as follows: *First*, the translation parameters $\{X_0, Y_0, Z_0\}$ are eliminated by differencing

$$\begin{bmatrix} f_{14} := f_1 - f_4 = x_1X_{12} - x_1cY_{12} + x_1bZ_{12} - a_{12} - cb_{12} + bc_{12} \\ f_{25} := f_2 - f_5 = x_1cX_{12} + x_1Y_{12} - x_1aZ_{12} + ca_{12} - b_{12} - ac_{12} \\ f_{39} := f_3 - f_9 = -x_1bX_{13} + x_1aY_{13} + x_1Z_{13} - ba_{13} + ab_{13} - c_{13} \\ f_{69} := f_6 - f_9 = -x_1bX_{23} + x_1aY_{23} + x_1Z_{23} - ba_{23} + ab_{23} - c_{23}, \end{bmatrix} \tag{17.16}$$

where

$$\left.\begin{matrix} X_{ij} = X_i - X_j,\ Y_{ij} = Y_i - Y_j,\ Z_{ij} = Z_i - Z_j \\ \\ a_{ij} = a_i - a_j,\quad b_{ij} = b_i - b_j,\quad c_{ij} = c_i - c_j \end{matrix}\right] \mid i, j \in \{1, 2, 3\}, i \neq j.$$

The reduced Groebner basis of (17.16) is then obtained using (4.38) on p. 45 by

$$GroebnerBasis[\{f_{14}, f_{25}, f_{37}, f_{67}\}, \{x_1, a, b, c\}, \{a, b, c\}].$$

This gives only the elements of the *Groebner basis* in which the variables $\{a, b, c\}$ have been eliminated and only the scale factor x_1 is left. The scale parameter is then given by the following *quartic* polynomial:

$$a_4x_1^4 + a_3x_1^3 + a_2x_1^2 + a_1x_1 + a_0 = 0, \tag{17.17}$$

with the coefficients as in [23, Boxes 2-2] or [37, Appendix A]. Once the admissible value of scale parameter $x_1 \in \mathbb{R}^+$ has been chosen from the four roots in (17.17), the elements of the skew-symmetric matrix $\mathbf{S}$ in (17.4) can then be obtained via the linear functions in [23, Boxes 2-3] or [37, Appendix B]. Substituting the skew-symmetric matrix $\mathbf{S}$ in (17.1) gives the rotation matrix $\mathbf{X}_3$, from which the Cardan rotation angles are deduced using (17.7) in Solution 17.1. The translation elements $\mathbf{x}_2$ can then be computed by substituting the scale parameter x_1 and the rotation matrix $\mathbf{X}_3$ in (17.1). Three sets of translation parameters are then obtained, from which their mean is taken.

Example 17.1 (Computation of transformation parameters using Groebner basis algorithm). Cartesian coordinates of seven stations in Stuttgart Germany are given in the local and global system (WGS-84) in *Tables* 17.1 and 17.2 respectively. Desired are the seven parameters of datum transformation. Using explicit solutions in [23, Boxes 2-2 and 2-3] or [37, Appendices A and B], the 7 transformation parameters are computed and presented in *Table* 17.3. These parameters are then used to transform the three points involved in the computations from the local reference system in *Table* 17.1 to the WGS-84, as shown in Table 17.4.

Table 17.1. Coordinates for system A (local system)

Station Name	$X(m)$	$Y(m)$	$Z(m)$
Solitude	4157222.543	664789.307	4774952.099
Buoch Zeil	4149043.336	688836.443	4778632.188
Hohenneuffen	4172803.511	690340.078	4758129.701
Kuehlenberg	4177148.376	642997.635	4760764.800
Ex Mergelaec	4137012.190	671808.029	4791128.215
Ex Hof Asperg	4146292.729	666952.887	4783859.856
Ex Kaisersbach	4138759.902	702670.738	4785552.196

Table 17.2. Coordinates for system B (WGS-84)

Station Name	$X(m)$	$Y(m)$	$Z(m)$
Solitude	4157870.237	664818.678	4775416.524
Buoch Zeil	4149691.049	688865.785	4779096.588
Hohenneuffen	4173451.354	690369.375	4758594.075
Kuehlenberg	4177796.064	643026.700	4761228.899
Ex Mergelaec	4137659.549	671837.337	4791592.531
Ex Hof Asperg	4146940.228	666982.151	4784324.099
Ex Kaisersbach	4139407.506	702700.227	4786016.645

17-22 Dixon resultant solution

In this section we solve the $C_7(3)$ Helmert transformation with the Dixon resultant using the Kapur-Saxena-Yang (KSY) method (see Kapur et al. [239]), in the Computer Algebra System of *Mathematica*. This method was implemented into *Mathematica* by Nakos and Williams [309, 308]. In order to evaluate the performance of the Dixon resultant, we compared its performance with that of Groebner basis solution (see the previous section and Závoti and Jancso [433]).

As we have seen in the previous section, in order to solve the 7-parameter datum transformation problem, a *minimum* of three corresponding points in both systems

Table 17.3. Groebner basis' 7-transformation parameters

Transformation parameter	Value	unit
Scale $k-1$	-1.4343	[ppm]
Rotation $\mathbf{X}_1(a)$	0.32575149	[“]
Rotation $\mathbf{X}_2(b)$	-0.46037399	[“]
Rotation $\mathbf{X}_3(c)$	-0.00810606	[“]
Translation ΔX	643.0953	[m]
Translation ΔY	22.6163	[m]
Translation ΔZ	481.6023	[m]

Table 17.4. Transformed Cartesian coordinates of System A (*Table* 17.1) into System B (*Table* 17.2) using the parameters in *Table* 17.3

Site	$X(m)$	$Y(m)$	$Z(m)$
System A: Solitude	4157222.5430	664789.3070	4774952.0990
System B	4157870.2370	664818.6780	4775416.5240
Transformed value	4157870.3070	664818.6742	4775416.5240
Residual	- 0.0700	0.0038	0.0000
System A: Buoch Zeil	4149043.3360	688836.4430	4778632.1880
System B	4149691.0490	688865.7850	4779096.5880
Transformed value	4149691.1190	688865.7812	4779096.5880
Residual	- 0.0700	0.0038	0.0000
System A: Hohenneuffen	4172803.5110	690340.0780	4758129.7010
System B	4173451.3540	690369.3750	4758594.0750
Transformed value	4173451.2141	690369.3826	4758594.0750
Residual	0.1399	-0.0076	0.0000

are required. For these three points, nine equations $(f_1, f_2, \ldots, f_9)$ can be written (see 17.15). In order to obtain a solvable system of equations, seven equations should be chosen. In the previous section, a combination was chosen arbitrarily as $\{f_1, f_2, f_3, f_4, f_5, f_6, f_9\}$. However, there exists $\binom{9}{7} = 36$ possible combinations of selecting seven equations from the nine original equations in (17.15). From the selected 7 equations, the *translation parameters* X_0, Y_0, Z_0 are eliminated in a differential mode (see 17.16) reducing the seven equations to a system of *four* equations with four unknown parameters (i.e., scale x_1 and rotation elements a, b, c). After selecting a combination, the possible configurations to eliminate the translation parameters are:

$$
\begin{array}{ccc}
X_0 & Y_0 & Z_0 \\
f_1 - f_4 & f_2 - f_5 & f_3 - f_6 \\
f_1 - f_7 & f_2 - f_8 & f_3 - f_9 \\
f_4 - f_7 & f_5 - f_8 & f_6 - f_9
\end{array}
\tag{17.18}
$$

Considering now a different combination than used in the previous section, e.g., $(f_2, f_3, f_4, f_5, f_6, f_8, f_9)$, the eliminations $(f_i - f_j)$ whose indices (i, j) cover the sequence $(2, 3, 4, 5, 6, 8, 9)$ are

$$\begin{aligned} r_1 &= f_2 - f_8; \\ r_2 &= f_2 - f_5; \\ r_3 &= f_3 - f_9; \\ r_4 &= f_6 - f_9; \end{aligned} \tag{17.19}$$

leading to

$$\begin{aligned} r_1 &= -ca_1 + ca_3 - cx_1X_1 + cx_1X_3 + b_1 - b_3 - x_1Y_1 + x_1Y_3 + ac_1 - ac_3 + ax_1Z_1 - ax_1Z_3; \\ r_2 &= -ca_1 + ca_2 - cx_1X_1 + cx_1X_2 + b_1 - b_2 - x_1Y_1 + x_1Y_2 + ac_1 - ac_2 + ax_1Z_1 - ax_1Z_2; \\ r_3 &= ba_1 - ba_3 + bx_1X_1 - bx_1X_3 - ab_1 + ab_3 - ax_1Y_1 + ax_1Y_3 + c_1 - c_3 - x_1Z_1 + x_1Z_3; \\ r_4 &= ba_2 - ba_3 + bx_1X_2 - bx_1X_3 - ab_2 + ab_3 - ax_1Y_2 + ax_1Y_3 + c_2 - c_3 - x_1Z_2 + x_1Z_3; \end{aligned} \tag{17.20}$$

This system can be simplified using the same new variables as in the previous section, instead of the original coordinate differences. The simplified system, with these new variables which we call *relative coordinates* is

$$\begin{aligned} r_1 &= -ca_{13} - cx_1X_{13} + b_{13} - x_1Y_{13} + ac_{13} + ax_1Z_{13}; \\ r_2 &= -ca_{12} - cx_1X_{12} + b_{12} - x_1Y_{12} + ac_{12} + ax_1Z_{12}; \\ r_3 &= ba_{13} + bx_1X_{13} - ab_{13} - ax_1Y_{13} + c_{13} - x_1Z_{13}; \\ r_4 &= ba_{23} + bx_1X_{23} - ab_{23} - ax_1Y_{23} + c_{23} - x_1Z_{23}; \end{aligned} \tag{17.21}$$

Equation (17.21) can now be solved by eliminating the variables a, b and c to obtain a polynomial for the scale parameter (x_1) via the Dixon resultant or Groebner basis in *Mathematica*[1].

First, the Dixon prompt should be called by typing,

$$<< Resultant`Dixon`$$

then we can compute the Dixon resultant by

$$DixonResultant[\{r_1, r_2, r_3, r_4\}, \{a, b, c\}, \{\sigma_1, \sigma_2, \sigma_3\}]$$

where $\{\sigma_1, \sigma_2, \sigma_3\}$ are the auxiliary variables of the Dixon polynomial, $r_i\{i = 1, \ldots, 4\}$ the equations, and $\{a, b, c\}$ the variables to be eliminated from the equation system. This results in a *quartic* polynomial for the scale factor (x_1), similar to the Groebner basis solution (see 17.17). The Dixon resultant takes 0.016 seconds to solve this system of equations on a HP xw 4100 workstation with Windows XP operation system, 3GHz P4 Intel processor and 1 GB RAM. Let us compare the results now with those of the Groebner basis solution. The same result can be achieved by typing in *Mathematica*,

$$GroebnerBasis[\{r_1, r_2, r_3, r_4\}, \{x_1, a, b, c\}, \{a, b, c\}]$$

where $r_i\{i = 1, \ldots, 4\}$ are the equations, $\{x_1, a, b, c\}$ the variables in the system, and $\{a, b, c\}$ the variables to be eliminated from the equation system, where the

[1] *Remark:* With the Dixon resultant the system of Eq. (17.20), without using the relative coordinates, also provided a solution, which was un-achievable with Groebner basis (see Paláncz et al. [319]).

order means the elimination order of the variables. However, using the Groebner basis method requires 3017.69 seconds!

We have chosen different 7-equation combinations from the original 9 equations and we changed also the monomial order during the processing (see Table 17.5 for a few chosen sequences). The Dixon resultant solution was indifferent to these changes, the running time always being ca. 0.016 seconds. However, the running time for the reduced Groebner basis solution was dependent on the chosen 7-equation combination and changes also with the order of variables to be eliminated (see Table 17.5). In case of four chosen sequences (from the 36 possible choices), one notices considerable change in the required computational time, see for example the last column in Table 17.5. For example, for the combination of $(f_2, f_3, f_4, f_5, f_6, f_8, f_9)$, the computation time was 50 minutes using the monomial order $\{a, b, c\}$ and only 0.484 seconds if we changed the monomial order to $\{b, c, a\}$! With Dixon resultant, the running time was only 0.016 seconds, independent from the order of variables. The Groebner basis approach is clearly affected by both

(a) the combinatorial sequence and
(b) the monomial order.

These two factors are undesirable since users are not often privy to the optimal sequence and order during data processing. However, there is a *third* important factor having strong influence on the performance of the Groebner basis computation: this is the *elimination order*. In general, using MonomialOrder − > EliminationOrder can ensure the best behavior, (Lichtblau, Priv. Comm.).

Table 17.5. Running times (seconds) for cases of different sequences and order of the variables to be eliminated using reduced Groebner basis with relative coordinates

Sequence Order of variables	$f_1, f_2, f_3, f_4,$ f_6, f_7, f_8	$f_1, f_2, f_3, f_4,$ f_5, f_8, f_9	$f_1, f_2, f_3, f_4,$ f_5, f_7, f_9	$f_2, f_3, f_4, f_5,$ f_6, f_8, f_9
a,b,c	0.219	0.688	0.985	3017.69
a,c,b	0.36	35.921	0.672	2601.75
b,a,c	0.25	0.922	0.765	3172.48
b,c,a	163.765	0.547	33.985	0.484
c,a,b	0.719	52.328	0.562	1831.42
c,b,a	174.547	0.75	47.797	0.532

The Dixon resultant therefore proved to be faster in this case, and very robust in that it is insensitive to the order of variables, unlike the Groebner basis. This feature can be very important from a practical point of view, because in the case of Groebner basis, the user should find the proper combinatorial sequence and monomial order via a trial-error method. For the sequence $(f_2, f_3, f_4, f_5, f_6, f_8, f_9)$ in Table 17.5 for example, only two orders can provide a solution in a reasonable time from the six possible orders of the variables.

As a test, three Hungarian points in the ETRS89 system $(x_1, y_1, z_1, \ldots, z_3)$ and in the local Hungarian system HD72 (Hungarian Datum 1972) $(X_1, Y_1, Z_1, \ldots,$

Z_3) listed in *Table* 17.6 were used. Both Dixon and Groebner basis yielded identical results. These examples are available on *Mathematica* in notebook format[2].

Table 17.6. Hungarian points in the ETRS89 and HD72 datums

Point	x(m)	y(m)	z(m)	X(m)	Y(m)	Z(m)
1	4171409.677	1470823.777	4580140.907	4171352.311	1470893.887	4580150.178
2	4146957.889	1277033.850	4659439.264	4146901.301	1277104.509	4659448.287
3	3955632.880	1611863.197	4720991.316	3955575.649	1611933.124	4721000.952

17-23 Gauss-Jacobi combinatorial transformation

When more than three points in both systems are given and the transformation parameters desired, the Gauss-Jacobi combinatorial algorithm is applied. In such a case, the dispersion matrix has to be obtained via the nonlinear error propagation law/variance-covariance propagation law. From the algebraic system of equations (17.15), the Jacobi matrices are given (using e.g., (7.30) and (7.31) on p. 94) as

$$\mathbf{J}_x = \begin{bmatrix} \frac{\partial f_1}{\partial x_1} & \frac{\partial f_1}{\partial a} & \frac{\partial f_1}{\partial b} & \frac{\partial f_1}{\partial c} & \frac{\partial f_1}{\partial X_0} & \frac{\partial f_1}{\partial Y_0} & \frac{\partial f_1}{\partial Z_0} \\ \frac{\partial f_2}{\partial x_1} & \frac{\partial f_2}{\partial a} & \frac{\partial f_2}{\partial b} & \frac{\partial f_2}{\partial c} & \frac{\partial f_2}{\partial X_0} & \frac{\partial f_2}{\partial Y_0} & \frac{\partial f_2}{\partial Z_0} \\ \frac{\partial f_3}{\partial x_1} & \frac{\partial f_3}{\partial a} & \frac{\partial f_3}{\partial b} & \frac{\partial f_3}{\partial c} & \frac{\partial f_3}{\partial X_0} & \frac{\partial f_3}{\partial Y_0} & \frac{\partial f_3}{\partial Z_0} \\ \frac{\partial f_4}{\partial x_1} & \frac{\partial f_4}{\partial a} & \frac{\partial f_4}{\partial b} & \frac{\partial f_4}{\partial c} & \frac{\partial f_4}{\partial X_0} & \frac{\partial f_4}{\partial Y_0} & \frac{\partial f_4}{\partial Z_0} \\ \frac{\partial f_5}{\partial x_1} & \frac{\partial f_5}{\partial a} & \frac{\partial f_5}{\partial b} & \frac{\partial f_5}{\partial c} & \frac{\partial f_5}{\partial X_0} & \frac{\partial f_5}{\partial Y_0} & \frac{\partial f_5}{\partial Z_0} \\ \frac{\partial f_6}{\partial x_1} & \frac{\partial f_6}{\partial a} & \frac{\partial f_6}{\partial b} & \frac{\partial f_6}{\partial c} & \frac{\partial f_6}{\partial X_0} & \frac{\partial f_6}{\partial Y_0} & \frac{\partial f_6}{\partial Z_0} \\ \frac{\partial f_7}{\partial x_1} & \frac{\partial f_7}{\partial a} & \frac{\partial f_7}{\partial b} & \frac{\partial f_7}{\partial c} & \frac{\partial f_7}{\partial X_0} & \frac{\partial f_7}{\partial Y_0} & \frac{\partial f_7}{\partial Z_0} \end{bmatrix}, \tag{17.22}$$

and

[2]http://library.wolfram.com/infocenter/MathSource/6654/

$$
\mathbf{J}_y = \begin{bmatrix}
\frac{\partial f_1}{\partial a_1} & \frac{\partial f_1}{\partial b_1} & \frac{\partial f_1}{\partial c_1} & \frac{\partial f_1}{\partial a_2} & \frac{\partial f_1}{\partial b_2} & \frac{\partial f_1}{\partial c_2} & \frac{\partial f_1}{\partial a_3} & \cdots\cdots\cdots & \frac{\partial f_1}{\partial Z_3} \\
\frac{\partial f_2}{\partial a_1} & \frac{\partial f_2}{\partial b_1} & \frac{\partial f_2}{\partial c_1} & \frac{\partial f_2}{\partial a_2} & \frac{\partial f_2}{\partial b_2} & \frac{\partial f_2}{\partial c_2} & \frac{\partial f_2}{\partial a_3} & \cdots\cdots\cdots & \frac{\partial f_2}{\partial Z_3} \\
\frac{\partial f_3}{\partial a_1} & \frac{\partial f_3}{\partial b_1} & \frac{\partial f_3}{\partial c_1} & \frac{\partial f_3}{\partial a_2} & \frac{\partial f_3}{\partial b_2} & \frac{\partial f_3}{\partial c_2} & \frac{\partial f_3}{\partial a_3} & \cdots\cdots\cdots & \frac{\partial f_3}{\partial Z_3} \\
\frac{\partial f_4}{\partial a_1} & \frac{\partial f_4}{\partial b_1} & \frac{\partial f_4}{\partial c_1} & \frac{\partial f_4}{\partial a_2} & \frac{\partial f_4}{\partial b_2} & \frac{\partial f_4}{\partial c_2} & \frac{\partial f_4}{\partial a_3} & \cdots\cdots\cdots & \frac{\partial f_4}{\partial Z_3} \\
\frac{\partial f_5}{\partial a_1} & \frac{\partial f_5}{\partial b_1} & \frac{\partial f_5}{\partial c_1} & \frac{\partial f_5}{\partial a_2} & \frac{\partial f_5}{\partial b_2} & \frac{\partial f_5}{\partial c_2} & \frac{\partial f_5}{\partial a_3} & \cdots\cdots\cdots & \frac{\partial f_5}{\partial Z_3} \\
\frac{\partial f_6}{\partial a_1} & \frac{\partial f_6}{\partial b_1} & \frac{\partial f_6}{\partial c_1} & \frac{\partial f_6}{\partial a_2} & \frac{\partial f_6}{\partial b_2} & \frac{\partial f_6}{\partial c_2} & \frac{\partial f_6}{\partial a_3} & \cdots\cdots\cdots & \frac{\partial f_6}{\partial Z_3} \\
\frac{\partial f_7}{\partial a_1} & \frac{\partial f_7}{\partial b_1} & \frac{\partial f_7}{\partial c_1} & \frac{\partial f_7}{\partial a_2} & \frac{\partial f_7}{\partial b_2} & \frac{\partial f_7}{\partial c_2} & \frac{\partial f_7}{\partial a_3} & \cdots\cdots\cdots & \frac{\partial f_7}{\partial Z_3}
\end{bmatrix}, \tag{17.23}
$$

where the doted points in $\mathbf{J}_y$ represent the partial derivatives of (17.15) with respect to

$$\{b_3, c_3, X_1, Y_1, Z_1, X_2, Y_2, Z_2, X_3, Y_3\}.$$

From the dispersion $\mathbf{\Sigma}_y$ of the vector of observations $\mathbf{y}$ and with (17.22) and (17.23) forming $\mathbf{J} = \mathbf{J}_x^{-1}\mathbf{J}_y$, the dispersion matrix $\mathbf{\Sigma}_x$ is then obtained using (7.31). Finally, we obtained the dispersion matrix $\mathbf{\Sigma}$ from (7.33) on p. 94. The solution is performed stepwise as discussed on p. 259. There exist two possibilities of using the combinatorial algorithm. These are:

(1) Forming combinations of the given coordinates, each minimal set comprising 3 points. Given n number of points in both systems, combinations can be formed from (7.28) on p. 93, each set containing $m = 3$ points. For each combination, the desired transformation parameters are computed using the explicit formulae in [23, Boxes 2-2 and 2-3] or [37, Appendices A and B]. The resulting combinatorial solutions are then adjusted using the special linear Gauss-Markov model.

(2) Alternatively, instead of forming combinations from points alone and solving as in (1) above, combinations are formed both for the points and also from the 9 equations in (17.15). In this case, each minimal combinatorial in points will have three stations from which a further combinatorial in terms of equations are formed. From the 9 equations in (17.15), combinations are formed with a minimum of seven equations per set. The solution of the seven equations of each combinatorial set delivers equations of the form in [23, Boxes 2-2 and 2-3]. Once the solution is completed for a minimum combinatorial set for three points, the procedure is repeated for other points until all the combinations have been solved. The resulting combinatorial solutions are then adjusted using

the special linear Gauss-Markov model as already explained. This approach is labour intensive, but may offer improved accuracy as compared to the approach in (1) as all the available information is exploited. We leave it as an exercise for an interested reader.

Example 17.2 (Computation of transformation parameters using Gauss-Jacobi combinatorial algorithm). We repeat Example 17.1 by computing the 7 transformation parameters for the overdetermined case using the combinatorial algorithm. All 7 points of *Tables* 17.1 and 17.2 are used, unlike in *Example* 17.1 where only three points were used (e.g., the minimal case). The computed transformation parameters are presented in *Table* 17.7. In order to check the accuracy of these parameters, they are used to transform the Cartesian coordinates from the local reference system in *Table* 17.1 to WGS-84. *Table* 17.8 gives the residuals computed by subtracting the transformed values from the actual GPS coordinates of *Table* 17.2. *Table* 17.9 gives for comparison purposes the residuals obtained using least squares method. The residuals from both procedures are of the same magnitude. We also compute the residual norm (square root of the sum of squares of residuals) and present them in *Table* 17.10. The computed norms from the combinatorial solutions are somewhat better than those of the linearized least squares solutions. Figure 17.1 presents the scatter of the computed 36 minimal combinatorial solutions of scale indicated by doted points ($\bullet$) around the adjusted value indicated by a line ($-$). Figures 17.2 and 17.3 plot the scatter of the computed 36 minimal combinatorial solutions of the translation and rotation parameters indicated by doted points ($\bullet$) around the adjusted values indicated by stars ($\star$).

Table 17.7. Gauss-Jacobi combinatorial's 7 transformation parameters

Transformation parameter	Value	Root-mean-square	unit
Scale $k-1$	4.92377597	0.350619414	[ppm]
Rotation $\mathbf{X}_1(a)$	-0.98105498"	0.040968549	["]
Rotation $\mathbf{X}_2(b)$	0.68869774"	0.047458707	["]
Rotation $\mathbf{X}_3(c)$	0.96671738"	0.044697434	["]
Translation ΔX	639.9785	2.4280	[m]
Translation ΔY	68.1548	3.0123	[m]
Translation ΔZ	423.7320	2.7923	[m]

Table 17.8. Residuals of the transformed Cartesian coordinates of System A (Table 17.1) into System B using the parameters in *Table* 17.7

Site	$X(m)$	$Y(m)$	$Z(m)$
Solitude	0.0739	0.1381	0.1397
Buoch Zeil	0.0328	-0.0301	0.0095
Hohenneuffen	-0.0297	-0.0687	-0.0020
Kuelenberg	0.0246	-0.0347	-0.0793
Ex Mergelaec	-0.1405	0.0228	-0.0148
Ex Hof Asperg	-0.0477	0.0116	-0.0599
Ex Keisersbach	-0.0673	0.0335	-0.0070

Table 17.9. Residuals of the transformed Cartesian coordinates of System A (*Table* 17.1) into System B using the parameters computed by a least squares method

Site	$X(m)$	$Y(m)$	$Z(m)$
Solitude	0.0940	0.1351	0.1402
Buoch Zeil	0.0588	-0.0497	0.0137
Hohenneuffen	-0.0399	-0.0879	-0.0081
Kuelenberg	0.0202	-0.0220	-0.0874
Ex Mergelaec	-0.0919	0.0139	-0.0055
Ex Hof Asperg	-0.0118	0.0065	-0.0546
Ex Keisersbach	-0.0294	0.0041	-0.0017

Table 17.10. Computed residual norms

Method	$X(m)$	$Y(m)$	$Z(m)$
Linearized Least Squares Solution	0.1541	0.1708	0.1748
Gauss-Jacobi Combinatorial	0.1859	0.1664	0.1725

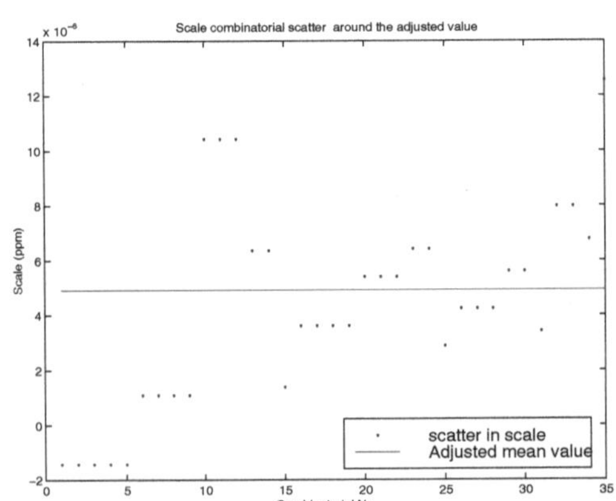

Figure 17.1. Scatter of the computed 36 minimal combinatorial values of scale around the adjusted value

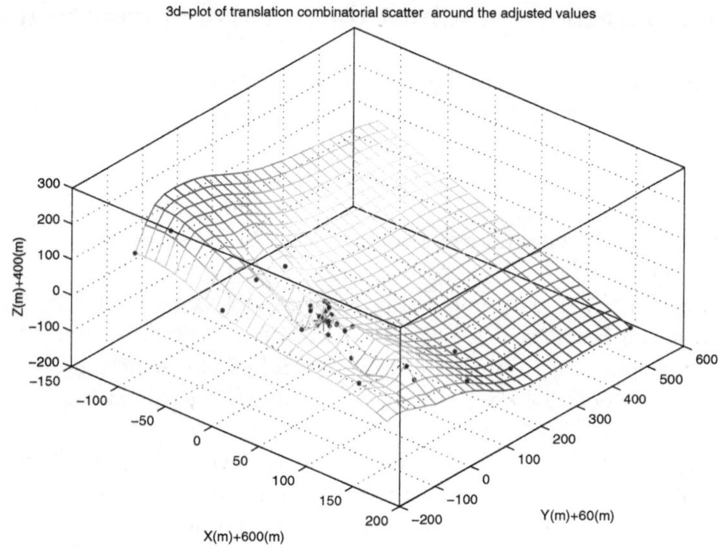

Figure 17.2. Scatter of the 36 computed translations around the adjusted values

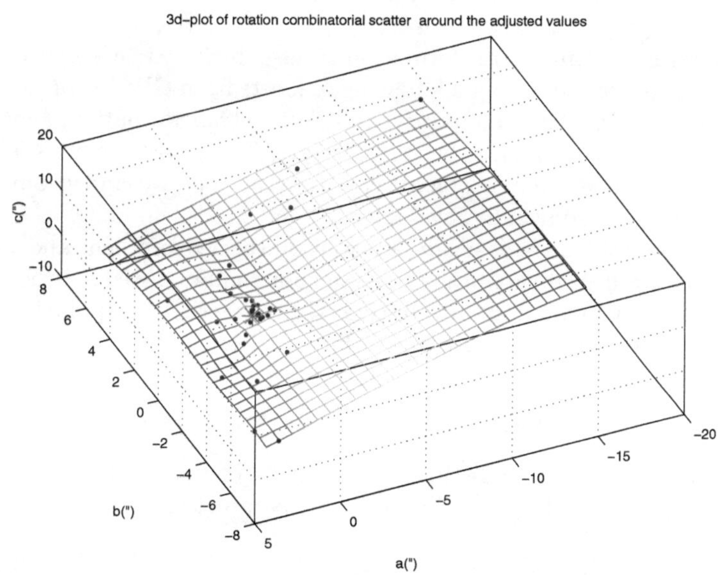

Figure 17.3. Scatter of the 36 computed rotations around the adjusted values (dots) and (stars)

17-3 The 9-parameter (affine) datum transformation

Due to the distortions between the traditional terrestrial and GPS derived networks, the 7-parameter similarity transformations in some cases may not offer satisfactory precision. For example, transforming GPS global coordinates to the local Hungarian system with similarity transformation gives 0.5 m maximal residuals, see, e.g., Papp and Szűcs [326]. To reduce the remaining residuals, other transformation models with more parameters can be used.

The 9-parameter affine transformation is not only a logical extension but even a generalization of the 7-parameter similarity model. This transformation is the modification of the Helmert $C_7(3,3)$ transformation, where 3 different scales are used in the corresponding coordinate axes instead of one scale factor. In the case of the 3 scale parameters being equal, the model reverts back to the similarity transformation.

The solution of the 9-parameter model was achieved by Späth [368] using the numerical minimization technique of the residuum vector, as well as by Papp and Szűcs [326] who used the linearized *least squares* method. Watson [403] pointed out that the *Gauss-Newton method* or its variants can be easily implemented for the 9-parameter problem using separation of variables and iteration with respect to the rotation parameters alone, while other parameters can be calculated via a simple linear least squares solution. The Watson [403] method is analogous to other methods for separated least square problems, which goes back at least to Golub and Pereyra [152].

The 9-parameter affine transformation is also included in some coordinate-transformation software developed following requests from GPS users (see e.g. [295, 143]). To determine the 9 parameters of the 3D affine transformation, a minimum of 3 points with known coordinates $(x_i, y_i, z_i, X_i, Y_i, Z_i)$ in both systems is required. This is the so-called 3-point problem. However, in geodesy and geoinformatics, $N > 3$ known points are usually available.

The N-point problem is basically an overdetermined problem and because of the size of real world problems, the Gauss-Jacobi combinatorial solution becomes inadequate. In what follows, an alternative ALESS method is applied to solve for the 9 parameters of the *overdetermined affine* model.

Definition of the problem

The 3D affine transformation is the generalization of the $C_7(3,3)$ Helmert transformation, using three different scale (s_1, s_2, s_3) parameters instead of a single one. Consider (17.24)

$$\begin{pmatrix} x_i \\ y_i \\ z_i \end{pmatrix} = \mathbf{W}\mathbf{R} \begin{pmatrix} X_i \\ Y_i \\ Z_i \end{pmatrix} + \begin{pmatrix} X_0 \\ Y_0 \\ Z_0 \end{pmatrix}, \tag{17.24}$$

where $\mathbf{W}$ is the scale matrix, X_0, Y_0, Z_0 the translation parameters, $\mathbf{R}$ the rotation matrix, and x_i, y_i, z_i and X_i, Y_i, Z_i are the coordinates of the points in the two coordinate systems. For such a case (cf x_1 in Eqn. 17.1), the scale factors can be modeled by a diagonal matrix

$$\mathbf{W} = \begin{pmatrix} s_1 & 0 & 0 \\ 0 & s_2 & 0 \\ 0 & 0 & s_3 \end{pmatrix}. \tag{17.25}$$

Instead of using the traditionally simplified Cardan angles representation of $\mathbf{R}$ (see, e.g., Papp and Szűcs [326]), we will use the same rotation matrix as in the case of the 7-parameter transformation, expressed by a skew-symmetric matrix, facilitating the symbolic-numeric solution of the problem without simplifications. The rotation matrix

$$\mathbf{R} = (\mathbf{I}_3 - \mathbf{S})^{-1}(\mathbf{I}_3 + \mathbf{S}), \tag{17.26}$$

where $\mathbf{S}$ is the skew-symmetric matrix was given in 17.4 and parameterized using $a,\ b$ and c as

$$\mathbf{R} = \begin{pmatrix} \frac{1+a^2-b^2-c^2}{1+a^2+b^2+c^2} & \frac{2(ab-c)}{1+a^2+b^2+c^2} & \frac{2(b+ac)}{1+a^2+b^2+c^2} \\ \frac{2(ab+c)}{1+a^2+b^2+c^2} & \frac{1-a^2+b^2-c^2}{1+a^2+b^2+c^2} & -\frac{2(a-bc)}{1+a^2+b^2+c^2} \\ \frac{2(-b+ac)}{1+a^2+b^2+c^2} & \frac{2(a+bc)}{1+a^2+b^2+c^2} & \frac{1-a^2-b^2+c^2}{1+a^2+b^2+c^2} \end{pmatrix}, \tag{17.27}$$

for which the orthogonality relation holds such that $\mathbf{R}\mathbf{R}^T$ is an identity matrix. The axial rotation angles (Cardan angles) can be calculated also from the rotation matrix $\mathbf{R}$ in 17.7. Furthermore, instead of the scale parameters (s_1, s_2, s_3), the inverse values of the scale parameters $(\sigma_1, \sigma_2, \sigma_3)$ can be used to obtain simplified equations.

Let us call $\sigma_i = 1/s_i$ and introduce $\boldsymbol{\Omega} = \mathbf{W}^{-1}$

$$\boldsymbol{\Omega} = \begin{pmatrix} \sigma_1 & 0 & 0 \\ 0 & \sigma_2 & 0 \\ 0 & 0 & \sigma_3 \end{pmatrix}. \tag{17.28}$$

Expressing the rotation matrix with the skew-symmetric matrix and using the inverse of the scale matrix ($\boldsymbol{\Omega}$), Eq. (17.24) can be written in the form,

$$\begin{pmatrix} x_i \\ y_i \\ z_i \end{pmatrix} = \boldsymbol{\Omega}^{-1}(\mathbf{I}_3 - \mathbf{S})^{-1}(\mathbf{I}_3 + \mathbf{S}) \begin{pmatrix} X_i \\ Y_i \\ Z_i \end{pmatrix} + \begin{pmatrix} X_0 \\ Y_0 \\ Z_0 \end{pmatrix}. \tag{17.29}$$

Multiplying both sides by $(\mathbf{I}_3 - \mathbf{S})\boldsymbol{\Omega}$ and rearranging the equation leads to

$$\begin{pmatrix} 0 \\ 0 \\ 0 \end{pmatrix} = (\mathbf{I}_3 - \mathbf{S})\boldsymbol{\Omega} \begin{pmatrix} x_i \\ y_i \\ z_i \end{pmatrix} - (\mathbf{I}_3 + \mathbf{S}) \begin{pmatrix} X_i \\ Y_i \\ Z_i \end{pmatrix} - (\mathbf{I}_3 - \mathbf{S})\boldsymbol{\Omega} \begin{pmatrix} X_0 \\ Y_0 \\ Z_0 \end{pmatrix}. \tag{17.30}$$

Finally, substituting $\boldsymbol{\Omega}$ and $\mathbf{S}$ in to Eq. (17.30) gives,

$$\begin{pmatrix} 0 \\ 0 \\ 0 \end{pmatrix} = \begin{pmatrix} 1 & c & -b \\ -c & 1 & a \\ b & -a & 1 \end{pmatrix} \begin{pmatrix} \sigma_1 & 0 & 0 \\ 0 & \sigma_2 & 0 \\ 0 & 0 & \sigma_3 \end{pmatrix} \begin{pmatrix} x_i \\ y_i \\ z_i \end{pmatrix} - \\ \begin{pmatrix} 1 & -c & b \\ c & 1 & -a \\ -b & a & 1 \end{pmatrix} \begin{pmatrix} X_i \\ Y_i \\ Z_i \end{pmatrix} - \begin{pmatrix} 1 & c & -b \\ -c & 1 & a \\ b & -a & 1 \end{pmatrix} \begin{pmatrix} \sigma_1 & 0 & 0 \\ 0 & \sigma_2 & 0 \\ 0 & 0 & \sigma_3 \end{pmatrix} \begin{pmatrix} X_0 \\ Y_0 \\ Z_0 \end{pmatrix}, \tag{17.31}$$

where the unknown parameters are the *three rotation parameters* (a, b, c), the *three translation parameters* (X_0, Y_0, Z_0) and the *three inverse scale parameters* $(\sigma_1, \sigma_2, \sigma_3)$.

17-4 Algebraic solution of the 9-parameter transformation

17-41 The 3-point affine transformation problem

For the determination of the 9 parameters (a, b, c, X_0, Y_0, Z_0, s_1, s_2, s_3) of the 3D affine transformation, we need 9 equations. According to Eq. (17.31), for one point with known coordinates in both coordinate systems, we can write three equations. For the determination of all parameters therefore, 3 non-collinear points with known coordinates are needed. In this case the 9 equations, which give a nonlinear system ($f_i = 0$) (see Eq.(17.32)) are written as

$$\begin{aligned}
f_1 &= -X_1 + cY_1 - bZ_1 + x_1\sigma_1 - X_0\sigma_1 + cy_1\sigma_2 - cY_0\sigma_2 - bz_1\sigma_3 + bZ_0\sigma_3 \\
f_2 &= -cX_1 - Y_1 + aZ_1 - cx_1\sigma_1 + cX_0\sigma_1 + y_1\sigma_2 - Y_0\sigma_2 + az_1\sigma_3 - aZ_0\sigma_3 \\
f_3 &= bX_1 - aY_1 - Z_1 + bx_1\sigma_1 - bX_0\sigma_1 - ay_1\sigma_2 + aY_0\sigma_2 + z_1\sigma_3 - Z_0\sigma_3 \\
f_4 &= -X_2 + cY_2 - bZ_2 + x_2\sigma_1 - X_0\sigma_1 + cy_2\sigma_2 - cY_0\sigma_2 - bz_2\sigma_3 + bZ_0\sigma_3 \\
f_5 &= -cX_2 - Y_2 + aZ_2 - cx_2\sigma_1 + cX_0\sigma_1 + y_2\sigma_2 - Y_0\sigma_2 + az_2\sigma_3 - aZ_0\sigma_3 \\
f_6 &= bX_2 - aY_2 - Z_2 + bx_2\sigma_1 - bX_0\sigma_1 - ay_2\sigma_2 + aY_0\sigma_2 + z_2\sigma_3 - Z_0\sigma_3 \\
f_7 &= -X_3 + cY_3 - bZ_3 + x_3\sigma_1 - X_0\sigma_1 + cy_3\sigma_2 - cY_0\sigma_2 - bz_3\sigma_3 + bZ_0\sigma_3 \\
f_8 &= -cX_3 - Y_3 + aZ_3 - cx_3\sigma_1 + cX_0\sigma_1 + y_3\sigma_2 - Y_0\sigma_2 + az_3\sigma_3 - aZ_0\sigma_3 \\
f_9 &= bX_3 - aY_3 - Z_3 + bx_3\sigma_1 - bX_0\sigma_1 - ay_3\sigma_2 + aY_0\sigma_2 + z_3\sigma_3 - Z_0\sigma_3 \,.
\end{aligned} \tag{17.32}$$

To solve this system of equations, different symbolical and numerical methods can be used. In what follows, symbolic solution of this problem using the *Dixon resultant* and *Groebner basis* are presented, as was done previously for the 7-parameter problem.

17-411 Simplifications for the symbolic solution

Similar to the $\mathbb{C}_7(3,3)$ problem, we can reduce the system of equation (17.32) by differencing the equations. In this way, the translation parameters (X_0, Y_0, Z_0)

can be eliminated. The elimination of the translation parameters can be done by subtractions

$$
\begin{aligned}
g_1 = f_1 - f_7 =& - X_1 + X_3 + (x_1 - x_3)\sigma_1 + c(Y_1 - Y_3 + (y_1 - y_3)\sigma_2) \\
& - b(Z_1 - Z_3 + (z_1 - z_3)\sigma_3) \\
g_2 = f_4 - f_7 =& - X_2 + X_3 + (x_2 - x_3)\sigma_1 + c(Y_2 - Y_3 + (y_2 - y_3)\sigma_2) \\
& - b(Z_2 - Z_3 + (z_2 - z_3)\sigma_3) \\
g_3 = f_2 - f_8 =& - Y_1 + Y_3 - c(X_1 - X_3 + (x_1 - x_3)\sigma_1) + (y_1 - y_3)\sigma_2 \\
& + a(Z_1 - Z_3 + (z_1 - z_3)\sigma_3) \\
g_4 = f_5 - f_8 =& - Y_2 + Y_3 - c(X_2 - X_3 + (x_2 - x_3)\sigma_1) + (y_2 - y_3)\sigma_2 \\
& + a(Z_2 - Z_3 + (z_2 - z_3)\sigma_3) \\
g_5 = f_3 - f_9 =& - Z_1 + Z_3 + b(X_1 - X_3 + (x_1 - x_3)\sigma_1) - a(Y_1 - Y_3 \\
& + (y_1 - y_3)\sigma_2) + (z_1 - z_3)\sigma_3 \\
g_6 = f_6 - f_9 =& - Z_2 - Z_3 + b(X_2 - X_3 + (x_2 - x_3)\sigma_1) - a(Y_2 - Y_3 \\
& + (y_2 - y_3)\sigma_2) + (z_2 - z_3)\sigma_3 .
\end{aligned}
\tag{17.33}
$$

In the remaining 6 equations ($g_i = 0$) there are only 6 unknown parameters (a, b, c, σ_1, σ_2, σ_3). This system can be simplified by the introduction of the *relative coordinates* as was the case for the 7-parameter transformation.

$$
\left.\begin{aligned}
x_{ij} = x_i - x_j, y_{ij} = y_i - y_j, z_{ij} = z_i - z_j \\
X_{ij} = X_i - X_j, Y_{ij} = Y_i - Y_j, Z_{ij} = Z_i - Z_j .
\end{aligned}\right\} \rightarrow i, j \in \{1, 2, 3\}, i \neq j \tag{17.34}
$$

Then our equation system becomes

$$
\begin{aligned}
g_1 &= -X_{13} + cY_{13} - bZ_{13} + x_{13}\sigma_1 + cy_{13}\sigma_2 - bz_{13}\sigma_3 \\
g_2 &= -X_{23} + cY_{23} - bZ_{23} + x_{23}\sigma_1 + cy_{23}\sigma_2 - bz_{23}\sigma_3 \\
g_3 &= -cX_{13} - Y_{13} + aZ_{13} - cx_{13}\sigma_1 + y_{13}\sigma_2 + az_{13}\sigma_3 \\
g_4 &= -cX_{23} - Y_{23} + aZ_{23} - cx_{23}\sigma_1 + y_{23}\sigma_2 + az_{23}\sigma_3 \\
g_5 &= +bX_{13} - aY_{13} - Z_{13} + bx_{13}\sigma_1 - ay_{13}\sigma_2 + z_{13}\sigma_3 \\
g_6 &= +bX_{23} - aY_{23} - Z_{23} + bx_{23}\sigma_1 - ay_{23}\sigma_2 + z_{23}\sigma_3 .
\end{aligned}
\tag{17.35}
$$

17-412 Symbolic solution with Dixon resultant

Symbolic solution of the system g_i means the reduction of the multivariate polynomial system via computer algebra to a single *univariate polynomial* and *computing its roots*. Then the other unknowns can be computed backwards, as done in the *Gauss elimination method* for solving linear systems of equations. In the case of the affine transformation, the system can be reduced with different methods to a univariate polynomial containing only one scale parameter σ_i. This can be done by Dixon resultant with *Kapur-Saxena-Yang* and with the Early Discovery Factors

(EDF) method also, see Paláncz et al. [325] and Zaletnyik and Paláncz [431]. Employing the Kapur-Saxena-Yang method, using pairwise-elimination step by step, a univariate polynomial of degree 29 for σ_1 is obtained, see. e.g., [430, 432]. To find the proper solution of this polynomial of degree 29, a good initial value is needed.

The univariate polynomial for σ_1 can also be computed by employing the accelerated Dixon resultant by the *EDF algorithm*, which was suggested and implemented in the computer algebra system *Fermat* by Lewis [267, 269]. Using this method one can get the results in the following form

$$\prod_{i=1}^{5} \varphi_i(\sigma_1)^{K_i} \tag{17.36}$$

where $\varphi_i(\sigma_1)$ are irreducible polynomials with low degrees, but their powers, K_i are very large positive integer numbers. Expanding this expression would result in millions of terms. Consequently, we shall consider $K_i = 1$, for $i = 1, \ldots, 5$, namely

$$\prod_{i=1}^{5} \varphi_i(\sigma_1) \tag{17.37}$$

as the Dixon resultant. These polynomials are as follows

$$\begin{aligned}
\varphi_1 =& y_{13}\, z_{23} - y_{23}\, z_{13};\\
\varphi_2 =& x_{13}^2\, y_{23}\, z_{23}\, \sigma_1^2 - x_{13}\, x_{23}\, y_{13}\, z_{23}\, \sigma_1^2 - x_{13}\, x_{23}\, y_{23}\, z_{13}\, \sigma_1^2 + x_{23}^2\, y_{13}\, z_{13}\, \sigma_1^2\\
& - Z_{13}^2\, y_{23}\, z_{23} - Y_{13}^2\, y_{23}\, z_{23} - X_{13}^2\, y_{23}\, z_{23} + Z_{13}\, Z_{23}\, y_{13}\, z_{23}\\
& + Y_{13}\, Y_{23}\, y_{13}\, z_{23} + X_{13}\, X_{23}\, y_{13}\, z_{23} + Z_{13}\, Z_{23}\, y_{23}\, z_{13} + Y_{13}\, Y_{23}\, y_{23}\, z_{13}\\
& + X_{13}\, X_{23}\, y_{23}\, z_{13} - Z_{23}^2\, y_{13}\, z_{13} - Y_{23}^2\, y_{13}\, z_{13} - X_{23}^2\, y_{13}\, z_{13};\\
\varphi_3 =& x_{13}\, y_{23}\, \sigma_1 - x_{23}\, y_{13}\, \sigma_1 + X_{13}\, y_{23} - X_{23}\, y_{13};\\
\varphi_4 =& Z_{13}\, x_{13}\, x_{23}\, z_{23}\, \sigma_1^2 - Z_{23}\, x_{13}^2\, z_{23}\, \sigma_1^2 - Z_{13}\, x_{23}^2\, z_{13}\, \sigma_1^2 + Z_{23}\, x_{13}\, x_{23}\, z_{13}\, \sigma_1^2\\
& + X_{13}\, Z_{13}\, x_{23}\, z_{23}\, \sigma_1 - 2\, X_{13}\, Z_{23}\, x_{13}\, z_{23}\, \sigma_1 + X_{23}\, Z_{13}\, x_{13}\, z_{23}\, \sigma_1\\
& + X_{13}\, Z_{23}\, x_{23}\, z_{13}\, \sigma_1 - 2\, X_{23}\, Z_{13}\, x_{23}\, z_{13}\, \sigma_1 + X_{23}\, Z_{23}\, x_{13}\, z_{13}\, \sigma_1\\
& - X_{13}^2\, Z_{23}\, z_{23} + X_{13}\, X_{23}\, Z_{13}\, z_{23} + X_{13}\, X_{23}\, Z_{23}\, z_{13} - X_{23}^2\, Z_{13}\, z_{13};\\
\varphi_5 =& Z_{13}\, z_{23} - Z_{23}\, z_{13};
\end{aligned} \tag{17.38}$$

From these *factor polynomials*, one which provides the good (positive and real) solution is selected. In the case of more than one real, positive solution, an initial value can help in deciding the correct polynomial.

Fortunately for the value of σ_1, a very good estimation can be given as $\sigma_1 = 1/s_1$, where s_1 (the first scale parameter) can be estimated by dividing the sum of distances from the center of gravity in both systems, see e.g., Albertz and Kreiling [8].

The factor polynomial providing the proper root (φ_4), has degree 2, therefore its solution can be expressed in an analytical form (only one of the roots is correct, the positive σ_1). Similarly, we can also get simple explicit forms for σ_2 and σ_3.

The results of the *Dixon-EDF method* is not only faster and more elegant, but also more precise than that of the *Dixon-KSY method*. However, one should still check the solutions of all polynomials with degrees 1 and 2 in order to choose the proper results.

17-413 Symbolic solution with reduced Groebner basis

The same results for σ_i, $i = 1, 2, 3$ can be achieved by using the *reduced Groebner basis* built in to *Mathematica*, where the monomial order should be defined as the elimination order (Lichtblau[3], Private Communication). With Groebner basis, the solutions of σ_1, σ_2 and σ_3 as a second order univariate polynomial can be obtained in one step.

For σ_1 for instance, one writes in *Mathematica*:

$$\begin{aligned}\sigma_1 =& GroebnerBasis[\{g_1, g_2, g_3, g_4, g_5, g_6\}, \sigma_1, \{a, b, c, \sigma_2, \sigma_3\},\\ & MonomialOrder \to EliminationOrder]\end{aligned}$$

Solving the problem with Groebner basis has the advantage that the solution for σ_i can be obtained in one step without the need to select the proper solution using good initial values. For all σ_i, the resulting univariate polynomials are of degree 2 where only one solution is positive. Its solution can therefore be expressed in analytical form as expressed by equations (17.39)-(17.41), i.e.,

$$\begin{aligned}\sigma_1 =& \left[X_{23}^2 y_{13} z_{13} + y_{13} Y_{23}^2 z_{13} + X_{13}^2 y_{23} z_{23} + Y_{13}^2 y_{23} z_{23} + y_{23} Z_{13}^2 z_{23}\right.\\ & - X_{13} X_{23}(y_{23} z_{13} + y_{13} z_{23}) - Y_{13} Y_{23}(y_{23} z_{13} + y_{13} z_{23}) - y_{23} z_{13} Z_{13} Z_{23} -\\ & \left. y_{13} Z_{13} z_{23} Z_{23} + y_{13} z_{13} Z_{23}^2\right]^{1/2}\\ & / \left[(x_{23} y_{13} - x_{13} y_{23})(x_{23} z_{13} - x_{13} z_{23})\right]^{1/2}\end{aligned} \tag{17.39}$$

$$\begin{aligned}\sigma_2 =& \left[-X_{13}^2 x_{23} z_{23} + X_{13} X_{23}(x_{23} z_{13} + x_{13} z_{23})\right.\\ & + x_{23}\left(Y_{13} Y_{23} z_{13} - Y_{13}^2 z_{23} - Z_{13}^2 z_{23} + z_{13} Z_{13} Z_{23}\right)\\ & \left. - x_{13}\left(X_{23}^2 z_{13} + Y_{23}^2 z_{13} - Y_{13} Y_{23} z_{23} - Z_{13} z_{23} Z_{23} + z_{13} Z_{23}^2\right)\right]^{1/2}\\ & / \left[(x_{13} y_{23} - x_{23} y_{13})(y_{13} z_{23} - y_{23} z_{13})\right]^{1/2}\end{aligned} \tag{17.40}$$

and

$$\begin{aligned}\sigma_3 =& \left[X_{13}^2 x_{23} y_{23} - X_{13} X_{23}(x_{23} y_{13} + x_{13} y_{23})\right.\\ & + x_{23}\left(Y_{13}^2 y_{23} - y_{13} Y_{13} Y_{23} + y_{23} Z_{13}^2 - y_{13} Z_{13} Z_{23}\right)\\ & \left. + x_{13}\left(X_{23}^2 y_{13} - Y_{13} y_{23} Y_{23} + y_{13} Y_{23}^2 - y_{23} Z_{13} Z_{23} + y_{13} Z_{23}^2\right)\right]^{1/2}\\ & / \left[(x_{23} z_{13} - x_{13} z_{23})(y_{23} z_{13} - y_{13} z_{23})\right]^{1/2}\end{aligned} \tag{17.41}$$

---

[3] researcher of Wolfram Research

Knowing $\sigma_1, \sigma_2, \sigma_3$ the equation system g_i (Eq.(17.35)) will be linear and a, b, c can be expressed easily analytically by

$$\begin{aligned} a = &-[-X_{23}Y_{13} + X_{13}Y_{23} + (X_{23}y_{13} - X_{13}y_{23})\sigma_2 \\ &+ \sigma_1(-x_{23}Y_{13} + x_{13}Y_{23} + (x_{23}y_{13} - x_{13}y_{23})\sigma_2)]/ \\ &[X_{23}Z_{13} - X_{13}Z_{23} + (X_{23}z_{13} - X_{13}z_{23})\sigma_3 \\ &+ \sigma_1(x_{23}Z_{13} - x_{13}Z_{23} + (x_{23}z_{13} - x_{13}z_{23})\sigma_3)] \end{aligned}, \tag{17.42}$$

$$\begin{aligned} b = &-[-Y_{23}Z_{13} + Y_{13}Z_{23} + (Y_{23}z_{13} - Y_{13}z_{23})\sigma_3 \\ &+ \sigma_2(-y_{23}Z_{13} + y_{13}Z_{23} + (y_{23}z_{13} - y_{13}z_{23})\sigma_3)]/ \\ &[-X_{23}Y_{13} + X_{13}Y_{23} + (-X_{23}y_{13} + X_{13}y_{23})\sigma_2 \\ &+ \sigma_1(-x_{23}Y_{13} + x_{13}Y_{23} + (-x_{23}y_{13} + x_{13}y_{23})\sigma_2)] \end{aligned}, \tag{17.43}$$

and

$$\begin{aligned} c = &-[X_{23}Z_{13} - X_{13}Z_{23} + (X_{23}z_{13} - X_{13}z_{23})\sigma_3 \\ &+ \sigma_1(-x_{23}Z_{13} + x_{13}Z_{23} + (-x_{23}z_{13} + x_{13}z_{23})\sigma_3)]/ \\ &[-Y_{23}Z_{13} + Y_{13}Z_{23} + (-Y_{23}z_{13} + Y_{13}z_{23})\sigma_3 \\ &+ \sigma_2(-y_{23}Z_{13} + y_{13}Z_{23} + (-y_{23}z_{13} + y_{13}z_{23})\sigma_3)] \end{aligned}. \tag{17.44}$$

Substituting (a, b, c, σ_1, σ_2, σ_3) in to the original equation (Eq. (17.24)), the translation parameters (X_0, Y_0, Z_0) can be calculated using

$$\begin{aligned} \begin{pmatrix} X_0 \\ Y_0 \\ Z_0 \end{pmatrix} &= \begin{pmatrix} x_1 \\ y_1 \\ z_1 \end{pmatrix} - \mathbf{WR} \begin{pmatrix} X_1 \\ Y_1 \\ Z_1 \end{pmatrix} = \begin{pmatrix} x_1 \\ y_1 \\ z_1 \end{pmatrix} - \mathbf{\Omega}^{-1}\mathbf{R} \begin{pmatrix} X_1 \\ Y_1 \\ Z_1 \end{pmatrix} \\ &= \begin{pmatrix} x_1 \\ y_1 \\ z_1 \end{pmatrix} - \begin{pmatrix} 1/\sigma_1 & 0 & 0 \\ 0 & 1/\sigma_2 & 0 \\ 0 & 0 & 1/\sigma_3 \end{pmatrix} \frac{1}{1+a^2+b^2+c^2} \\ &\begin{pmatrix} 1+a^2-b^2-c^2 & 2(ab-c) & 2(b+ac) \\ 2(ab+c) & 1-a^2+b^2-c^2 & 2(-a+bc) \\ 2(-b+ac) & 2(a+bc) & 1-a^2-b^2+c^2 \end{pmatrix} \begin{pmatrix} X_1 \\ Y_1 \\ Z_1 \end{pmatrix} \end{aligned} \tag{17.45}$$

As we have seen in this section, the *Dixon-KSY method* as well as the *Dixon-EDF method* required the estimation of the initial values to select the proper solution from different solutions. With *reduced Groebner basis*, we get the same solution in one step as the proper solution of the *Dixon-EDF method.* With this method, a fully analytic solution requiring neither initial conditions nor iterations can be given.

The computer algebra method, namely the *accelerated Dixon resultant* with the technique of *Early Discovery Factors* as well as the *reduced Groebner basis*, provides a very simple, elegant symbolic solution for the 3-points problem. The main advantages of the symbolic solutions originate from its *iteration-free* feature, very short- practically zero-computation time, and the independence of the value of the actual numerical data.

The symbolic solution of the 3-points problem can be used for the N-points problem also. One possibility is to use the solutions of the 3-points problem as initial guess value for solving the N-points problem with some local numerical method. Another possibility is just to use the solutions of the different triplets in the *Gauss-Jacobi combinatorial method.*

17-42 The N-points problem

For local numerical methods the initial values can be computed from the 3-points model, but one need to be cautious and not to compute it blindly. In this section, we shall discuss how one can properly select 3-points from the N ones, in order to compute good initial values ensuring fast convergence of the *Newton-type methods* employed for the solution of the N-points problem.

We will examine a numerical example with 81 first order Hungarian stations, with coordinates in both the global system of ETRS89 and the local Hungarian system HD72 (Hungarian Datum 1972). Let us choose two different sets each containing 3 points from the local datum data set (see Fig. 17.4) and calculate using symbolic solution the parameters of the coordinate transformation between the global WGS84 and the local Hungarian system. The results for the two triplets

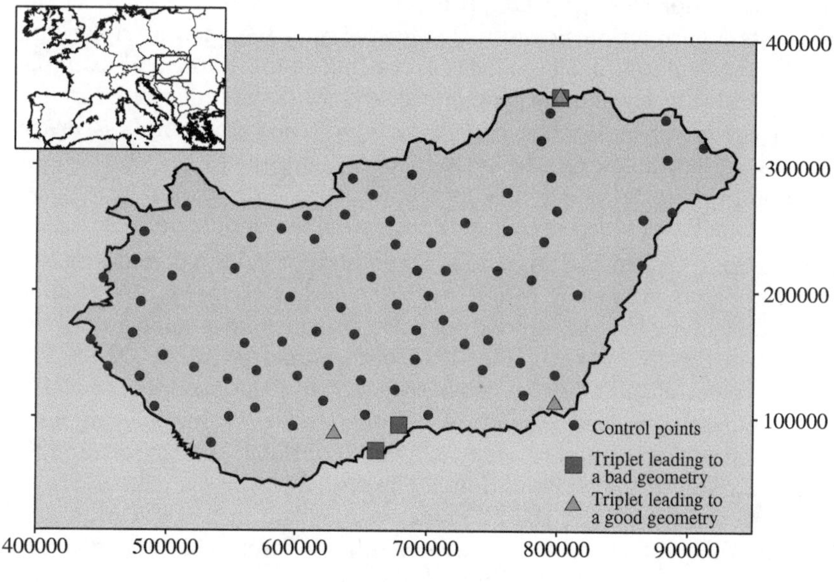

Figure 17.4. Map of Hungary showing the 81 points in the local coordinate system together with the chosen triplets

are quite different, as shown in *Table* 17.11.

Using the parameters of the *1st set* of *Table* (17.11) as initial values for the *Newton-Raphson method* to solve the N-points problem (here, all the N=81 points

Table 17.11. Calculated coordinate transformation parameters from two different sets for the case of Hungarian Datum (81 points) to ETRS89 coordinate system

	1st set	2nd set
X_0	-77.523	-496.192
Y_0	+90.366	+124.797
Z_0	+25.151	+543.061
a	$+1.520 \cdot 10^{-6}$	$+5.470 \cdot 10^{-6}$
b	$+1.526 \cdot 10^{-6}$	$+27.692 \cdot 10^{-6}$
c	$-0.543 \cdot 10^{-6}$	$-2.415 \cdot 10^{-6}$
σ_1	0.999998865	0.999957317
σ_2	1.000001182	0.999989413
σ_3	1.000001642	1.000069623

of Hungary), the method converges rapidly after 4 iteration steps. On replacing the values of the 1st set with those of the *2nd set*, the method does not converge even after 100 iteration steps (see Paláncz et al. [325]). This signifies the importance of properly selecting the 3 points from the N-points to calculate symbolically the *initial guess values* for the N-points problem.

There exists a correlation between the geometry of the chosen triplet and the goodness of the calculated initial values. According to our numerical example, we get the *best* initial values when the geometry of the triplet is similar to an *equilateral triangle*, and the *worst* case when the geometry of the three points is nearly on a *line*. A geometrical index can be introduced to represent the geometry of the selected 3 points to avoid the solutions which provide disadvantageous starting values. This geometrical index is the sine of the minimum angle in the triangle, it's maximal value is $\frac{\sqrt{3}}{2}$ when the triangle is an equilateral triangle, and around zero when the three points are nearly *collinear*. In the earlier examples, the geometrical index in the first case (Fig. 17.4, good geometry), which gave good initial values, was 0.529 and in the second case (Fig. 17.4, bad geometry) was 0.002.

To check the correlation between the geometry and the goodness of the initial values, we calculated the transformation parameters for all 3-point combinations from the 81 points, giving a total of 85320 combinations. We then examined all the resulting Z_0 values for combinations. The real Z_0 value calculated using the *Newton-Raphson method* from the 81 points was 50.342 m, but the values calculated from the different triplets can be very different from this, for example the maximum value for Z_0 was 24 679 629 m! The geometrical index of this extreme triplet was 0.003, meaning that these 3 points were almost collinear.

In Figure 17.5, the calculated Z_0 values are presented as a function of the geometrical index for all 85320 combinations (for a better representation, $|Z_0| > 1000$ values are not represented since they are too large, e.g., the minimum Z_0 is -1 798 501 and the maximum is 24 679 629). The true value of Z_0 based on the 81 data points is also represented by a line.

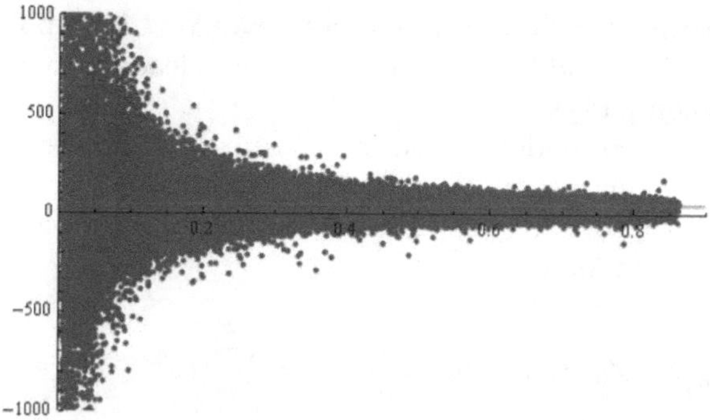

Figure 17.5. Values of Z_0 as function of the geometrical index for all combinations of triplets from the 81 point Hungarian data set (see Fig. 17.4)

The triplets that give extremely different solutions from the adjusted value all have geometrical indices less then 0.1. In this range, the round-off error is dominating. For higher values of this index, the improper fitting of the measurement data to the 9 parameter model causes the deviations in value of Z_0. In general, the more similar the geometry of the selected three points is to an equilateral triangle, the better is the initial value for of the N-points problem. It is therefore shown by this example that it is very important to examine the geometry of the selected three points for calculating the initial values, and to avoid nearly collinear triplets (see Zaletnyik [430]).

17-421 ALESS approach to overdetermined cases

In the cases where more than 3-points with known coordinates in both coordinate systems are known, as is usually the case, there are more independent equations than unknown variables leading to an overdetermined system. These equations generally are inconsistent due to inevitable *stochastic* observation and model errors. In such a situation, the 3 nonlinear equations can be written for a single point (according to Eq. (17.31) as:

$$\begin{aligned}
vx_i =& -X_i + cY_i - bZ_i + x_i\sigma_1 - X_0\sigma_1 + cy_i\sigma_2 - cY_0\sigma_2 - bz_i\sigma_3 + bZ_0\sigma_3 \\
vy_i =& -cX_i - Y_i + aZ_i - cx_i\sigma_1 + cX_0\sigma_1 + y_i\sigma_2 - Y_0\sigma_2 + az_i\sigma_3 - aZ_0\sigma_3 \\
vz_i =& bX_i - aY_i - Z_i + bx_i\sigma_1 - bX_0\sigma_1 - ay_i\sigma_2 + aY_0\sigma_2 + z_i\sigma_3 - Z_0\sigma_3,
\end{aligned} \tag{17.46}$$

where vx_i, vy_i and vz_i are 0 when we have the minimally required 3 homologous points to determine the 9 unknown parameters. However, when we have to solve an overdetermined system, vx_i, vy_i and vz_i are not zero, because of the inevitable observation and model errors.

For $N > 3$ homologous points, the nonlinear system to be solved leads to a system of $3N$ polynomial equations, minimizing $\sum v_i^2$ where $v_i^2 = (vx_i^2 + vy_i^2 + vz_i^2)$. In our numerical example, we will use 1138 points from the Hungarian OGPSH database, with coordinates in the global system of ETRS89 and in the local Hungarian system HD72 (Hungarian Datum 1972) (using the ellipsoidal coordinates without height parameter). For example, we have $N = 1138$ points which means we have 3414 equations and 9 unknown parameters, leading to an overdetermined multivariate polynomial system.

This overdetermined model can be transformed into a determined one by employing symbolic evaluation of the objective function ($\Delta = \sum_{i=1}^{N} v_i^2$), and its symbolic derivation, providing the necessary condition of its minimum (see Sect. 7-2). The objective function

$$\Delta(a, b, c, X_0, Y_0, Z_0, \sigma_1, \sigma_2, \sigma_3) = \sum_{i=1}^{N} v_i^2 = \sum_{i=1}^{N} (vx_i^2 + vy_i^2 + vz_i^2). \tag{17.47}$$

can be created easily with Computer Algebra Systems as

$$\begin{aligned}\Delta = \sum \Big(& X_i^2 + b^2 X_i^2 + c^2 X_i^2 - 2abX_iY_i + Y_i^2 + a^2 Y_i^2 + c^2 Y_i^2 - 2acX_iZ_i - \\ & 2bcY_iZ_i + Z_i^2 + a^2 Z_i^2 + b^2 Z_i^2 - 2x_iX_i\sigma_1 + 2b^2 x_iX_i\sigma_1 + 2c^2 x_iX_i\sigma_1 + \\ & 2X_0X_i\sigma_1 - 2b^2 X_0X_i\sigma_1 - 2c^2 X_0X_i\sigma_1 - 2abx_iY_i\sigma_1 + 4cx_iY_i\sigma_1 + \\ & 2abX_0Y_i\sigma_1 - 4cX_0Y_i\sigma_1 - 4bx_iZ_i\sigma_1 - 2acx_iZ_i\sigma_1 + 4bX_0Z_i\sigma_1 + \\ & 2acX_0Z_i\sigma_1 + x_i^2\sigma_1^2 + b^2 x_i^2\sigma_1^2 + c^2 x_i^2\sigma_1^2 - 2x_iX_0\sigma_1^2 - 2b^2 x_iX_0\sigma_1^2 - \\ & 2c^2 x_iX_0\sigma_1^2 + X_0^2\sigma_1^2 + b^2 X_0^2\sigma_1^2 + c^2 X_0^2\sigma_1^2 - 2abX_iy_i\sigma_2 - \\ & 4cX_iy_i\sigma_2 + 2abX_iY_0\sigma_2 + 4cX_iY_0\sigma_2 - 2y_iY_i\sigma_2 + 2a^2 y_iY_i\sigma_2 + \\ & 2c^2 y_iY_i\sigma_2 + 2Y_0Y_i\sigma_2 - 2a^2 Y_0Y_i\sigma_2 - 2c^2 Y_0Y_i\sigma_2 + 4ay_iZ_i\sigma_2 - \\ & 2bcy_iZ_i\sigma_2 - 4aY_0Z_i\sigma_2 + 2bcY_0Z_i\sigma_2 - 2abx_iy_i\sigma_1\sigma_2 + \\ & 2abX_0y_i\sigma_1\sigma_2 + 2abx_iY_0\sigma_1\sigma_2 - 2abX_0Y_0\sigma_1\sigma_2 + y_i^2\sigma_2^2 + \\ & a^2 y_i^2\sigma_2^2 + c^2 y_i^2\sigma_2^2 - 2y_iY_0\sigma_2^2 - 2a^2 y_iY_0\sigma_2^2 - 2c^2 y_iY_0\sigma_2^2 + \\ & Y_0^2\sigma_2^2 + a^2 Y_0^2\sigma_2^2 + c^2 Y_0^2\sigma_2^2 + 4bX_iz_i\sigma_3 - 2acX_iz_i\sigma_3 - 4aY_iz_i\sigma_3 - \\ & 2bcY_iz_i\sigma_3 - 4bX_iZ_0\sigma_3 + 2acX_iZ_0\sigma_3 + 4aY_iZ_0\sigma_3 + 2bcY_iZ_0\sigma_3 - \\ & 2z_iZ_i\sigma_3 + 2a^2 z_iZ_i\sigma_3 + 2b^2 z_iZ_i\sigma_3 + 2Z_0Z_i\sigma_3 - 2a^2 Z_0Z_i\sigma_3 - \\ & 2b^2 Z_0Z_i\sigma_3 - 2acx_iz_i\sigma_1\sigma_3 + 2acX_0z_i\sigma_1\sigma_3 + 2acx_iZ_0\sigma_1\sigma_3 - \\ & 2acX_0Z_0\sigma_1\sigma_3 - 2bcy_iz_i\sigma_2\sigma_3 + 2bcY_0z_i\sigma_2\sigma_3 + 2bcy_iZ_0\sigma_2\sigma_3 - \\ & 2bcY_0Z_0\sigma_2\sigma_3 + z_i^2\sigma_3^2 + a^2 z_i^2\sigma_3^2 + b^2 z_i^2\sigma_3^2 - 2z_iZ_0\sigma_3^2 - \\ & 2a^2 z_iZ_0\sigma_3^2 - 2b^2 z_iZ_0\sigma_3^2 + Z_0^2\sigma_3^2 + a^2 Z_0^2\sigma_3^2 + b^2 Z_0^2\sigma_3^2 \Big) \end{aligned} \tag{17.48}$$

The necessary conditions of the minimum for the objective function are

$$\begin{aligned} & \frac{\partial \Delta}{\partial a} = 0, \frac{\partial \Delta}{\partial b} = 0, \frac{\partial \Delta}{\partial c} = 0, \frac{\partial \Delta}{\partial X_0} = 0, \frac{\partial \Delta}{\partial Y_0} = 0, \frac{\partial \Delta}{\partial Z_0} = 0, \\ & \frac{\partial \Delta}{\partial \sigma_1} = 0, \frac{\partial \Delta}{\partial \sigma_2} = 0, \frac{\partial \Delta}{\partial \sigma_3} = 0. \end{aligned} \tag{17.49}$$

Considering the necessary conditions for the minimum, we have 9 equations and 9 variables. However, in this case the determined system are of a higher order and more complex polynomial system than the original overdetermined one, leading to most of these equations having many thousands of terms. It is therefore useful to collect terms corresponding to the same multivariate expression via computer algebra, see Zaletnyik [430]. Here, just as an illustration, let us see the first equation resulting from the derivation of the objective function with respect to the variable a (the whole system $(F_a, F_b, \ldots, F_{\sigma_3})$ can be found in Appendix A-4),

$$\begin{aligned}
&F_a\left(a,b,c,\sigma_1,\sigma_2,\sigma_3,X_0,Y_0,Z_0\right)=\frac{\partial\Delta}{\partial a}=\\
&-bNX_0Y_0\sigma_1\sigma_2+aNY_0^2\sigma_2^2-cNX_0Z_0\sigma_1\sigma_3+aNZ_0^2\sigma_3^2+bY_0\sigma_1\sigma_2\sum_{i=1}^N x_i\\
&+cZ_0\sigma_1\sigma_3\sum_{i=1}^N x_i+bY_0\sigma_2\sum_{i=1}^N X_i+cZ_0\sigma_3\sum_{i=1}^N X_i+aY_0\sigma_2^2\sum_{i=1}^N(-2y_i)\\
&+bX_0\sigma_1\sigma_2\sum_{i=1}^N y_i+b\sigma_1\sigma_2\sum_{i=1}^N(-x_iy_i)+b\sigma_2\sum_{i=1}^N(-X_iy_i)+a\sigma_2^2\sum_{i=1}^N y_i^2\\
&+aY_0\sigma_2\sum_{i=1}^N(-2Y_i)+bX_0\sigma_1\sum_{i=1}^N Y_i+Z_0\sigma_3\sum_{i=1}^N 2Y_i+b\sigma_1\sum_{i=1}^N(-x_iY_i)\\
&+b\sum_{i=1}^N(-X_iY_i)+a\sigma_2\sum_{i=1}^N 2y_iY_i+aZ_0\sigma_3^2\sum_{i=1}^N(-2z_i)+cX_0\sigma_1\sigma_3\sum_{i=1}^N z_i\\
&+c\sigma_1\sigma_3\sum_{i=1}^N(-x_iz_i)+c\sigma_3\sum_{i=1}^N(-X_iz_i)+\sigma_3\sum_{i=1}^N(-2Y_iz_i)+a\sigma_3^2\sum_{i=1}^N z_i^2\\
&+Y_0\sigma_2\sum_{i=1}^N(-2Z_i)+aZ_0\sigma_3\sum_{i=1}^N(-2Z_i)+cX_0\sigma_1\sum_{i=1}^N Z_i+c\sigma_1\sum_{i=1}^N(-x_iZ_i)\\
&+c\sum_{i=1}^N(-X_iZ_i)+\sigma_2\sum_{i=1}^N 2y_iZ_i+a\sigma_3\sum_{i=1}^N 2z_iZ_i+a\sum_{i=1}^N\left(Y_i^2+Z_i^2\right)=0.
\end{aligned} \tag{17.50}$$

In this way we get 9 polynomial equations (see also [323, 430]) with the 9 unknown parameters $(a, b, c, X_0, Y_0, Z_0, \sigma_1, \sigma_2, \sigma_3)$.

17-422 Homotopy solution of the ALESS-determined model

Different possibilities exist for solving the determined system created using ALESS method. One can use *global methods* to find all of the solutions of the *determined system*, and then select only the good solution which provides the least value of the *objective function* of ALESS. Another approach is to use *local methods*, like the *extended Newton-Raphson* method or Newton type homotopy, in the case of good initial values (see Chap. 6).

To solve such a complicated system as the 9-equation system created by ALESS in symbolic form is a difficult problem which, until now, has had no solution. The global homotopy solution, calculating automatically the start systems is also problematic. The highest order term in every equation is 5, therefore the degree of every equation is $d_i = 5, \quad i = 1, \ldots, 9$. This implies that the upper bound of the number of the solutions of this system is $5^9 = 1953125$, which means that to use the homotopy solution with the automatically calculated start system for polynomial systems, we would need to track nearly 2 millions paths[4].

[4] *Remark:* Tracking millions of paths is not unrealistic on supercomputers, clusters of workstations or even modern multiprocessor, multi-core desktop computers, see Blum et al. [84].

We have seen in Sect. 6-42, that one of the most simple ways to define a start system is by employing Newton-type homotopy, see Eq. (6.26). This type of homotopy requires a guess value for the solution of the original (target) system. Now seemingly, we have arrived back to our original problem of the missing proper initial value for *local methods* like Newton-Raphson. However, the situation is not so bad. On the one hand, we do not need a proper initial value, because homotopy is much more robust than local methods, and enlarges the domain of convergence. On the other hand, there are natural ways to compute the initial value, which is good enough for homotopy.

One possibility is to use the result of the symbolic solution of the 3-point problem. As we have seen in the previous section, the values calculated from the different triplets can differ significantly, depending on their geometrical configurations. A solution for one of the geometrically ill-posed triplets is illustrated in *Table* 17.12. In this Table, we represent the calculated parameters in their traditional geodetic forms, i.e., representing the rotation matrix with the three rotation angles (α, β, γ) in seconds, instead of a, b, c, and the deviations of the scale parameters from one ($k_i = s_i - 1 = 1/\sigma_i - 1$), in ppm (part per million)), instead of the inverse scale parameters $(\sigma_1, \sigma_2, \sigma_3)$.

Table 17.12. The start values computed from a symbolical 3-points solution

Variables	Start values	in geodetic form
$\tilde{a}$	−0.00002	$\tilde{\alpha} = +8.2508''$
$\tilde{b}$	0.00002	$\tilde{\beta} = -8.2504''$
$\tilde{c}$	0.00002	$\tilde{\gamma} = -8.2508''$
$\tilde{X}_0$	−243	$\tilde{X}_0 = -243$ m
$\tilde{Y}_0$	−227	$\tilde{Y}_0 = -227$ m
$\tilde{Z}_0$	337	$\tilde{Z}_0 = +337$ m
$\tilde{\sigma}_1$	1	$\tilde{k}_1 = 0$ ppm
$\tilde{\sigma}_2$	1	$\tilde{k}_2 = 0$ ppm
$\tilde{\sigma}_3$	1	$\tilde{k}_3 = 0$ ppm

Another way to calculate a guess value is to delete the nonlinear terms from the original overdetermined system, e.g., from Eqs. (17.46).

$$
\begin{aligned}
Lvx_i &= -X_i + cY_i - bZ_i + x_i\sigma_1 \\
Lvy_i &= -cX_i - Y_i + aZ_i + y_i\sigma_2 \\
Lvz_i &= bX_i - aY_i - Z_i + z_i\sigma_3
\end{aligned}
\qquad (17.51)
$$

The least square solution of this linear system $i = 1, \ldots, 1138$ can be computed via pseudoinverse with 1138 Hungarian data points. The result is presented in *Table* 17.13 which indicates that the two different methods gave fairly different values and both are far from the desired solution, see e.g., *Table* 17.14.

Employing Newton homotopy, the start system is,

Table 17.13. The start values computed from Eqs.(17.51) via pseudoinverse

Variables	Start values	in geodetic form
$\tilde{a}$	$-5.14934436480436 \cdot 10^{-6}$	$\tilde{\alpha} = +2.1244$"
$\tilde{b}$	$9.147743927965785 \cdot 10^{-6}$	$\tilde{\beta} = -3.7736$"
$\tilde{c}$	0.0000472128892111371	$\tilde{\gamma} = -19.4767$"
$\tilde{X}_0$	$3.78794980367417 \cdot 10^{-17}$	$\tilde{X}_0 = 0.000$ m
$\tilde{Y}_0$	0	$\tilde{Y}_0 = 0.000$ m
$\tilde{Z}_0$	0	$\tilde{Z}_0 = 0.000$ m
$\tilde{\sigma}_1$	1.0000011141420198	$\tilde{k}_1 = -1.114$ ppm
$\tilde{\sigma}_2$	1.0001180993441208	$\tilde{k}_2 = -118.085$ ppm
$\tilde{\sigma}_3$	0.9999810252524541	$\tilde{k}_3 = +18.975$ ppm

$$G(\chi) = F(\chi) - F(\chi_0) \tag{17.52}$$

with $F = (F_a(\chi), F_b(\chi), F_c(\chi), F_{X_0}(\chi), F_{Y_0}(\chi), F_{Z_0}(\chi), F_{\sigma_1}(\chi), F_{\sigma_2}(\chi), F_{\sigma_3}(\chi))$ as the target system, $\chi = (a, b, c, X_0, Y_0, Z_0, \sigma_1, \sigma_2, \sigma_3)$ as the unknown variables and $\chi_0 = (\tilde{a}, \tilde{b}, \tilde{c}, \tilde{X}_0, \tilde{Y}_0, \tilde{Z}_0, \tilde{\sigma}_1, \tilde{\sigma}_2, \tilde{\sigma}_3)$ as the start (initial) values for the homotopy function. The homotopy function is then given by (see, e.g., Eq. (6.28).

$$H(\chi, \lambda) = F(\chi) - (1 - \lambda)F(\chi_0). \tag{17.53}$$

The solution for all parameters are presented in *Table* 17.14. The Newton homotopy solution was successful with both initial values. As an illustration, Fig. 17.6 shows the paths of the homotopy solution in the complex plain for the parameter a using the two different initial values.

Table 17.14. Homotopy solution for the 1138 Hungarian points

Variables	Homotopy solutions	in geodetic form
a	$-8.389143102087336 \cdot 10^{-6}$	$\alpha = +3.4591$"
b	0.00011415902600771965	$\beta = -47.094$"
c	-0.00003482245241765358	$\gamma = +14.3649$"
X_0	-2298.5887892237693	$X_0 = -2298.589$ m
Y_0	526.5088259777841	$Y_0 = +526.509$ m
Z_0	2143.7159886648537	$Z_0 = +2143.716$ m
σ_1	0.9997302444546577	$k_1 = +269.828$ ppm
σ_2	1.000188709705781	$k_2 = -188.674$ ppm
σ_3	1.000245242648286	$k_3 = -245.183$ ppm

Table 17.15 shows the results of the different methods in case of the different initial guess values. It can be seen that the traditional *Newton-Raphson method* failed in both cases, while its modification, the *Newton-Krylov method*, had a singularity in the first case, and was slowed down by the increasing number of iterations

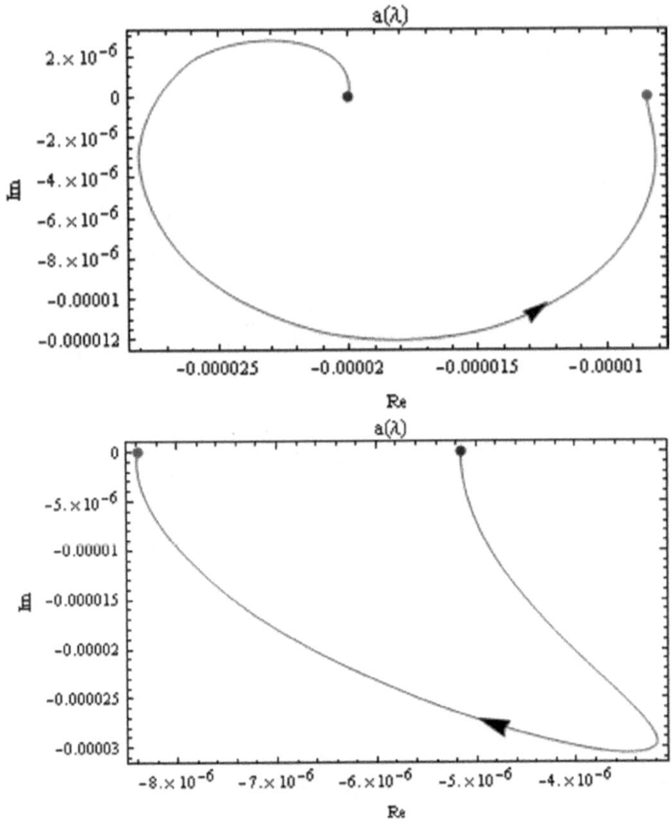

Figure 17.6. Paths of the homotopy solution in the complex plain for the parameter a, with two different initial values

required to ensure acceptable precision in the second case. In this example, homotopy solution seemed to be more robust as well as faster than the traditional Newton's type methods. Fig. 17.6 demonstrates clearly how large the domain of convergency of the homotopy method is.

Table 17.15. Comparing different methods for the case of 1138 Hungarian points for 3D affine transformation

Method	Computation time [s] with initial values in	
	Table 17.12	Table 17.13
Newton-Raphson	fails to converge	converging to a wrong solution
Newton-Krylov	singularity	3.5
Newton-Homotopy	0.72	0.73

17-43 Procrustes solution

Example 17.3 (Computation of 7-parameter transformation incorporating weights). We consider Cartesian coordinates of seven stations given in the local and global system (WGS-84) as in Tables 17.1 and 17.2 on pp. 309. Desired are the 7-datum transformation parameters; scale x_1, the translation vector $\mathbf{x}_2 \in \mathbb{R}^{3\times 1}$ and the rotation matrix $\mathbf{R} \in \mathbb{R}^{3\times 3}$. In addition to these 7 datum transformation parameters, we compute for control purposes the residual (error matrix) $\mathbf{E}$ upon which the mean error norm (9.55) is determined as a scalar measure of error of types $\mathbf{W}$-LESS. A two step procedure is carried out as follows:

In the first step, we computed the 7 transformation parameters using I-LESS (with weight matrix as identity) from Corollary 9.4 on p. 127. The computed values of scale x_1 and the rotation matrix $\mathbf{R} \in \mathbb{R}^{3\times 3}$ are used in (9.94) to obtain the dispersion of the error matrix $\mathbf{E}$. In-order to obtain the dispersions $\boldsymbol{\Sigma}_{vec\mathbf{Y}_1^*}$ and $\boldsymbol{\Sigma}_{vec\mathbf{Y}_1^*}$ of the pseudo-observations in the local and global systems respectively, we make use of "positional error sphere" for each point (position) in both systems. Here, the positional error sphere refers to the average of the variances ($\sigma_i^2 = \sqrt{(\sigma_x^2+\sigma_y^2+\sigma_z^2)/3}$) for the $i = 7$ points involved so as to achieve the isotropic condition.

The identity matrices are multiplied by these positional error spheres so as to obtain the dispersion matrices $\boldsymbol{\Sigma}_{vec\mathbf{Y}_1^*}$ and $\boldsymbol{\Sigma}_{vec\mathbf{Y}_2^*}$ which fulfill the isotropic condition. One obtains therefore the dispersion matrices $\boldsymbol{\Sigma}_{vec\mathbf{Y}_1^*}$ and $\boldsymbol{\Sigma}_{vec\mathbf{Y}_2^*}$ as being diagonal block matrices with each block corresponding to the variance-covariance matrices of the respective position. For points 1 and 2 in the local system for instance, assuming no correlation between the two points, one obtains

$$\boldsymbol{\Sigma}_{vec\mathbf{Y}_1^*} = \begin{bmatrix} \sigma_1^2 & & & & & \\ & \sigma_1^2 & & & & \\ & & \sigma_1^2 & & & \\ & & & \sigma_2^2 & & \\ & & & & \sigma_2^2 & \\ & & & & & \sigma_2^2 \end{bmatrix}, \tag{17.54}$$

where $\{\sigma_1^2, \sigma_2^2\}$ are positional error spheres for points 1 and 2, respectively. This is also performed for $\boldsymbol{\Sigma}_{vec\mathbf{Y}_2^*}$ and the resulting dispersion matrices used in (9.94) to obtain the dispersion matrix of the error matrix $\mathbf{E}$.

Since the obtained block diagonal error matrix $\mathbf{E}$ is a $3n \times 3n$ matrix, the $n \times n$ matrix is extracted by taking the trace of the block diagonal matrices of $\mathbf{E}$. Adopting such a matrix from [169, Table 7] as

$$\mathbf{W} = \begin{bmatrix} 1.8110817 & 0 & 0 & 0 & 0 & 0 & 0 \\ 0 & 2.1843373 & 0 & 0 & 0 & 0 & 0 \\ 0 & 0 & 2.1145291 & 0 & 0 & 0 & 0 \\ 0 & 0 & 0 & 1.9918578 & 0 & 0 & 0 \\ 0 & 0 & 0 & 0 & 2.6288452 & 0 & 0 \\ 0 & 0 & 0 & 0 & 0 & 2.1642460 & 0 \\ 0 & 0 & 0 & 0 & 0 & 0 & 2.359370 \end{bmatrix}. \tag{17.55}$$

One proceeds via the general Procrustes algorithm on *Solution* 9.3 in p. 128 to compute the transformation parameters which are presented in *Tables* 17.16 and 17.17. In these tables, the results of the **I**-LESS (step 1) and **W**-LESS (step 2) Procrustes transformations are presented, namely the 7-datum transformation parameters; the scale, rotation matrix and the translation parameters. We also present the residual (error) matrix and the norms of the error matrices. The computed residuals can be compared with those of linearized least squares procedure in *Table* 17.9 on p. 316.

Table 17.16. Results of the I-LESS Procrustes transformation

	Values			
Rotation Matrix $\mathbf{X}_3 \in \mathbb{R}^{3\times3}$	1.00000	-4.33276e-6	4.81463e-6	
	-4.81465e-6	1.00000	-4.84085e-6	
	4.33274e-6	4.84087e-6	1.00000	
Translation $\mathbf{x}_2 \in \mathbb{R}^{3\times1}(m)$	641.8804			
	68.6553			
	416.3982			
Scale $x_1 \in \mathbb{R}$	1.00000558251985			
Residual matrix $\mathbf{E}(m)$	Site	$X(m)$	$Y(m)$	$Z(m)$
	Solitude	0.0940	0.1351	0.1402
	Buoch Zeil	0.0588	-0.0497	0.0137
	Hohenneuffen	-0.0399	-0.0879	-0.0081
	Kuelenberg	0.0202	-0.0220	-0.0874
	Ex Mergelaec	-0.0919	0.0139	-0.0055
	Ex Hof Asperg	-0.0118	0.0065	-0.0546
	Ex Keisersbach	-0.0294	0.0041	0.0017
Error matrix norm (m) $\lvert\lvert\lvert E_l \rvert\rvert\rvert_{\mathbf{W}} := \sqrt{tr(\mathbf{E}_l^*\mathbf{E}_l)}$	0.2890			
Mean error matrix norm (m) $\lvert\lvert\lvert E_l \rvert\rvert\rvert_{\mathbf{W}} := \sqrt{tr(\mathbf{E}_l^*\mathbf{E}_l)/3n}$	0.0631			

Table 17.17. Results of the W-LESS Procrustes transformation

	Values			
Rotation Matrix $\mathbf{X}_3 \in \mathbb{R}^{3\times3}$	1.00000	4.77976e-6	-4.34410e-6	
	-4.77978e-6	1.00000	-4.83730e-6	
	4.34408e-6	4.83731e-6	1.00000	
Translation $\mathbf{x}_2 \in \mathbb{R}^{3\times1}(m)$	641.8377			
	68.4743			
	416.2159			
Scale $x_1 \in \mathbb{R}$	1.00000561120732			
Residual matrix $\mathbf{E}(m)$	Site	$X(m)$	$Y(m)$	$Z(m)$
	Solitude	0.0948	0.1352	0.1407
	Buoch Zeil	0.0608	-0.0500	0.0143
	Hohenneuffen	-0.0388	-0.0891	-0.0072
	Kuelenberg	0.0195	-0.0219	-0.0868
	Ex Mergelaec	-0.0900	0.0144	-0.0052
	Ex Hof Asperg	-0.0105	0.0067	-0.0542
	Ex Keisersbach	-0.0266	0.0036	0.0022
Error matrix norm (m) $\lvert\lvert\lvert E_l \rvert\rvert\rvert_{\mathbf{W}} := \sqrt{tr(\mathbf{E}_l^*\mathbf{W}\mathbf{E}_l)}$	0.4268			
Mean error matrix norm (m) $\lvert\lvert\lvert E_l \rvert\rvert\rvert_{\mathbf{W}} := \sqrt{tr(\mathbf{E}_l^*\mathbf{W}\mathbf{E}_l)/3n}$	0.0930			

Example 17.4 (Computation of 9-parameter transformation for 82-common known stations in both Australian Geodetic Datum (AGD 84) and Geocentric Datum Australia (GDA 94) using the ABC algorithm (see Sect. 9-51 on p. 9-51)). Awange et al. [42] computed 9-parameter transformation for Western Australia based on 82 stations common in both AGD 84 and GDA 94 (Fig. 17.7) using the ABC algorithm presented in Sect. 9-51. The AGD is defined by the ellipsoid with a semimajor axis of 6378160 m and a flattening of 0.00335289, while the GDA is defined by an ellipsoid of semi-major axis 6378137 m and a flattening of 0.00335281 (see, e.g., Kinneen and Featherstone [240]). The results of the 9-parameter solution were found to give a marginal 1.4% improvement when compared to the results obtained by the 7-parameter transformation. The results of the transformation parameters are given in *Table* 17.18.

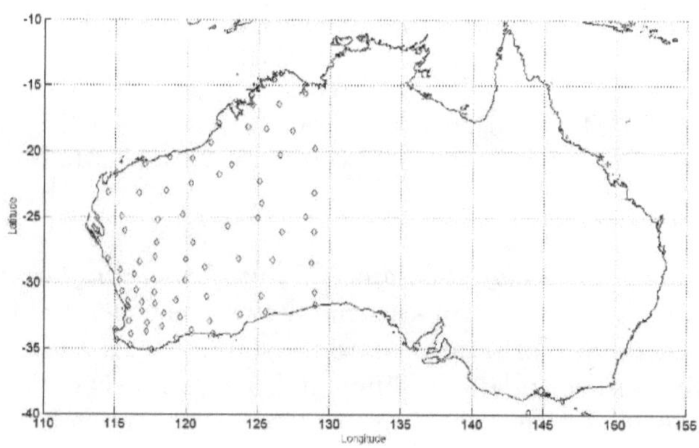

Figure 17.7. Locations of the 82 stations in WA, Australia (AGD84 and GDA94)-Source: Awange et al. [42]

Example 17.5 (Computation of 9-parameter transformation for a network with mild *anisotropy using the PZ algorith (see Sect. 9-52, p. 131)).* As pointed out in Sect. 9-52, the ABC algorithm used in the example above worlks well only when these scale factors do not differ from each other considerably. In this example and the next, we apply the PZ-algorthm to cure this problem. Considering the transformation problem of the AGD 84 to GDA 94 in *Example* 17.4, in addition to the ABC-algorithm, the problem is solved using (see results in *Table* 17.18):

1. Direct numerical solution via Global Minimization (GM). This approach solves a system of 246 equations for the unknown parameters a, b, c, s, X_0, Y_0, Z_0. The objective function can be constructed by considering the least square residuals as objective function (see, e.g., Paláncz et al. 2008). To carry out global minimization via genetic algorithm, the built -in function **NMinimize** in Mathematica is employed.

2. Application of the Helmert transformation (i.e., 7-parameter transformation) with original Procrustes approach.
3. Application of PZ method.

Table 17.18. Results of the different methods in case of network with mild anisotropy

Method	Time (sec)	Error (m)	Scale s_1, s_2, s_3	Translation (m)
Procrustes	0.062	6.867	1.00000368981	−115.838 −48.373 144.760
ABC	0.282	6.788	1.00000396085 1.00000354605 1.00000345982	−115.062 −47.676 144.096
PZ Method	0.687	6.642	1.00000416838 1.00000310067 1.00000346938	−112.169 −44.047 144.311
GM	3.562	6.642	1.00000416842 1.00000310066 1.00000346943	−112.169 −44.046 144.312

Example 17.6 (Computation of 9-parameter transformation for a network with strong *anisotropy).* For this test an artificial network is generated from Hungarian Datum points. The angles of rotations are not small values and the values of the 3 scale parameters are considerably different from each other. We have $N = 81$ points in both systems. Using the four approaches above leads to the results of Table 17.19.

Table 17.19. Results of the different methods in case of network with strong anisotropy

Method	Time (sec)	Error (m)	Scale s_1, s_2, s_3	Translation (m)
Procrustes	0.047	117285.16	1.2335164851	$-4.826 * 10^6$ 253120.162 $6.356 * 10^6$
ABC	0.281	93431.630	1.099389091087 1.287089145364 1.136301910900	$-4.074 * 10^6$ 219870.673 $6.642 * 10^6$
PZ Method	0.906	1.3581	0.6200004792118 1.3000001184524 1.86999586512189	1339.036 −236.061 152.711
GM	4.625	1.23799	0.62000047769 1.30000011779 1.86999590054	1339.072 −236.047 152.461

Example 17.7 (Testing different initial values for the PZ algorithm). In this example, we are interested in testing the starting values of the proposed PZ-Method. We consider four scenarios: (i) Using identity matrix $\mathbf{I}_3$ as starting values in which case ($s_{O1} = 1, s_{O2} = 1, s_{O2} = 1$); (ii) the PZ Approximation (9.84); (iii) the ABC-method without iteration (i.e., the results of the first run); and (iv) The ABC-method with the solutions after iterations. The results are presented in Table 17.20.

Table 17.20. Running time of PZ method with initial values (s_1, s_2, s_3) estimated with different methods in case of network with strong anisotropy

Time (sec)	s_1	s_2	s_3	Method
1.11	1	1	1	-
0.86	0.508673877	0.711886073089	2.1502674697508	PZ Method
0.70	0.538035777	1.290521383334	1.6274715796179	ABC without iteration
1.00	1.099389091	1.287089145364	1.1363019109003	ABC solution

17-5 Concluding remark

This chapter has illustrated how the algebraic technique of Groebner basis explicitly solves the nonlinear 7-parameter datum transformation equations once they have been converted into algebraic (polynomial) form. In particular, the algebraic tool of Groebner basis provides symbolic solutions; showing the scale parameter to fulfill a quartic polynomial and the rotation parameters are given by linear functions in scale. It has also been demonstrated how overdetermined versions of the problem can be solved using the Gauss-Jacobi combinatorial algorithm and the general Procrustes algorithm. Both approaches incorporate the stochasticity of both systems involved in the transformation problem.

Although computationally intensive, the Gauss-Jacobi combinatorial algorithm solves the weighted transformation problem without any assumptions. The general Procrustes algorithm functions well with the isotropic assumptions, i.e., all three coordinates $\{X_i, Y_i, Z_i\}$ of a point i are given the same weight. The weights are further assumed to be inhomogeneous, i.e., the weights of a point i differ from those of point j. Both of these assumptions are ideal and may not necessarily hold in practice.

The chapter further gave a symbolic-numeric solution to compute the parameters of a 3D affine transformation model. In the case of the 3-point problem, a fully symbolic solution was given to solve the 3D affine datum transformation problem. We demonstrated that elimination techniques, namely the enhanced and classical Dixon resultant and the reduced Groebner bases, can be used in the elimination process to determine the parameters. The symbolic reduction can be carried out off line, therefore it does not influence the computing time. This feature can be very useful, when one solves the 3 points problem many times, for example in the case of the Gauss-Jacobi solution of N-points problem. However, Gauss-Jacobi combinatorial solutions can be used only for cases where the number of known points is small. The symbolically calculated parameters for the 3-points problem can also be used in the case of the N-points problem, as initial values. Criteria for selecting an appropriate triplet from data points for initial values are also given.

The N-points problem is solved by a symbolic-numeric algorithm. *First* the *overdetermined system* is transformed into a *determined system* using the ALESS method symbolically via computer algebra, then the solution of the determined system is solved by Newton-type homotopy using different initial values. This method is *fast*, *robust* and has a *very low complexity*, according to its independence from the number of equations in the original *overdetermined system*. The homotopy solution can enlarge the convergence region and provide solution regardless of initial values, when local methods like standard Newton-Raphson fail.

For the 9-parameter Procrustes solution, in case of mild anisotropy of the network, the ABC method gives better approximation than the general Procrustes method employing Helmert transformation model, while, the PZ method provides precise, geometrically correct solution. The ABC method is about 2 times faster than the PZ method, and the later is roughly 5 times faster than the global optimization method applying to 3D affine model as we saw in Table 17.18. In case of strong anisotropy, use of the ABC method to solve 3D affine transformation completely fails, while PZ method provides correct solutions. However, even in that case, ABC method is still useful for computing proper initial values for PZ method in order to increase its efficiency as demonstrated in Table 17.20.

The subject of transformation in general is still an active area of research as evident in the works of [2, 3, 6, 14, 16, 23, 26, 37, 38, 51, 57, 60, 72, 82, 131, 168, 169, 174, 184, 185, 186, 189, 190, 202, 213, 238, 248, 251, 312, 313, 315, 380, 391, 406, 417, 425].

Appendix

Appendix A-1: Definitions

To enhance the understanding of the theory of Groebner bases presented in Chap. 4, the following definitions supplemented with examples are presented.

Three commonly used monomial ordering systems are; *lexicographic ordering, graded lexicographic ordering* and *graded reverse lexicographic ordering.* First we define the monomial ordering before considering the three types.

Definition .1 (Monomial ordering). *A monomial ordering on* $k[x_1, \ldots, x_n]$ *is any relation* $>$ *on* $\mathbb{Z}^n_{\geq 0}$ *or equivalently any relation on the set* x^α, $\alpha \in \mathbb{Z}^n_{\geq 0}$ *satisfying the following conditions:*
(a) is total (or linear) ordering on $\mathbb{Z}^n_{\geq 0}$
(b) If $\alpha > \beta$ *and* $\gamma \in \mathbb{Z}^n_{\geq 0}$*, then* $\alpha + \gamma > \beta + \gamma$
(c)$<$ *is a well ordering on* $\mathbb{Z}^n_{\geq 0}$.
This condition is satisfied if and only if every strictly decreasing sequence in $\mathbb{Z}^n_{\geq 0}$ *eventually terminates.*

Definition .2 (Lexicographic ordering). *This is akin to the ordering of words used in dictionaries. If we define a polynomial in three variables as* $P = k[x, y, z]$ *and specify an ordering* $x > y > z$*, i.e.,* x *comes before* y *and* y *comes before* z*, then any term with* x *will supersede that of* y *which in tern supersedes that of* z*. If the powers of the variables for respective monomials are given as* $\alpha = (\alpha_1, \ldots, \alpha_n)$ *and* $\beta = (\beta_1, \ldots, \beta_n), \alpha, \beta \in \mathbb{Z}^n_{\geq 0}$*, then* $\alpha >_{lex} \beta$ *if in the vector difference* $\alpha - \beta \in \mathbb{Z}^n$*, the most left non-zero entry is positive. For the same variable (e.g.,* x*) this subsequently means* $x^\alpha >_{lex} x^\beta$.

Example .1. $x > y^5 z^9$ is an example of lexicographic ordering. As a second example, consider the polynomial $f = 2x^2y^8 - 3x^5yz^4 + xyz^3 - xy^4$, we have the *lexicographic order*; $f = -3x^5yz^4 + 2x^2y^8 - xy^4 + xyz^3 \,|x > y > z$.

Definition .3 (Graded lexicographic ordering). *In this case, the total degree of the monomials is taken into account. First, one considers which monomial has the highest total degree before looking at the lexicographic ordering. This ordering looks at the left most (or largest) variable of a monomial and favours the largest power. Let* $\alpha, \beta \in \mathbb{Z}^n_{\geq 0}$*, then* $\alpha >_{grlex} \beta$ *if* $|\alpha| = \sum\limits_{i=1}^{n} \alpha_i > |\beta| = \sum\limits_{i=1}^{n} \beta_i$ or $|\alpha| = |\beta|$, *and* $\alpha >_{lex} \beta$*, in* $\alpha - \beta \in \mathbb{Z}^n$*, the most left non zero entry is positive.*

Example .2. $x^8y^3z^2 >_{grlex} x^6y^2z^3 \,|(8,3,2) >_{grlex} (6,2,3)$, since $|(8,3,2)| = 13 > |(6,2,3)| = 11$ and $\alpha - \beta = (2,1,-1)$. Since the left most term of the difference (2) is positive, the ordering is graded lexicographic. As a second example, consider the polynomial $f = 2x^2y^8 - 3x^5yz^4 + xyz^3 - xy^4$, we have the graded lexicographic order; $f = -3x^5yz^4 + 2x^2y^8 - xy^4 + xyz^3 \,|x > y > z$.

J.L. Awange et al., *Algebraic Geodesy and Geoinformatics*, 2nd ed.,
DOI 10.1007/978-3-642-12124-1,

Definition .4 (Graded reverse lexicographic ordering). *In this case, the total degree of the monomials is taken into account as in the case of graded lexicographic ordering. First, one considers which monomial has the highest total degree before looking at the lexicographic ordering. In contrast to the graded lexicographic ordering, one looks at the right most (or largest) variable of a monomial and favours the smallest power. Let* $\alpha, \beta \in \mathbb{Z}^n_{\geq 0}$*, then* $\alpha >_{grevlex} \beta$ *if* $|\alpha| = \sum_{i=1}^{n} \alpha_i > |\beta| = \sum_{i=1}^{n} \beta_i$ or $|\alpha| = |\beta|$*, and* $\alpha >_{grevlex} \beta$*, and in* $\alpha - \beta \in \mathbb{Z}^n$ *the right most non zero entry is negative.*

Example .3. $x^8y^3z^2 >_{grevlex} x^6y^2z^3 \,|(8,3,2) >_{grevlex} (6,2,3)$ since $|(8,3,2)| = 13 > |(6,2,3)| = 11$ and $\alpha - \beta = (2,1,-1)$. Since the right most term of the difference (-1) is negative, the ordering is *graded reverse lexicographic*. As a second example, consider the polynomial $f = 2x^2y^8 - 3x^5yz^4 + xyz^3 - xy^4$, we have the *graded reverse lexicographic order*: $f = 2x^2y^8 - 3x^5yz^4 - xy^4 + xyz^3 \,|x > y > z$.

If we consider a non-zero polynomial $f = \sum_\alpha a_\alpha x^\alpha$ in $k\,[x_1, \ldots, x_n]$ and fix the monomial order, the following additional terms can be defined:

Definition .5.
Multidegree of f *: Multideg (* f *)*$=\max(\alpha \in \mathbb{Z}^n_{\geq 0} \,|a_\alpha \neq 0)$
Leading Coefficient of f *: LC (* f *)*$=a_{\text{multideg}(f)} \in k$
Leading Monomial of f *: LM (* f *)*$=x^{\text{multideg}(f)}$ *(with coefficient 1)*
Leading Term of f *: LT (* f *)=LC (* f *) LM (* f *)*

Example .4. Consider the polynomial $f = 2x^2y^8 - 3x^5yz^4 + xyz^3 - xy^4$ with respect to lexicographic order $\{x > y > z\}$, we have
Multideg (f)=(5,1,4)
LC (f)= -3
LM (f)= x^5yz^4
LT (f)= $-3x^5yz^4$

The definitions of polynomial ordering above have been adopted from [117, pp. 52–58].

Appendix A-2: C. F. Gauss combinatorial formulation

CARL FRIEDRICH GAUSS

WERKE

NEUNTER BAND.

HERAUSGEGEBEN

VON DER

KÖNIGLICHEN GESELLSCHAFT DER WISSENSCHAFTEN

ZU

GÖTTINGEN.

IN COMMISSION BEI B. G. TEUBNER IN LEIPZIG.

1903.

BESTIMMUNG

DES

BREITENUNTERSCHIEDES

ZWISCHEN DEN

STERNWARTEN VON GÖTTINGEN UND ALTONA

DURCH

BEOBACHTUNGEN AM RAMSDENSCHEN ZENITHSECTOR

VON

CARL FRIEDRICH GAUSS,

RITTER DES GUELPHEN- UND DANNEBROG-ORDENS; K. GROSSBR. HANNOVERSCHEM HOFRATH; PROFESSOR DER ASTRONOMIE UND DIRECTOR DER STERNWARTE IN GÖTTINGEN; MITGLIED DER AKADEMIEN UND SOCIETÄTEN VON BERLIN, COPENHAGEN, EDINBURG, GÖTTINGEN, LONDON, MÜNCHEN, NEAPEL, PARIS, PETERSBURG, STOCKHOLM, DER AMERIKANISCHEN, ITALIENISCHEN, KURLÄNDISCHEN, LONDONER ASTRONOMISCHEN U. A.

GÖTTINGEN,

BEI VANDENHOECK UND RUPRECHT.

1828.

NACHLASS.

[1.]

Endresultat für den Ort eines Punktes in einer Ebene, der von drei bekannten aus angeschnitten ist.

Es bedeuten 10, 20, 30 die drei beobachteten Richtungen [nach P] und α, β, γ die entsprechenden Entfernungen.

Die drei einzelnen Resultate aus den Combinationen 2—3, 1—3, 1—2 seien A, B, C, zugleich die Winkel des durch jene gebildeten Dreiecks; die ihnen gegenüber stehenden Seiten a, b, c.

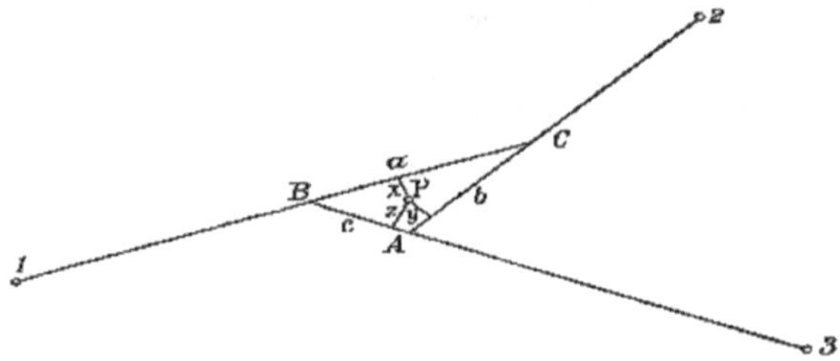

Perpendikel von dem gesuchten Orte auf a, b, c seien x, y, z. S doppelter Flächeninhalt des Dreiecks.

Es sind dann

$$\frac{x}{\alpha}, \quad \frac{y}{\beta}, \quad \frac{z}{\gamma}$$

die übrig bleibenden Fehler, also

$$\frac{xx}{\alpha\alpha} + \frac{yy}{\beta\beta} + \frac{zz}{\gamma\gamma} \quad \text{Minimum}$$

und

$$ax + by + cz = S.$$

Also werden x, y, z proportional den Grössen $\alpha\alpha a$, $\beta\beta b$, $\gamma\gamma c$;

$$x = \frac{\alpha\alpha a S}{\alpha\alpha aa + \beta\beta bb + \gamma\gamma cc},$$
etc.

[Bezeichnet (ABC) die Fläche des Dreiecks ABC, u. s. f., so ist

$$S = 2(ABC) = (\alpha\alpha aa + \beta\beta bb + \gamma\gamma cc)k$$
$$2(BPC) = \alpha\alpha aak$$
$$2(APC) = \beta\beta bbk$$
$$2(APB) = \gamma\gamma cck,$$

wo k die Correlate der Bedingungsgleichung ist. P ist der durch die Perpendikel x, y, z bestimmte Punkt.

Folglich wird, wenn A, B, C, P die complexen Grössen bedeuten, denen die Eckpunkte des Dreiecks ABC und der Punkt P entsprechen:

$$(\alpha\alpha aa + \beta\beta bb + \gamma\gamma cc)P = \alpha\alpha aaA + \beta\beta bbB + \gamma\gamma ccC.]$$

Es folgt hieraus, dass das Endresultat*)

$$\frac{\alpha\alpha aaA + \beta\beta bbB + \gamma\gamma ccC}{\alpha\alpha aa + \beta\beta bb + \gamma\gamma cc}$$

lso ein Mittel aus den drei partiellen Resultaten A, B, C ist, indem man iesen die Gewichte

$$\alpha\alpha aa, \quad \beta\beta bb, \quad \gamma\gamma cc$$

eilegt, oder

$$\alpha\alpha \sin A^2, \quad \beta\beta \sin B^2, \quad \gamma\gamma \sin C^2.$$

Offenbar ist hier A zugleich der Winkel zwischen 20 und 30, u. s. f.

Appendix A-3: Linear homotopy

In *Mathematica* one can solve the system (6.31)-(6.32) in the following way:
1) Preprocessing:
the equations:
$\boldsymbol{f_1[x\_, y\_] = x^2 + y^2 - 1}$
$\boldsymbol{f_2[x\_, y\_] = x^3 + y^3 - 1}$
the system
$\boldsymbol{F = \{f_1[x, y], f_2[x, y]\}}$
the variables: **X={x,y};**
computing start system and its solutions:
start=StartingSystem[F, X];
the start system:
$\mathbf{G{=}start[[1]]{=}}\ \boldsymbol{\{(0.673116 + 0.739537i)\left((-0.933825 + 0.35773i) + x^2\right),}$
$\boldsymbol{(-0.821746 - 0.569853i)\left((-0.957532 - 0.288325i) + y^3\right)\}}$
the solutions of the start system:
X0=start[[2]]={{0.983317- 0.1819 i, -0.413328 -0.910582 i},
{ 0.983317- 0.1819 i, 0.995251+ 0.0973382 i},
{0.983317- 0.1819 i, -0.581923 + 0.813244 i},
{-0.983317 + 0.1819 i, -0.413328 - 0.910582 i},
{-0.983317 + 0.1819 i, 0.995251+ 0.0973382 i},
{-0.983317 + 0.1819 i, -0.581923 + 0.813244 i}}
2) Processing:
computing homotopy path by solving the corresponding differential equation system Eq.(6.50) with initial values **X0**
γ **={1,1};**
in case the start system is not complex, it contains n random complex numbers
P = 0;
in case of computation with very high precision (20 digits or more)
P = 1
computing homotopy paths:
hpath = LinearHomotopyNDS01[X,F,G,X0,γ**, P]**
the solution of the target system:
Sol=hpath[[1]]={{0,1},{0,1},{1,0},{1,0}, {-1-0.707107 i,-1-0.707107 i},
{-1+0.707107 i,-1+0.707107 i}}
the interpolation functions of the paths, the trajectories of the solutions:
Traject=hpath[[2]]=
={x[λ**]→ InterpolatingFunction[][**λ**], y[**λ**]→ InterpolatingFunction[][**λ**]}**
3) Postprocessing:
these paths can be visualized by function: **Path[X,Traject,X0],** see Fig. 6.8.
More details of the computation and functions applied here can be found in Palancz —citePalancz08.

Appendix A-4: Determined system of the 9-parameter transformation N point problem

$$F_a(a,b,c,\sigma_1,\sigma_2,\sigma_3,X_0,Y_0,Z_0)=\frac{\partial\Delta}{\partial a}=$$

$$-bNX_0Y_0\sigma_1\sigma_2+aNY_0^2\sigma_2^2-cNX_0Z_0\sigma_1\sigma_3+aNZ_0^2\sigma_3^2+bY_0\sigma_1\sigma_2\sum_{i=1}^N x_i$$

$$+cZ_0\sigma_1\sigma_3\sum_{i=1}^N x_i+bY_0\sigma_2\sum_{i=1}^N X_i+cZ_0\sigma_3\sum_{i=1}^N X_i+aY_0\sigma_2^2\sum_{i=1}^N(-2y_i)$$

$$+bX_0\sigma_1\sigma_2\sum_{i=1}^N y_i+b\sigma_1\sigma_2\sum_{i=1}^N(-x_iy_i)+b\sigma_2\sum_{i=1}^N(-X_iy_i)+a\sigma_2^2\sum_{i=1}^N y_i^2$$

$$+aY_0\sigma_2\sum_{i=1}^N(-2Y_i)+bX_0\sigma_1\sum_{i=1}^N Y_i+Z_0\sigma_3\sum_{i=1}^N 2Y_i+b\sigma_1\sum_{i=1}^N(-x_iY_i)$$

$$+b\sum_{i=1}^N(-X_iY_i)+a\sigma_2\sum_{i=1}^N 2y_iY_i+aZ_0\sigma_3^2\sum_{i=1}^N(-2z_i)+cX_0\sigma_1\sigma_3\sum_{i=1}^N z_i$$

$$+c\sigma_1\sigma_3\sum_{i=1}^N(-x_iz_i)+c\sigma_3\sum_{i=1}^N(-X_iz_i)+\sigma_3\sum_{i=1}^N(-2Y_iz_i)+a\sigma_3^2\sum_{i=1}^N z_i^2$$

$$+Y_0\sigma_2\sum_{i=1}^N(-2Z_i)+aZ_0\sigma_3\sum_{i=1}^N(-2Z_i)+cX_0\sigma_1\sum_{i=1}^N Z_i+c\sigma_1\sum_{i=1}^N(-x_iZ_i)$$

$$+c\sum_{i=1}^N(-X_iZ_i)+\sigma_2\sum_{i=1}^N 2y_iZ_i+a\sigma_3\sum_{i=1}^N 2z_iZ_i+a\sum_{i=1}^N\left(Y_i^2+Z_i^2\right)=0.$$

$$F_b(a,b,c,\sigma_1,\sigma_2,\sigma_3,X_0,Y_0,Z_0)=\frac{\partial\Delta}{\partial b}=$$

$$bnX_0^2\sigma_1^2-anX_0Y_0\sigma_1\sigma_2-cnY_0Z_0\sigma_2\sigma_3+bnZ_0^2\sigma_3^2+bX_0\sigma_1^2\sum_{i=1}^n -2x_i+$$

$$aY_0\sigma_1\sigma_2\sum_{i=1}^n x_i+b\sigma_1^2\sum_{i=1}^n x_i^2+bX_0\sigma_1\sum_{i=1}^n -2X_i+Z_0\sigma_3\sum_{i=1}^n -2X_i+$$

$$aY_0\sigma_2\sum_{i=1}^n X_i+b\sigma_1\sum_{i=1}^n 2x_iX_i+aX_0\sigma_1\sigma_2\sum_{i=1}^n y_i+cZ_0\sigma_2\sigma_3\sum_{i=1}^n y_i+$$

$$a\sigma_1\sigma_2\sum_{i=1}^n -x_iy_i+a\sigma_2\sum_{i=1}^n -X_iy_i+aX_0\sigma_1\sum_{i=1}^n Y_i+cZ_0\sigma_3\sum_{i=1}^n Y_i+$$

$$a\sigma_1\sum_{i=1}^n -x_iY_i+a\sum_{i=1}^n -X_iY_i+bZ_0\sigma_3^2\sum_{i=1}^n -2z_i+cY_0\sigma_2\sigma_3\sum_{i=1}^n z_i+$$

$$\sigma_3\sum_{i=1}^n 2X_iz_i+c\sigma_2\sigma_3\sum_{i=1}^n -y_iz_i+c\sigma_3\sum_{i=1}^n -Y_iz_i+b\sigma_3^2\sum_{i=1}^n z_i^2+$$

$$bZ_0\sigma_3\sum_{i=1}^n -2Z_i+cY_0\sigma_2\sum_{i=1}^n Z_i+X_0\sigma_1\sum_{i=1}^n 2Z_i+\sigma_1\sum_{i=1}^n -2x_iZ_i+$$

$$c\sigma_2\sum_{i=1}^n -y_iZ_i+c\sum_{i=1}^n -Y_iZ_i+b\sigma_3\sum_{i=1}^n 2z_iZ_i+b\sum_{i=1}^n\left(X_i^2+Z_i^2\right)=0$$

$$F_c\left(a,b,c,\sigma_1,\sigma_2,\sigma_3,X_0,Y_0,Z_0\right)=\frac{\partial\Delta}{\partial c}=$$

$$cnX_0^2\sigma_1^2+cnY_0^2\sigma_2^2-anX_0Z_0\sigma_1\sigma_3-bnY_0Z_0\sigma_2\sigma_3+cX_0\sigma_1^2\sum_{i=1}^{n}-2x_i+$$

$$aZ_0\sigma_1\sigma_3\sum_{i=1}^{n}x_i+c\sigma_1^2\sum_{i=1}^{n}x_i^2+cX_0\sigma_1\sum_{i=1}^{n}-2X_i+aZ_0\sigma_3\sum_{i=1}^{n}X_i+Y_0\sigma_2\sum_{i=1}^{n}2X_i+$$

$$c\sigma_1\sum_{i=1}^{n}2x_iX_i+cY_0\sigma_2^2\sum_{i=1}^{n}-2y_i+bZ_0\sigma_2\sigma_3\sum_{i=1}^{n}y_i+\sigma_2\sum_{i=1}^{n}-2X_iy_i+$$

$$c\sigma_2^2\sum_{i=1}^{n}y_i^2+X_0\sigma_1\sum_{i=1}^{n}-2Y_i+cY_0\sigma_2\sum_{i=1}^{n}-2Y_i+bZ_0\sigma_3\sum_{i=1}^{n}Y_i+\sigma_1\sum_{i=1}^{n}2x_iY_i+$$

$$c\sigma_2\sum_{i=1}^{n}2y_iY_i+c\sum_{i=1}^{n}\left(X_i^2+Y_i^2\right)+aX_0\sigma_1\sigma_3\sum_{i=1}^{n}z_i+bY_0\sigma_2\sigma_3\sum_{i=1}^{n}z_i+$$

$$a\sigma_1\sigma_3\sum_{i=1}^{n}-x_iz_i+a\sigma_3\sum_{i=1}^{n}-X_iz_i+b\sigma_2\sigma_3\sum_{i=1}^{n}-y_iz_i+b\sigma_3\sum_{i=1}^{n}-Y_iz_i+$$

$$aX_0\sigma_1\sum_{i=1}^{n}Z_i+bY_0\sigma_2\sum_{i=1}^{n}Z_i+a\sigma_1\sum_{i=1}^{n}-x_iZ_i+a\sum_{i=1}^{n}-X_iZ_i+b\sigma_2\sum_{i=1}^{n}-y_iZ_i+$$

$$b\sum_{i=1}^{n}-Y_iZ_i=0$$

$$F_{X_0}\left(a,b,c,\sigma_1,\sigma_2,\sigma_3,X_0,Y_0,Z_0\right)=\frac{\partial\Delta}{\partial X_0}=$$

$$nX_0\sigma_1^2+b^2nX_0\sigma_1^2+c^2nX_0\sigma_1^2-abnY_0\sigma_1\sigma_2-acnZ_0\sigma_1\sigma_3+$$

$$\sigma_1^2\sum_{i=1}^{n}-x_i+b^2\sigma_1^2\sum_{i=1}^{n}-x_i+c^2\sigma_1^2\sum_{i=1}^{n}-x_i+b^2\sigma_1\sum_{i=1}^{n}-X_i+c^2\sigma_1\sum_{i=1}^{n}-X_i+$$

$$\sigma_1\sum_{i=1}^{n}X_i+ab\sigma_1\sigma_2\sum_{i=1}^{n}y_i+c\sigma_1\sum_{i=1}^{n}-2Y_i+ab\sigma_1\sum_{i=1}^{n}Y_i+ac\sigma_1\sigma_3\sum_{i=1}^{n}z_i+$$

$$ac\sigma_1\sum_{i=1}^{n}Z_i+b\sigma_1\sum_{i=1}^{n}2Z_i=0$$

$$F_{Y_0}\left(a,b,c,\sigma_1,\sigma_2,\sigma_3,X_0,Y_0,Z_0\right)=\frac{\partial\Delta}{\partial Y_0}=$$

$$-abnX_0\sigma_1\sigma_2+nY_0\sigma_2^2+a^2nY_0\sigma_2^2+c^2nY_0\sigma_2^2-bcnZ_0\sigma_2\sigma_3+$$

$$ab\sigma_1\sigma_2\sum_{i=1}^{n}x_i+ab\sigma_2\sum_{i=1}^{n}X_i+c\sigma_2\sum_{i=1}^{n}2X_i+\sigma_2^2\sum_{i=1}^{n}-y_i+a^2\sigma_2^2\sum_{i=1}^{n}-y_i+$$

$$c^2\sigma_2^2\sum_{i=1}^{n}-y_i+a^2\sigma_2\sum_{i=1}^{n}-Y_i+c^2\sigma_2\sum_{i=1}^{n}-Y_i+\sigma_2\sum_{i=1}^{n}Y_i+bc\sigma_2\sigma_3\sum_{i=1}^{n}z_i+$$

$$a\sigma_2\sum_{i=1}^{n}-2Z_i+bc\sigma_2\sum_{i=1}^{n}Z_i=0$$

$$F_{Z_0}(a,b,c,\sigma_1,\sigma_2,\sigma_3,X_0,Y_0,Z_0)=\frac{\partial\Delta}{\partial Z_0}=$$
$$-acnX_0\sigma_1\sigma_3-bcnY_0\sigma_2\sigma_3+nZ_0\sigma_3^2+a^2nZ_0\sigma_3^2+b^2nZ_0\sigma_3^2+$$
$$ac\sigma_1\sigma_3\sum_{i=1}^{n}x_i+b\sigma_3\sum_{i=1}^{n}-2X_i+ac\sigma_3\sum_{i=1}^{n}X_i+bc\sigma_2\sigma_3\sum_{i=1}^{n}y_i+bc\sigma_3\sum_{i=1}^{n}Y_i+$$
$$a\sigma_3\sum_{i=1}^{n}2Y_i+\sigma_3^2\sum_{i=1}^{n}-z_i+a^2\sigma_3^2\sum_{i=1}^{n}-z_i+b^2\sigma_3^2\sum_{i=1}^{n}-z_i+a^2\sigma_3\sum_{i=1}^{n}-Z_i+$$
$$b^2\sigma_3\sum_{i=1}^{n}-Z_i+\sigma_3\sum_{i=1}^{n}Z_i=0$$

$$F_{\sigma_1}(a,b,c,\sigma_1,\sigma_2,\sigma_3,X_0,Y_0,Z_0)=\frac{\partial\Delta}{\partial\sigma_1}=$$
$$nX_0^2\sigma_1+b^2nX_0^2\sigma_1+c^2nX_0^2\sigma_1-abnX_0Y_0\sigma_2-acnX_0Z_0\sigma_3+$$
$$X_0\sigma_1\sum_{i=1}^{n}-2x_i+b^2X_0\sigma_1\sum_{i=1}^{n}-2x_i+c^2X_0\sigma_1\sum_{i=1}^{n}-2x_i+abY_0\sigma_2\sum_{i=1}^{n}x_i+$$
$$acZ_0\sigma_3\sum_{i=1}^{n}x_i+\sigma_1\sum_{i=1}^{n}x_i^2+b^2\sigma_1\sum_{i=1}^{n}x_i^2+c^2\sigma_1\sum_{i=1}^{n}x_i^2+b^2X_0\sum_{i=1}^{n}-X_i+$$
$$c^2X_0\sum_{i=1}^{n}-X_i+X_0\sum_{i=1}^{n}X_i+\sum_{i=1}^{n}-x_iX_i+b^2\sum_{i=1}^{n}x_iX_i+c^2\sum_{i=1}^{n}x_iX_i+$$
$$abX_0\sigma_2\sum_{i=1}^{n}y_i+ab\sigma_2\sum_{i=1}^{n}-x_iy_i+cX_0\sum_{i=1}^{n}-2Y_i+abX_0\sum_{i=1}^{n}Y_i+$$
$${}_{i=1}^{n}-2Y_i+abX_0\sum_{i=1}^{n}Y_i+$$
$$ab\sum_{i=1}^{n}-x_iY_i+c\sum_{i=1}^{n}2x_iY_i+acX_0\sigma_3\sum_{i=1}^{n}z_i+ac\sigma_3\sum_{i=1}^{n}-x_iz_i+acX_0\sum_{i=1}^{n}Z_i+$$
$$bX_0\sum_{i=1}^{n}2Z_i+b\sum_{i=1}^{n}-2x_iZ_i+ac\sum_{i=1}^{n}-x_iZ_i=0$$

$$F_{\sigma_2}(a,b,c,\sigma_1,\sigma_2,\sigma_3,X_0,Y_0,Z_0)=\frac{\partial\Delta}{\partial\sigma_2}=$$
$$-abnX_0Y_0\sigma_1+nY_0^2\sigma_2+a^2nY_0^2\sigma_2+c^2nY_0^2\sigma_2-bcnY_0Z_0\sigma_3+$$
$$abY_0\sigma_1\sum_{i=1}^{n}x_i+abY_0\sum_{i=1}^{n}X_i+cY_0\sum_{i=1}^{n}2X_i+Y_0\sigma_2\sum_{i=1}^{n}-2y_i+$$
$$a^2Y_0\sigma_2\sum_{i=1}^{n}-2y_i+c^2Y_0\sigma_2\sum_{i=1}^{n}-2y_i+abX_0\sigma_1\sum_{i=1}^{n}y_i+bcZ_0\sigma_3\sum_{i=1}^{n}y_i+$$
$$ab\sigma_1\sum_{i=1}^{n}-x_iy_i+c\sum_{i=1}^{n}-2X_iy_i+ab\sum_{i=1}^{n}-X_iy_i+\sigma_2\sum_{i=1}^{n}y_i^2+a^2\sigma_2\sum_{i=1}^{n}y_i^2+$$
$$c^2\sigma_2\sum_{i=1}^{n}y_i^2+a^2Y_0\sum_{i=1}^{n}-Y_i+c^2Y_0\sum_{i=1}^{n}-Y_i+Y_0\sum_{i=1}^{n}Y_i+\sum_{i=1}^{n}-y_iY_i+$$
$$a^2\sum_{i=1}^{n}y_iY_i+c^2\sum_{i=1}^{n}y_iY_i+bcY_0\sigma_3\sum_{i=1}^{n}z_i+bc\sigma_3\sum_{i=1}^{n}-y_iz_i+aY_0\sum_{i=1}^{n}-2Z_i+$$
$$bcY_0\sum_{i=1}^{n}Z_i+bc\sum_{i=1}^{n}-y_iZ_i+a\sum_{i=1}^{n}2y_iZ_i=0$$

$$F_{\sigma_3}(a, b, c, \sigma_1, \sigma_2, \sigma_3, X_0, Y_0, Z_0) = \frac{\partial \Delta}{\partial \sigma_3} =$$

$$- acnX_0 Z_0 \sigma_1 - bcnY_0 Z_0 \sigma_2 + nZ_0^2 \sigma_3 + a^2 nZ_0^2 \sigma_3 + b^2 nZ_0^2 \sigma_3 +$$

$$acZ_0 \sigma_1 \sum_{i=1}^{n} x_i + bZ_0 \sum_{i=1}^{n} -2X_i + acZ_0 \sum_{i=1}^{n} X_i + bcZ_0 \sigma_2 \sum_{i=1}^{n} y_i + bcZ_0 \sum_{i=1}^{n} Y_i +$$

$$aZ_0 \sum_{i=1}^{n} 2Y_i + Z_0 \sigma_3 \sum_{i=1}^{n} -2z_i + a^2 Z_0 \sigma_3 \sum_{i=1}^{n} -2z_i + b^2 Z_0 \sigma_3 \sum_{i=1}^{n} -2z_i +$$

$$acX_0 \sigma_1 \sum_{i=1}^{n} z_i + bcY_0 \sigma_2 \sum_{i=1}^{n} z_i + ac\sigma_1 \sum_{i=1}^{n} -x_i z_i + ac \sum_{i=1}^{n} -X_i z_i +$$

$$- x_i z_i + ac \sum_{i=1}^{n} -X_i z_i +$$

$$b \sum_{i=1}^{n} 2X_i z_i + bc\sigma_2 \sum_{i=1}^{n} -y_i z_i + a \sum_{i=1}^{n} -2Y_i z_i + bc \sum_{i=1}^{n} -Y_i z_i + \sigma_3 \sum_{i=1}^{n} z_i^2 +$$

$$a^2 \sigma_3 \sum_{i=1}^{n} z_i^2 + b^2 \sigma_3 \sum_{i=1}^{n} z_i^2 + a^2 Z_0 \sum_{i=1}^{n} -Z_i + b^2 Z_0 \sum_{i=1}^{n} -Z_i + Z_0 \sum_{i=1}^{n} Z_i +$$

$$\sum_{i=1}^{n} -z_i Z_i + a^2 \sum_{i=1}^{n} z_i Z_i + b^2 \sum_{i=1}^{n} z_i Z_i = 0$$

References

1. Abel JS, Chaffee JW (1991) Existence and uniqueness of GPS solutions. IEEE Transactions on Aerospace and Electronic Systems 27: 952–956
2. Abd-Elmotaal H, El-Tokhey M (1995) Effect of spherical approximation on datum transformation. Manuscripta Geodaetica 20: 469–474
3. Abusali PAM, Schutz BE, Tapley BD (1994) Transformation between SLR/VLBI and WGS-84 frames. Bulletin Geodesique 69: 61–72
4. Aduol FWO (1987) Detection of outliers in geodetic networks using principal component analysis and bias parameter estimation. Institute of Geodesy, University of Stuttgart, Technical Report No. 2, Stuttgart
5. Aduol FWO (1994) Robust geodetic parameter estimation through iterative weighting. Survey Review 32: 359–367
6. Aduol FWO, Gacoki TG (2002) Transformation between GPS coordinates and local plane UTM coordinates using the excel spreadsheet. Survey Review 36: 449–462
7. Aduol FWO, Schaffrin B (1986) On outlier identification in geodetic networks using principal component analysis. Conference on Influential Data Analysis, University of Sheffield
8. Albertz J, Kreiling W (1975) Photogrammetric Guide, Herbert Wichmann Verlag, Karlsruhe, pp. 58-60
9. Allgower EL, Georg K (1990) Numerical Continuation Methods. An Introduction. SCM 13. Springer-Verlag
10. Altamimi Z, Collilieux X, Legrand J, Garayt B, Boucher C (2007) ITRF2005: A new release of the International Terrestrial Reference Frame based on time series of station positions and Earth Orientation Parameters, J. Geophys. Res., 112, B09401, doi:10.1029/2007JB004949.
11. Ansermet A (1910) Eine Auflösung des Rückwärtseinschneidens. Zeitschrift des Vereins Schweiz. Konkordatsgeometer, Jahrgang 8, pp. 88–91
12. Anthes R (2003): The Constellation Observing system for Meteorology Ionosphere and Climate (COSMIC). International Workshop on GPS Meteorology, 14th-17th January 2003, Tsukuba, Japan
13. Anthes RA (2004) Application of GPS Remote Sensing to Meteorology and Related Fields, Journal of Meteorological Society of Japan, Vol. 82, No. 1B
14. Awange JL (2002a) Groebner bases, multipolynomial resultants and the Gauss-Jacobi combinatorial algorithms-adjustment of nonlinear GPS/LPS observations. Ph.D. thesis, Department of Geodesy and GeoInformatics, Stuttgart University, Germany. Technical Reports, Report Nr. 2002 (1)
15. Awange JL (2002b) Groebner basis solution of planar resection. Survey Review 36: 528–543
16. Awange JL (2003a): Partial procrustes solution of the threedimensional orientation problem from GPS/LPS observations. In: Grafarend EW, Krumm FW, Schwarze VS (eds) Geodesy - the Challenge of the 3rd Millennium. Springer, Heidelberg pp.277–286
17. Awange JL (2003b) Buchberger algorithm applied to planar lateration and intersection problems. Survey Review 37: 319–329
18. Awange JL (2005): Diagnosis of Outlier of type Multipath in GPS Pseudo-range observations. Survey Review, 38: 177-189
19. Awange JL, Aduol FWO (1999) An evaluation of some robust estimation techniques in the estimation of geodetic parameters. Survey Review 35: 146–162

20. Awange JL, Aduol FWO (2002) An evaluation of some robust estimation techniques in the estimation of geodetic parameters-part II. Survey Review 36: 380–389
21. Awange JL, Fukuda Y (2003) On possible use of GPS-LEO satellite for flood forecasting. Accepted to the International Civil Engineering Conference on Sustainable Development in the 21st Century "The Civil Engineer in Development" 12 - 16 August 2003 Nairobi, Kenya
22. Awange JL, Grafarend EW (2002a) Sylvester resultant solution of planar ranging problem. Allgemeine Vermessungs-Nachrichten 108: 143–146
23. Awange JL, Grafarend EW (2002b) Linearized least squares and nonlinear Gauss-Jacobi combinatorial algorithm applied to 7-parameter datum transformation. Zeitschrift für Vermessungswessen 127: 109–117
24. Awange JL, Grafarend EW (2002c) Algebraic solution of GPS pseudo-ranging equations. Journal of GPS Solutions 4: 20–32
25. Awange JL, Grafarend EW (2002d) Nonlinear adjustment of GPS observations of type pseudo-range. Journal of GPS Solutions 4: 80–93
26. Awange JL, Grafarend EW (2003a) Closed form solution of the overdetermined nonlinear 7-parameter datum transformation. Allgemeine Vermessungs Nachrichtern 109: 130–148
27. Awange JL, Grafarend EW (2003b) Groebner basis solution of the three-dimensional resection problem (P4P). Journal of Geodesy, 77: 327–337
28. Awange JL, Grafarend EW (2003c) Multipolynomial resultant solution of the three-dimensional resection problem (P4P). Bollettino di Geodesia e Science Affini 62: 79–102
29. Awange JL, Grafarend EW (2003d) Explicit solution of the overdetermined three-dimensional resection problem. Journal of Geodesy 76: 605–616
30. Awange JL, Grafarend EW (2005) From space angles to point position using Sylvester resultant. Allgemeine Vermessungs-Nachrichten, 112: 265-269
31. Awange JL, Fukuda Y, Takemoto S, Grafarend EW (2003a) Direct Polynomial approach to nonlinear distance (ranging) problems. Earth, Planets and Space 55: 231–241
32. Awange JL, Fukuda Y, Takemoto S, Ateya I, Grafarend EW (2003b): Ranging algebraically with more observations than unknowns. Earth, Planets and Space 55 (2003) 387-394.
33. Awange JL, Grafarend EW, Fukuda Y (2003c) Closed form solution of the triple three-dimensional intersection problem. Zeitschrift für Geodaesie, Geoinfornation und Landmanagement 128: 395–402
34. Awange JL, Fukuda Y, Takemoto S, Grafarend EW (2003d) Resultants approach to the triple three-dimensional intersection problem. Journal of Geodetic Society of Japan 49: 243–256
35. Awange JL, Fukuda Y, Takemoto S (2004a) B. Strumfel's resultant solution of planar resection problem. Allgemeine Vermessungs-Nachrichten 111(6): 214–219
36. Awange JL, Fukuda Y, Takemoto S, Wickert J, Aoyama Y (2004) Analytic solution of GPS atmospheric sounding refraction angles. Earth, Planets and Space 56: 573–587
37. Awange JL, Grafarend EW, Fukuda Y (2004) Exact solution of the nonlinear 7-parameter datum transformation by Groebner basis. Bollettino di Geodesia e Scienze Affini 63: 117-127
38. Awange JL, Grafarend EW, Fukuda Y, Takemoto S (2005) The role of algebra in Modern day Geodesy. In: F. Sanso, A window on future in the future of Geodesy, Springer, Heidelberg 2005, pp. 524-529

39. Awange JL, Grafarend EW, Fukuda Y, Takemoto S (2005) The application of commutative algebra to Geodesy: two examples. Journal of Geodesy, 79: 93-102
40. Awange JL, Grafarend EW, Fukuda Y (2004) A combinatorial scatter approach to the overdetermined three-dimensional intersection problem. Bollettino di Geodesia e Scienze Affini 63: 235-248.
41. Awange JL, Grafarend EW (2005) Solving Algebraic Computational Problems in Geodesy and Geoinformatics. Springer, Berlin
42. Awange JL, Bae K-H, Claessens S (2008) Procrustean solution of the 9 -parameter transformation problem. Earth, Planets and Space, 60: 529-537
43. Awange JL (2010) GNSS environmental monitoring. Springer Verlag. Berlin
44. Baarda W (1967a) A generalization of the concept strength of the figure. Publications on Geodesy, New Series 2(4), Delft
45. Baarda W (1967b) Statistical concepts in geodesy. The Netherlands geodetic commission, Publication in geodesy, New Series 2, No. 4 , Delft
46. Baarda W (1968a) Statistics - A compass for the land surveyor. Computing centre of the Delft Geodetic Institute, Delft
47. Baarda W (1968b) A testing procedure for use in geodetic networks. Publication in geodesy, New Series 2, No. 5 , Delft
48. Baarda W (1973) S-transformation and criterion matrices. Netherlands Geodetic Commission. Publications on Geodesy, New Series 5(1), Delft
49. Bajaj C, Garity T, Waren J (1988) On the applications of multi-equational resultants. Department of Computer Science, Purdue University, Technical Report CSD-TR-826, pp. 1-22
50. Baker HC, Dodson AH, Penna NT, Higgins M, Offiler D (2001) Ground-based GPS water vapour estimation: Potential for meteorological forecasting, Journal of Atmospheric and Solar-Terrestrial Physics 63(12): 1305–1314.
51. Baki IzH, Chen YQ (1999) Tailored datum transformation model for locally distributed data. Journal of Surveying Engineering 125: 25–35
52. Balodimos DD, Korakitis R, Lambrou E, Pantazis G. (2003) Fast and accurate determination of astronomical coordinates $\boldsymbol{\Theta}, \boldsymbol{\Lambda}$ and azimuth using total station and GPS receiver. Survey Review 37: 269–275
53. Bancroft S (1985) An algebraic solution of the GPS equations. IEEE Transaction on Aerospace and Electronic Systems AES-21: 56–59.
54. Bancroft S (1985) An algebraic solution of the GPS equations. IEEE Transaction on Aerospace and Electronic Systems AES-21: 56–59
55. Barbeau EJ (2003) Polynomials. Problem books in mathematics. Springer, New York, Berlin
56. Bard Y (1974) Nonlinear Parameter Estimation, Academic Press, New York.
57. Barsi A (2001) Performing coordinate transformation by artificial neural network. Allgemeine Vermessungs-Nachrichten 108: 134–137
58. Bartelme N, Meissl P (1975) Ein einfaches, rasches und numerisch stabiles Verfahren zur Bestimmung des kürzesten Abstandes eines Punktes von einem sphäroidischen Rotationsellipsoid. Allgemeine Vermessungs-Nachrichten 82: 436–439
59. Bates D, Hauenstein J, Sommese A, Wampler C (2008) Software for numerical algebraic geometry: a paradigm and progress towards its implementation. Software for algebraic geometry, 1-14, IMA Vol. Math. Appl. 148, Springer, New York
60. Bazlov YA, Galazin VF, Kaplan BL, Maksimov VG, Rogozin VP (1999) Propagating PZ 90 and WGS 94 transformation parameters. GPS Solutions 3: 13–16
61. Bähr HG (1988) A quadratic approach to the non-linear treatment of non-redundant observations. Manuscripta Geodaetica 13: 191–197

62. Bähr HG (1991) Einfach überbestimmtes ebenes Einschneiden, differentialgeometrisch analysiert. Zeitschrift für Vermessungswesen 116: 545–552
63. Becker T, Weispfenning V (1993) Gröbner bases. A computational approach to commutative algebra. Graduate Text in Mathematics 141, Springer, New York
64. Becker T, Weispfenning V (1998) Gröbner bases. A computational approach to commutative algebra. Graduate Text in Mathematics 141, 2nd Edition, Springer, New York
65. Beinat A, Crosilla F (2001) Generalised Procrustes analysis for size and shape 3-D object reconstructions. Optical 3-D measurement techniques, Wien 1-4 October, V, pp. 345–353
66. Beinat A, Crosilla F (2003) Generalised Procrustes algorithm for the conformal updating of a cadastral map. Zeitschrift für Geodasie, Geoinformation und Landmanagement 128: 341-349
67. Bevis M, Businger S, Chiswell S, Herring TA, Anthes RA, Rocken C, Ware RH (1994) GPS Meteorology: Mapping zenith wet delays onto precipitable water. Journal of Applied Meteorology. 33: 379–386
68. Ben-Israel A (1966) A Newton - Raphson Method for the solution of systems of equations. J.Math.Anal.Applic. 15. pp .243 - 252
69. Ben Israel A, Greville TNE (1974) Generalized Inverse Matrices, Wiley - Interscience, New York.
70. Benning W (1974) Der kürzeste Abstand eines in rechtwinkligen Koordinaten gegebenen Außenpunktes vom Ellipsoid. Allgemeine Vermessungs-Nachrichten 81: 429–433
71. Benning W (1987) Iterative ellipsoidische Lotfußpunktberechung. Allgemeine Vermessungs-Nachrichten 94: 256–260
72. Beranek M (1997) Uberlegungen zur Normalverteilung und theoretische Analyse der 3- und 4-Parameter-Transformation. Allgemeine Vermessungs-Nachrichten 104: 137–141
73. Bernstein DS (2009) Matrix Mathematics, 2nd Edition, Princeton University Press, Princeton
74. Bernstein DN (1975) The number of roots of a system of equations. Funct. Anal.Appl. 9:183-185
75. Bernstein DN, Kushnirenko AG, Khovanskii AG (1976) Newton polyhedra. Uspehi Mat. Nauk 31:201-202
76. Bernstein DS (2005) Matrix Mathematics, Princeton University Press, Princeton, p.142
77. Biagi L, Sanso F (2004) Sistemi di riferimento in geodesia: algebra e geometria dei minimi quadrati per un modello con deficienza di rango (parte seconda). Bollettino di Geodesia e Science Affini 63: 29–52
78. Bil WL (1992) Sectie en Projectie. NGT (Dutch Geodetic Magazine) Geodesia 92-10: 405–411
79. Bingham C, Chang T, Richards D (1992) Approximating the matrix fischer and Bingham distributions: Applications to spherical regression and procrustes analysis. Journal of Multivariate Analysis 41:314–337
80. Binous H (2007a) Homotopy Continuation Method to Solve a System of Nonlinear Algebraic Equations. Wolfram Library Archive, MathSource, http://library.wolfram.com/infocenter/MathSource/6710/ [Accessed 1 Dec 2008]
81. Binous H (2007b) Homotopy Continuation Method to Find All Roots of a Polynomial Equation I - II. Wolfram Library Archive, MathSource, http://library.wolfram.com/infocenter/MathSource/6717/ [Accessed 1 Dec 2008]

82. Birardi G (1996) The future global geodetic network and the transformation of local onto WGS coordinates. Bullettino di Geodesia e Scienze Affini 55: 49–56
83. Blaha G, Besette RP (1989) Nonlinear least squares method via an isomorphic geometrical setup. Bulletin Geodesique 63: 115–138
84. Blum L, Cucker F, Shub M, Smale S (1998) Complexity and Real Computation, Springer-Verlag
85. Bock W (1959) Mathematische und geschichtliche Betrachtungen zum Einschneiden. Schriftenreihe Niedersaechsisches Landesvermessungsamt. Report 9, Hannover
86. Bojanczyk AW, Lutoborski A (2003) The procrustes problem for orthogonal Kronecker products. SIAM Journal of Scientific Computation 25: 148–163
87. Borg I, Groenen P (1997) Modern multidimensional scaling. Springer, New York
88. Borkowski KM (1987) Transformation of geocentric to geodetic coordinates without approximation. Astrophys. Space. Sci. 139: 1–4
89. Borkowski KM (1989) Accurate algorithm to transform geocentric to geodetic coordinates. Bull. Geod. 63: 50–56
90. Bowring BR (1976) Transformation from spatial to geographical coordinates. Survey Review 23: 323–327
91. Bowring BR (1985) The accuracy of geodetic latitude and height equations. Survey Review 28: 202–206
92. Brandstätter G (1974) Notiz zur analytischen Lösung des ebenen Rückwärtsschnittes. Österreichische Zeitschrift für Vermessungswesen 61: 34–136
93. Brokken FB (1983) Orthogonal Procrustes rotation maximizing congruence. Psychometrika 48: 343–352
94. Brunner FK (1979) On the analysis of geodetic networks for the determination of the incremental strain tensor. Survey Review 25: 146–162
95. Buchberger B (1965) An algorithm for finding a basis for the residue class ring of a zero dimensional polynomial ideal (German). Ph.D. thesis, Institute of Mathematics, University of Innsbruck (English Translation: J. of Symbolic Computation, Special Issue on Logic, Mathematics, and Computer Science: Interactions. Volume 41, Number 3-4, Pages 475-511, 2006).
96. Buchberger B (1970) Ein algorithmisches Kriterium für die Lösbarkeit eines algebraischen Gleichungsystems (An Algorithmic Criterion for the Solvability of Algebraic Systems of Equations).. Aequationes Mathematicae 4/3: 374–383 (English translation in: B. Buchberger, F. Winkler (eds.), Groebner Bases and Applications, Proceedings of the International Conference "33 Years of Groebner Bases", 1998, RISC, Austria, London Mathematical Society Lecture Note Series, Vol. 251, Cambridge University Press, 1998, pp. 535 -545.)
97. Buchberger B (1979) A criterion for detecting unnecessary reductions in the construction of Groebner bases. Proceedings of the 1979 European Symposium on Symbolic and Algebraic computation, Springer lecture notes in Computer Science 72, pp. 3-21, Springer, Berlin-Heidelberg-New York
98. Buchberger B (2001) Gröbner bases. A short introduction for system theorists. In: Moreno-Diaz R et al. (eds): EUROCAST 2001, LNCS 2178, pp. 1–19
99. Buchberger B, Winkler F (1998) Groebner bases and applications. London mathematical society lecture note series 251, Cambridge university press
100. Canny JF (1988) The complexity of robot motion planning. ACM doctoral dissertation award, MIT Press

101. Canny JF, Kaltofen E, Yagati L (1989) Solving systems of nonlinear polynomial equations faster. Proceedings of the International Symposium on Symbolic and Algebraic Computations ISSAC, July 17-19, Portland, Oregon, pp.121–128
102. Canny JF, Emiris IZ (2000) A subdivision based algorithm for the sparse resultant. Journal of the ACM 47(3), pp.417-451
103. Cattani E, Dickenstein A, Sturmfels B (1998) Residues and resultants. J. Math. Sci. University of Tokyo 5: 119–148
104. Cayley A (1865) On the theory of elimination. Cambridge and Dublin Mathematical Journal, III., pp. 210-270
105. Chaffee JW, Abel JS (1994) On the exact solutions of the pseudorange equations. IEEE Transactions on Aerospace and Electronic Systems 30: 1021–1030
106. Chapra SC, Canale RP (1998) Numerical Methods for Engineers, With Programming and Software Applications, Third Edition, McGraw-Hill, Boston, New York, London
107. Chen G, Herring TA (1997) Effects of atmospheric azimuthal asymmetry on the analysis of apace geodetic data. Journal of Geophysical Research 102(B9): 20489–20502
108. Cheng C-L, Van Ness JW (1999) Statistical regression with measurement error. Oxford University Press, 198 Madison Avenue, New York
109. Choi SH, Book NL (1991) Unreachable Roots for Global Homotopy Continuation Methods, AIChE Journal, 37., 7: 1093-1095.
110. Chu MT, Driessel R (1990) The projected gradient method for least squares matrix approximations with spectral constraints. SIAM J. Numer. Anal. 27(4): 1050–1060
111. Chu MT, Trendafilov NT (1998) Orthomax rotation problem. A differential equation approach. Behaviormetrika 25(1): 13–23
112. Chrystal G (1964) Textbook of Algebra (Vol. 1). Chelsea, New York
113. Chtcherba AD, Kapur D (2004) Construction Sylvester-type resultant matrices using the Dixon formulation, Journal of Symbolic Computation 38, pp.777-814
114. Chtcherba AD, Kapur D, Minimair M (2005) Cayley-Dixon construction of Resultants of Multi-Univariate Composed Polynomials, University of New Mexico, Department of Computer Science, Technical Report TR-CS-2005-15, April
115. Cox DA (1998) Introduction to Gröbner bases. Proceedings of Symposia in Applied Mathematics 53: 1–24
116. Cox TF, Cox MA (1994) Multidimensional scaling. St. Edmundsbury Press, St. Edmunds, Suffolk
117. Cox D, Little J, O'Shea D (1997) Ideals, Varieties, and Algorithms. An introduction to computational algebraic geometry and commutative algebra, Springer, New York
118. Cox D, Little J, O'Shea D (1998) Using algebraic geometry. Graduate Text in Mathematics 185. Springer, New York
119. Croceto N (1993) Point projection of topographic surface onto the reference ellipsoid of revolution in geocentric Cartesian coordinates. Survey Review 32: 233–238
120. Crosilla F (1983a) A criterion matrix for a second order design of control networks. Bull. Geod. 57: 226–239
121. Crosilla F (1983b) Procrustean transformation as a tool for the construction of a criterion matrix for control networks. Manuscripta Geodetica 8: 343–370
122. Crosilla F, Beinat A (2003) Procrustes analysis and geodetic sciences. In: Grafarend EW, Krumm FW, Schwarze VS (eds) Geodesy - the Challenge of the 3rd Millennium. Springer, Heidelberg pp. 277–286

123. Crosilla F (2003) Use of generalised Procrustes analysis for the photogrammetric block adjustement by independent models. Journal of Photogrammetric & Remote sensing 56: 195–209
124. Dach R (2000) Einfluß von Auflasteffekten auf Präzise GPS-Messungen, DGK, Reihe C, Heft Nr. 519
125. Davenport JH, Siret Y, Tournier E (1988) Computer algebra. Systems and algorithms for algebraic computation. Academic Press Ltd., St. Edmundsbury, London
126. Davis JL, Herring TA, Shapiro II, Rogers AE, Elgered G (1985) Geodesy by radio interferometry: Effects of atmospheric modeling errors on estimates of baseline length. Radio Sci 20: 1593–1607
127. Decarolis F, Mayer R, Santamaria M (2002) Homotopy Continuation Methods. http://home.uchicago.edu/∼fdc/H-topy.pdf [Accessed 1 Dec 2008]
128. Dixon AL (1908) The elimination of three quantics in two independent variables. Proc. London Mathematical Society series 2 6: 468–478
129. Drexler FJ (1977) Eine Methode zur Berechnung sämtlicher Lösungen von Polynomgleichungssystemen. Numer. Math., 29:45-58.
130. Dryden IL (1998) General shape and registration analysis. In: Barndorff-Nielsen O, Kendall WS, van Lieshout MNM (eds) Stochastic Geometry: likelihood and computation. Chapman and Hall, London pp: 333–364
131. Featherstone WE (1997) A comparison of existing co-ordinate transformation models and parameters in Australia. Cartography 26: 13–25
132. Featherstone WE, Lichti DD (2008) Fitting gravimetric geoid models to vertical deflections. Journal of Geodesy 83:583-589, DOI: 10.1007/s00190-008-0263-4
133. Finsterwalder S, Scheufele W (1937) Das Rückwartseinschneiden im Raum. Sebastian Finsterwalder zum 75 Geburtstage, pp. 86–100, Verlag Hebert Wichmann, Berlin
134. Fischbach FF (1965) A satellite method for pressure and temperature below 24km. Bull. Am. Meteorol. 46: 528–532
135. Fischler MA, Bolles RC (1981) Random sample consensus: A paradigm for modell fitting with application to image analysis and automated cartography. Communications of the ACM 24: 381–395
136. Fitzgibbon A, Pilu M, Fisher RB (1999) Direct least squares fitting of ellipses. IEEE Transactions on Pattern Analysis and Machine Intelligence 21: 476–480
137. Fletcher R(1970) Generalized inverses for nonlinear eqautions and optimization. In Rabinowitz, P. ed. Numerical ethods for Nonlinear Algebraic Equations, pp .75 - 85, Gordon and Breach, London
138. Flores A, Ruffini G, Rius A (2000) 4D Tropospheric Tomography Using GPS Slant Wet Delay. Ann. Geophys. 18: 223–234
139. Fotiou A (1998) A pair of closed expressions to transform geocentric to geodetic coordinates. Zeitschrift für Vermessungswesen 123: 133–135
140. Foulds LR (1984) Combinatorial optimization for undergraduates. Springer, New York
141. Francesco D, Mathien PP, Senechal D (1997) Conformal field theory. Springer, Heidelberg, New York
142. Fröhlich H, Hansen HH (1976) Zur Lotfußpunktrechnung bei rotationsellipsoidischer Bezugsfläche. Allgemeine Vermessungs-Nachrichten 83: 175–179
143. Fröhlich H, Bröker G (2003) Trafox version 2.1. 3d-kartesische Helmert-Transformation. http://www.koordinatentransformation.de/data/trafox.pdf [Accessed 27 August 2008]

144. Fukushima T (1999) Fast transform from geocentric to geodetic coordinates. Journal of Geodesy 73: 603–610
145. Gander W, Golub GH, Strebel R (1994) Least-Squares fitting of circles and ellipses. BIT No. 43: 558–578
146. Garcia CB, Zangwill WI (1979) Determining all solutions to certain systems of nonlinear equations. Math. Operations Res., 4:1-14.
147. Garcia CB, Zangwill WI (1981) Pathways to Solutions, Fixed Points and Equilibria, Prentice Hall, Englewood Cliffs. NJ
148. Gelfand IM, Kapranov MM, Zelevinsky AV (1990) Generalized Euler Integrals and A-Hypergeometry Functions. Advances in Mathematics 84: 255–271
149. Gelfand IM, Kapranov MM, Zelevinsky AV (1994) Discriminants, resultants and multidimensional determinants. Birkhäuser, Boston
150. Gelfand MS, Mironor AA, Perzner PA (1996) Gene recognition via spliced sequence alignment. Proc. Natl. Aead. Sci. USA 93: 9061–9066
151. Golub GH (1987) Least squares, singular values and matrix approximation. Aplikace matematiky 13: 44–51
152. Golub GH, Pereyra V (1973) The differentiation of pseudo-inverses and nonlinear least squares problems whose variables separate. *SIAM J. Num. Anal.*, **10**, 413-432.
153. Goodall C (1991) Procrustes methods in statistical analysis of shape. J. Royal Statistical Soc. B53: 285–339
154. Gordon SJ, Lichti DD (2004) Terrestrial laser scanners with a narrow field of view: the effect on 3D resection solutions. Survey Review 37:448–468
155. Gotthardt E (1940) Zur Unbestimmtheit des räumlichen Rückwärtseinschnittes, Mitteilungen der Ges. f. Photogrammetry e.V., Jänner 1940, Heft 5
156. Gotthardt E (1974) Ein neuer gefährlicher Ort zum räumlichen Rückwärtseinschneiden, Bildm. u. Luftbildw.
157. Gower JC (1975) Generalized procrustes analysis. Psychometrika 40(1): 33–51
158. Grafarend EW (1975) Three dimensional Geodesy 1. The holonomity problem. Zeitschrift für Vermessungswesen 100: 269–280
159. Grafarend EW (1981) Die Beobachtungsgleichungen der dreidimensionalen Geodäsie im Geometrie- und Schwereraum. Ein Beitrag zur operationellen Geodäsie. Zeitschrift für Vermessungswesen 106: 411–429
160. Grafarend EW (1985) Variance-Covariance component estimation; theoretical results and geodetic applications. Statistics & Decisions, Supplement Issue No. 2, 407–441
161. Grafarend EW (1988) Azimuth transport and the problem of orientation within geodetic traverses and geodetic networks. Vermessung, Photogrammetrie, Kulturtechnik 86: 132–150
162. Grafarend EW (1989) Photogrammetrische Positionierung. Festschrift für Prof. Dr.-Ing. Dr. h.c Friedrich Ackermann zum 60. Geburtstag, Institut für Photogrammetrie, Univerität Stuttgart, Heft 14, pp.44-55, Stuttgart
163. Grafarend EW (1990) Dreidimensionaler Vorwaertschnitt. Zeitschrift für Vermessungswesen 115: 414–419
164. Grafarend EW (1991) Application of Geodesy to Engineering. In: Eds. Linkwitz K, Eisele V, Mönicke HJ, IAG-Symposium No. 108, Springer, Berlin-Heidelberg-New York
165. Grafarend EW (2000) Gaußsche flächennormale Koordinaten im Geometrie- und Schwereraum. Erste Teil: Flächennormale Ellipsoidkoordinaten. Zeitschrift für Vermessungswesen 125: 136–139

166. Grafarend EW (2000) Gaußsche flächennormale Koordinaten im Geometrie- und Schwereraum. Erste Teil: Flächennormale Ellipsoidkoordinaten. Zeitschrift für Vermessungswesen 125: 136–139
167. Grafarend EW, Ardalan A (1999) World geodetic datum 2000. Journal of Geodesy 73: 611–623
168. Grafarend EW, Awange JL (2000) Determination of vertical deflections by GPS/LPS measurements. Zeitschrift für Vermessungswesen 125: 279–288
169. Grafarend EW, Awange JL (2003) Nonlinear analysis of the three-dimensional datum transformation (conformal group $\mathbb{C}_7(\mathbf{3})$. Journal of Geodesy 77: 66–76
170. Grafarend EW, Keller W (1995) Setup of observational functionals in gravity space as well as in geometry space. Manuscripta Geodetica 20: 301–325
171. Grafarend EW, Kunz J (1965) Der Rückwärtseinschnitt mit dem Vermessungskreisel. Bergbauwissenschaften 12: 285–297
172. Grafarend EW, Lohse P (1991) The minimal distance mapping of the topographic surface onto the (reference) ellipsoid of revolution. Manuscripta Geodaetica 16: 92–110
173. Grafarend EW, Mader A (1993) Robot vision based on an exact solution of the threedimensional resection-intersection. Applications of Geodesy to Engineering. In K. Linkwitz, V. Eisele and H-J Moenicke, Symposium No. 108, Spinger Berlin-Heidelberg-Newyork-London-Paris-Tokyo-HongKong-Barcelona-Budapest.
174. Grafarend EW, Okeke F (1998) Transformation of conformal coordinates of type Mercator from global datum (WGS 84) to local datum (regional, national). Marine Geodesy 21: 169–180
175. Grafarend EW, Richter B (1977) Generalized Laplace condition. Bull. Geod. 51: 287–293
176. Grafarend EW, Sanso F (1985) Optimization and design of geodetic networks. Springer, Berlin-Heidelberg-New York-Tokyo
177. Grafarend EW, Schaffrin B (1974) Unbiased Freenet adjustment. Survey Review 22: 200–218
178. Grafarend EW, Schaffrin B (1989) The geometry of nonlinear adjustment-the planar trisection problem-. In: Kejlso E, Poder K, Tscherning CC (eds) Festschrift to T. Krarup, pp. 149-172, Denmark
179. Grafarend EW, Schaffrin B (1991) The planar trisection problem and the impact of curvature on non-linear least -squares estimation. Computational statistics & data analysis 12: 187–199
180. Grafarend EW, Schaffrin B (1993) Ausgleichungsrechnung in Linearen Modellen. B. I. Wissenschaftsverlag, Mannheim
181. Grafarend EW, Shan J (1996) Closed-form solution of the nonlinear pseudo-ranging equations (GPS). ARTIFICIAL SATELLITES, Planetary Geodesy 31: 133–147
182. Grafarend EW, Shan J (1997a) Closed-form solution of P4P or the three-dimensional resection problem in terms of Möbius barycentric coordinates. Journal of Geodesy 71: 217–231
183. Grafarend EW, Shan J (1997b) Closed form solution of the twin P4P or the combined three dimensional resection-intersection problem in terms of Möbius barycentric coordinates. Journal of Geodesy 71: 232–239
184. Grafarend EW, Syffus R (1997) Strip transformation of conformal coordinates of type Gauss-Kruger and UTM. Allgemeine Vermessungs-Nachrichten 104: 184–190
185. Grafarend EW, Syffus R (1998) Transformation of conformal coordinates of type Gauss-Krüger or UTM from local datum (regional, national, European) to global

datum (WGS 84) part 1: The transformation equations. Allgemeine Vermessungs-Nachrichten 105: 134–141
186. Grafarend EW, Hendricks A, Gilbert A (2000) Transformation of conformal coordinates of type Gauss-Kruger or UTM from a local datum (Regional, National, European) to a global datum (WGS 84, ITRF 96). Allgemeine Vermessungs-Nachrichten 107: 218–222
187. Grafarend EW, Lohse P, Schaffrin B (1989) Dreidimensionaler Rückwärtsschnitt. Zeitschrift für Vermessungswesen 114: 61–67,127–137,172–175,225–234,278–287
188. Grafarend EW, Knickmeyer EH, Schaffrin B (1982) Geodätische Datumtransformationen. Zeitschrift für Vermessungswesen 107: 15–25
189. Grafarend EW, Krumm F, Okeke F (1995) Curvilinear geodetic datum transformation. Zeitschrift für Vermessungswesen 120: 334–350
190. Grafarend EW, Syffus R, You RJ (1995) Projective heights in geometry and gravity space. Allgemeine Vermessungs-Nachrichten 102: 382–402
191. Green B (1952) The orthogonal approximation of an oblique stracture in factor analysis. Psychometrika 17: 429–440
192. Grewal MS, Weill LR, Andrews AP (2001) Global Positioning Systems, Inertial Navigation and Integration, John Wiley & Sons, New York
193. Gritton KS, Seader JD, Lin WJ (2001) Global homotopy continuation procedures for seeking all roots of a nonlinear equation. Computers and Chem. Engn. 25:1003-1019
194. Grunert JA (1841) Das Pothenotsche Problem in erweiterter Gestalt; nebst Bemerkungen über seine Anwendungen in der Geodäsie. Grunerts Archiv für Mathematik und Phsyik 1 pp. 238–241
195. Guckenheimer J, Myers M, Sturmfels B (1997) Computing Hopf birfucations. SIAM J. Numerical Analysis 34: 1–21
196. Gui Q, Zhang J (1998) Robust biased estimation and its applications in geodetic adjustments. Journal of Geodesy 72: 430–435
197. Gulliksson M (1995a) The partial Procrustes problem - A first look. Department of Computing Science, Umea University, Report UMINF-95.11, Sweden
198. Gulliksson M (1995b) Algorithms for the partial Procrustes problem. Department of Industrial Technology, Mid Sweden University s-891 18, Report 1995:27, Ornskoldsvik, Sweden
199. Gullikson M, Söderkvist I (1995) Surface fitting and parameter estimation with nonlinear least squares. Zeitschrift für Vermessungswesen 25: 611–636
200. Gunji T, Kim S, Kojima M, Takeda A, Fujisawa K, Mizutani T (2004) “PHoM” - a Polyhedral Homotopy Continuation Method for Polynomial Systems. Computing 73(1):57-77
201. Guolin L (2000) Nonlinear curvature measures of strength and nonlinear diagnosis. Allgemein Vermessungs-Nachrichten 107: 109–111
202. Guo J, Jin F (2001) A new model of digitizing coordinate transformation and its nonlinear solution. Allgemeine Vermessungs-Nachrichten 108: 311–317
203. Gurbunov ME, Gurvich AS, Bengtsson L (1996) Advanced algorithms of inversion of GPS/MET satellite data and their application to the reconstruction of temperature and humidity, Rep. No. 211, Max-Plunk-Institut für Meteorologie, Hamburg, Germany
204. Han SC, Kwon JH, Jekeli C (2001) Accurate absolute GPS positioning through satellite clock error estimation. Journal of Geodesy 75: 33–43
205. Hampel FR, Ronchetti EM, Rousseeuw P, Stahel WA (1986) Robust Statistic - the approach based non influence Functions. John Wiley & Sons, New York

206. Hammer E (1896) Zur graphischen Ausgleichung beim trigonometrischen Einschneiden von Punkten. Optimization methods and softwares 5: 247–269
207. Hanselman D, Littlefield B (1997) The student edition of Matlab. Prentice-Hall, New Jersey
208. Haneberg WC (2004) Computational Geosciences with *Mathematica*, Springer, Berlin, Heidelberg, New York
209. Hanssen RF, Weckwerth TM, Zebker HA, Klees R (1999) High-Resolution Water Vapor Mapping from Interferometric Radar Measurements. Science 283: 1297–1299
210. Haralick RM, Lee C, Ottenberg K, Nölle M (1991) Analysis and solution of the three point perspective pose estimation problem. Proc. IEEE Org. on Computer Vision and Pattern Recognition, pp. 592–598
211. Haralick RM, Lee C, Ottenberg K, Nölle M (1994) Review and analysis of solution of the three point perspective pose estimation problem. International Journal of Computer Vision 13 3: 331–356
212. Harper D, Wooff C, Hodgkinson D (1991) A guide to computer algebra system. John Wiley & Sons, New York
213. Harvey BR (1986) Transformation of 3D coordinates. The Australian Surveyor 33: 105–125
214. Haselgrove CB (1961) The solution of nonlinear equations and of diferntial equations with two - point boundary conditions. Computing J. 4: 255 - 259
215. Hazaveh K, Jeffrey DJ, Reid GJ, Watt SM, Wittkopf AD (2003) An exploration of homotopy solving in Maple. http://www.apmaths.uwo.ca/~djeffrey/Offprints/ascm2003.pdf [Accessed 27 Aug 2008]
216. Healey S, Jupp A, Offiler D, Eyre J (2003) The assimilation of radio occultation measurements. In Reigber C, Lühr H, Schwintzer P (eds), First CHAMP mission results for gravity, magnetic and atmospheric studies, Springer, Heidelberg
217. Heck B (1987) Rechenverfahren und Auswertemodelle der Landesvermessung. Wichmann Verlag, Karlsruhe, Germany
218. Heikkinen M (1982) Geschlossene Formeln zur Berechnung räumlicher geodätischer Koordinaten aus rechtwinkligen Koordinaten. Zeitschrift für Vermessungswesen 107: 207–211
219. Heindl G (1982) Experiences with non-statistical method of detecting outliers. International symposium on geodetic network and computations of the I. A. G. Munich, Aug. 30th to Sept. 5, 5: 19–28
220. Heiskanen WA, Moritz H (1967) Physical Geodesy, Freeman and Company, London
221. Hirvonen R, Moritz H (1963) Practical computation of gravity at high altitudes, Institute of Geodesy, Photogrammetry and Cartography. Ohio State University, Report No. 27, Ohio
222. Hirt C, Bürki B (2002) The Digital Zenith Camera - A New High-Precision and Economic Astrogeodetic Observation System for Real-Time Measurement of Deflections of the Vertical. Proc. of the 3rd Meeting of the International Gravity and Geoid Commission of the International Association of Geodesy, Thessaloniki, Greece (ed. I. Tziavos): 161-166.
223. Hirt C, Seeber G (2002) Astrogeodätische Lotabweichungsbestimmung mit dem digitalen Zenitkamerasystem TZK2-D. Zeitschrift für Vermessungswesen 127: 388-396.
224. Hirt C (2003) The Digital Zenith Camera TZK2-D - A Modern High Precision Geodetic Instrument for Automatic Geographic Positioning in Real-Time. Astro-

nomical Data Analysis Software and Systems XII, Astronomical Society of the Pacific Conference Series Vol. 295, San Francisco: 155-159.

225. Hirt C, Seeber G (2008): Accuracy Analysis of vertical deflection data observed with the Hannover Digital Zenith Camera System TZK2-D. Journal of Geodesy 82: 231-248. DOI: 10.1007/s00190-007-0184-7
226. Hirt C, Bürki B, Somieski A, Seeber G (2009). Modern Determination of vertical deflections using digital zenith cameras. Journal Surveying Engineering, accepted.
227. Hofman-Wellenhof B, Lichtenegger H, Collins J (2001) Global Positioning System: Theory and practice, 5th Edition, Springer, Wien
228. Hofman-Wellenhof B, Lichtenegger H, Wasle E (2008) GNSS Global Navigation Satellite System: GPS, GLONASS; Galileo and more, Springer, Wien
229. Horaud R, Conio B, Leboulleux O (1989) An analytical solution for the perspective 4-point problem. Computer Vision, Graphics and Image Processing 47: 33–44
230. Hornoch AT (1950) Über die Zurückführung der Methode der kleinsten Quadrate auf das Prinzip des arithmetischen Mittels. Zeitschrift für Vermessungswesen 38: 13–18
231. Huber PJ (1964) Robust estimation of a location parameter. Annals of Mathematical Statistics 35: 73–101
232. Huber PJ (1972) Robust Statistics; A review. Annals of Mathematical Statistics 43: 1041–1067
233. Huber PJ (1981) Robust Statistics. John Wiley & Sons, New York
234. Ireland K, Rosen M (1990) A classical introduction to modern number theory. Springer, New York
235. Irving RS (2004) Integers, polynomials, and rings. Springer, New York
236. Jacobi CGI (1841) Deformatione et proprietatibus determinantum, Crelle's Journal für die reine und angewandte Mathematik, Bd. 22
237. Kahmen H, Faig W (1988) Surveying. Walter de Gruyter, Berlin
238. Kampmann G (1996) New adjustment techniques for the determination of transformation parameters for Cadastral and Engineering purposes. Geomatica 50: 27–34
239. Kapur D, Saxena T, Yang L (1994) Algebraic and geometric reasoning using Dixon resultants. In ACM ISSAC 94, International Symposium on Symbolic and Algebraic Computation, Oxford, England, July 1994, 99-107.
240. Killian K (1990) Der gefährliche Ort des überbestimmten räumlichen Rückwärtseinschneidens. Öst. Zeitschrift für Vermessungswesen und Photogrammetry 78: 1–12
241. Kinneen RW, Featherstone WE (2004) Empirical evaluation of published transformation parameters from the Australian Geodetic Datums (AGD66 and AGD84) to the Geocentric Datum of Australia (GDA94), J. Spat. Sci., 49(2), 1–31.
242. Kleusberg A (1994) Die direkte Lösung des räumlichen Hyperbelschnitts. Zeitschrift für Vermessungswesen 119: 188-192
243. Kleusberg A (2003) Analytical GPS navigation solution. In: Grafarend EW, Krumm FW, Schwarze VS (eds) Geodesy - the Challenge of the 3rd Millennium. Springer, Heidelberg pp.93–96
244. Koch KR (1999) Parameter estimation and hypothesis testing in linear models. Springer, Berlin, Heidelberg
245. Koch KR (2001) Bermekung zu der Veröffentlichung "Zur Bestimmung eindeutiger transformationparameter". Zeitschrift für Vermessungswesen 126: 297
246. Koch KR, Yang Y (1998a) Konfidenzbereiche und Hypothesenteste für robuste Parameterschätzungen. ZfV 123: 20–26

247. Koch KR, Yang Y (1998b) Robust Kalman filter for rank deficient observation models. Journal of Geodesy 72: 436–441
248. Koch KR, Fröhlich H, Bröker G (2000) Transformation räumlicher variabler Koordinaten. Allgemeine Vermessungs-Nachrichten 107: 293–295
249. Kotsireas IS (2001) Homotopies and polynomial systems solving I: Basic Principles, ACM SIGSAM Bulletin, 35., 1:19-32
250. Kuo Y.-H, Sokolovski SV, Anthens RA, Vandenberghe F (2000) Assimilation of the GPS radio occultation data for numerical weather prediction. Terrestrial, Atmospheric and Oceanic Science, 11: 157–186
251. Krarup T (1979) S transformation or how to live without the generalized inverse - almost. Geodetisk Institut, Charlottenlund, Denmark
252. Krarup T (1982) Nonlinear adjustment and curvature. In: Forty years of thought, Delft, pp. 145–159
253. Krause LO (1987) A direct solution of GPS-type navigation equations. IEEE Transactions on Aerospace and Electronic Systems 23: 225–232
254. Krishna S, Manocha D (1995) Numerical algorithms for evaluating one-dimensional algebraic sets. Proceedings of the International Symposium on Symbolic and Algebraic Computation ISSAC, July 10-12, pp. 59–67, Montreal, Canada
255. Kubik KK (1967) Iterative Methoden zur Lösunge des nichtlinearen Ausgleichungsproblemes. Zeitschrift für Vermessungswesen 91: 145–159
256. Kursinski ER, Hajj GA, Schofield JT, Linfield RP, Hardy KR (1997) Observing Earth's atmosphere with radio occultation measurements using Global Positioning System. J. Geophy. Res. 102: 23429–23465
257. Kurz S (1996) Positionierung mittels Rückwartsschnitt in drei Dimensionen. Studienarbeit, Geodätisches Institut, University of Stuttgart, Stuttgart
258. Lam TY (2003) Exercises in classical ring theory. Springer, New York, Tokyo
259. Lannes A, Durand S (2003) Dual algebraic formulation of differential GPS. Journal of Geodesy 77: 22–29
260. Lapaine M (1990) A new direct solution of the transformation problem of Cartesian into ellipsoidal coordinates. In: Rapp RH. Sanso F, Determination of the geoid: Present and future. pp. 395-404. Springer, New York
261. Larson LW (1996) Destructive water: Water-caused natural disasters, their abetment and control. IAHS conference, Anaheim California, June 24-28
262. Lauritzen N (2003) Concrete abstract algebra. From numbers to Gröbner bases. Cambridge University Press, UK
263. Lee TL, Li TY, Tsai CH (2008) HOM4PS-2.0: A software package for solving polynomial systems by the polyhedral homotopy continuation method. Computing 83(2-3):109-133.
264. Leick A (2003) GPS satellite surveying, 3rd Edition, John Wiley & Sons, New York
265. Lenzmann E, Lenzmann L (2001a) Zur Bestimmung eindeutiger transformationparameter. Zeitschrift für Vermessungswesen 126: 138–142
266. Lenzmann E, Lenzmann L (2001b) Erwiderung auf die Anmerkung von Jörg Reinking und die Bermekungen von Karl-Rudolf Koch zu unserem Meitrag "Zur Bestimmung eindeutiger transformationparameter". Zeitschrift für Vermessungswesen 126: 298–299
267. Lewis RH (2002) Using the Dixon resultant on big problems. in CBMS Conference, Texas A&M University, http://www.math.tamu.edu/conferences/cbms/abs.html [Accessed 27 Aug 2008]

268. Lewis RH, 2004: Exploiting symmetry in a polynomial system with the Dixon resultant. in *Int. Conf. on Appl. of Computer Algebra (ACA)*, Lamar University, Texas, July 2004.
269. Lewis RH (2008) Heuristics to Accelerate the Dixon Resultant. Mathematics and Computers in Simulation, 77, 4:400-407.
270. Leykin A, Verschelde J (2004) PHCmaple: A Maple Interface to the Numerical Homotopy Algorithms in PHCpack, http://www.ima.umn.edu/~leykin/papers/PHCmaple.pdf [Accessed 1 Dec 2008]
271. Lictbbau D (2009) Cylinders through five points: Computational Algebra and Geometry. http://library.wolfram.com/infocenter/Conferences/7521/ [Accessed 13/10/2009]
272. Lichtenegger H (1995) Eine direkte Lösung des räumlichen Bogenschnitts. Österreichische Zeitschrift für Vermessung und Geoinformation 83: 224–226
273. Lidl R, Pilz G (1998) Applied abstract algebra. 2nd edition, Springer, New York
274. Lin KC, Wang J (1995) Transformation from geocentric to geodetic coordinates using Newton's iteration. Bulletin Geodesique 69: 300–303
275. Linnainmaa S, Harwood D, Davis LS (1988) Pose determination of a three-dimensional object using triangle pairs. IEEE transaction on pattern analysis and Machine intelligence 105: 634–647
276. Lohse P (1990) Dreidimensionaler Rückwärtsschnit. Ein Algorithmus zur Streckenberechnung ohne Hauptachsentransformation. Zeitschrift für Vermessungswesen 115: 162–167
277. Lohse P (1994) Ausgleichungsrechnung in nichtlinearen Modellen. DGK, Reihe C, Heft Nr. 429
278. Loskowski P (1991) Is Newton's iteration faster than simple iteration for transformation between geocentric and geodetic coordinates? Bulletin Geodesique 65: 14–17
279. Lyubeznik G (1995) Minimal resultant system. Journal of Algebra 177: 612–616
280. Macaulay F (1902) On some formulae in elimination. Proceeding in London Mathematical Society, pp. 3-27
281. Macaulay F (1916) The algebraic theory of modular systems. Cambridge Tracts in Mathematics 19, Cambridge University Press, Cambridge
282. Macaulay F (1921) Note on the resultant of a number of polynomials of the same degree. Proceeding in London Mathematical Society 21: 14–21
283. Mackenzie FT (2003) Our changing planet; an introduction to Earth system science and global environmental change. 3rd edition, Prentice Hall, New Jersey
284. Manocha D (1992) Algebraic and numeric techniques for modeling and robotics. Ph.D. thesis, Computer Science Division, Department of Electrical Engineering and Computer Science, University of California, Berkeley
285. Manocha D (1993) Efficient algorithms for multipolynomial resultant. The Computer Journal 36: 485–496
286. Manocha D (1994a) Algorithms for computing selected solutions of polynomial equations. Extended abstract appearing in the proceedings of the ACM ISSAC 94.
287. Manocha D (1994b) Computing selected solutions of polynomial equations. Proceedings of the International Sypmosium on Symbolic and Algebraic Computations ISSAC, July 20-22, pp.1-8, Oxford
288. Manocha D (1994c) Solving systems of polynomial equations. IEEE Computer Graphics and application 14: 46–55
289. Manocha D (1998) Numerical methods for solving polynomial equations. Proceedings of Symposia in Applied Mathematics 53: 41–66

290. Manocha D, Canny J (1991) Efficient techniques for multipolynomial resultant algorithms. Proceedings of the International Symposium on Symbolic Computations, July 15-17, 1991, pp. 86–95, Bonn
291. Manocha D, Canny J (1992) Multipolynomial resultant and linear algebra. Proceedings of the International Symposium on Symbolic and Algebraic Computations ISSAC, July 27-29, pp. 158-167, Berkeley
292. Manocha D, Canny J (1993) Multipolynomial resultant algorithms. Journal of Symbolic Computations 15: 99–122
293. Mardia K (1978) Some properties of classical multidimensional scaling. Commun. Statist.-Theory Meth. A7(13): 1233–1241
294. Mathar R (1997) Multidimensionale Skalierung. B. G. Teubner Verlag, Stuttgart
295. Mathes A (2002) EasyTrans Pro-Edition, Professionelle Koordinatentransformation für Navigation, *Vermessung und GIS*, ISBN 978-3-87907-367-2, CD-ROM mit Benutzerhandbuch
296. Mathes A (1998) GPS und GLONASS als Teil eines hybrid Meßsystems in der Geodäsie am Beispiel des Systems HIGGINS, Dissertationen, DGK, Reihe C, Nr. 500
297. Mautz R (2001) Zur Lösung nichtlinearer Ausgleichungsprobleme bei der Bestimmung von Frequenzen in Zeitreihen. DGK, Reihe C, Nr. 532
298. McCoy NH, Janusz GJ (2001) Introduction to abstract algebra. Harcout Academic Press, San Diego
299. Meissl P (1982) Least squares adjustment. A modern approach, Mitteilungen der geodätischen Institut der Technischen Universität Craz, Folge 43. 17
300. Melbourne WG, Davis ES, Duncan CB, Hajj GA, Hardy K, Kursinski R, Mechan TK, Young LE, Yunck TP (1994) The application of spaceborne GPS to atmospheric limb sounding and global change monitoring. JPL Publication 94-18
301. Merritt EL (1949) Explicit Three-point resection in space. Phot. Eng. 15: 649–665
302. Mittermayer E (1972) A generalization of least squares adjustment of free networks. Bull. Geod. 104: 139–155
303. Monhor D (2001) The real linear algebra and linear programming. Müszaki Könyvkiadó, Budapest
304. Monhor D (2002) Clarification of and complements to the concept of outlier. Geodezia es Kartografia 12: 21–27
305. Morgan AP (1992) Polynomial continuation and its relationship to the symbolic reduction of polynomial systems. In Symbolic and Numerical Computations for Artificial Intelligence, pp. 23–45
306. Mukherjee K(1996) Robust estimation in nonlinear regression via minimum distance method. Mathematical methods of statistics, Vol 5, No. 1, Allerton Press. Inc., New York
307. Müller FJ (1925) Direkte (Exakte) Lösungen des einfachen Rückwarscnittseinschneidens im Raum. 1 Teil. Zeitschrift für Vermessungswesen 37: 249–255, 265–272, 349–353, 365–370, 569–580
308. Nakos G, Williams R (2002) A fast algorithm implemented in Mathematica provides one-step elimination of a block of unknowns from a system of polynomial equations, Wolfram Library Archive, MathSource, http://library.wolfram.com/infocenter/MathSource/2597/ [Accessed 27 Aug 2008]
309. Nakos G, Williams R (1997) Elimination with the Dixon resultant. Mathematica in Education and Research, 6, pp.11- 21
310. Nicholson WK (1999) Introduction to abstract algebra. Second Edition, John Wiley & Sons, New York-Chichester-weinheim-Brisbane-Singapore

311. Niell AE (1996) Global Mapping Functions for the Atmosphere Delay at Radio Wavelengths. Journal of Geophysical Research 101(B2): 3227–3246
312. Newsome G, Harvey BR (2003) GPS coordinate transformation parameters for Jamaica. Survey Review 37: 218–233
313. Nitschke M, Knickmeyer EH (2000): Rotation parameters - a survey of techniques. Journal of Surveying Engineering 126: 83–105
314. Ojika T (1987) Modified deflation algorithm for the solution of singular problems. I. A system of nonlinear algebraic equations. J.Math.Anal.Appl. 123: 199 - 221
315. Okeke FI (1998) The curvilinear datum transformation model, DGK, Reihe C, Heft Nr. 481
316. Ozone MI (1985) Non-iterative solution of the ϕ equations. Surveying and Mapping 45: 169–171
317. Paláncz B., 2006: GPS N-points Problem, Mathematica in Education and Research, 11., 2:153-177.
318. Paláncz B, Awange JL, Grafarend EW (2007) Computer algebra Solution of the GPS N-points problem, GPS Solutions, 11., 4:1080, Springer, Heidelberg, http://www.springerlink.com/content/75rk6171520gxq72/ [Accessed 1 Dec 2008]
319. Paláncz B , Zaletnyik P, Awange JL, Grafarend EW(2008) Dixon resultant's solution of systems of geodetic polynomial equations, Journal of Geodesy 82:(8) pp. 505-511. (2008)
320. Paláncz B (2008) Introduction to Linear Homotopy, Wolfram Library Archive, MathSource, http://library.wolfram.com/infocenter/MathSource/7119/ [Accessed 27 Aug 2008]
321. Paláncz B, Zaletnyik P, Awange JL (2008a) Extension of Procrustes algorithm for 3D affine coordinate transformation, Wolfram Library Archive, MathSource, http://library.wolfram.com/infocenter/MathSource/7171/ [Accessed 27 Aug 2008]
322. Paláncz B, Lewis RH, Zaletnyik P, Awange JL (2008b) Computational Study of the 3D Affine Transformation, Part I. 3-point problem, Wolfram Library Archive, MathSource, http://library.wolfram.com/infocenter/MathSource/7090/ [Accessed 27 Aug 2008]
323. Paláncz B, Zaletnyik P, Lewis RH, Awange JL (2008c) Computational Study of the 3D Affine Transformation, Part II. N-point problem,Wolfram Library Archive, MathSource, http://library.wolfram.com/infocenter/MathSource/7121/ [Accessed 27 Aug 2008]
324. Paláncz B, Zaletnyik P, Awange LJ (2008d) Extension of Procrustes algorithm for 3D affine coordinate transformation http://library.wolfram.com/infocenter/MathSource/7171/
325. Paláncz B, Zaletnyik P, Awange JL, Lewis R (in press) Computational Study of the 3D Affine Transformation, In Peter G. Chareton (ed): Computational Mathematics: Theory, Methods and Applications. Nova Science Publishers, New York. 30 pages. to appear. ISBN: 978-1-60876-271-2
326. Papp E, Szucs L (2005) Transformation Methods of the Traditional and Satellite Based Networks (in Hungarian with English abstract), Geomatikai Közlemenyek VIII. 85-92.
327. Paul MK (1973) A note on computation of geodetic coordinates from geocentric (Cartesian) coordinates. Bull. Geod. No. 108: 135–139
328. Penev P (1978) The transformation of rectangular coordinates into geographical by closed formulas. Geo. Map. Photo 20: 175–177
329. Perelmuter A (1979) Adjustment of free networks. Bull. Geod. 53: 291–295

330. Pick M (1985) Closed formulae for transformation of Cartesian coordinates into a system of geodetic coordinates. Studia geoph. et geod. 29: 653–666
331. Pistone G, Wynn HP (1996) Generalized confounding with Gröbner bases. Biometrika 83: 112–119
332. Pope A (1982) Two approaches to non-linear least squares adjustments. The Canadian Surveyor 28: 663–669
333. Preparata FP, Shamos MI (1985) Computational geometry. An Introduction. Springer, New York-Berlin-Heidelberg-London-Paris-Tokyo-Hong Kong-Barcelona-Budapest
334. Press WH, Teukolsky SA, Vetterling WT, Flannery BP (1992) Numerical recipes in Fortran 77: The art of scientific computing, 2nd edition, Cambridge University Press
335. Prestel A, Delzell CN (2001) Positive polynomials: from Hilbert's 17th problem to real algebra. Springer, Berlin
336. Quoc-Nam Tran (1994) Extended Newton' s method for finding the roots of an arbitrary system of nonlinear equations. In Hamza, M.H. ed. Proc. 12 th IASTED Int. Conf.on Applied Informatics, IASTED, Anaheim, CA.
337. A Symbolic - Numerical Method for Finding a Real Solution of an Arbitrary System of Nonlinear Algebraic Equations, J.Symbolic Computation 26. pp. 739 - 760
338. Rao CR (1967) Least squares theory using an estimated dispersion matrix and its application to measurement of signals. Procedure of the Fifth Barkeley Symposium, Barkeley
339. Rao CR (1971) Estimation of variance and covariance components - MINQUE theory. Journal of Multivariate Analysis 1: 257–275
340. Rao CR (1973) Representation of the best linear unbiased estimators in the Gauss-Markov model with singular dispersion matrix. Journal of multivariate analysis 3: 276–292
341. Rao CR (1978) Choice of the best linear estimators in the Gauss-Markov model with singular dispersion matrix. Comm. Stat. Theory Meth. A7 (13): 1199–1208
342. Rao CR, Kleffe J (1979) Variance and covariance components estimation and applications. Technical Report No. 181, Ohio State University, Dept. of Statistics, Columbus, Ohio
343. Reigber C, Lühr H, Schwintzer P (2003) First CHAMP mission results for gravity, magnetic and atmospheric studies, Springer, Heidelberg
344. Reinking J (2001) Anmerkung zu "Zur Bestimmung eindeutiger transformationparameter". Zeitschrift für Vermessungswesen 126: 295–296
345. Richter B (1986) Entwurf eines nichtrelativistischen geodätisch-astronomischen Bezugssystems, DGK, Reihe C, Heft Nr. 322
346. Rinner K (1962) Über die Genauigkeit des räumlichen Bogenschnittes. Zeitschrift für Vermessungswesen 87: 361–374
347. Ritt JF (1950) Differential algebra. AMS colloquium publications 18
348. Rocken C, Anthes R, Exner M, Hunt D, Sokolovski S, Ware R, Gorbunov M, Schreiner S, Feng D, Hermann B, Kuo Y.-H, Zou X (1997) Analysis and validation of GPS/MET data in the neutral atmosphere. J. Geophys. Res. 102: 29849–29860
349. Runge C (1900) Graphische Ausgleichung beim Rückwätseinchneiden. Zeitschrift für Vermessungswesen 29: 581–588
350. Saito T (1973) The non-linear least squares of condition equations. Bull. Geod. 110: 367–395
351. Saleh J (2000) Robust estimation based on energy minimization principles. Journal of Geodesy 74: 291–305

352. Salmon G (1876) Lessons Introductory to modern higher algebra. Hodges, Foster and Co., Dublin
353. Schaffrin B (1983) Varianz-Kovarianz-Komponenten-Schätzung bei der Ausgleichung heterogener Wiederholungsmessungen, DGK, Reihe C, Heft Nr.282
354. Schek HJ, Maier P (1976) Nichtlineare Normalgleichungen zur Bestimmung der Unbekannten und deren Kovarianzmatrix. Zeitschrift für Vermessungswesen 101: 140–159
355. Schönemann PH (1966) Generalized solution of the orthogonal Procrustes problem. Psychometrika 31: 1–10
356. Schönemann PH, Caroll RM (1970) Fitting one matrix to another under choice of a certain dilatation and rigid motion. Psychometrika 35(2): 245–255
357. Schottenloher M (1997) A mathematical introduction to conformal field theory. Springer, Berlin, Heidelberg, New York
358. Schram TG (1988) Properties of gravitational lens mapping. In: Kaiser N, Lasenby AN (eds) The post-recombination universe. Kluwer Academic Publishers pp. 319–321
359. Schram TG (1998) Computer algebra system in engineering education. Global Journal of Engng. Educ. 2: 187–194
360. Shut GH (1958/59) Construction of orthogonal matrices and their application in analytical Photogrammetrie. Photogrammetria XV: 149–162
361. Schwarze VS (1995) Satellitengeodätische Positionierung in der relativistischen Raum-Zeit, DGK, Reihe C, Heft Nr.449
362. Shut GH (1958/59) Construction of orthogonal matrices and their application in analytical Photogrammetrie. Photogrammetria XV: 149-162
363. Singer P, Ströbel D, Hördt R, Bahndorf J, Linkwitz K (1993) Direkte Lösung des räümlichen Bogenschnitts. Zeitschrift für Vermessungswesen 118: 20–24
364. Sjöberg LE (1999) An efficient iterative solution to transform rectangular geocentric coordinates to geodetic coordinates. Zeitschrift für Vermessungswesen 124: 295–297
365. Soler T, Hothem LD (1989) Important parameters used in geodetic transformations. Journal of Surveying Engineering 115: 414–417
366. Sommese AJ, Verschelde J, Wampler CW (2003) Introduction to Numerical Algebraic Geometry. in "Solving Polynomial Equations. Foundations, Algorithms and Applications" , edited by Alicia Dickenstein and Ioannis Z. Emiris, volume 14 of "Algorithms and Computation in Mathematics", pages 301-337, Springer-Verlag
367. Sommese AJ, Wampler CW (2005) The Numerical solution of systems of polynomials arising in engineering and science, World Scientific
368. Späth H (2004) A numerical method for determining the spatial Helmert transformation in case of different scale factors, *Zeitschrift für Geodäsie, Geoinformation und Landmanagement*, **129**, 255-257.
369. Steiner AK (1998) High resolution sounding of key climate variables using the radio occultation technique. Dissertation, Institute for Meteorology and Geophysics, University of Graz, No. 3
370. Steiner AK, Kirchengast G, Foelsche U, Kornblueh L, Manzini E, Bengtsson L (2001) GNSS occultation sounding for climate monitoring. Phys. Chem. Earth (A) 26: 113–124
371. Stillwell J (2003) Elements of number theory, Springer, New York
372. Strang G, Borre K (1997) Linear Algebra, Geodesy and GPS, Wellesley Cambridge Press, Wellesley
373. Sturmfels B (1994) Multigraded resultant of Sylvester type. Journal of Algebra 163: 115–127

374. Sturmfels B (1996) Gröbner bases and convex polytopes. American Mathematical Society, Providence
375. Sturmfels B (1998) Introduction to resultants. Proceedings of Symposia in Applied Mathematics 53: 25–39
376. Sünkel H (1999) Ein nicht-iteratives Verfahren zur Transformation geodätischer Koordinaten. Öster. Zeitschrift für Vermessungswesen 64: 29–33
377. Sylvester JJ (1853) On a theory of syzygetic relations of two rational integral functions, comprising an application to the theory of Strum's functions and that of the greatest algebraic common measure. Philos.Trans. 143, pp.407-548
378. Ten Berge, JMF (1977) Orthogonal procrustes rotation for two or more matrices. Psychometrika 42: 267–276
379. Teunissen PJG (1990) Nonlinear least squares. Manuscripta Geodaetica 15: 137–150
380. Teunissen PJG (1988) The non-linear 2d symmetric Helmert transformation: an exact nonlinear least squares solution. Bull. Geod. 62: 1–15
381. Teunissen PJG, Knickmeyer EH (1988) Non-linearity and least squares. CISM Journal ASCGC 42: 321–330
382. Thompson EH (1959a) A method for the construction of orthogonal matrices. Photogrammetria III: 55–59
383. Thompson EH (1959b) An exact linear solution of the absolute orientation. Photogrammetria XV: 163–179
384. Torge W (1991) Geodesy. 2nd Edition. Walter de Gruyter, Berlin
385. Trefethen LN, Bau D (1997) Numerical linear algebra. SIAM, Philadelphia
386. Trott M (2005) Mathematica Guide Book for Symbolic Computation, Springer, New York
387. Tsuda T, Heki K, Miyazaki S, Aonashi K, Hirahara K, Tobita M, Kimata F, Tabei T, Matsushima T, Kimura F, Satomura M, Kato T, Naito I (1998) GPS meteorology project of Japan-Exploring frontiers of geodesy-. Earth Planets Space, 50(10): i–v
388. Tsuda T, Hocke K (2004) Application of GPS occultation for studies of atmospheric waves in the Middle Atmosphere and Ionosphere. In Anthens et al. (eds). Application of GPS Remote Sensing to Meteorology and Related Fields, Journal of Meteorological Society of Japan, Vol. 82, No. 1B, pp. 419–426
389. Van Mierlo J (1988) Rückwärtschnitt mit Streckenverhältnissen. Algemain Vermessungs Nachrichten 95: 310–314
390. Vanicek P, Krakiwski EJ (1982) Geodesy: The concepts. North-Holland Publishing Company. Amsterdam. New York-Oxford
391. Vanicek P, Steeves RR (1996) Transformation of coordinates between two horizontal geodetic datums. Journal of Geodesy 70: 740–745
392. Vasconcelos WV (1998) Computational methods in commutative algebra and algebraic geometry, Springer, Berlin-Heidelberg
393. Verschelde J (1999) Algorithm 795: PHCpack: a general-purpose solver for polynomial systems by homotopy continuation. ACM Trans. Math. Softw. 25(2):251-276
394. Verschelde J (2007) Homotopy Methods for Solving Polynomial Systems tutorial at ISSAC' 05, Beijing, China. http://www.math.uic.edu/~jan/tutorial.pdf [Accessed 1 Dec 2008]
395. Vincenty T (1978) Vergleich zweier Verfahren zur Berechnung der geodätischen Breite und Höhe aus rechtwinkligen koorninaten. Allgemeine Vermessungs-Nachrichten 85: 269–270
396. Vincenty T (1980) Zur räumlich-ellipsoidischen Koordinaten-Transformation. Zeitschrift für Vermessungswesen 105: 519–521

397. Voigt C (1998) Prokrustes Transformationen. Geodätisches Institut, Stuttgart
398. Voigt C, Denker H, Hirt C (2009) Regional Astrogeodetic Validation of GPS/Levelling Data and Quasigeoid Models. In: M.G. Sideris (ed.), Observing our Changing Earth, International Association of Geodesy Symposia 133, pp. 413-420.
399. Von zur Gathen J, Gerhard J (2003) Modern computer algebra. 2nd edition, Cambridge University Press, UK
400. Vorob'ev, VV, Krasil'nikova TG (1994) Estimation of the accuracy of atmospheric refractive index recovery from Doppler shift measurements at frequencies used in the NAVSTAR system. Phys. of Atmos. and Oceans 29: 602–609
401. Van Der Waerden BL (1950) Modern Algebra. 3rd Edition, F. Ungar Publishing Co., New York
402. Watson LT, Sosonkina M, Melville RC, Morgan AP, Walker HF (1997) Algorithm 777: HOMPACK90: a suite of Fortran 90 codes for globally convergent homotopy algorithms. ACM Trans. Math. Softw. 23(4):514-549
403. Watson GA (2006) Computing Helmert transformations, *Journal of Computational and Applied Mathematics* 197: 387-395.
404. Weiss J (1993) Resultant methods for the inverse kinematics problem. In: Angeles et al. (eds.) Computational Kinematics, Kluwer Academic Publishers, Netherlands
405. Wellisch S (1910) Theorie und Praxis der Ausgleichsrechnung. Bd. II: Probleme der Ausgleichsrechnung
406. Welsch WM (1993) A general 7-parameter transformation for the combination, comparison and accuracy control of the terrestrial and satellite network observations. Manuscripta Geodaetica 18: 295–305
407. Ware H, Fulker D, Stein S, Anderson D, Avery S, Clerk R, Droegmeier K, Kuettner J, Minster B, Sorooshian S (2000) SuomiNet: A real time national GPS network for atmospheric research and education. Bull. Am. Meteorol. Soc. 81: 677–694
408. Werkmeister P (1916) Trigonometrische Punktbestimmung durch einfaches Einschneiden mit Hilfe von Vertikalwinkeln. Zeitschrift für Vermessungswesen 45: 248–251
409. Werkmeister P (1920) Über die Genauigkeit trigonometrischer Punktbestimmungen. Zeitschrift für Vermessungswesen 49: 401–412,433–456
410. Werner D (1913) Punktbestimmung durch Vertikalwinkelmessung. Zeitschrift für Vermessungswesen 42: 241–253
411. Wester MJ (1999) Computer algebra system- a practical guide. John Wiley & Sons, Chichester, United Kingdom
412. Wickert J, Michalak G, Schmidt T, Beyerle G, Cheng CZ, Healy SB, et al (2008) GPS radio occultation: Results from CHAMP, GRACE and FORMOSAT-3/COSMIC. Terrestrial, Atmospheric and Oceanic Sciences. In print.
413. Wickert J (2002) Das CHAMP-Radiookkultationsexperiment: Algorithmen, Prozessierungssystem und erste Ergebnisse. Dissertation. Scientific Technical Report STR02/07, GFZ Potsdam
414. Wild F (2001) Test an der Geschlossenen Lösung des "Twin P4P-Problems": Dreidimensionaler Vorwärts- und Rückwärtsschnitt. Studienarbeit, Geodetic Institute, Stuttgart University
415. Winkler F (1996) A polynomial algorithm in computer algebra. Springer, Wien
416. Wieser A, Brunner FK, 2002 Short static GPS sessions: Robust estimation results. Journal of GPS Solutions 5: 70–79
417. Wolfrum O (1992) Merging terestrial and satellite networks by a ten-parameter transformation model. Manuscripta Geodaetica 17: 210–214

418. Wu W (1984) On the decision problem and mechanization of the theorem proving elementary geometry. Scientia Sinica 21: 150–172
419. Xu G (2003) GPS. Theory, algorithms and applications, Springer, Berlin Heidelberg
420. Xu P (1987) A test method for many outliers. I. T. C. Journal 4: 314–317
421. Xu P (1989a) Statistical criteria for robust methods. I. T. C. Journal 1: 37–40
422. Xu P (1989b) On robust estimation with correlated observations. Bull. Geod. 63: 237–252
423. Xu P (2002) A hybrid global optimization method: the one-dimensional case. Journal of Computation and Applied mathematics 147: 301–314
424. Xu P (2003) A hybrid global optimization method: the multi-dimensional case. Journal of Computation and Applied mathematics 155: 423–446
425. Yang Y (1999) Robust estimation of geodetic datum transformation. Journal of Geodesy 73: 268–274
426. Yang Y, Cheng MK, Shum CK, Tapley BD (1999) Robust estimation of systematic errors of satellite laser range. Journal of Geodesy 73: 345–349
427. You RJ (2000) Transformation of Cartesian to geodetic coordinates without iterations. Journal of Surveying Engineering 126: 1–7
428. Youcai H, Mertikas SP (1995) On the design of robust regression estimators. Man. Geod. 20: 145–160
429. Yunck TP (2003): The promise of spaceborne GPS for Earth remote sensing. International Workshop on GPS Meteorology, 14th-17th January 2003, Tsukuba, Japan
430. Zaletnyik, P (2008) Application of Computer Algebra and Neural Networks to Solve Coordinate Transformation Problems. PhD. Thesis at Department of Geodesy and Surveying [in Hungarian], Budapest University of Technology and Economics, Hungary
431. Zaletnyik P, Paláncz B (2009): The symbolic solution of the 3D affin transformation in case of three known points (in Hungarian with English abstract), Geomatikai Közlemények XII, 35-45.
432. Zaletnyik P, Paláncz B, Awange JL, Grafarend EW(2008): Application of Computer Algebra System to Geodesy, In: Michael G Sideris (ed), Observing our Changing Earth. Springer-Verlag, 2008. pp. 803-808. (International Association of Geodesy Symposia; 133.)
433. Závoti J, Jancso T (2006) The solution of the 7-parameter datum transformation problem with- and without the Groebner basis, Acta Geod. Geoph. Hung.,41(1), pp.87-100
434. Zhang S (1994) Anwendung der Drehmatrix in Hamilton normierten Quaternionenen bei der Bündelblock Ausgleichung. Zeitschrift für Vermessungswesen 119: 203-211
435. Zhao A (2007) Newton' s Method with Deflation for Isolated Singularities of Polynomial Systems, Ph.D. Thesis. University of Illinois at Chicago, USA
436. Zippel R (1993) Effective polynomial computation, Kluwer Academic Publications, Boston 1993

Index

3d-orientation, 140

Abel's inversion, 270
Abelian axioms, 15
Abelian group, 14, 15, 20
additive associativity, 14
additive commutativity, 14
additive distributivity, 15
additive identity, 10, 14
additive inverse, 10, 11, 14
aerosol, 266
algebra, 13
  definition, 14
  rings, 16
  sub-algebra, 15
algebraic, 1
algorithms
  Ansermet, 245
  Awange-Grafarend, 179
  Buchberger, 35, 36, 42
  Dixon's resultant, 304
  Euclidean, 20, 34
  Gauss-Jacobi combinatorial, 84, 97, 189, 304, 315
  Gauss-Newton's , 79
  Grafarend-Shan, 245
  Groebner basis, 301, 304, 308
  least squares, 189
  Newton's, 79, 181, 273, 278
  Newton-Raphson, 79
  partial Procrustes solution, 116
  polynomial resultants, 49, 301
  Procrustes, 123, 128, 304
  Procrustes II, 301
  Ritt-Wu, 62
  Steepest-descent, 79
angles, 9, 13
associativity, 14, 16
astrogeodetic observations, 151
astronomical
  latitude, 144, 147, 149
  longitude, 144, 147, 149
atmosphere, 139, 266, 269
atmospheric refractions errors, 290
azimuth, 142

Baarda doctrine, 249
base vectors, 143
Bezout's method, 56, 57
bifurcation, 24
Bjerhammer sphere, 159
BLUUE, 87, 199
Buchberger's optimization criterion, 44, 229

C. F. Gauss, 84, 86, 158
C. G. I. Jacobi, 86
C/A-code, 141, 269
cadastral maps, 112
cancer-causing gene, 115
Cardan angles, 305, 309
Cardano's formula, 25
Cayley, 12
Cayley's approach, 57
CCD cameras, 111
centering matrix, 124, 127, 128
Central Processing Unit (CPU), 3

CHAMP, 286
climatic changes, 265
coefficient matrix, 193
coefficients, 53
collinear, 326
combination, 93
combinatorial program, 97
commutativity, 14
complex numbers, 24
Computer Aided Design (CAD), 1
computer algebra, 24, 25, 27
conformal field theory, 303
conformal group, 120, 249
Conventional International Origin (CIO), 141
coordinates, 11
  canonical, 126
  Cartesian, 144, 161
  Möbius barycentric, 249
  maps', 111
  spherical, 142–144, 147
correlation matrix, 90
criterion matrices, 135

D. Hilbert, 38
Dedekind, 38
deflection of the vertical, 140
degenerate conics, 163
degenerate quadrics, 237
desertification, 265
determinant, 49, 51, 86, 232
determined system, 329, 338
digitizing table, 9
dilatation, 119, 129
Direct minimization, 101
directions, 13
  horizontal, 151, 226
  vertical, 151, 226
dispersion matrix, 87, 90, 93–95, 198, 211
distance ratios, 249, 262, 303
distances, 9, 11, 13, 21, 161, 173, 250
distributivity, 15
Dixon matrix, 55
Dixon Polynomial, 55
Dixon- KSY method, 323
Dixon-EDF method, 323
doppler shift, 270, 273, 274, 279

E. Noether, 38
Early Discovery Factors, 324
eccentricity, 168
eigenspace, 126
eigenvalues, 52, 162
eigenvectors, 52
elimination, 195
elimination order, 312
elimination theory, 51, 55
ellipsoid-of-revolution, 155
ellipsoidal
  anti-starshaped surface, 159
  coordinates, 156
  heights, 156, 159
  latitude, 149
  longitude, 149
  normal, 149
  starshaped surface, 159
environment, 139
environmental monitoring, 265, 286
equations
  algebraic, 19
  Grunert's, 145, 226, 227
  nonlinear systems, 1, 2, 9, 16
  polynomial, 17, 18, 38
  quadratic, 21
equivariant, 303
error
  matrix, 127–129
  norm, 189
  propagation, 92, 93, 95, 291, 313
  weight, 121
errors, 2, 289
Euclidean space, 150, 249
Euler, 11
Euler angles, 145, 306
Euler map, 307
Euler rotation matrix, 144, 145
Evariste Galois, 22
excess Doppler shift, 270
expectation, 94
explicit
  formulae, 2–4
  procedures, 3
Extended Newton-Raphson, 1
extension theorem, 45
exterior orientation, 244, 249
external relation, 14
extraneous factors, 53

F. Crosilla, 112
factor analysis, 111
factor polynomials, 322
Fermat, 61
Fermat's principle, 270
field extension, 22
fields (cf. rings), 14
first moments, 94
flooding, 265
four square identity, 12
frequency, 271
Frobenius error matrix, 121–123, 127, 128
Frobenius norm, 116–118

Gauss elimination, 34–36
Gauss elimination method, 321
Gauss-Krueger projection, 198
Gauss-Krulger Mapps, 155
Gauss-Newton method, 318
geodetic
  coordinates, 141, 156
  height, 141
  heights, 159
  latitude, 141
  longitude, 141
geoid, 155
geometry, 326
geopotential heights, 266, 273
GIS, 9
global method, 63
global methods, 329
global numerical method, 63
global warming, 265, 268, 286
GPS, 2
  applications, 139
  C/A-code, 139
  control segment, 141
  meteorology, 2, 139, 173, 266, 267
  P-code, 139
  PDOP, 181
  pseudo-ranges, 174, 290
  receivers, 266
  remote sensing, 139
  Selective Availability (SA), 139
  signals, 139
  space segment, 141
  user segment, 141
GPS meteorology
  ECMWF, 269
  EUMETSAT, 267
  GEONET, 272
  GRAS, 267
  IPWV, 267
  METEOSAT, 265
  NCEP, 269
  NOAA, 265
  NWP, 281
  NWP models, 265
  refraction angles
    Groebner basis solution, 276
    Sylvester resultant solution, 277
GPS system, 140
GRACE, 286
gravimetric, 24
gravitation lens, 49
gravity, 9, 13
gravity vector, 112, 142–145, 147, 151
greatest common divisors (gcd), 34
greenhouse gas, 265
Greenwich, 141
Groebner basis, 33, 51
  definition, 39, 42
  Maple computation, 46
  Mathematica computation, 45
group axioms, 15
Grunert's equations, 219

H. Hironaka, 33
Hamilton quaternion, 145, 237
Hamilton's letter, 12
heights, 90
Helmert transformation, 130
Helmert's projection, 156
Hilbert Basis Theorem, 38, 39
homogeneous, 121
Homotopy continuation method, 1

Ideal, 33, 36, 37, 164, 229
Ideal membership, 40
idempotent, 128
identity, 14
ill- conditioned, 101
impact parameter, 270, 278
injective function, 88
integers, 9, 10, 13, 15
internal relation, 14
international reference ellipsoid, 156, 158
intersection, 21, 249

planar
conventional solution, 250
Groebner basis solution, 251
three-dimension
closed form, 254
conventional solution, 255
Grafarend-Shan method, 262
Groebner basis solution, 255
overdetermined, 259
photogrammetric, 261
resultant solution, 256
invariant, 303
inverse, 14
ionosphere, 266, 269
ionospheric errors, 278, 290
isotropic, 121
isotropic dilatation, 116
isotropic error, 135
isozenithal, 158

J. A. Grunert, 218, 226
Jacobi matrix, 101, 108, 189
Jacobian determinant, 52, 54, 177, 178, 207, 233, 257

Kronecker, 38
Kronecker-Zehfu decomposition, 134
Kummer, 38

L1-signals, 281
L2-signals, 281
Lagrangean, 123, 161, 163
centralized, 124
multiplier, 36
reduced, 124
unconstraint, 123
Leading Coefficient (LC), 340
Leading Monomial (LM), 39, 229, 340
Leading Term (LT), 340
Least Common Multiple (LCM), 41, 229
least square, 79
optimization, 84
least squares, 49, 79, 101, 111, 122, 135, 181, 199, 245, 305, 315, 318
levelling, 24, 89, 90
lexicographic ordering, 36, 40–42, 44, 164, 176, 222, 228, 230, 252, 276, 339
graded, 44, 339
graded reverse, 340
linear algebra, 14, 49
linear Gauss-Markov model, 79, 135
local level system, 142
local methods, 101, 329
LPS systems, 140

manifold, 88
Maple, 36
Groebner basis computation, 46
Mathematica, 24, 25, 27, 36
Groebner basis computation, 45
Matlab, 24, 25, 27, 36
"det" command, 51
"polyfit" function, 279
"randn" command, 151
"roots" command, 36, 40, 51, 174, 196, 224, 232, 277
matrix properties, 117
maximum, 36
mean error matrix, 129, 132
minimal norm, 166
minimization problem, 80
minimum, 36
minimum distance mapping (MDM), 156
models
Gauss-Markov, 87
linear Gauss-Markov, 90
mixed, 87
non-stochastic, 87
stochastic, 87
monoid, 16
monomial ideal, 103
monomial ordering, 339
monomials, 17
multideg, 40, 340
multidimensional scaling, 114
multipath, 139, 290
multiplicative associativity, 14
multiplicative commutativity, 14
multiplicative distributivity, 15
multiplicative identity, 10, 14
multiplicative inverse, 11, 14
multipolynomial resultants, 49, 51

nadir-viewing microwave, 268
necessary condition, 53, 101, 162, 180
network densification, 3
Newton polytope, 31
Newton-Raphson method, 325

Newton-type methods, 101, 325
Niels Henrick Abel, 22
  impossibility theorem, 22
nonlinearity, 2
nonlinearity errors, 273
NSolve, 195
numbers
  complex, 9, 11, 13
  invention of zero, 10
  natural, 9, 10, 22
  prime, 21
  rational, 9, 11, 13, 15
  real, 9, 11, 13, 14
  ring, 22
numerical methods, 1

objective function, 80, 329
orientation unknown, 140, 144, 145
orthogonal group, 120
orthonormal matrix, 120, 126
outliers, 123, 289
overdetermined system, 338
overdetermined systems, 79, 101

P-code, 141, 269
P. J. Huber, 4, 290
partial Procrustes algorithm, 219
PDOP, 181, 182
permutation, 93
perspective center, 244, 262
phase delay, 271
photo coordinates, 9, 13
Pizetti's projection, 156
planar ranging, 21
polygon, 31
polynomials, 3, 9, 17, 18
  S–polynomial, 41, 42, 229
  critical degree, 52
  cubic, 25, 54, 230
  degree zero, 223
  equivalence, 40
  extraneous factor, 53
  factorization, 21
  homogeneous, 49, 53
  impossibility theorem, 22
  irreducible, 21
  Legendre, 1
  minimal, 22
  multivariate, 18, 19
  of degree zero, 50
  ordering
    lexicographic, 36
  quadratic, 21, 24, 54
  quartic, 26, 27, 219, 256, 275, 308
  reduced cubic, 25
  resultants, 33, 49
  roots, 2, 24, 25, 27, 36
  solvability test, 44
  Sylvester resultant, 274
  total degree, 52
  univariate, 3, 12, 19, 36, 163, 165, 232
polytope, 31
positional error sphere, 333
positional norm, 180
positive definite, 87, 90, 162
positive semi-definite, 123
Pothenot's problem, 219
precipitable water, 271
pressure, 139, 173, 266, 269, 273
Procrustes
  application in medicine, 114
  gene recognition, 114
  general solution, 116
  identification of malarial parasites, 115
  magic bed, 112
  partial solution, 111, 116
  software, 114
Procrustes algorithm, 101
projection lines, 160
Pythagorean length, 11
Pythagorean theorem, 275

quadratic formula, 24
quaternions, 9, 86
quotient, 11

radial distance, 180
radical solutions, 22
radio waves, 269
radiometers, 268
radiosondes, 265
Random Access Memory (RAM), 3
random errors, 86, 89
ranging, 173
  3d-ranging
    closed form, 203
    elimination approach-1, 204
    elimination approach-2, 205

Groebner approach, 206
ovedetermined case, 211
resultant approach, 207
GPS pseudo-ranging
Groebner basis solution, 179
Sturmfels' solution, 177
planar
conventional approach, 192
overdetermined, 198
reduced Groebner approach, 194
Sylvester approach, 193
reduced Groebner basis, 37, 45, 274, 276
redundant observations, 79
reference ellipsoid, 155
reference frames, 141
refraction angle, 269, 273
refractive index, 270
remote sensing, 173
resection, 217
3dimension
exact solution, 226
Grafarend et al. approach, 237
Groebner basis solution, 228
Grunert's equations, 226
overdetermined, 240
photogrammetric, 244
resultant solution, 230
planar
analytic solution, 220
Groebner basis approach, 222
Sturmfels' resultant approach, 223
scanner application, 217
residual, 101
residuals, 191, 290
rings, 9
commutative, 16, 19
congruence, 14
definition, 15
division, 16
fruit, 14
modular arithmetic, 14, 18
numbers, 9, 13, 18
polynomial, 9, 14, 16, 18, 19
with identity, 15, 16
robotics, 1
robust estimation, 4, 290
Rodrigues, 11
root-mean-square-errors, 191
roots, 24
rotation, 116, 129, 303
rotation matrix, 116, 119, 148, 262, 334
Runge-Kutta method, 70

satellites
CHAMP, 266
clock bias, 175
COSMIC, 266
ephemerids, 141
GALILEO, 111, 139, 140, 173, 267
geostationary, 265
GLONASS, 111, 139, 140, 173, 267
GNSS, 139, 173
GPS, 111, 139, 141, 173, 266
GRACE, 266
LEO, 269
occultations, 2, 278
polar, 265
receiver range bias, 175
remote sensing, 266
SAC-C, 266
velocities, 270, 274
scale factor, 119, 303, 334
second moments, 94
semi-group, 16
semi-major axis, 159, 168
semi-minor axis, 159, 168
semi-norm, 127, 128
set, 10, 11
shape analysis, 111
singular value decomposition (SVD), 118, 119, 123, 126, 128
singular values, 107
singularity, 102
skew-symmetric matrix, 305, 309
slant water, 272
smooth-linear bending, 115
Snell's law, 270
solutions
analytic, 3
closed form, 3
exact, 2
least squares, 19, 89
numerical, 3
space angles, 226, 249, 254, 303
Spatial Reference System, 173
spherical harmonics, 1
square matrix, 53
square root, 11

standard basis, 33
standard deviation, 152
standard errors, 152
starshaped surface, 159
stencil method, 237
stochastic, 327
stochasticity, 4, 97
stratosphere, 266, 268
sufficient condition, 53, 162, 180
surface normal, 159
surjective mapping, 120
Sylvester resultants, 49, 50, 207
symbolic solution, 321
Synthetic Aperture Radar (SAR), 268

Taylor points, 94
Taylor series, 102
Taylor series expansion, 189
TEC, 267
temperature, 139, 173, 265, 266, 269, 273, 286
Theseus, 113
topographical surface, 159
Topography, 155
total degree, 49, 53, 231, 339, 340
transformations
  7-parameter, 4, 111
  affine, 115
  Cartesian to ellipsoidal, 156, 167
  Gauss-Jacobi solution, 313
  Groebner solution, 305
  Karussel, 143
  orthogonal, 111
  orthogonal Procrustes, 120
  similarity, 120, 249, 262
translation, 116, 119, 129, 303, 334
trigonometric addition formulae, 274
tropopause, 268, 286
troposphere, 268

UCAR, 266
univariate polynomial, 321
unknown orientation, 143, 147, 220, 222, 223, 251

variance-covariance, 181, 211, 333
  matrix, 94
  propagation, 92, 93, 95, 291, 313
vectors, 14
vertical deflection, 112
vertical deflections, 140

W-semi-norm, 121
W. R. Hamilton, 12, 86
W. WU, 61
water vapour, 173, 265, 266, 269, 273, 286
weather, 139, 265
weight matrix, 97, 291
weighted arithmetic mean, 89
Weitzenböck space, 120, 249
World Geodetic System WGS-84, 140

zenith delay, 271

FSC
www.fsc.org
MIX
Papier | Fördert gute Waldnutzung
FSC® C083411

Kolibri 360 GmbH
Ferdinand-Jühlke-Straße 7
99095 Erfurt, Deutschland
produktsicherheit@kolibri360.de